KINGS FOR BID WINNING —BID PROPOSALS

得标为王—方案篇

2013-2014

上 册

龙志伟 编著
Edited by Long Zhiwei

| 度假休闲 | 商业建筑 | 酒店建筑 | 购物中心 | 文化艺术建筑 |

| Resort | Commercial Building | Hotel | Shopping Center | Culture & Art Building |

广西师范大学出版社
· 桂林 ·

图书在版编目(CIP)数据

得标为王:方案篇 2013－2014/龙志伟 编著. —桂林:广西师范大学出版社, 2014.4
ISBN 978－7－5495－2743－4

Ⅰ.①得… Ⅱ.①龙… Ⅲ.①建筑设计－作品集－世界－现代 Ⅳ.①TU206

中国版本图书馆 CIP 数据核字(2012)第 245615 号

出 品 人:刘广汉
责任编辑:王晨晖
装帧设计:龙志杰

广西师范大学出版社出版发行
(广西桂林市中华路 22 号　　邮政编码:541001
网址:http://www.bbtpress.com)
出版人:何林夏
全国新华书店经销
销售热线:021－31260822－882/883
上海锦良印刷厂印刷
(上海市普陀区真南路 2548 号 6 号楼　邮政编码:200331)
开本:646mm×960mm　　1/8
印张:113.5　　字数:50 千字
2014 年 4 月第 1 版　　2014 年 4 月第 1 次印刷
定价:898.00 元(全 3 册)

序 Preface

“物竞天择，适者生存”这句流传了几千年的警句在当今这个竞争激烈的时代已被奉为至理名言，“优胜劣汰”这一法则在建筑设计领域也同样发挥了指挥棒的作用。如何提高公司的业务水平及竞争力，使己方提出的设计方案赢得招标方及大众的青睐和认可进而得标，是每一个设计师和设计团队都必须深思的问题。一个能在众多方案中脱颖而出、独占鳌头的方案，不仅关系到提案的经济性，更与提案的原创性、创新性、合理性、实用性和完整性息息相关。一个富于想象又不脱离实际、富有创意又经济实用、彰显个性又贴近生活的独创性方案才是招标方心中的首选。

《得标为王——方案篇 2013-2014》是一本大型的设计方案集锦。该书收录了阿特金斯、斯蒂文·霍尔建筑师事务所、Jaspers-Eyers & Partners、3LHD、蓝天组、澳大利亚 SDG、C.F.Møller、UNStudio、深圳天方、北京殊舍等近百家国内外优秀建筑设计公司的知名设计方案。本书精选了度假休闲、商业建筑、酒店建筑、购物中心、文化艺术建筑、城市综合体、办公建筑、医疗建筑、学校建筑、交通建筑、住宅建筑等类别的方案近 200 个。

无论是理念的创新、思维方式及构思角度的转换，还是最新技术的运用、生态节能材料的使用，抑或是独特的造型与外观，它们既使该方案变得独特而唯一，也使之成为备受客户认可和推崇的新设计、新理念、新方案。本书指明了当今建筑设计领域智能、仿生、生态的设计新趋势，阐释了当今备受世人关注的绿色、低碳、以人为本的设计理念，反映了使用者生理上和心理上的需求。这些方案的实施，不仅将促使许多新形式、新类别的建筑的诞生，使人与社会、人与自然和谐发展，同时也将改善人们的生活方式。

“Survival of the Fittest in Natural Selection” has been claimed to be words of wisdom in this fiercely competitive age. “Survival of the Fittest” also plays a leading role in architectural design. How to boost competitiveness and improve professional skills, making the design scheme be favored and accepted by tenderers and the public so as to win the bid has become a thought-provoking issue to every designer and each design team as well. The design scheme, which stands out from numerous proposals, must be not only economical, but also originative, innovative, rational, practical and integrated. Only the most unique proposal that is imaginative, practical, creative, economical, characteristic and close to life can be the priority for tenderers.

Kings for Bid Winning – Bid Proposals 2013-2014 is a large collection of design schemes. Well-known design proposals from nearly 100 national and international renowned architectural design companies, such as Atkins, Steven Holl Architects, Jaspers-Eyers Architects, 3LHD, COOP HIMMELB(L)AU, Shine Design Group, C.F. Møller, UNStudio, Shenzhen TAF Architect and Beijing Shushe Architecture, are included. New projects of about 200 cases involve Resort, Commercial Building, Hotel, Shopping Center, Culture & Art Building, Urban Complex, Office Building, Medical Building, School Building, Transportation Building and Residence.

The innovative concept, thinking mode, design perspective, or the application of latest technology and ecological energy-saving materials, or distinct shape and appearance, make the proposal a special and unique new design, new concept, new scheme recognized and highly appraised by clients. The book implies the architectural design trend for intelligent, bionic and ecological design solutions, interprets a design concept of “Green, Low-carbon and People-oriented” and reflects physical and psychological demands of users. Implementing these proposals not only means the emerging of new forms and new types of buildings, a harmonious development between man and society, man and nature, but also improves people’s lifestyle.

目录 Contents

文化艺术建筑 234 Culture & Art Building

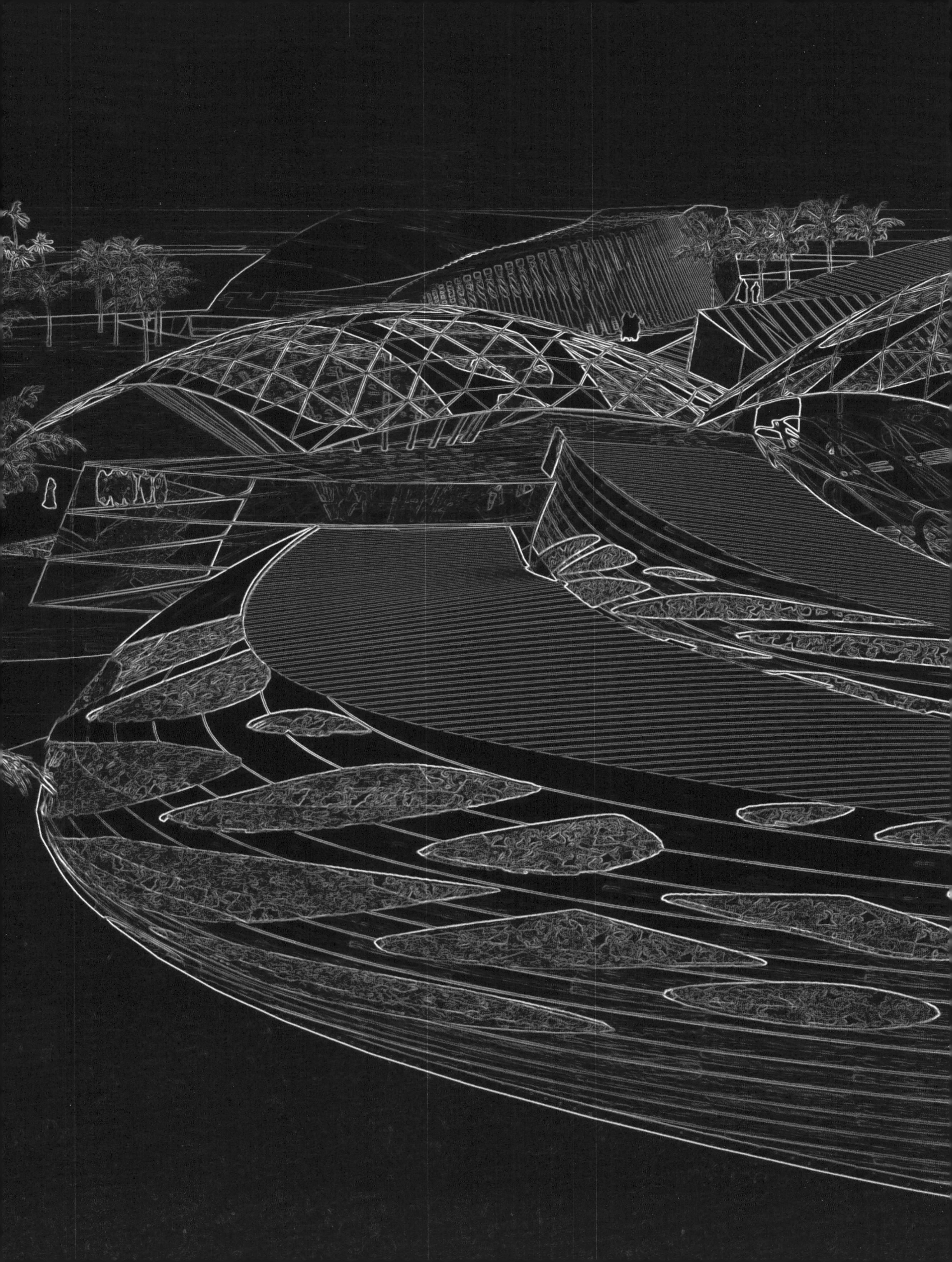

度假休闲 Resort

美国新泽西州大西洋城旅游区总体规划

Atlantic City Tourism District Master Plan

设计单位：捷得国际建筑师事务所

项目地址：美国新泽西州

Designed by: The Jerde Partnership

Location: New Jersey, USA

项目概况

大西洋城旅游区总体规划旨在通过构建一个现代主义的框架将大西洋城转变为美国东南部一个极具吸引力的滨海度假胜地。设计构建了一个整洁、绿色、安全的城市空间，而对周边独特的岛屿环境的利用，则使项目为不同年龄阶段的人们提供了多样的景点和体验。

规划设计

持续的革新举措将使这一滨海区成为极具发展潜力、充满生机和活力的空间。总体规划为核心地带的建筑提供了城市设计理念，包括对海滩和木栈道进行创新性的建设。

新的旅游区总体规划占地将近 6 879 655 平方米，这一规划将实现在该区引入创新型的、具有吸引力的新元素与通过贯彻实施基本的城市规划原则来整体改造大西洋城这两者之间的平衡。

该规划以“视觉体验”和“城市再生”两大概念为依据，通过基础城市规划，在不同的区域构建一个极具刺激性和凝聚力的、内部交通流线通畅的休闲空间。

Profile

The new Atlantic City Tourism District master plan sets out to attract more people through a realistic framework by transforming Atlantic City into the preferred coastal resort destination of the Northeastern United States. By creating a clean, green, safe city and taking advantage of its unique island setting, the new Atlantic City will offer a wide range of attractions and experiences for all ages.

Planning Design

The sustained renewal initiatives will allow the seaside destination to realize its full potential as a vibrant, exciting place to be and be seen. The master plan provides urban design principles for key improvements within the core areas including the Beach and Boardwalk during the initial phase of improvement initiatives.

Encompassing approximately 6,879,655 square meters of area, the new Tourism District Master Plan requires a careful balance of introducing innovative and inspirational new elements of attraction, while implementing fundamental urban planning principles to transform the perception of Atlantic City.

Two major concepts essential to the plan revolve around "Experiential Visioning" and "Urban Regeneration", creating excitement, identity, cohesion and connection as well as a leisure space with fluid internal traffic lines.

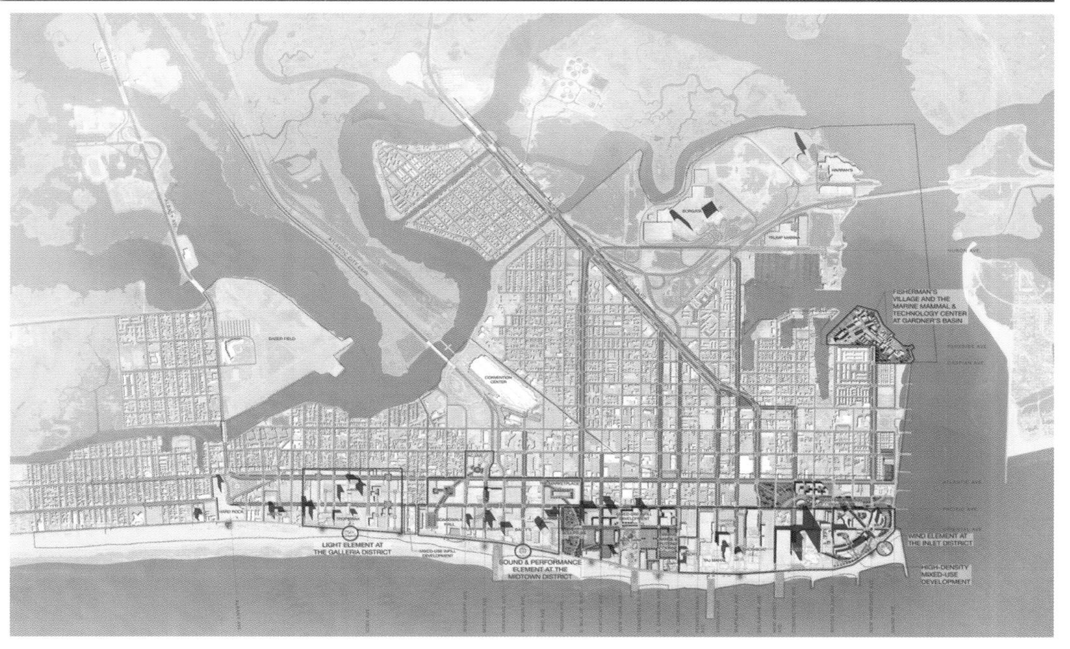
FISHERMAN'S
VILLAGE AND THE
MARINE MAMMAL &
TECHNOLOGY CENTER
AT GARDNER'S BASIN
LIGHT ELEMENT AT
THE GALLERIA DISTRICT
SOUND & PERFORMANCE
ELEMENT AT THE
MIDTOWN DISTRICT
WIND ELEMENT AT
THE INLET DISTRICT
HIGH-DENSITY
MIXED-USE
DEVELOPMENT

EASTER PARADE
King's Food & Gifts
Heaven
CUT L
AFTERTHOUGHTS

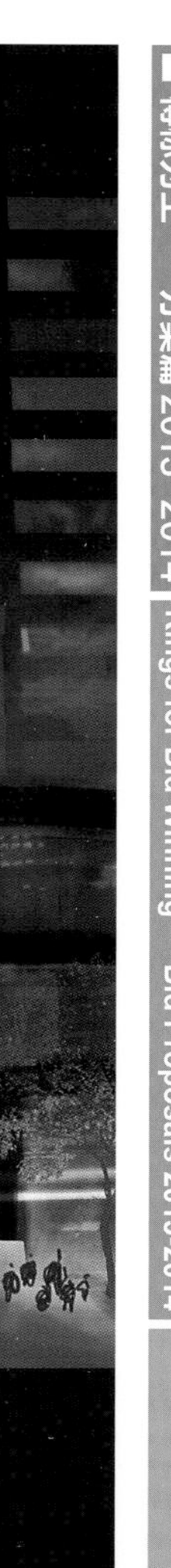

阿联酋迪拜 Isla Moda

Isla Moda

设计单位：Oppenheim Architecture+Design
项目地址：阿联酋迪拜
项目面积：104 330 ㎡

Designed by: Oppenheim Architecture + Design
Location: Dubai, UAE
Area: 104,330 m^2

项目概况

该项目是由世界时尚设计师 Karl Lagerfeld 和科尔酒店集团合作完成的。Isla Moda 是距离迪拜 20 千米的人工岛屿，这个人工打造的时尚岛，将包括 3 个豪华酒店和 150 多栋别墅。

建筑设计

项目的设计灵感来自于印度的漂浮宫殿，相当于现代的漂浮游轮。设计旨在让人们在这个人工模拟的世界中感受真实的体验。

这个虚拟的人工岛的设想源自对时尚与风格之间细微差别的理解以及对未知的、无限的生活体验的追求。设计将这个庞大的建筑体量融入所处的环境中，实现了建筑和抬高了的自然景观之间的平衡。建筑的形态遵循了完美的比例，是对“黄金分割”的物理体现。

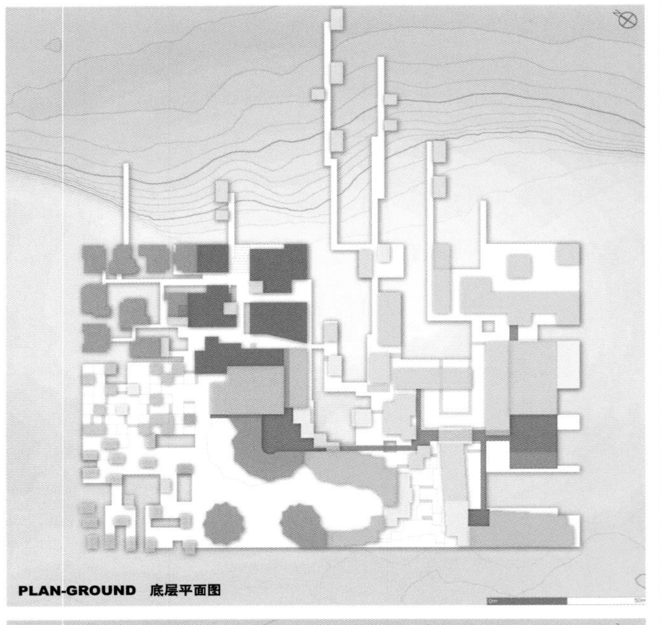
PLAN-GROUND 底层平面图

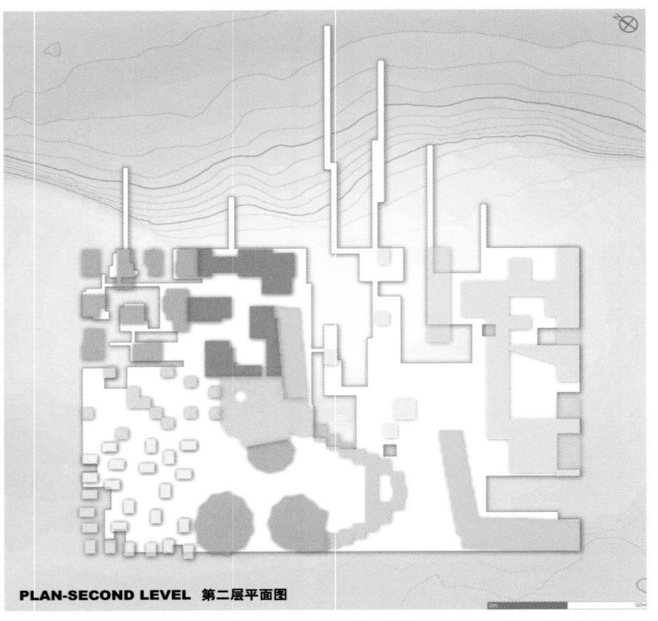
PLAN-SECOND LEVEL 第二层平面图

PLAN-THIRD LEVEL 第三层平面图

PLAN-FOURTH LEVEL 第四层平面图

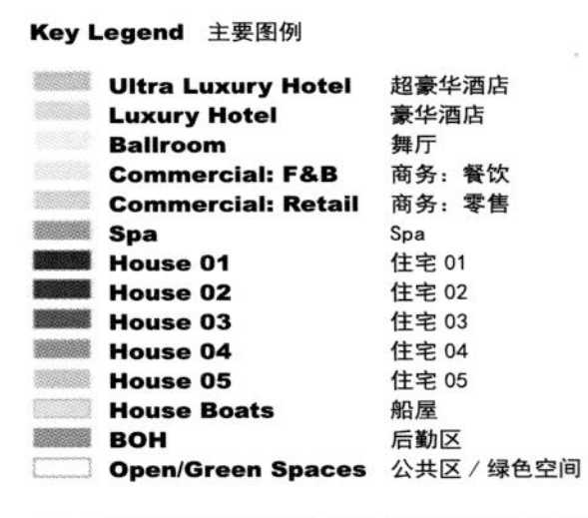
Key Legend 主要图例
Ultra Luxury Hotel 超豪华酒店
Luxury Hotel 豪华酒店
Ballroom 舞厅
Commercial: F&B 商务：餐饮
Commercial: Retail 商务：零售
Spa Spa
House 01 住宅 01
House 02 住宅 02
House 03 住宅 03
House 04 住宅 04
House 05 住宅 05
House Boats 船屋
BOH 后勤区
Open/Green Spaces 公共区 / 绿色空间

Total Area per Floor per Unit OR per House
总面积／层／单元或每个住宅

Program Legend 图例	Keys 关键	Area(gross) 总面积
Luxury Hotel-Lobby 豪华酒店 - 大厅		112 m²
Luxury Hotel-Standard 豪华酒店 — 标准	9	473 m²
Luxury Condo/Hotel-Studio Lock-off 豪华公寓／酒店 — 可分割工作室	6	300 m²
Luxury Condo/Hotel-1BR 豪华公寓／酒店 — 1BR	6	612 m²
Luxury Condo/Hotel-Circulation 豪华公寓／酒店 — 流通线路		871 m²
Ultra Luxury Condo/Hotel-Lobby 超豪华公寓／酒店 — 大厅		373 m²
Ultra Luxury Condo/Hotel-Jr.Suite 超豪华公寓／酒店 — 普通套间	11	650 m²
Ultra Luxury Condo/Hotel-2BR Lock-off 超豪华公寓／酒店 —2BR 可分割套房	11	715 m²
Ultra Luxury Condo/Hotel-4BR Villa 超豪华公寓／酒店 — 4BR 别墅	1	144 m²
Ultra Luxury Condo/Hotel-Circulation 超豪华公寓／酒店 — 流通线路		597 m²
Spa Lux Condo/Hotel-Lobby Spa 豪华公寓／酒店 — 大厅		505 m²
Spa Lux Condo/Hotel-Jr.Suite(Studio) Spa 豪华公寓／酒店 —普通套间（工作室）	1	63 m²
Spa Lux Condo/Hotel-1 BR Suite Spa 豪华公寓／酒店 —1 BR 套间	3	268 m²
Spa Lux Condo/Hotel-2 BR Suite Spa 豪华公寓／酒店 —2 BR 套间	-	
Spa Lux Condo/Hotel-3 BR Suite Spa 豪华公寓／酒店 —3 BR 套间	-	
Spa Lux Condo/Hotel-Circulation Spa 豪华公寓／酒店 — 流通线路		138 m²
Residential-House Winter 冬季住宅		1014 m²
Residential-House Cruise 巡航住宅		699 m²
House Cruise-Circulation 巡航住宅 — 流通线路		111 m²
Residential-House Spring 春季住宅	8x2br	1111 m²
House Spring-Circulation 春季住宅 — 流通线路		219 m²
Residential-House Summer 夏季住宅	5x2br, 1x3br,	1132 m²
House Summer-Circulation 夏季住宅 — 流通线路		435 m²
Residential-House Fall 秋季住宅	4x2br, 1x3br,	1078 m²
House Fall-Circulation 秋季住宅 — 流通线路		133 m²
Residential-Royal Villa 皇家别墅		209 m²
Royal Villa-Circulation 皇家别墅 — 流通线路		48 m²
Retail 零售		2018 m²
Spa Spa		64 m²
Meeting/Banquet/Event 会议／宴会／活动		
F+B 餐饮		934 m²
Food Service 餐饮服务		886 m²
Residential and Hotel Circulation 住宅和酒店流通线路		2615 m²
Mechanical/Electrical 机械／电气		8173 m²
BOH 后勤区		2463 m²
TOTAL CIRCULATION AT THIS LEVEL 本层总流通线路		

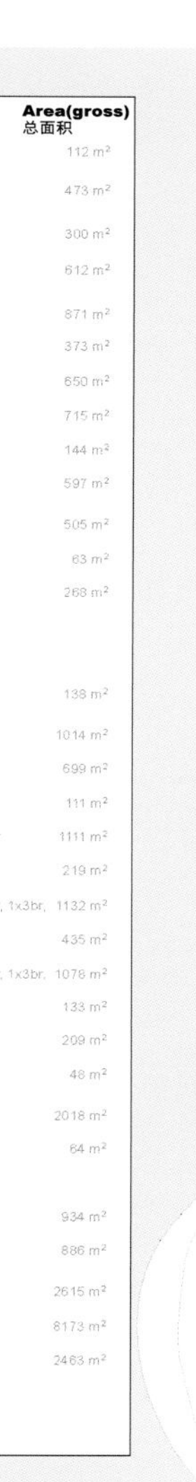

概念性建筑平面
CONCEPT ARCHITECTURAL PLANS

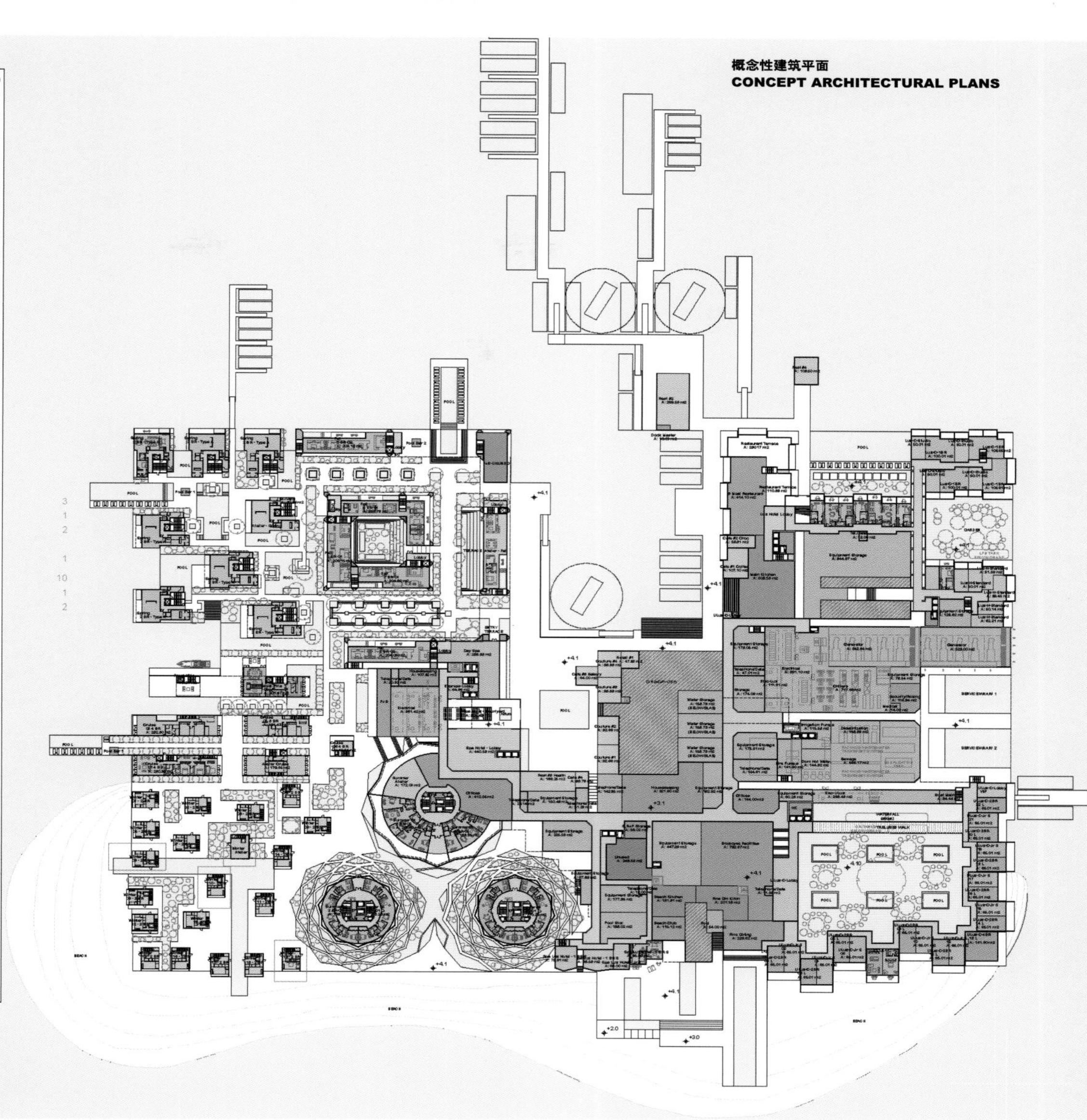

Profile

In collaboration with the world's most renowned fashion designer, Karl Lagerfeld and KOR Hotel Group, Isla Moda is a resort complex that comprises 3 hotels and 150 residential units on a manmade island 20 kilometers off the coast of Dubai.

Architectural Design

Inspired by the floating palaces of India, and the modern day equivalent—the cruise ship, Isla Moda distinguishes itself from other thematic reincarnations in "The World" by celebrating the notion of "manmade" and attempts authenticity within this simulacra of the real world.

The fabricated island evolves from an intrinsic understanding of the nuances of style and fashion, and their inextricable link to the possibilities of how life can and should be experienced. When woven into the lived environment, it reaches an apogee through a symphonic balance between architecture and an elevated natural environment. A monolithic volume—the physical embodiment of the Golden Rule—is intricately carved to establish maximum diversity of experience and typology.

SUSTAINABLE DESIGN OBJECTIVES
可持续设计目标

ECONOMIC 经济性

Reduce hauling & disposal costs
减少搬运和清理成本
Reduce annual energy and O&M costs
减少每年的能源消耗以及运行和维修成本
Increase productivity/reduce absenteeism
提高生产率 / 减少缺勤
Enhance asset value
提升资产价值

ENVIRONMENTAL 环保

Conserve limited resources (energy, water, materials)
节约有限的资源（能源、水、材料）
Reduce pollution (air, water)
减少污染（空气、水）
Preserve natural habitats
保护自然栖息地
Decrease emissions
减排

SOCIAL 社会效益

Improve occupant health
改善居住者的健康状况
Enhance occupant comfort
提升居住者的舒适感
Community benefits
社区福利
Boost public relations
促进公共关系
Improve guest experience
提升顾客体验

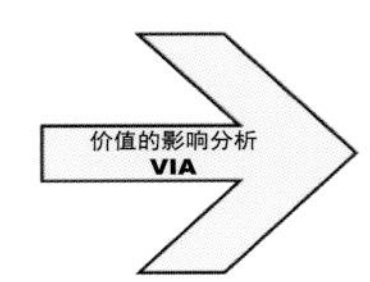
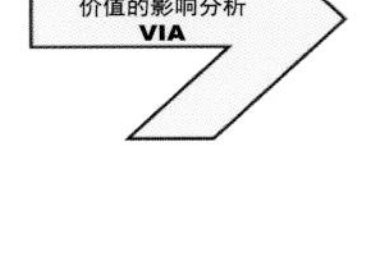

SUSTAINABLE DESIGN STRATEGIES
可持续设计策略

PASSIVE SYSTEMS 被动系统

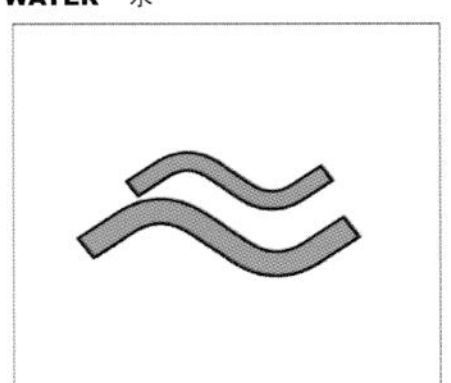

ENERGY 能量

WATER 水

WASTE 废弃物

TRANSPORTATION 交通

SEACOAST DESIGN 海岸设计

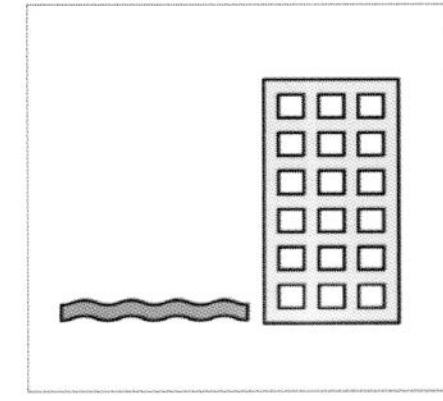

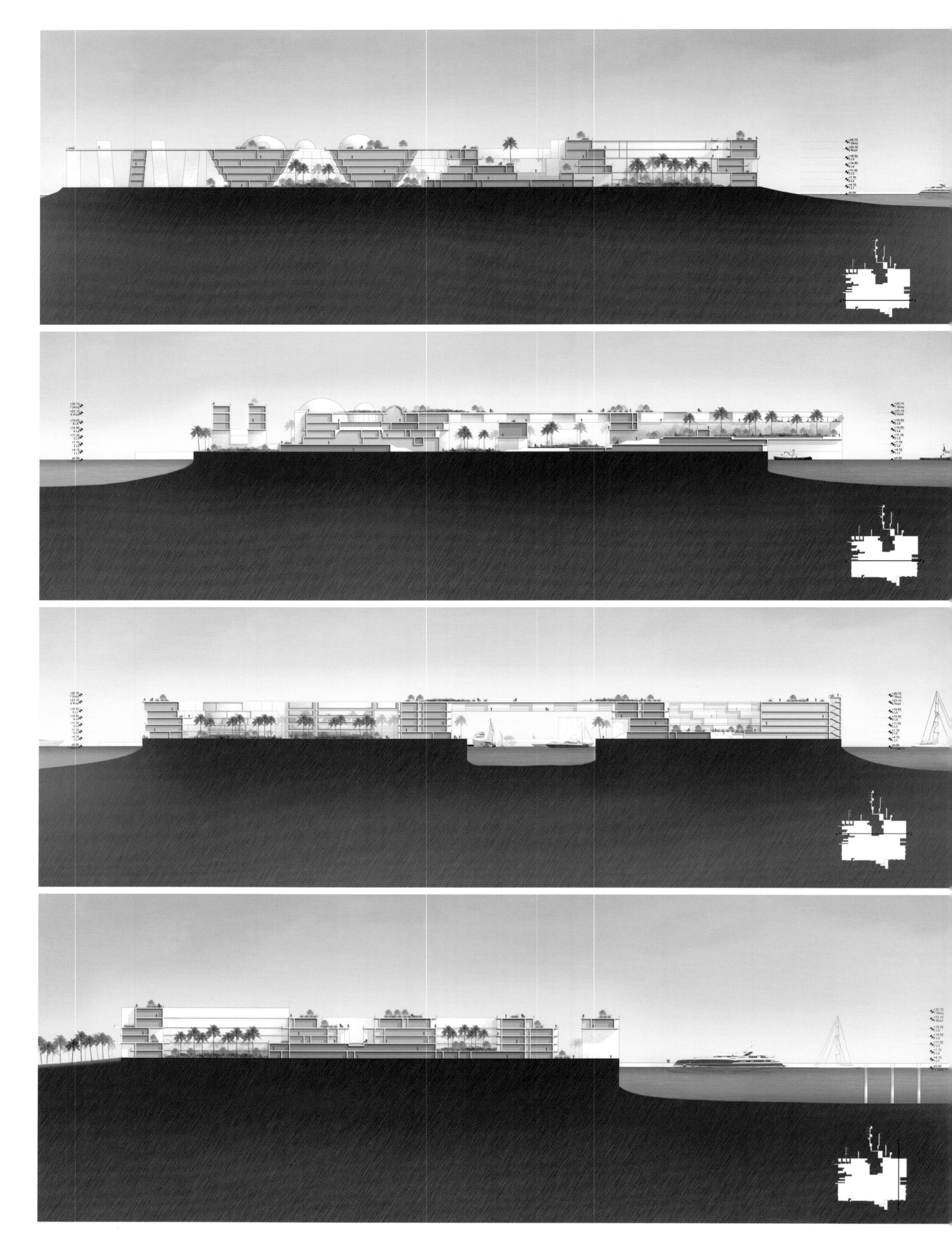

BIODEGRADABLE SOAPS:TO PERMIT GRAY WATER RE-USE WITH MINIMAL TREATMENT.
生物可降解肥皂：允许中水经最小化处理的情况下重新使用。

HIGH EFFICIENCY PLUMBING FIXTURES:FAUCETS,WATER CLOSETS,AND SHOWER HEADS WILL BE HIGH EFFICIENCY TO REDUCE WATER CONSUMPTION.
高效率浴间设备：高效率龙头、抽水马桶以及喷头能减少耗水量。

DUAL FLUSH TOILET:WATER CLOSET WILL HAVE TWO FLUSHING OPTIONS TO OPTIMIZE WATER CONSUMPTION BASED ON INTENDED USE.
双抽水马桶 抽水马桶将有两种冲水选择，根据预期用途最优化耗水量。

RECYCLED GLASS TILES:TILEWORK WITHIN WET AREAS WILL BE WITH 100% RECYCLED GLASS TILES.
回收玻璃瓦板：湿区瓦作将100%采用回收玻璃瓦板。

NON-VOC PAINT:PAINTS WILL CONTAIN NO VOLATILE ORGANIC COMPOUNDS THUS CONTRIBUTING TO A NON-TOXIC INTERIOR ENVIRONMENT.
非挥发性涂料：涂料将不含挥发性有机化合物，有助于形成无毒室内环境。

ORGANIC NATURAL FIBERS:RUGS, BEDDING,MATTRESS,AND PILLOWS WILL BE FABRICATED FROM ORGANIC NATURAL FIBERS SUCH AS COTTOM,HEMP,AND WOOL.
有机天然纤维：地毯、寝具、床垫及枕头采用有机天然纤维制造而成，例如棉花、大麻、羊毛。

LED LIGHT FIXTURES:LIGHT FIXTURES WITHIN ROOMS WILL USE THE LATEST ENERGY SAVING TECHNOLOGY.
LED灯具：室内灯具将采用最新节能工艺。

LOCAL ARTWORK:ART WITHIN ROOMS WILL BE FROM LOCAL ARTISTS AND ARTISANS USING RECYCLED MATERIAL.
当地艺术品：室内艺术品出自当地艺术家之手，且艺术品采用的是回收的材料。

GRAY WATER RE-USE:WATER FROM TOILETS AND FAUCETS WILL BE DIVERTED TO TREATMENT AND STORAGE TANKS TO BE RE-USED FOR IRRIGATION.
中水再利用：洗手间和龙头放出的水将进行处理并移送至贮水箱，用于灌溉。

FSC CERTIFIED WOOD: WOOD UTILIZED IN CASEWORK AND FURNITURE WILL BE CERTIFIED AS SUSTAINABLY HARVESTED BY THE FOREST STEWARDSHIP COUNCIL OR EQUIVALENT.
经森林管理委员会认证的木材使用：家具中采用的木材，将经过森林管理委员会或具有同等效力的机构的可持续性砍伐认证。

FSC

ENERGY STAR APPLIANCES: APPLIANCES WILL BE HIGHLY EFFICIENT AS CERTIFIED BY EPA'S ENERGY STAR LABELING SYSTEM.
节能电气：所有的电气都将具有高效率，需经过环境保护局能源之星标识系统的认证。

LOCAL STONE COUNTERTOP: COUNTER TOP WILL BE FABRICATED FROM INDIGENOUS STONE MATERIAL.
当地石材台面：案台将由当地石材制成。

RECYCLED WASTE:PAPER,PLASTIC, GLASS,AND ORGANIC WASTE WULL BE RECYCLED AND COMPOSTED.
废物回收：对纸、塑料、玻璃和有机废物进行回收和堆肥处理。

BIPV:GENERATES ELECTRICITY FROM SOUTHERN SUN EXSOURE.
光伏建筑一体化：利用南侧日照发电。

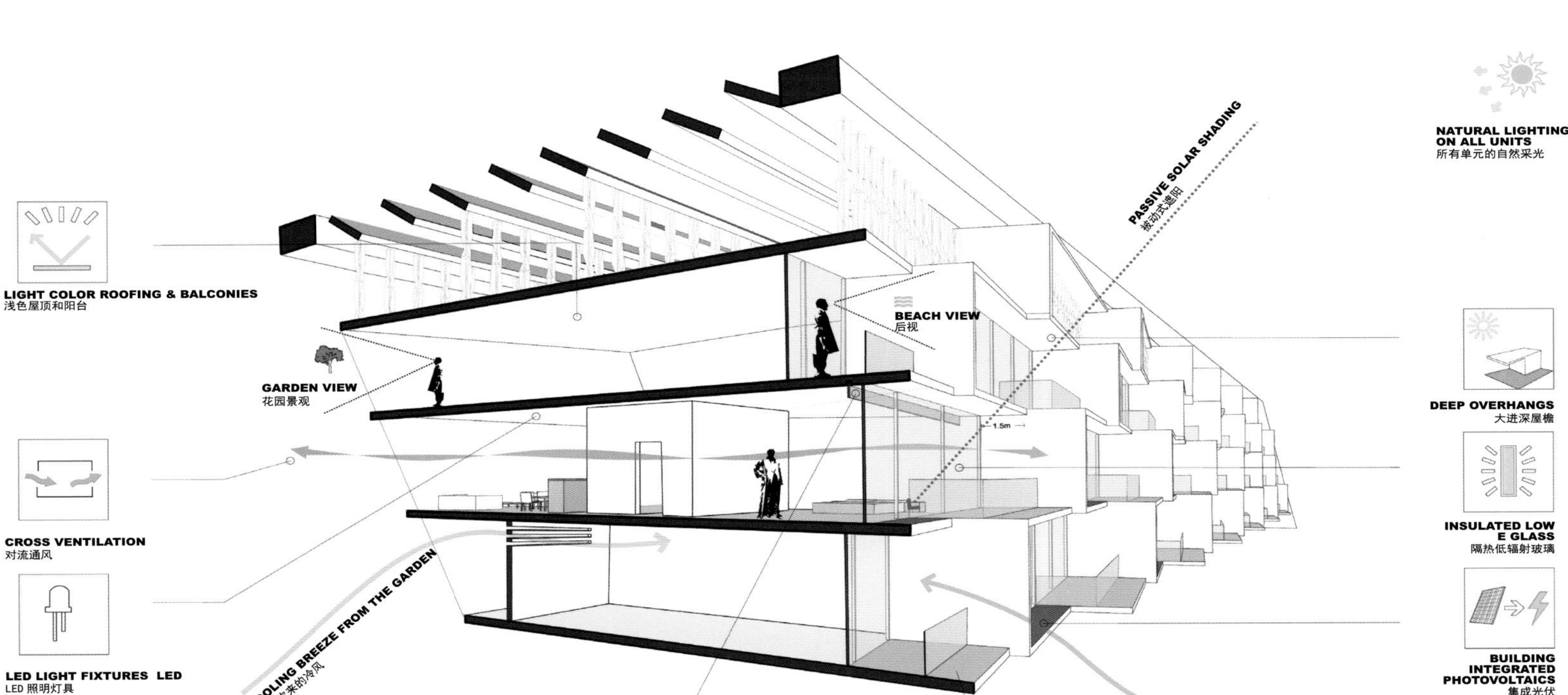

AUTOMATED SOLAR SHADES OCCUPANCY LIGHTING CONTROL
自动化遮阳及照明控制

XERISCAPE PLANTS(landscaping in ways that do not require supplemental irrigation)
节水型园艺植物（景观美化不需要额外的灌溉）

阿联酋阿布扎比女士俱乐部

Abu Dhabi Ladies Club

设计单位： Tony Owen Partners
合作单位： UPA Planning
项目地址： 阿联酋阿布扎比
项目面积： 30 000 ㎡

Designed by: Tony Owen Partners
Collaboration: UPA Planning
Location: Abu Dhabi, UAE
Area: 30,000 m^2

项目概况

Tony Owen Partners 和 UPA Planning 在阿联酋阿布扎比专门设计了一个女士俱乐部，这个独一无二的项目将提供保健、教育、儿童看护等方面的一系列服务，还将建设体育设施。

建筑设计

这个总面积为 30 000 平方米的建筑位于阿布扎比海滨，将会建设一个水上公园和大型的花园。俱乐部将包括有 1 000 个座位的大厅、会议厅、多功能厅、室内外体育馆、室内游泳池、水疗中心、妇女医疗设施等多种设施，可为该地区的女士提供文化、娱乐和教育等方面的设施以及小型的业务资源和培训服务。

设计特色

建筑由一系列形似翅膀的空间构成，这些空间围绕着中心带有顶棚的开放空间分布。建筑的外墙采用了电脑模拟的 L- 系统，弯曲的石墙上分布着树叶状的特色窗体图案。

建筑的一大特色是屋顶，屋顶的造型酷似大量交错弯曲的花瓣，其装饰图案由不透明和半透明的 PVC 材料构成，形成了明显的光影分区。

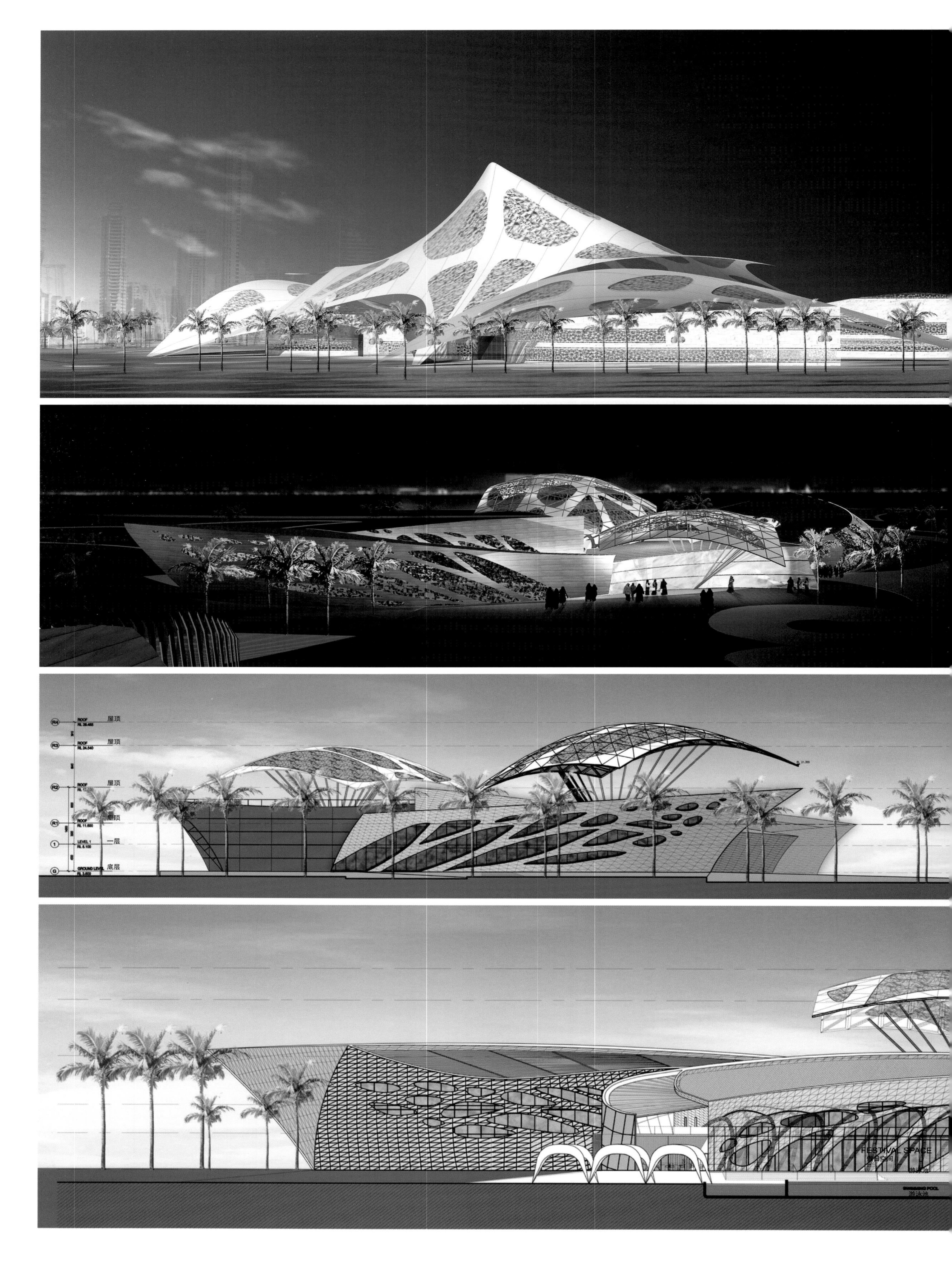

R4
ROOF 屋顶
RL 28.455
R3
ROOF 屋顶
RL 24.540
R2
ROOF 屋顶
R1
ROOF 屋顶
RL 11.600
1
LEVEL 1 一层
RL 8.100
G
GROUND LEVEL 底层
RL 3.600
FESTIVAL SPACE
游泳池
SWIMMING POOL
游泳池

Profile

Construction is now well advanced on the Abu Dhabi Ladies Club by Tont Owen Partners and UPA planning. This unique facility has been designed specifically for women in the UAE. It provides a range of health, education, child care and sporting facilities for women.

Architectural Design

The Ladies Club is a 30,000 square meters cultural facility located on the waterfront in Abu Dhabi city. The site will eventually contain a water park and extensive gardens. The Club will contain a 1,000 seat function hall, conference hall and accommodation, multi-purpose auditorium, indoor and outdoor sports facilities, indoor and outdoor swimming pools and a spa facility, women's health facilities, craft and education facilities and childrens creche and recreation areas. The facility will also provide small business resources and training for women in the UAE.

Design Feature

The complex consists of a series of wings arranged around a central covered festival space. The facades of the curved stone clad walls contain a distinctive window patterning. This pattern is based on the braching pattern of folliage and is derrived using computer generated "L-System" modelling. A feature of the design is a series of large fabric roof forms derived from the petals of a flower. These roofs are made from a unique pattern using two types of PVC fabric: a solid fabric and a transluscent lace fabric. This design allows for areas of shade and areas of natural light.

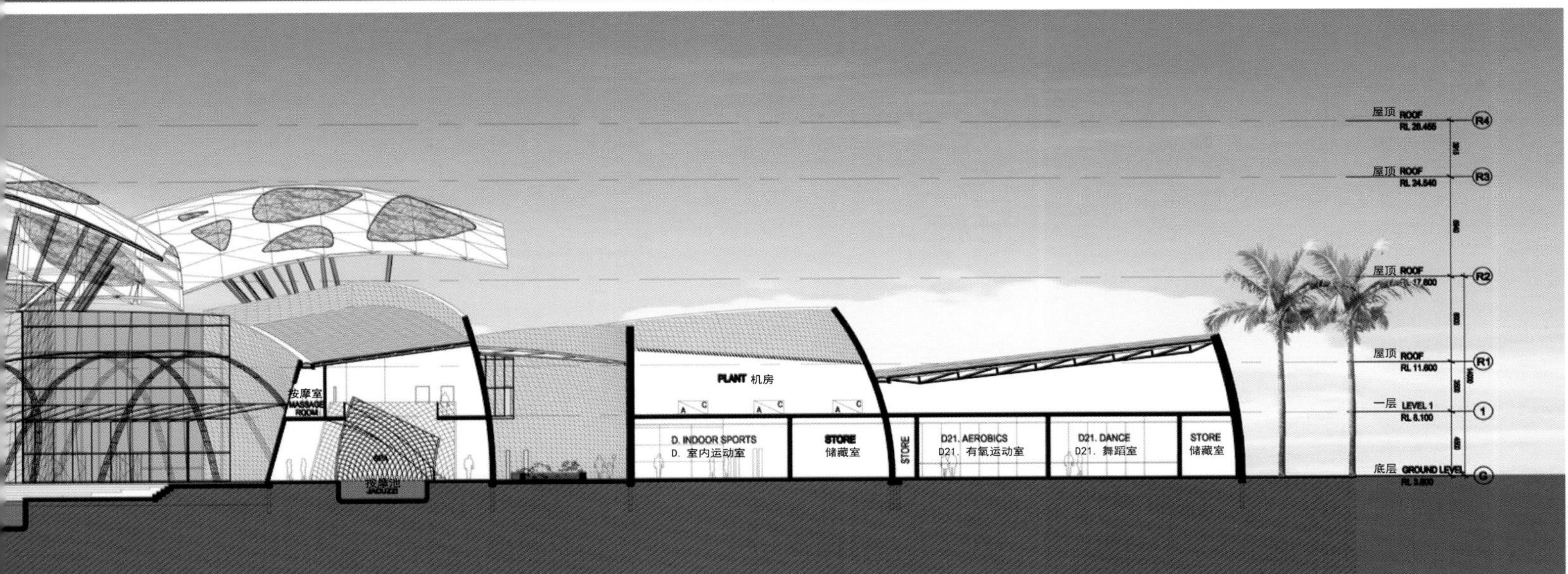

英国普罗维登西亚莱斯岛格雷斯湾“Caya”度假村

“Caya” by Grace Bay Resorts

设计单位：Oppenheim Architecture+Design
项目地址：英国普罗维登西亚莱斯岛
项目面积：17 187 m²

***Designed by:** Oppenheim Architecture + Design*
***Location:** Providenciales, Turks & Caicos Islands, UK*
***Area:** 17,187 m²*

项目概况

这个 17 187 平方米的度假村项目包括了多功能的酒店、餐厅、公寓、零售、海滩俱乐部等功能区间，它既展现了特克斯和凯科斯群岛的独有风情和精髓，同时也在普罗维登西亚莱斯岛为现代的、具有热带风情的度假村设计确立了新的标准。

建筑设计

“Caya”旨在在捕捉普罗维登西亚莱斯岛、特克斯和凯科斯群岛的自然风光的同时，在格雷斯湾设置高雅、独特的娱乐休闲设施。芳香的花园、宁静的水体景观、郁郁葱葱的热带植被，都成为这个简约的建筑形式的设计原型。

项目自然采光和通风条件良好，且行人流线流畅，设计通过对自然环境的利用，营造了轻松惬意的度假氛围。豪华度假村和私人酒店朝向格雷斯湾一览无遗的蓝绿色海面。两层的别墅可提供环绕着泳池庭院的私密住宅单元，有的住宅单元则带有私人花园。

海滩俱乐部可使游客欣赏到太阳从宁静的加勒比海上缓缓升起的美景，或于午后在有遮蔽的棕榈树下晒日光浴。置身在屋顶休息区，可欣赏特克斯和凯科斯群岛壮观的日落景观，感受微风拂面的舒适和惬意。

超越了传统的度假村给人的体验，“Caya”试图建立一个社区型的度假项目，同时确立这一度假村在普罗维登西亚莱斯岛的场所感。一个涵盖一系列的品牌商场和精品店的零售区将围绕公共广场设置，成为这个岛屿上一个现代的村落中心。

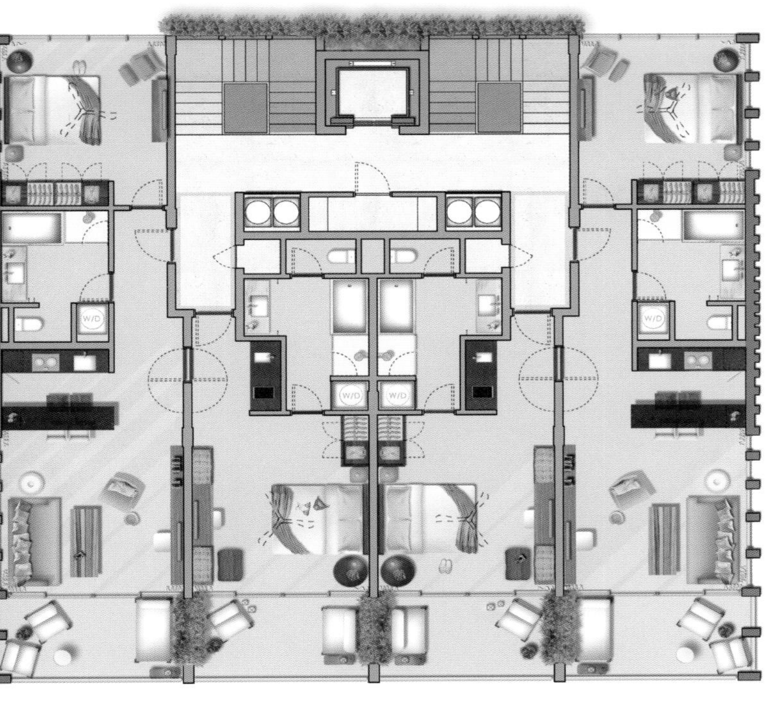

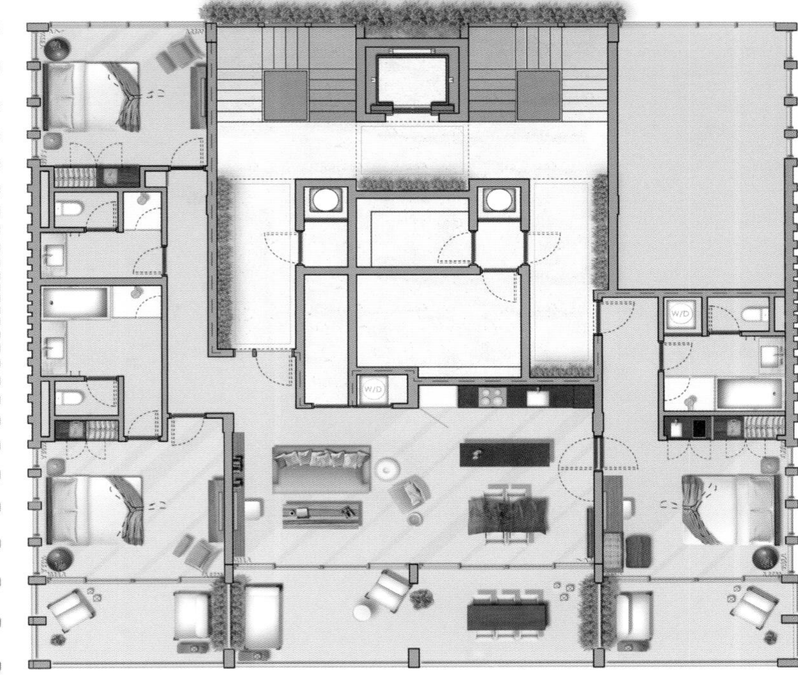

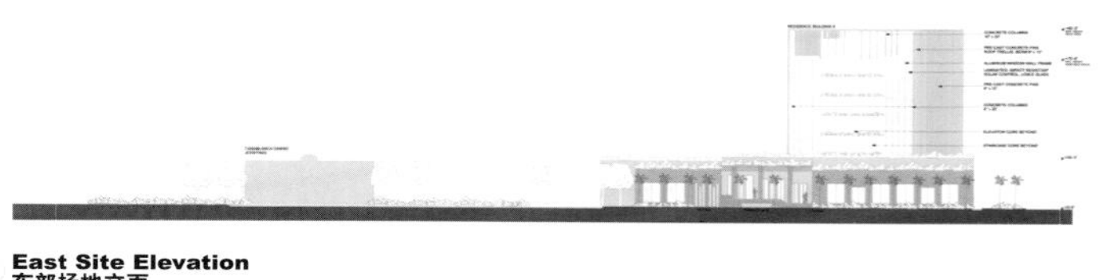

1 **East Site Elevation**
东部场地立面

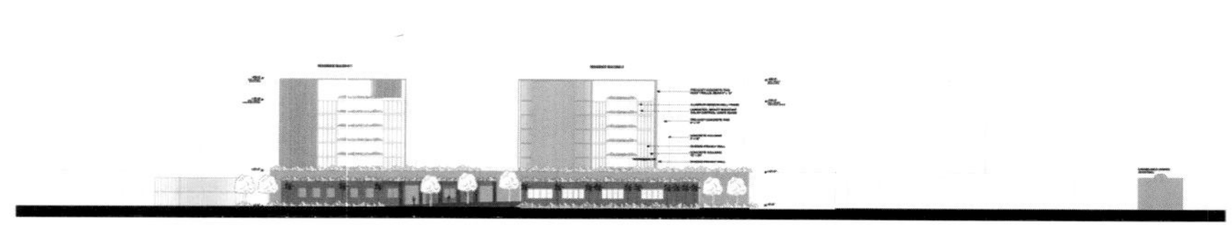

1 **South Site Elevation**
南部场地立面

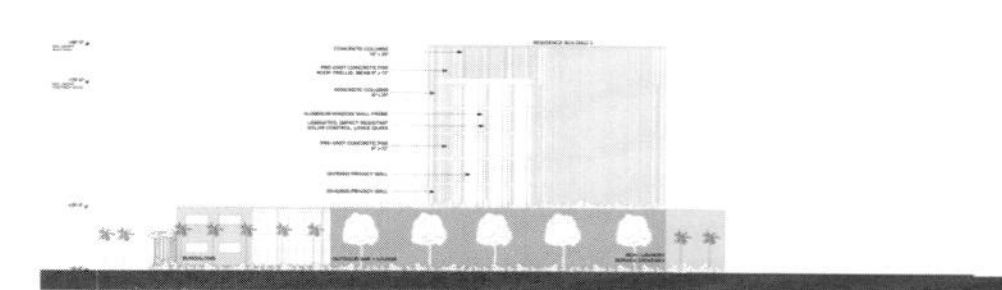

2 **West Site Elevation**
西部场地立面

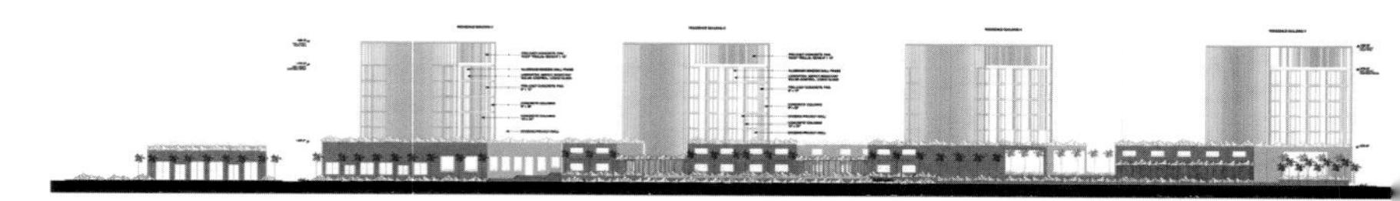

2 **North Site Elevation**
北部场地立面

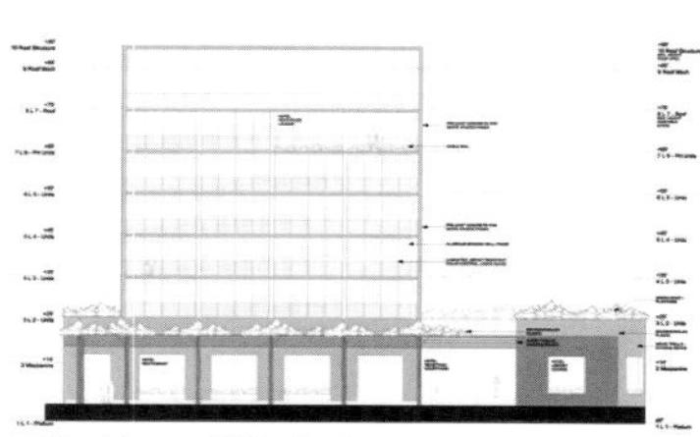

1 **Residence Bldg 1+Restaurant**
住宅建筑 1+ 餐厅

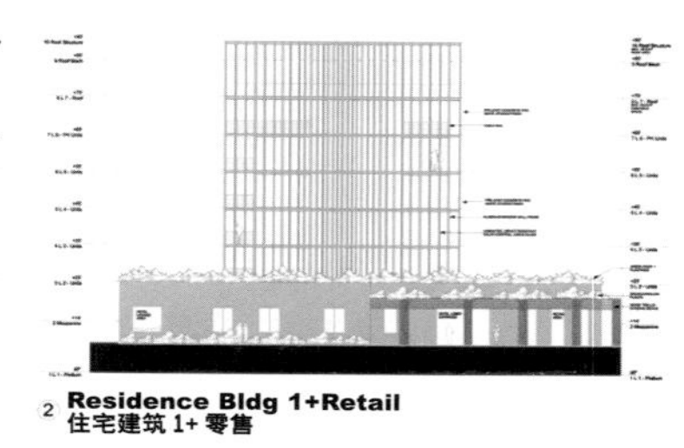

2 **Residence Bldg 1+Retail**
住宅建筑 1+ 零售

3 **Residence Bldg 1+Lobby/Retail**
住宅建筑 1+ 大厅 / 零售

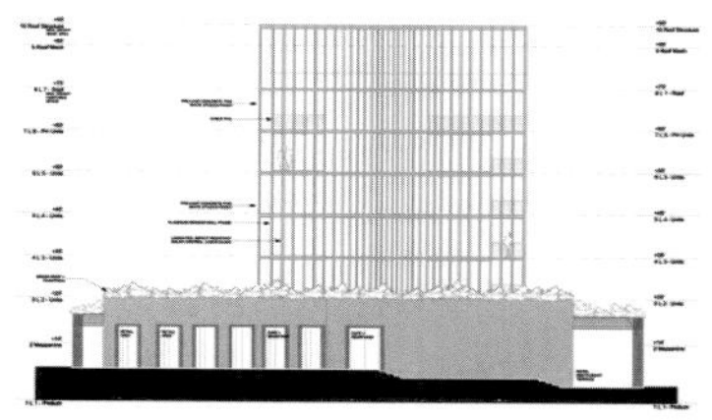

4 **Residence Bldg 1+Retail Area**
住宅建筑 1+ 零售区

Profile

The 17,187 square meters resorts consist of mixed-use hotel, restaurant, apartment, retail area and beach club. Caya portrays the pure essence of the Turks and Caicos while setting a new standard for modern, tropical design within Providenciales.

Architectural Design

Caya seeks to capture the natural beauty of Providenciales and the Turks and Caicos, while capitalizing on the casual elegance and exceptional service set forth by Grace Bay Resorts. Fragrant gardens, tranquil water features and vibrant tropical foliage guides the simplicity of the clean architectural forms.

The design itself is meant to be permeable to natural light, breezes and people, creating a relaxed resort atmosphere in tune with the natural environment. Luxury resort accommodations and private hotel residences are orientated towards sweeping views of the turquoise waters of Grace Bay. Two-story Bungalow units offer intimate accommodations around the main pool courtyards, some offering private gardens.

The Beach Club offers an opportunity to encounter the sun rising over the serene Caribbean Sea, or bathe in the afternoon sun under a gentle canopy of palm trees. Spectacular Turks and Caicos sunsets can be experienced and celebrated every evening from the rooftop lounge, capturing breezes from the trade winds.

Beyond the typical resort experience, Caya intends to create a sense of place and community in Providenciales. A retail component composed of luxury brands as well as boutiques will be organized around a public plaza, seeking to create a modern village center for the island.

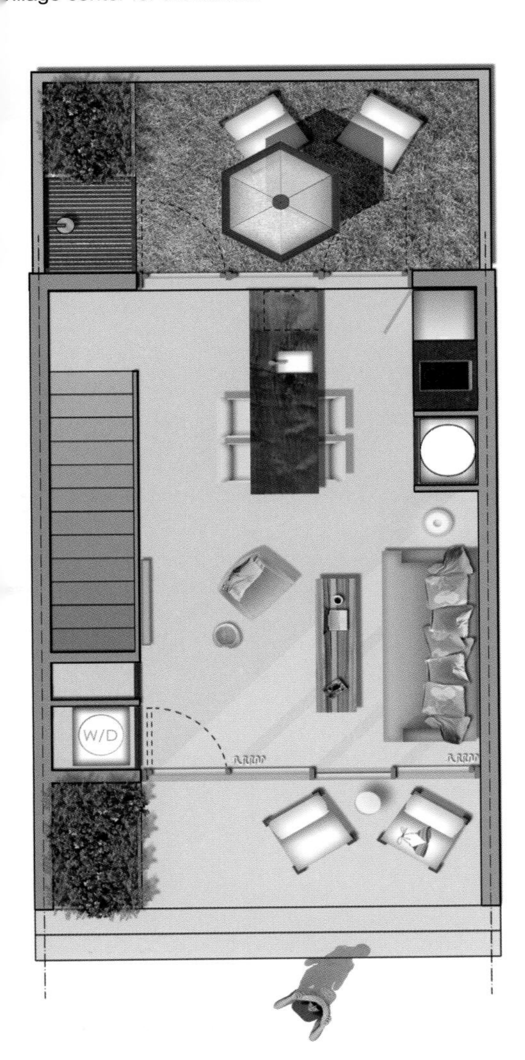

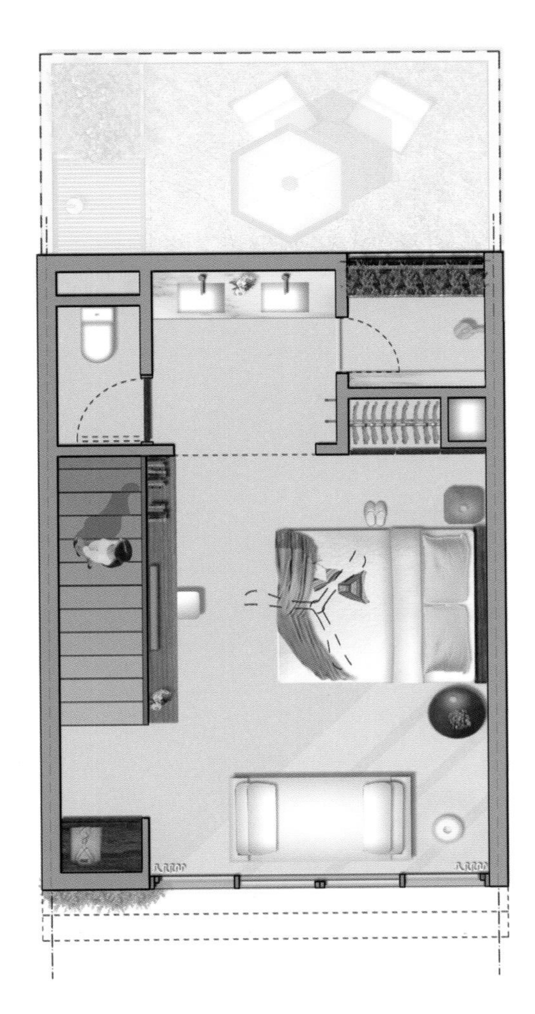

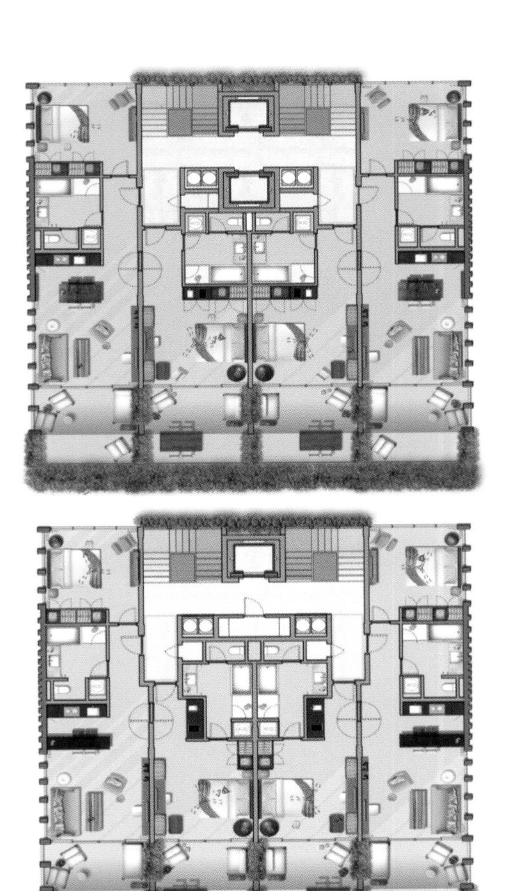

浙江杭州国际旅游中心

Shan-Shui Hangzhou

设计单位：斯蒂文·霍尔建筑师事务所
开发商：杭州旅游集团有限公司
项目地址：中国浙江省杭州市
用地面积：256 136 m²
建筑面积：275 000 m²

Designed by: Steven Holl Architects
Client: Hangzhou Tourism Group Co., Ltd.
Location: Hangzhou, Zhejiang, China
Site Area: 256,136 m²
Building Area: 275,000 m²

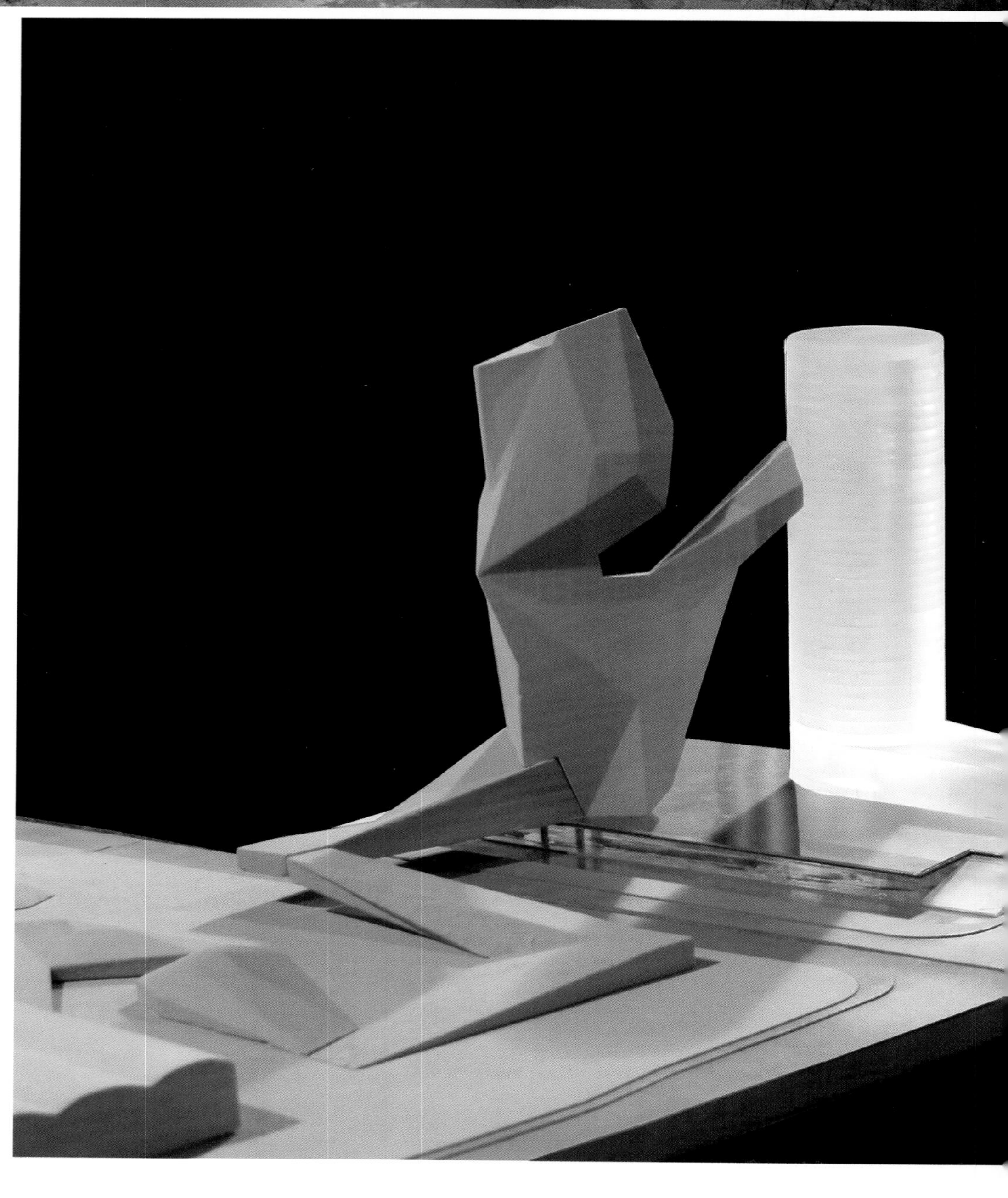

项目概况

杭氧、杭锅地块是杭州工业遗产规模最大、集聚度最强、含金量最高的区域，其品质和档次代表了杭州工业遗产的最高水平。斯蒂文·霍尔建筑师事务所借鉴中国传统绘画的技巧，以“山水杭州”为概念，从现有的历史风貌出发，选取并整合与之相似的元素，将之以城市的形态表现出来，赢得了这个杭州老工业基地改造和复兴项目的竞赛。

设计特色

整个场地的形状宛若蝴蝶领结，设计师将场地分为“水区”和“山区”两个部分。“水区”以公寓楼为主，圆形的塔楼从水池中升起，转折穿过主干道，向北分岔延伸成一个步行空间，形成“水之塔”的意象。“山区”将构建一个名为“3D 公园”的酒店，这是个低矮的、斜坡形结构的酒店，其屋顶将种植植物。与酒店相连的是一个现有的锅炉厂，将改建为一个运动场所。

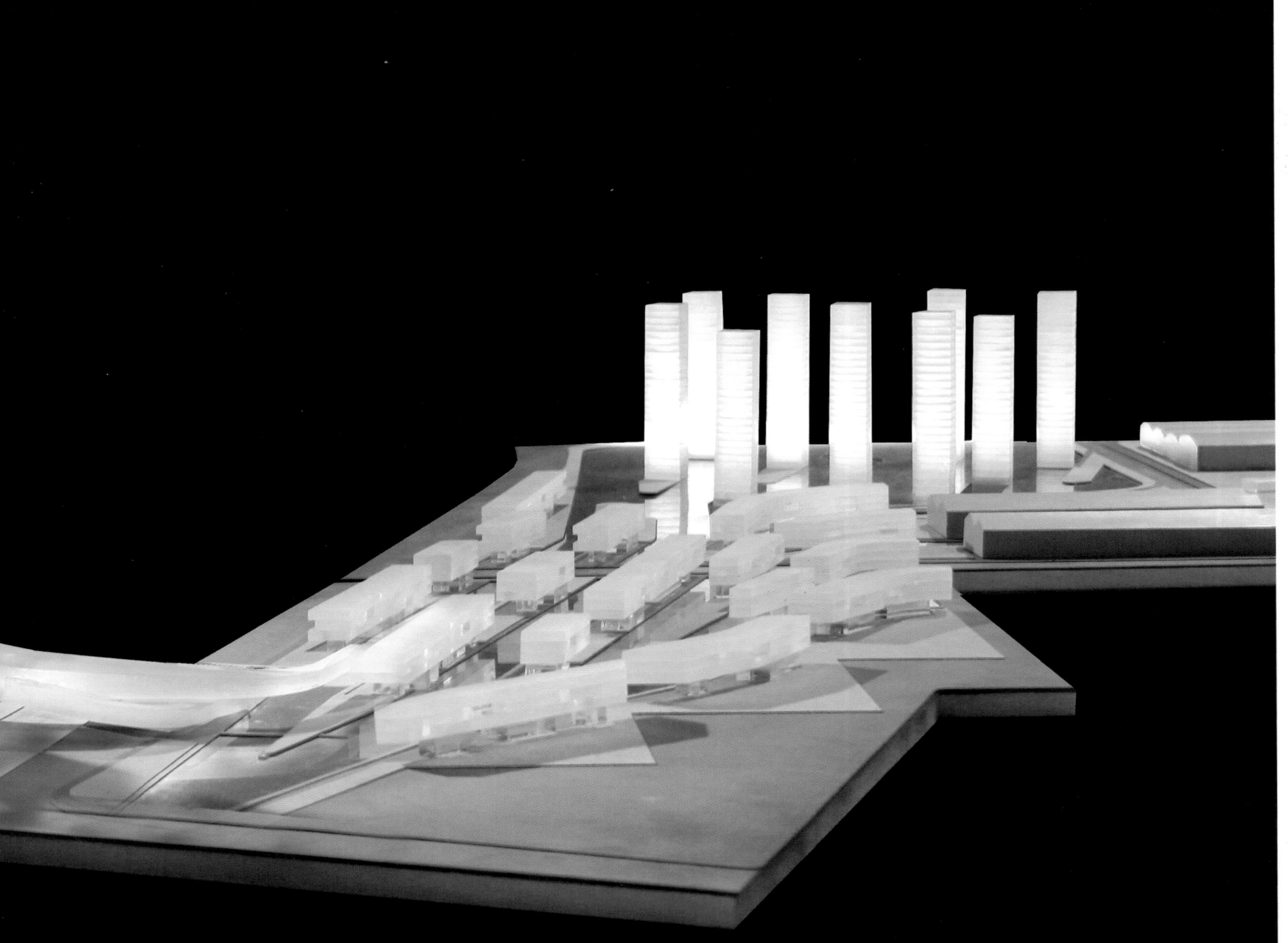

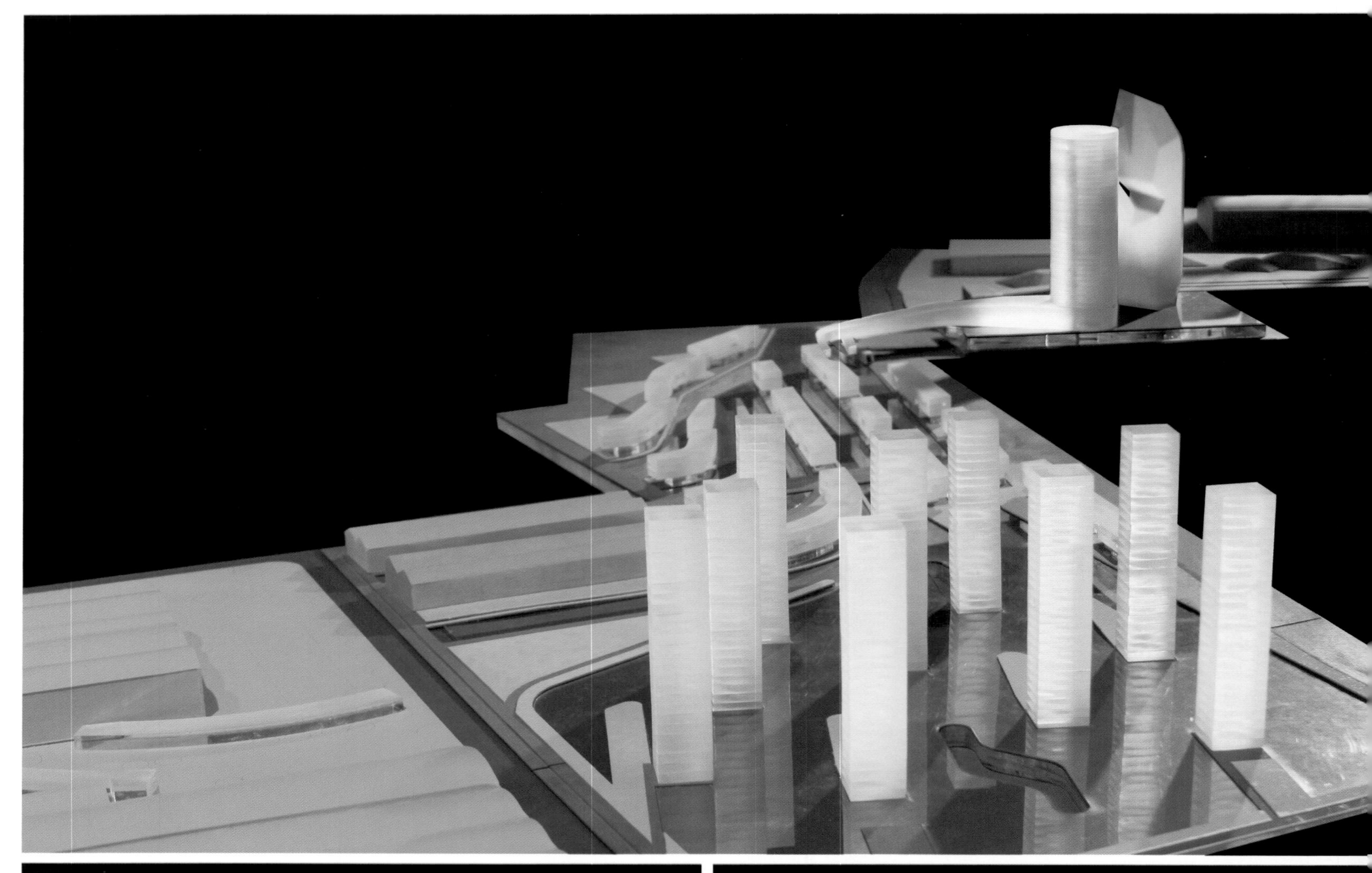

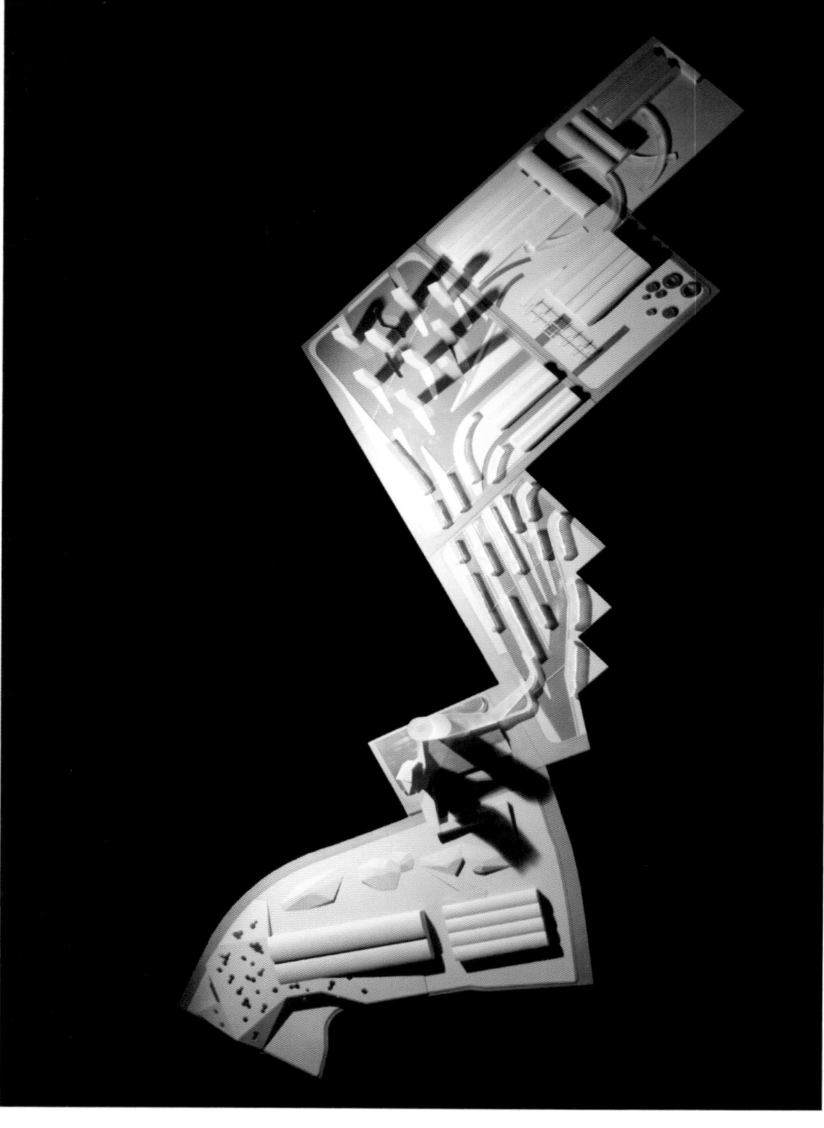

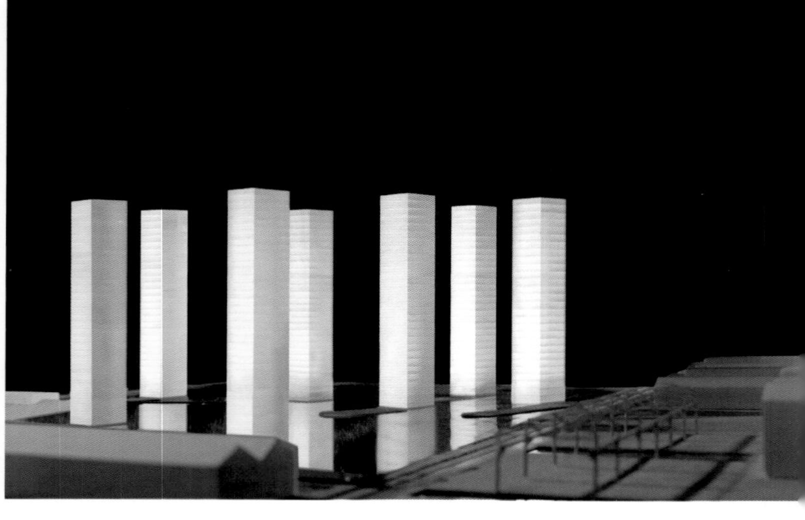

“水区”与“山区”的交会处设计有两栋高层建筑，两者通过中部的自动扶梯连接在一起，这也是该方案的独到之处。其一为圆柱形态的建筑，线条平滑流畅；另一栋棱角分明，仿若刀刻切削而成。这两栋形态迥然不同的建筑，一柔和似水，一刚硬如山，相得益彰。

生态设计

在对老厂房的改造上，设计师遵循了“低碳”的设计原则。设计师将锅炉厂一个巨型的复杂机器构建设计成楼梯入口、步道和电梯入口，破败的厂房则变身为摄影棚、马戏场、城市建筑博物馆等文化艺术空间。

整个项目的可再生能源利用率大于50%。所有建筑均采用土壤源热泵区域供热供冷系统，以经济和可持续的方式调节建筑温度，工厂屋顶安装太阳能光伏板，雨水和中水经处理后可用于园区内洗车、灌溉、道路喷洗、车库清洗等，这些措施的采用，极大地降低了建筑能耗。

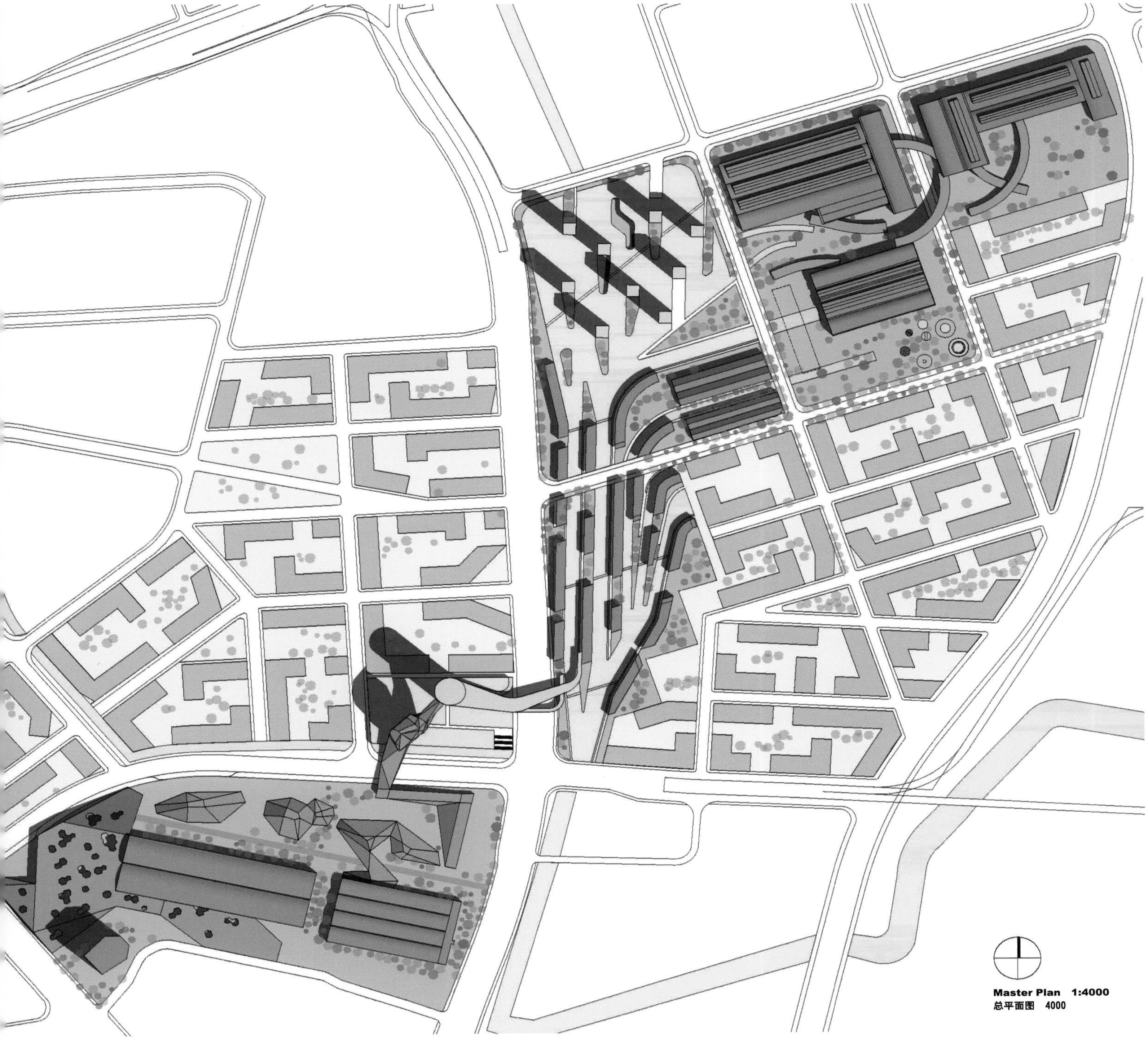

Master Plan 1:4000
总平面图 4000

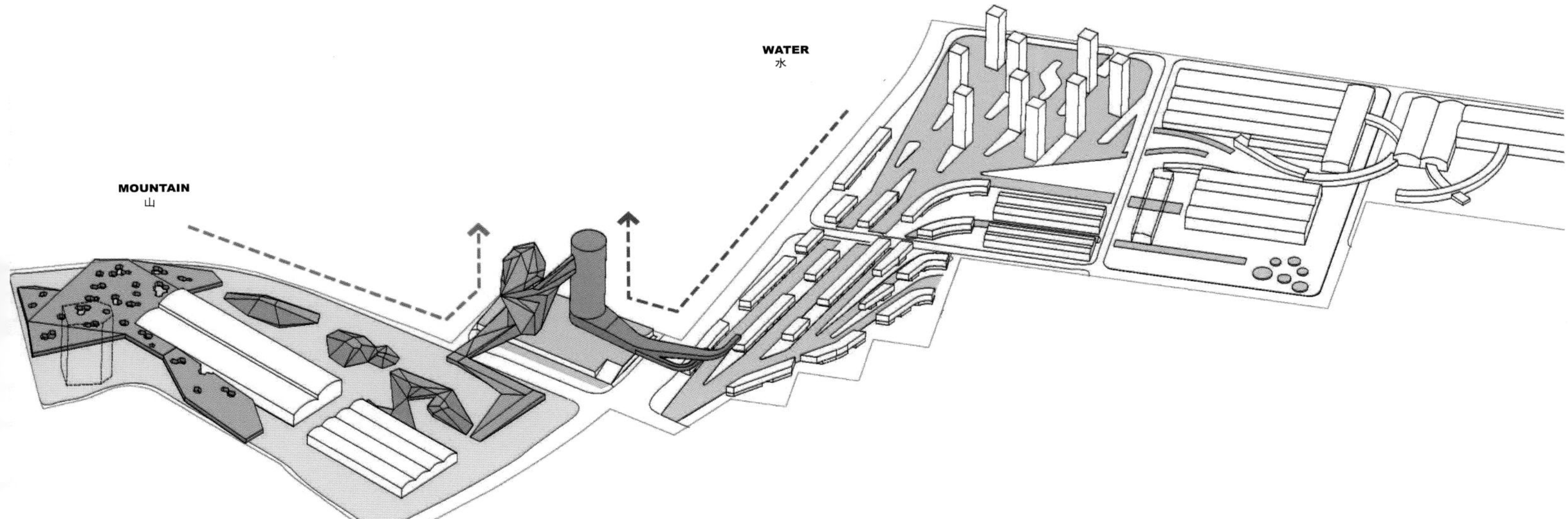

MOUNTAIN
山

WATER
水

Six large-scale elements which hover between landform and architecture

地形和建筑之间六个大型的元素

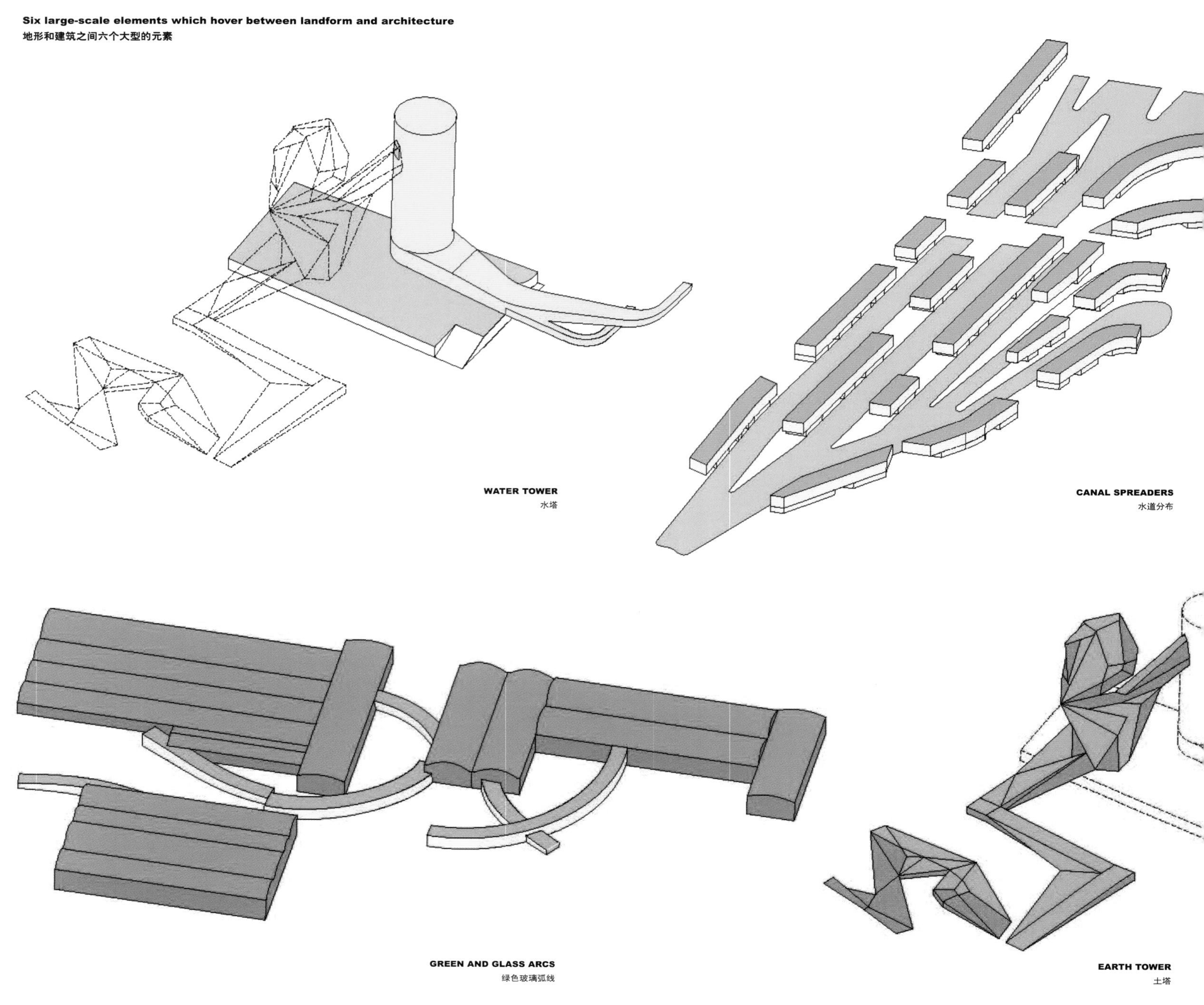

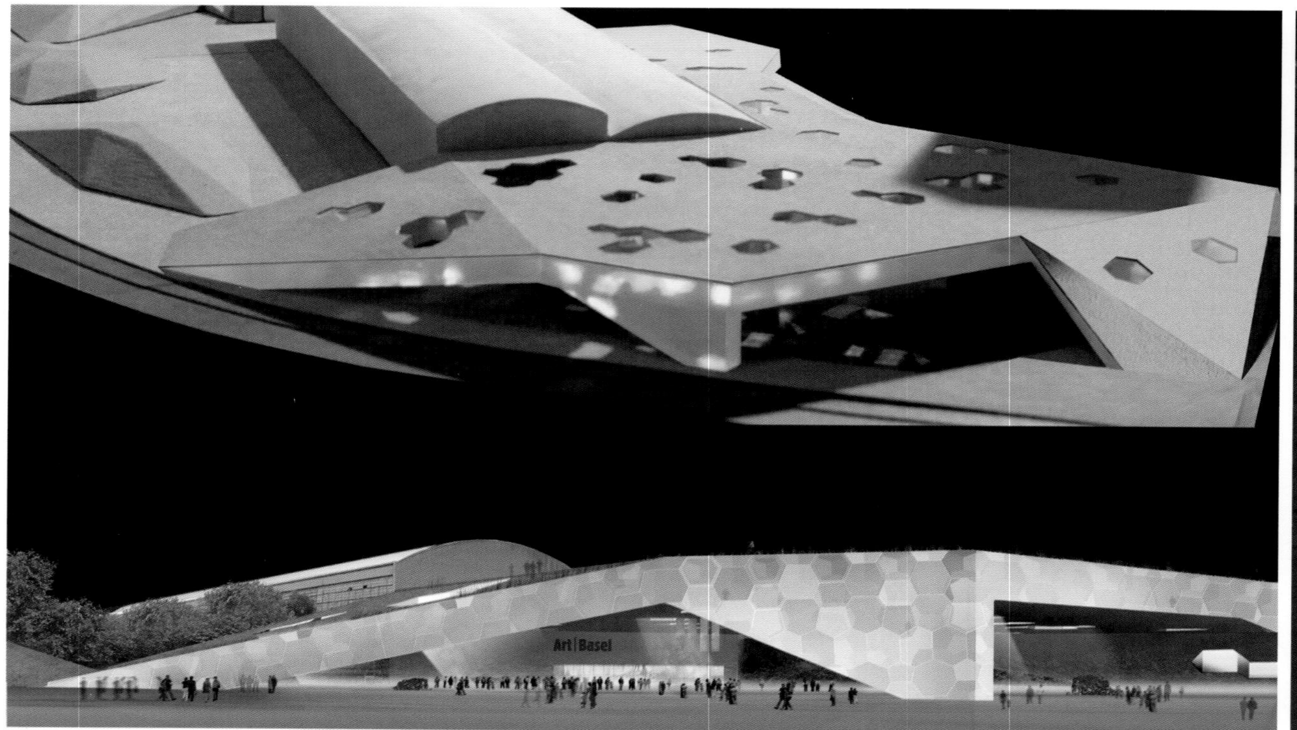

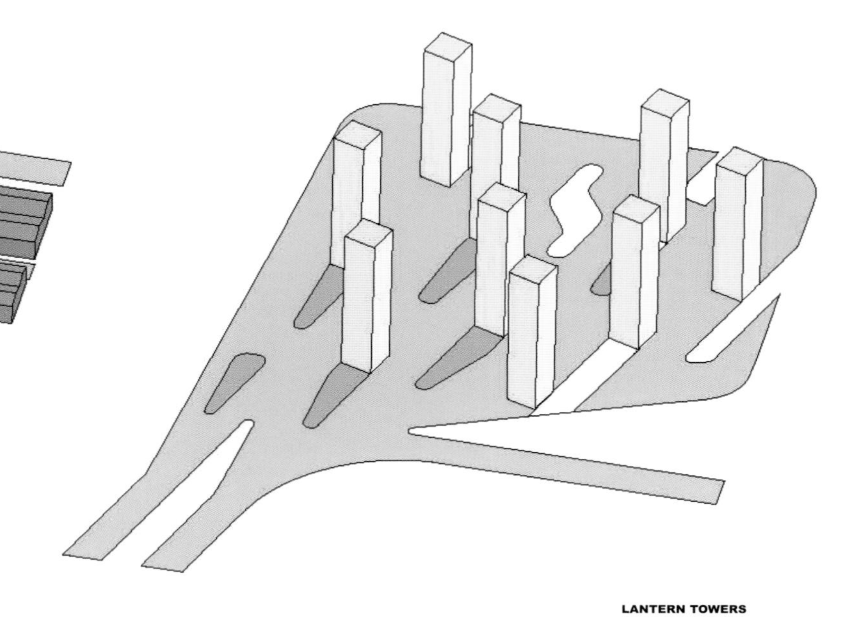

LANTERN TOWERS
灯塔

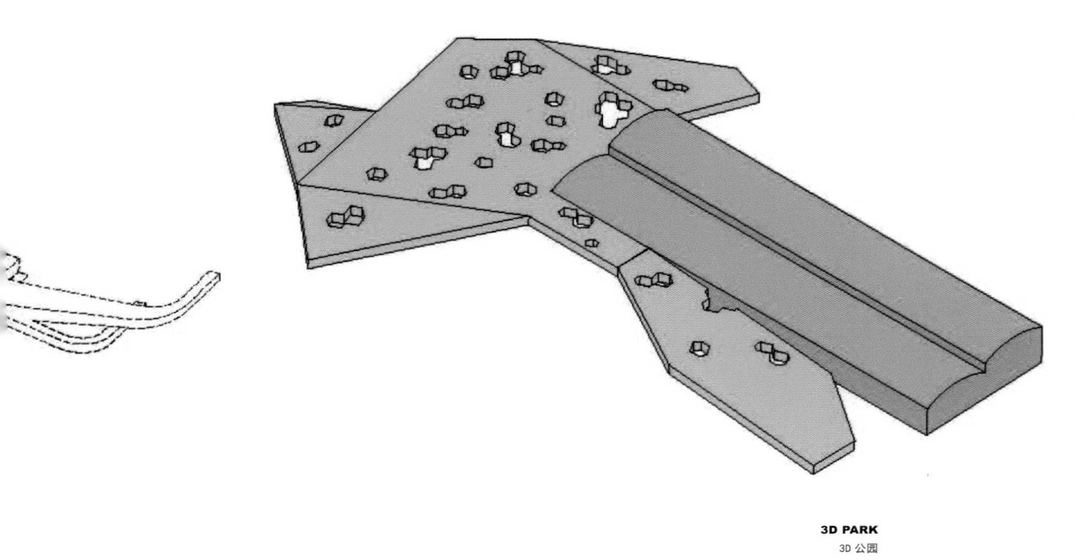

3D PARK
3D 公园

Profile

Oxygen sector and Boiler sector are the largest and densest heritage area in Hangzhou with the highest value. Its quality and level represent the top level of Hangzhou heritage. Steven Holl Architects refers to traditional drawing techniques of China which takes "Shan-Shui Hangzhou" as concept. Based on the existing historical features, they have selected and integrated similar historical elements to show city forms, which contributes to the bid winning of this old industrial base renovation and revitalization project.

Design Feature

Overall "Bow Tie" site is divided into "Water Area" and "Mountain Area". "Water Area" is mainly occupied by apartment buildings. A round tower rising from a water pond branches out to provide pedestrian circulation to the north, creating a "Water Tower" image. A "3D Park" hotel will be constructed in "Mountain Area". This is a low ramp-shaped hotel with vegetations planted on the roof. The existing boiler building connected with the Hotel will be transformed into a sports complex.

Two high rises at the intersection of "Water Area" and "Mountain Area" are connected to each other by an escalator in the middle. The cylindrical one has smooth lines while the other looks like cut and carved by sword. These two buildings, one is as gentle as water while the other as rigid as mountain, are perfectly complemented with each other.

Ecological Design

On the renovation of old plant, "Low Carbon" design principle is followed. A giant complex machine of boiler plant is utilized as entrance of staircase, entrance of passage and elevator. Old plants turn to photostudios, circuses, city architecture museum and other arts and cultural space.

Renewable energy applied occupies over 50%. Ground-source heat pump area in buildings acts as heating and cooling system, adjusting buildings temperature in an economical and sustainable way. Solar photovoltaic panels are installed on plants roof. Rainwater and recycled water after treatment can be used for car-washing, irrigation, road cleaning, garage cleaning etc.

澳门康科迪亚度假村

Concordia Macau Resort

设计单位： Moore Ruble Yudell Architects & Planners

合作单位： Andrew Lee King Fun & Associates Architects Limited

Archiplus International Limited

项目地址： 中国澳门

占地面积： 50 181 ㎡

摄影： Jim Simmons

Designed by: Moore Ruble Yudell Architects & Planners

Collaboration: Andrew Lee King Fun & Associates Architects Limited; Archiplus International Limited

Location: Macau, China

Site Area: 50,181 m^2

Photography: Jim Simmons

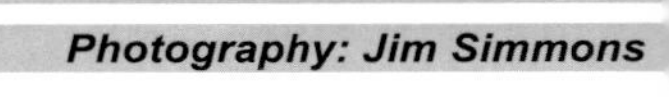

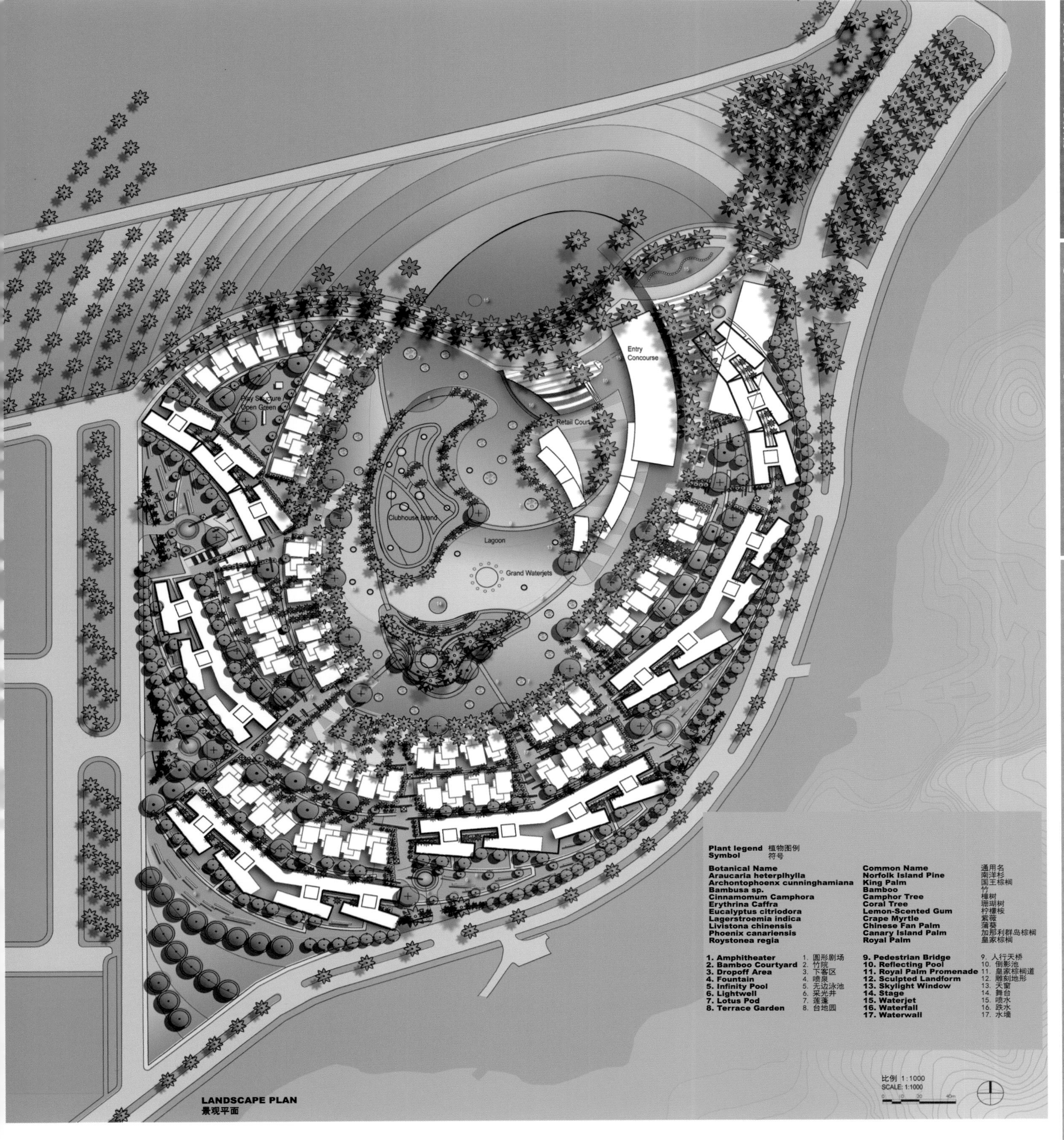

LANDSCAPE PLAN
景观平面

项目概况

澳门康科迪亚度假村为人们提供一个奢华的家庭住宅区的同时，还设置了酒店、零售店、休闲区等空间，以彰显其作为度假和娱乐胜地的独特身份。

建筑设计

在这个占地 50 181 平方米的空间中，公共场所、运动跑道、入口区的设置大大促进了居民之间的沟通和互动。中央潟湖是度假村的社交和娱乐中心，三个湖中浮岛为空间增添了无限幽远的意境，营造出浪漫、惬意的氛围，与度假村的特征相得益彰。中央潟湖被公共行人大道所环绕，并在景观、观景平台、桥梁和亭台楼阁的衬托下显得越发生动。

生态设计

设计始终贯穿了可持续发展和环保的理念，绿色草皮屋顶、集成光伏系统以及微型涡轮机发电系统将这一设计理念彰显得恰到好处。在节约用水方面，设计采用了生物过滤、废水管理、地面径流处理等全球性节水战略。郁郁葱葱的原生植物、棚架及檐篷则在热带气候环境下，为空间提供了有效的遮阳条件，并加强了自然通风。

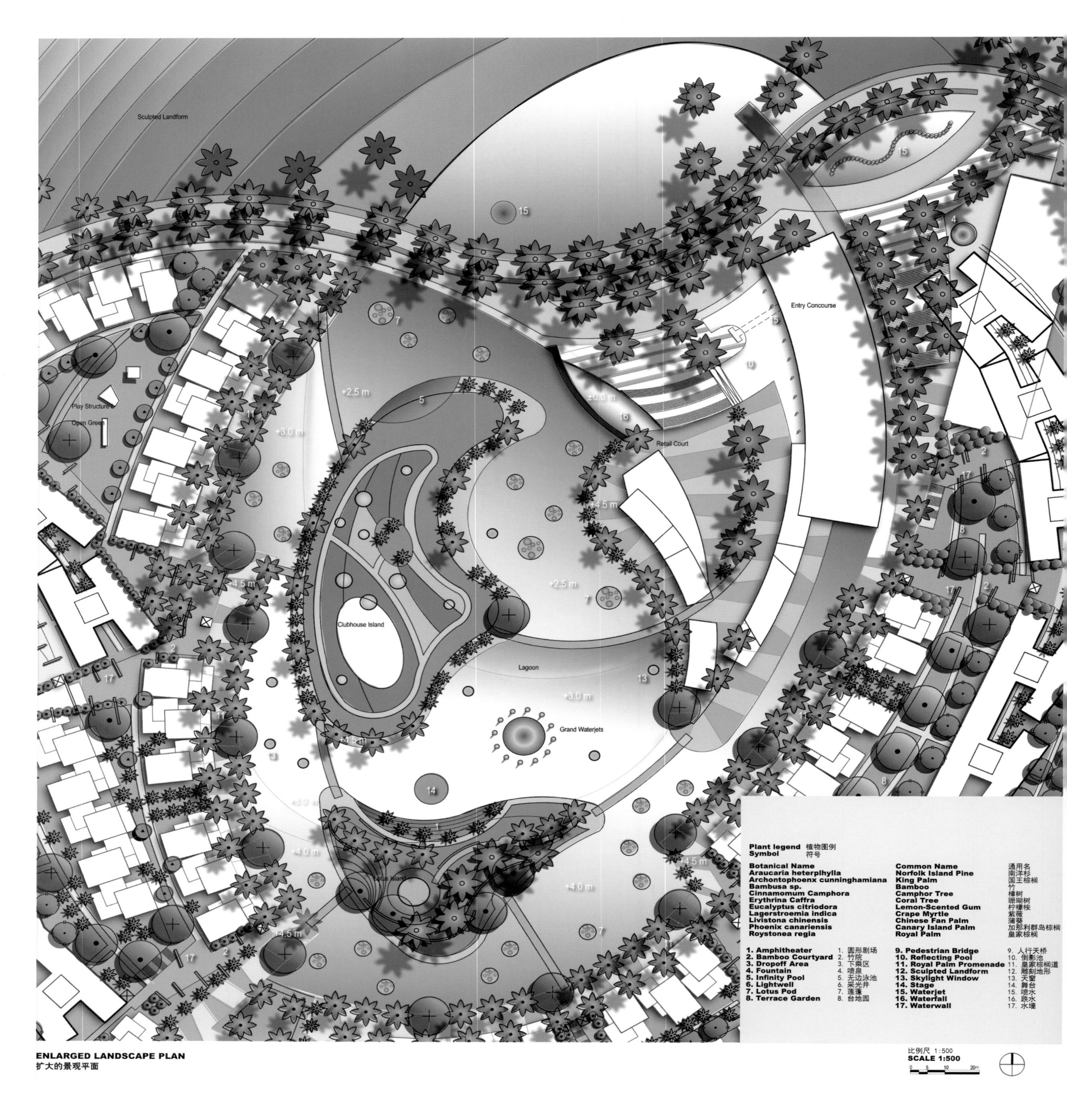

ENLARGED LANDSCAPE PLAN
扩大的景观平面

Profile

The Concordia Macau Resort creates a luxurious, family-oriented residential district while offering hotel, retail, and leisure facilities to celebrate the area's famed identity as a resort and entertainment destination.

Architectural Design

Public spaces, movement paths and residential entries work together on this spectacular 50,181 square meters site to encourage social interaction among residents. A Central Lagoon serves as the social and recreational heart of the project, with three floating "Islands in a Lake" which evoke a fantasy setting and enhance the project's resort atmosphere. A public pedestrian promenade encircles the Central Lagoon and is animated with landscaping, viewing terraces, bridges, and pavilions.

Ecological Design

Sustainable, ecologically-friendly features are optimized throughout the project, ranging from green sod roofs and integrated photovoltaic systems to micro-turbines that generate on-site power. Water conservation methods include bio-filtration, wastewater management, and a global strategy that utilizes waterfalls and grading to keep water in constant motion. Lush native plants, trellises and canopies provide shade, while screens facilitate natural ventilation in the hot and humid tropical climate.

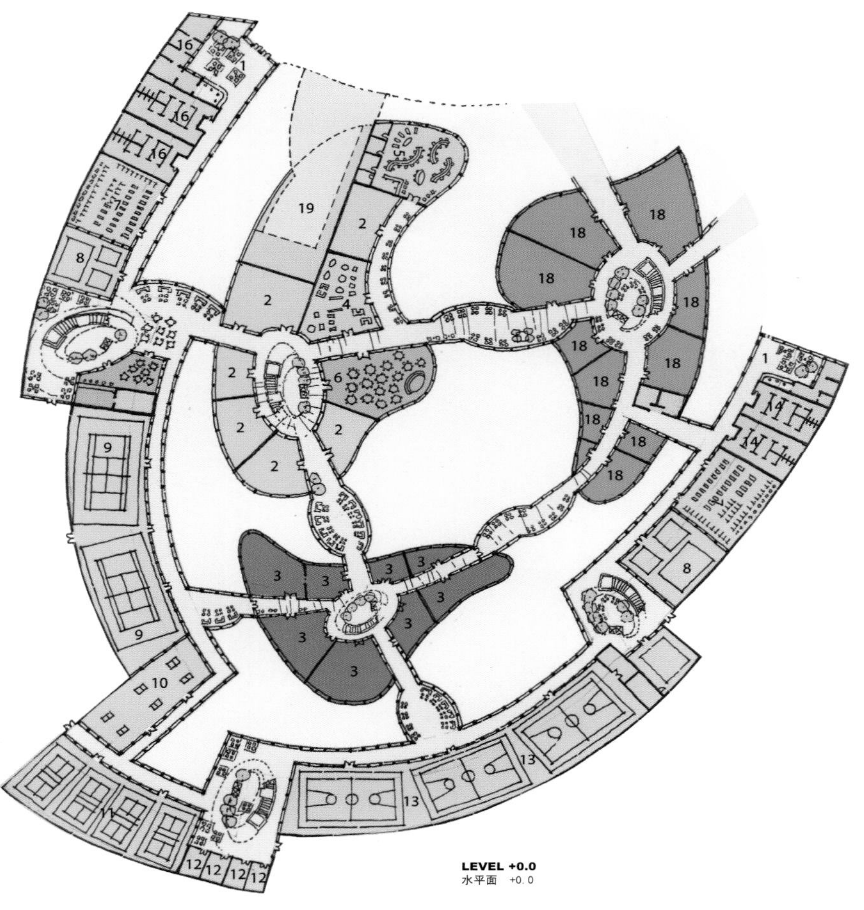

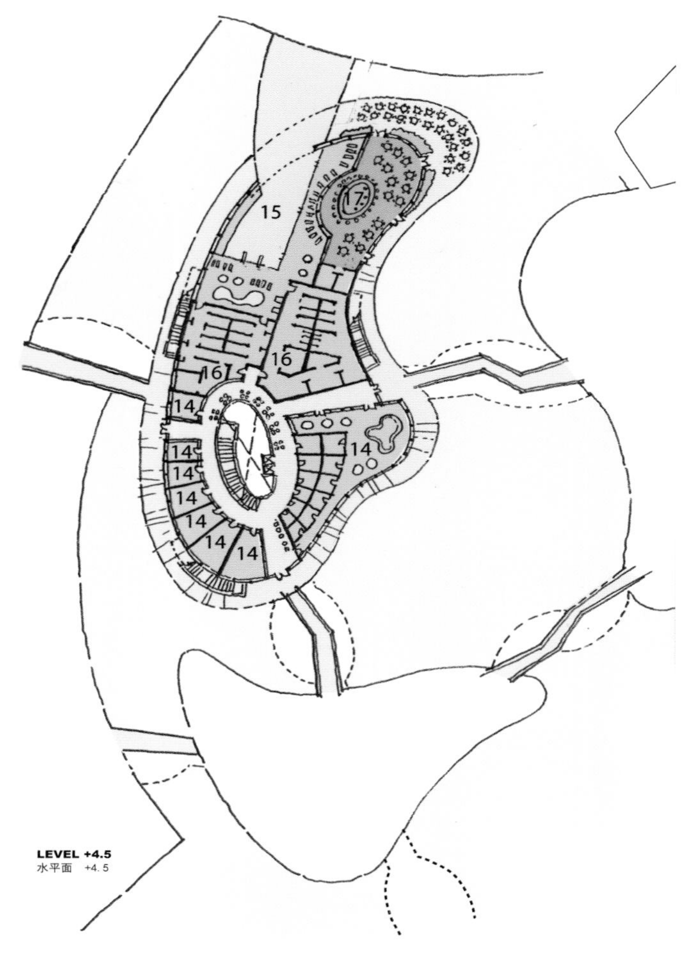

CLUBHOUSE PLAN 1/750
俱乐部平面图 1/750

1. COMUNITY CENTER	社区中心	**11. BADMINTON COURT**	羽毛球场
2. MULTIPURPOSE ROOM	多功能室	**12. RACQUET BALL COURT**	墙球厅
3. CHILDREN'S DISCOVERY CENTER	儿童探索中心	**13. BASKETBALL COURT**	篮球场
4. GAME/VIDEO ROOM	游戏／视听室	**14. SPA/HEALTH FITNESS**	Spa/ 健康中心
5. BEAUTY SALON	美容院	**15. INDOOR POOL**	室内游泳池
6. BANQUET ROOM	宴会厅	**16. LOCKER ROOM**	更衣室
7. WEGHTS ROOM	举重室	**17. CAFE/RESTAURANT**	咖啡厅／餐厅
8. GYM	健身房	**18. RETAIL**	零售
9. TENNIS COURT	网球场	**19. BOTTOM OF POOL&MECHANICAL**	池底 & 机械
10. TABLE TENNIS	乒乓球室		

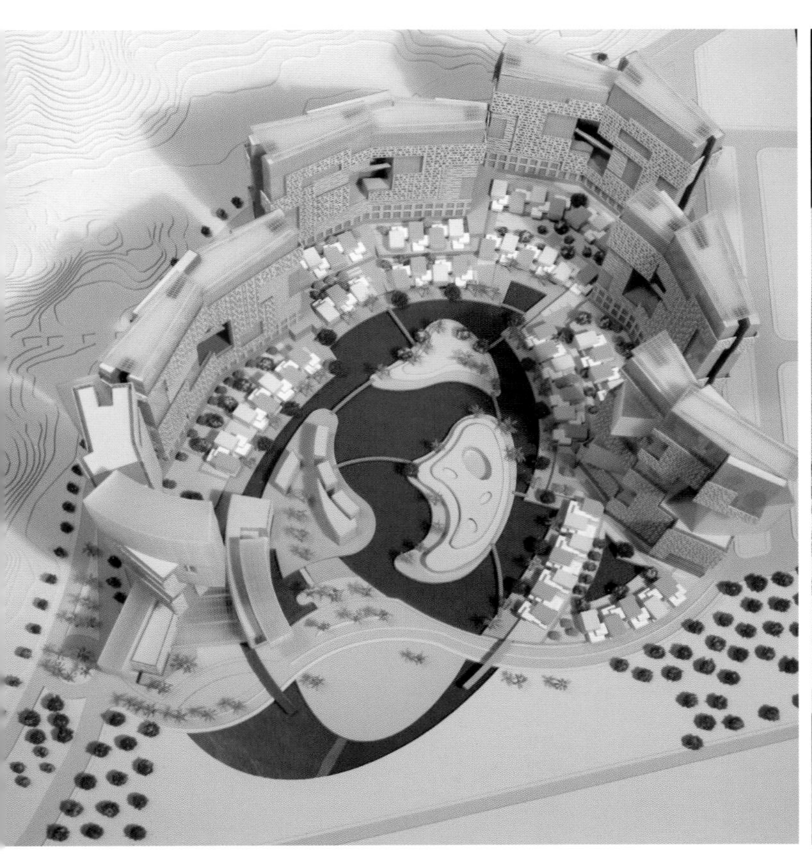

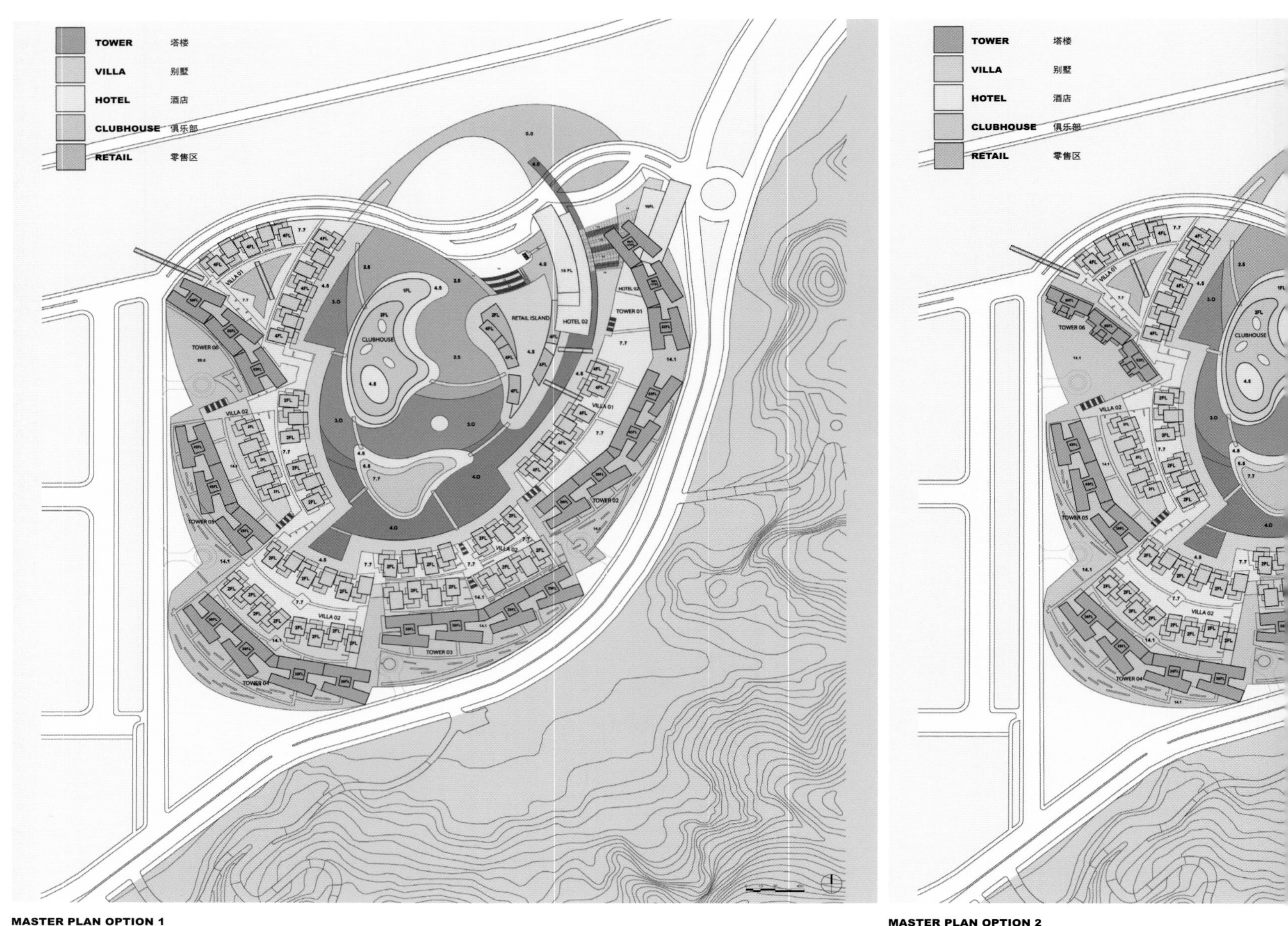

MASTER PLAN OPTION 1
总平面方案一

MASTER PLAN OPTION 2
总平面方案二

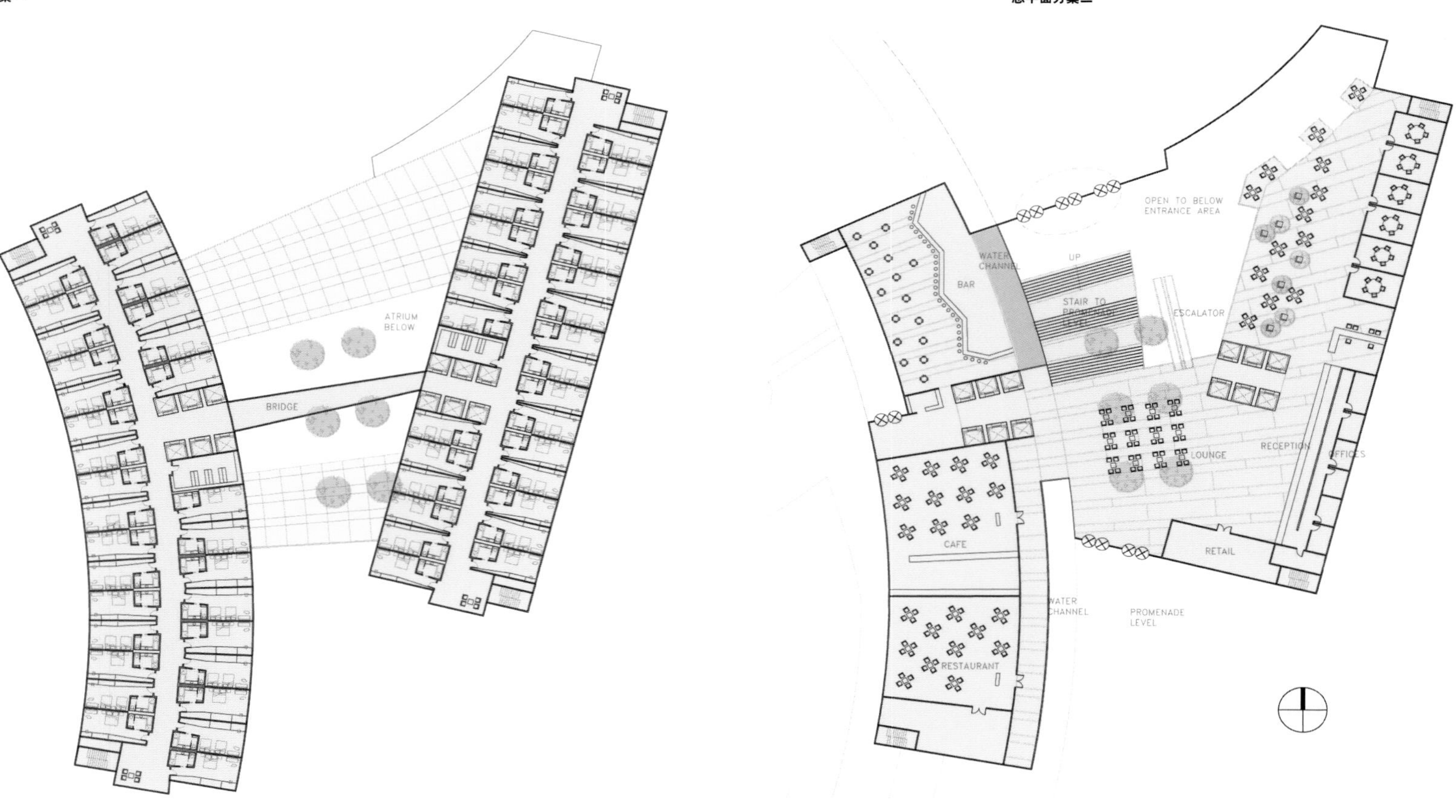

TYPICAL LEVEL PLAN
标准层平面

PROMENADE LEVEL PLAN
步道层平面

0 5 10 20m

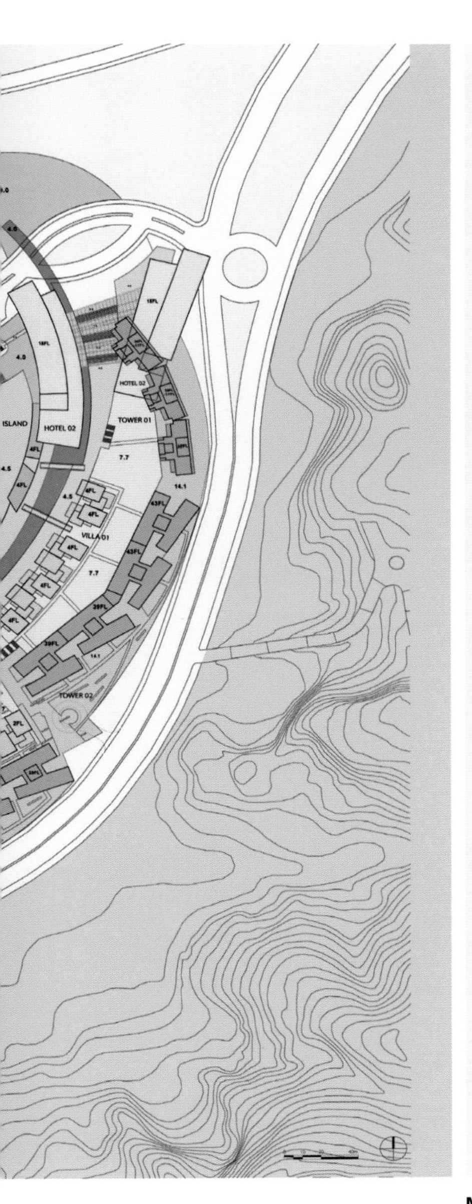

MASTER PLAN OPTION 3
总平面方案三

TYPICAL ROOM PLAN
典型房屋平面

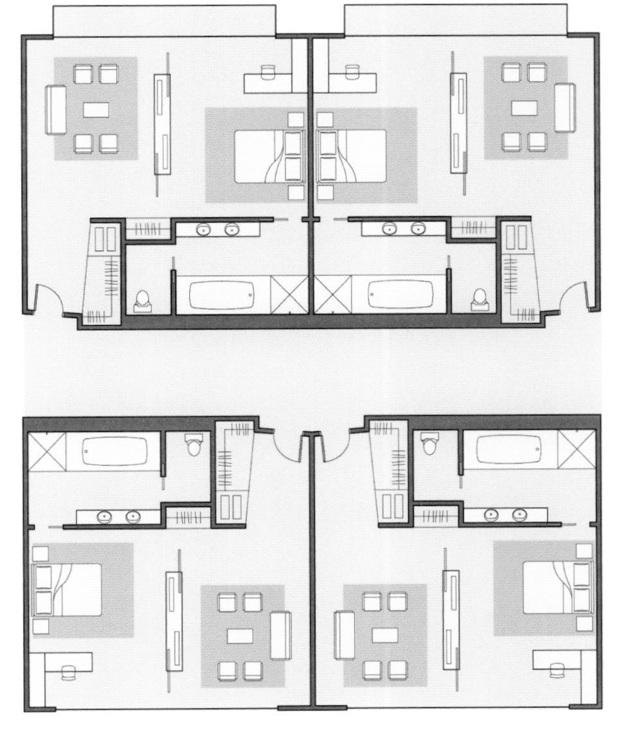

SUITE ROOM PLAN
套房平面

HOTEL PLANS
酒店平面

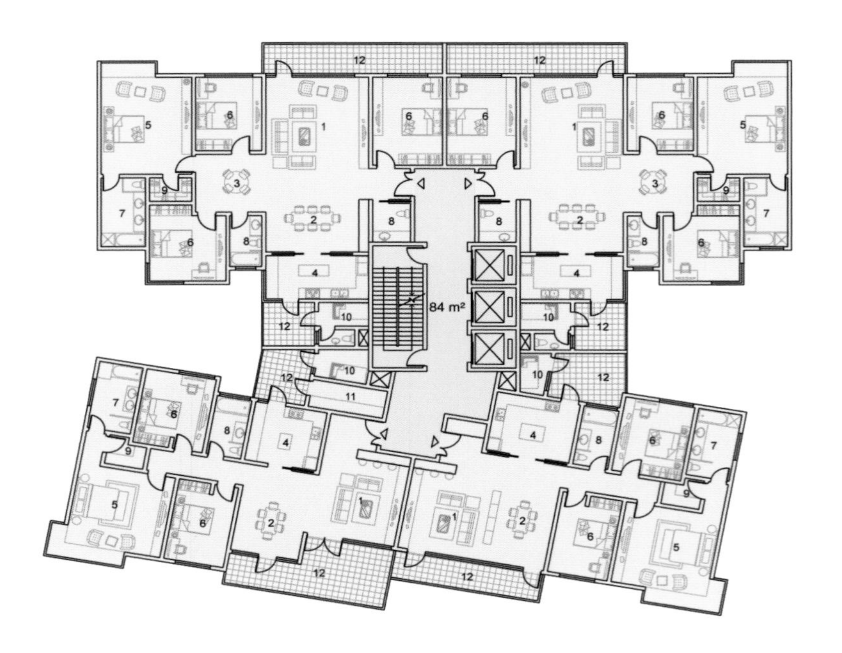

RESIDENTIAL TOWER 住宅大楼
TYPE1(4BR+3BRLX) 类型一（4BR+3BRLX）

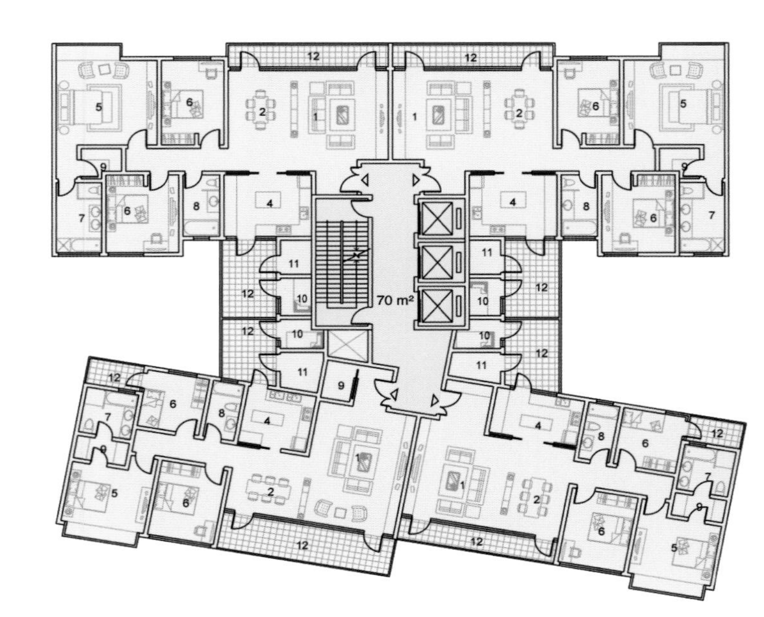

RESIDENTIAL TOWER 住宅大楼
TYPE2(3BRLX+3BR) 类型二（3BRLX+3BR）

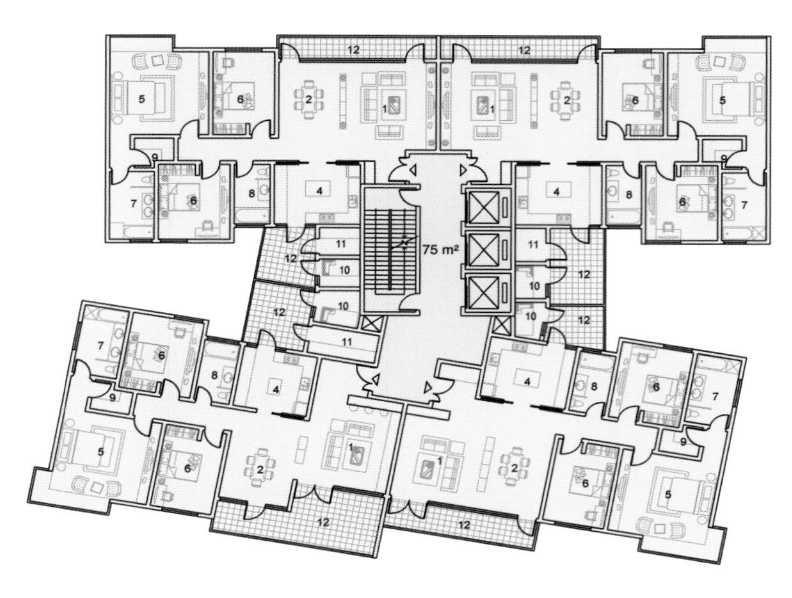

RESIDENTIAL TOWER 住宅大楼
TYPE3(3BRLX+3BRLX) 类型三（3BRLX+3BRLX）

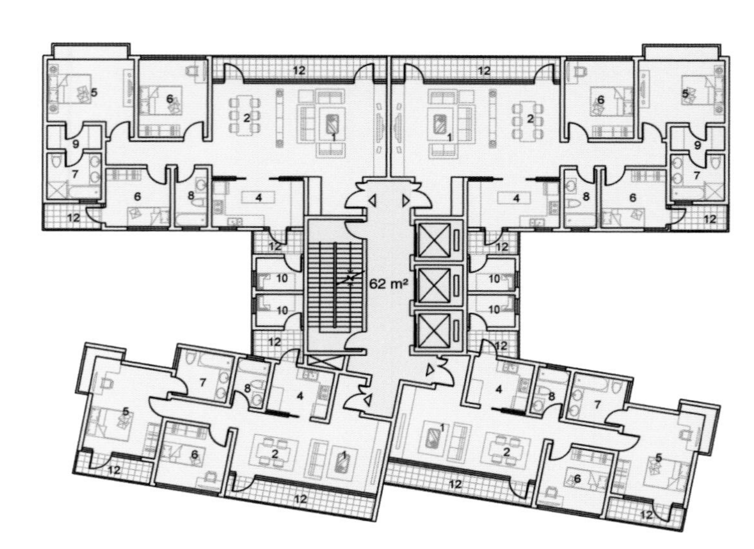

RESIDENTIAL TOWER 住宅大楼
TYPE4(3BR+2BR) 类型四（3BR+2BR）

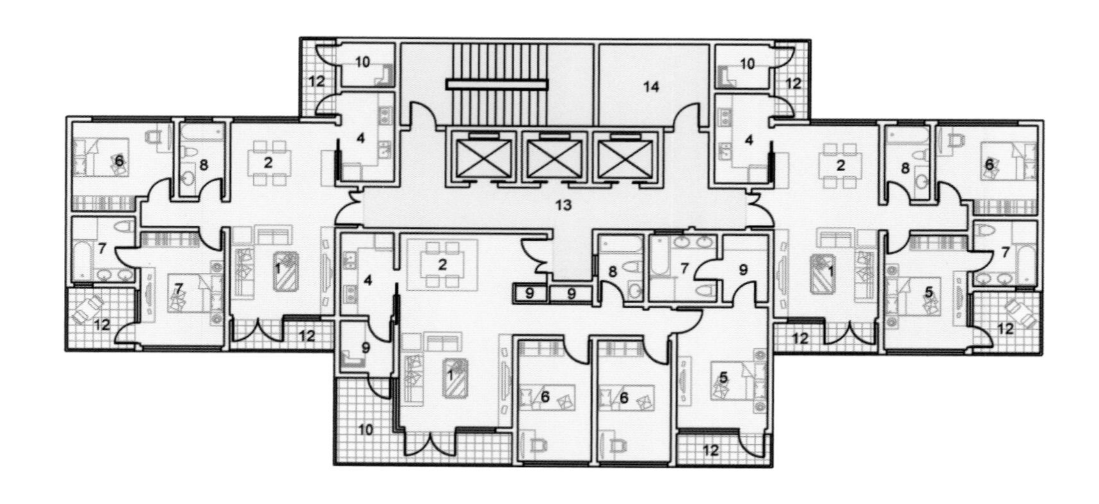

TOWER PLAN 1/200
大楼平面 1/200

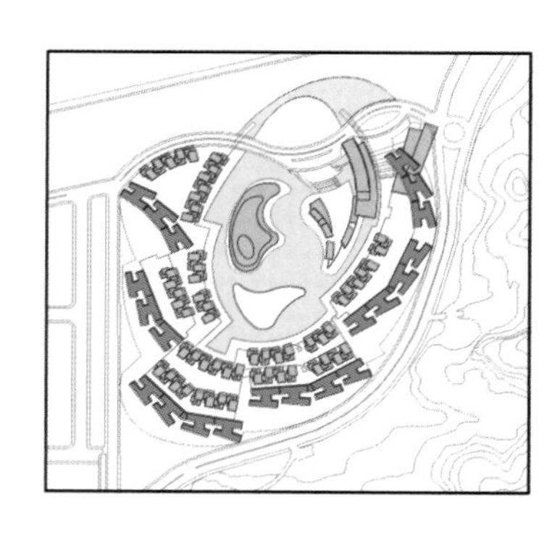

1.LIVING ROOM	1．客厅
2.DINING	2．餐饮区
3.FAMILY ROOM	3．娱乐室
4.KITCHEN	4．厨房
5.MASTER BEDROOM	5．主卧
6.BEDROOM	6．卧室
7.MASTER BATH ROOM	7．主浴室
8.BATH ROOM	8．浴室
9.CLOSET	9．壁橱
10.MAID ROOM	10．佣人房
11.UTILITY ROOM	11．杂物间
12.BALCONY	12．阳台
13.CORE	13．核心

ITE SECTION A 场地剖面图 A

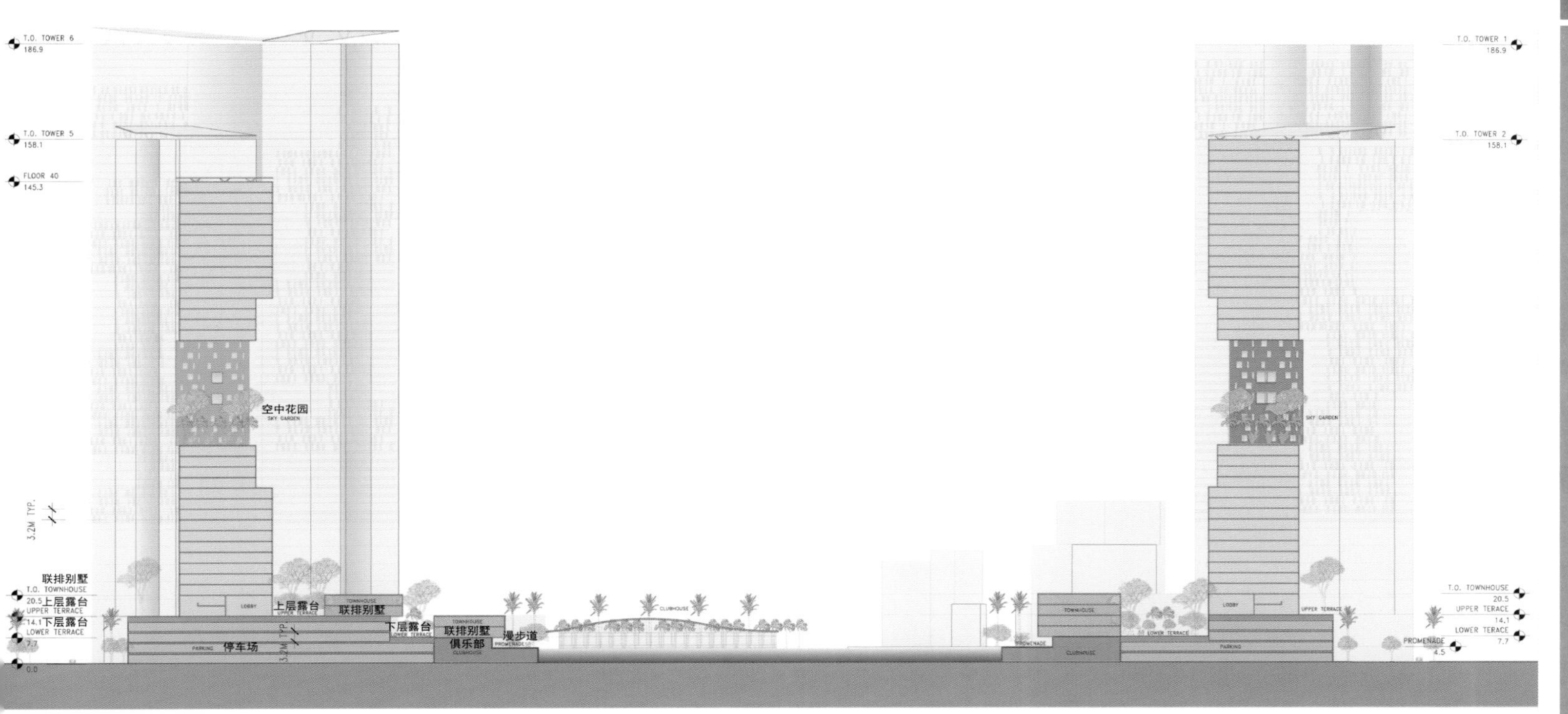

ITE SECTIONB 场地剖面图 B

越南岘港阿尔卡迪亚度假村

Arcadia Resort

设计单位：Campbell Shillinglaw and Partners
合作单位：UXC
开发商：Silver Shore Resort Stock Company
项目地址：越南岘港
场地面积：91 403 m²
建筑面积：22 888.65 m²
建筑密度：25.04%
容积率：1.17

Designed by: Campbell Shillinglaw and Partners
Collaboration: UXC
Developer: Silver Shore Resort Stock Company
Location: Da Nang, Vietnam
Site Area: 91,403 m²
Construction Area: 22,888.65 m²
Building Density: 25.04%
Plot Ratio: 1.17

项目概况

项目位于越南中部的港口城市岘港，是一个集五星级银岸赌场、酒店、住宅于一体的度假村。

建筑设计

项目包括46间豪华海景别墅、联房别墅、俱乐部，以及一栋涵盖了288间客房的酒店、住宅和商业区的综合大楼。住宅区位于大楼的15层，共有100套公寓。

设计充分利用了当地地形的高度特征，将建筑融入景观之中，构建了建筑与海景融为一体的别墅。同时，设计师通过将热带植被与水景结合，营造了清新、自然、郁郁葱葱的环境。

设计具有明显的地中海风格和意大利风格，弧形拱顶、凉廊、棚架和别墅外形流露出浓郁的地中海风情和意大利风情，给人们带来异国的情调和体验。奢华的装饰风格以及大理石、马赛克、宝石、木材等贵重材料的使用，构建了一个豪华的度假胜地，给人置身“王室宫殿”之感。

Profile

The project is located in Da Nang, the middle port city of Vietnam. It is a resort integrating 5 stars silver shore casino, hotel, villas and apartments.

Architectural Design

The design includes 46 luxury ocean view villas and club house, townhouses, a complex building with a hotel (288 rooms), apartments and commercial areas. The apartments are located on 15 floors of this building with a sum of 100 sets.

The landscape of the project is particularly designed to create ocean views for all the villas through a land elevation. It is also considered the tropical vegetation and water features to create a fresh and lush environment.

The design has a distinctive Italian and Mediterranean style. Arcs, loggias, pergolas and the shapes of the villas exert rich Italian and Mediterranean style which brings people exotic feelings and experience. Other design features as decorated columns, precious materials (marble, mosaics, stones, wood) create a luxurious experience, giving the sensation of being a “king in your palace”.

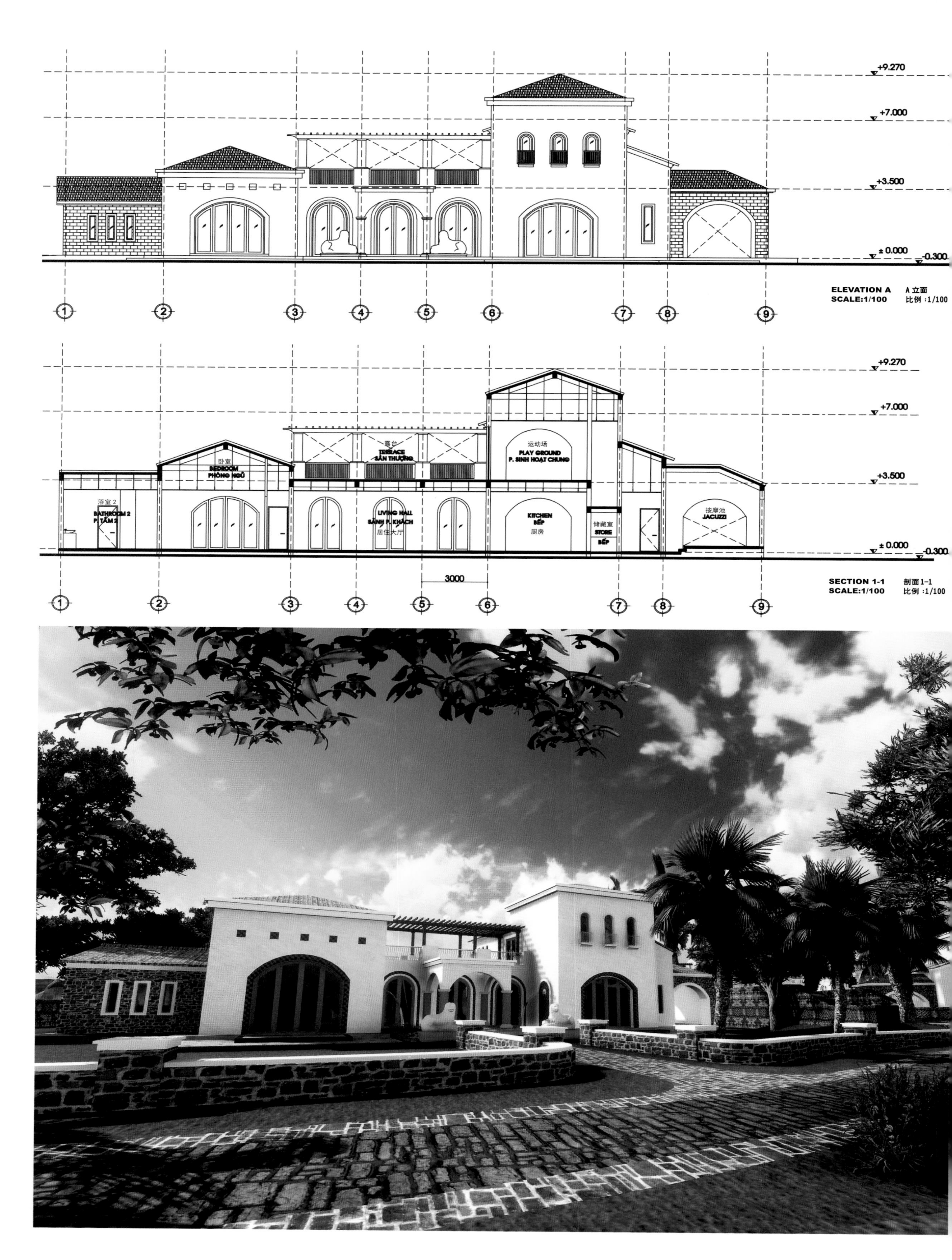
+9.270
+7.000
+3.500
± 0.000
-0.300
ELEVATION A
SCALE:1/100
A立面
比例 :1/100
1
2
3
4
5
6
7
8
9
露台
TERRACE
SÂN THƯỢNG
运动场
PLAY GROUND
P. SINH HOẠT CHUNG
卧室
BEDROOM
PHÒNG NGỦ
浴室 2
BATHROOM 2
P. TẮM 2
LIVING HALL
SẢNH P. KHÁCH
居住大厅
KITCHEN
BẾP
厨房
储藏室
STORE
BẾP
按摩池
JACUZZI
3000
SECTION 1-1
SCALE:1/100
剖面1-1
比例 :1/100

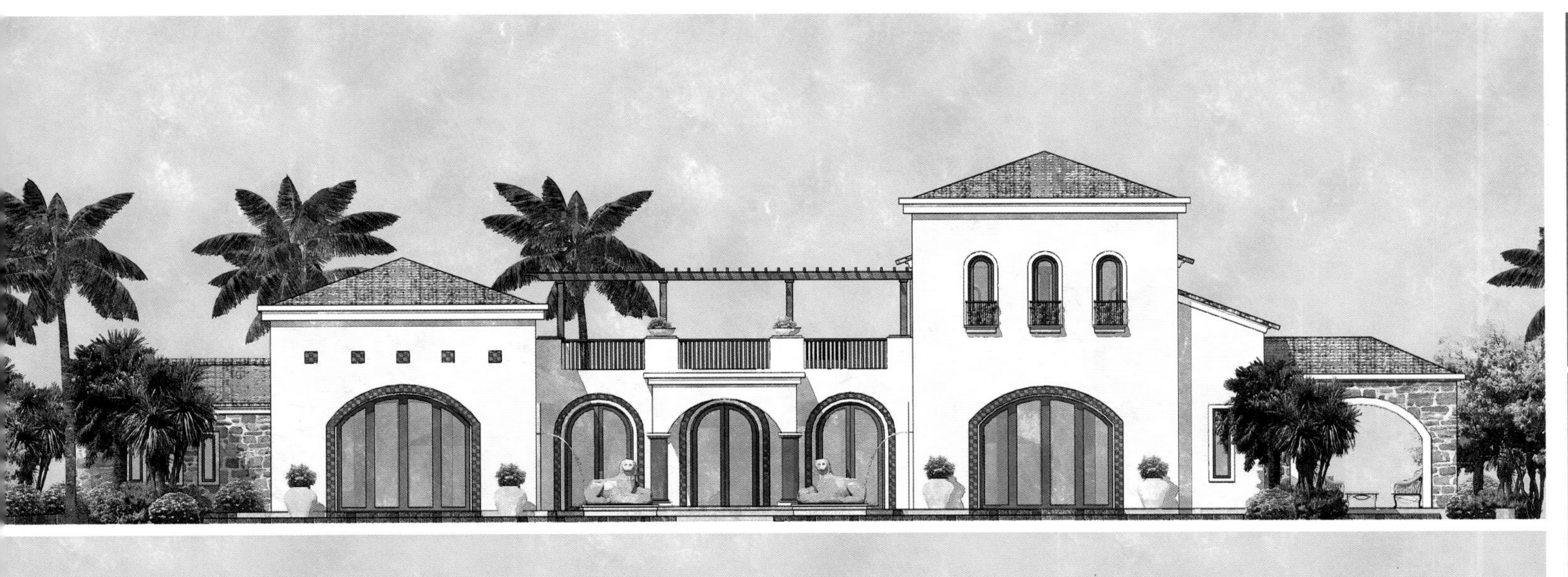

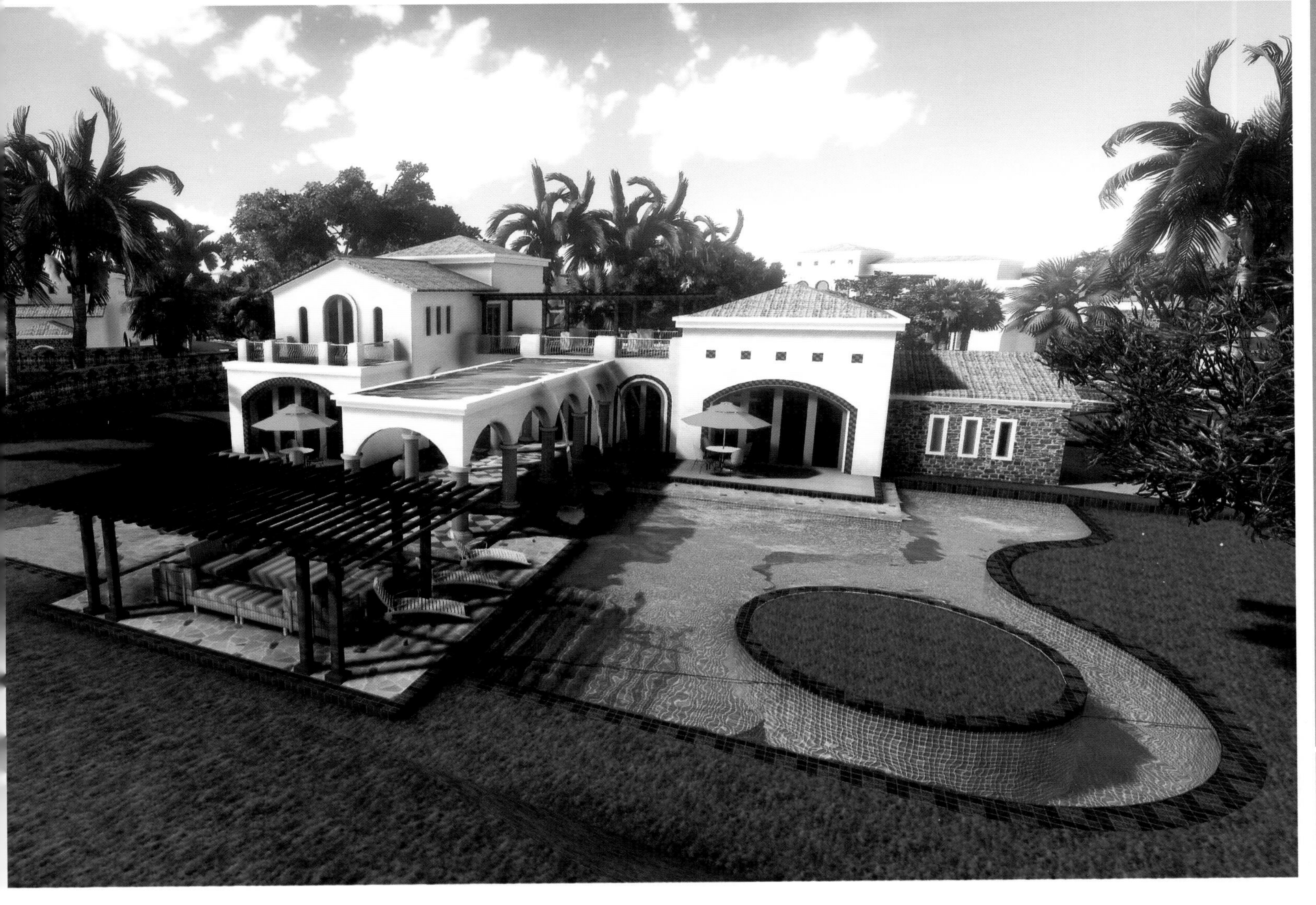

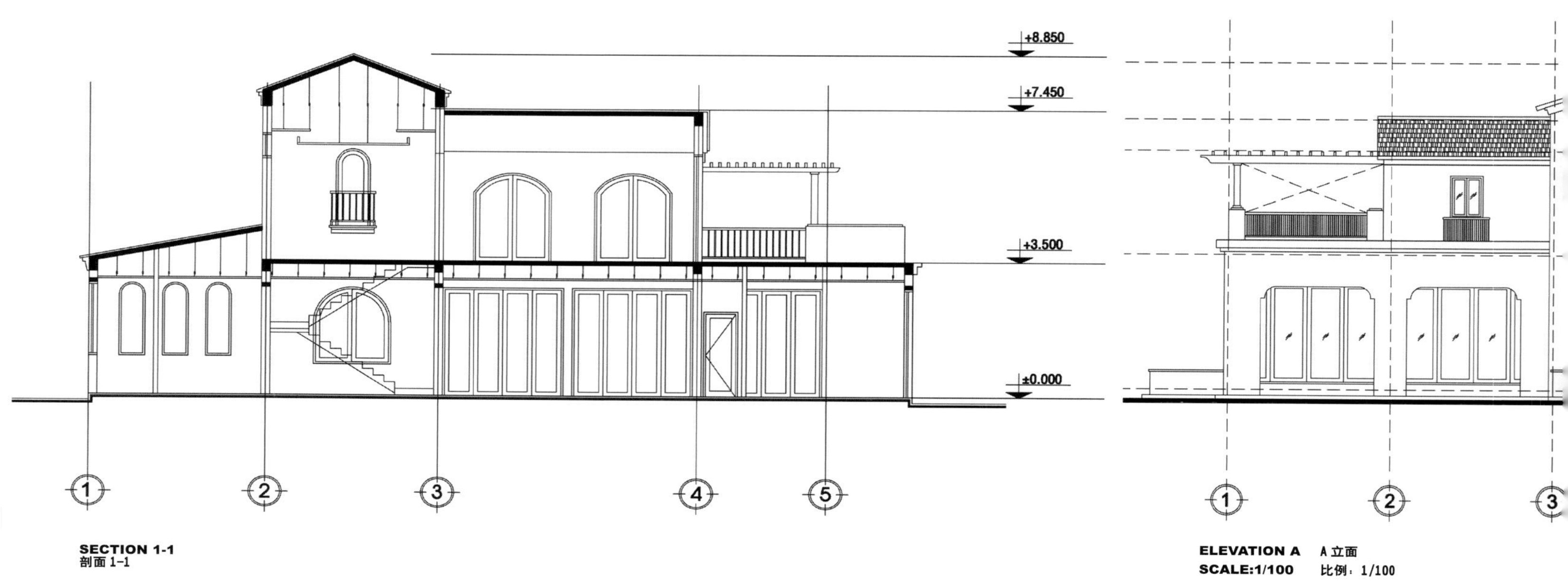

SECTION 1-1
剖面 1-1

ELEVATION A A立面
SCALE:1/100 比例：1/100

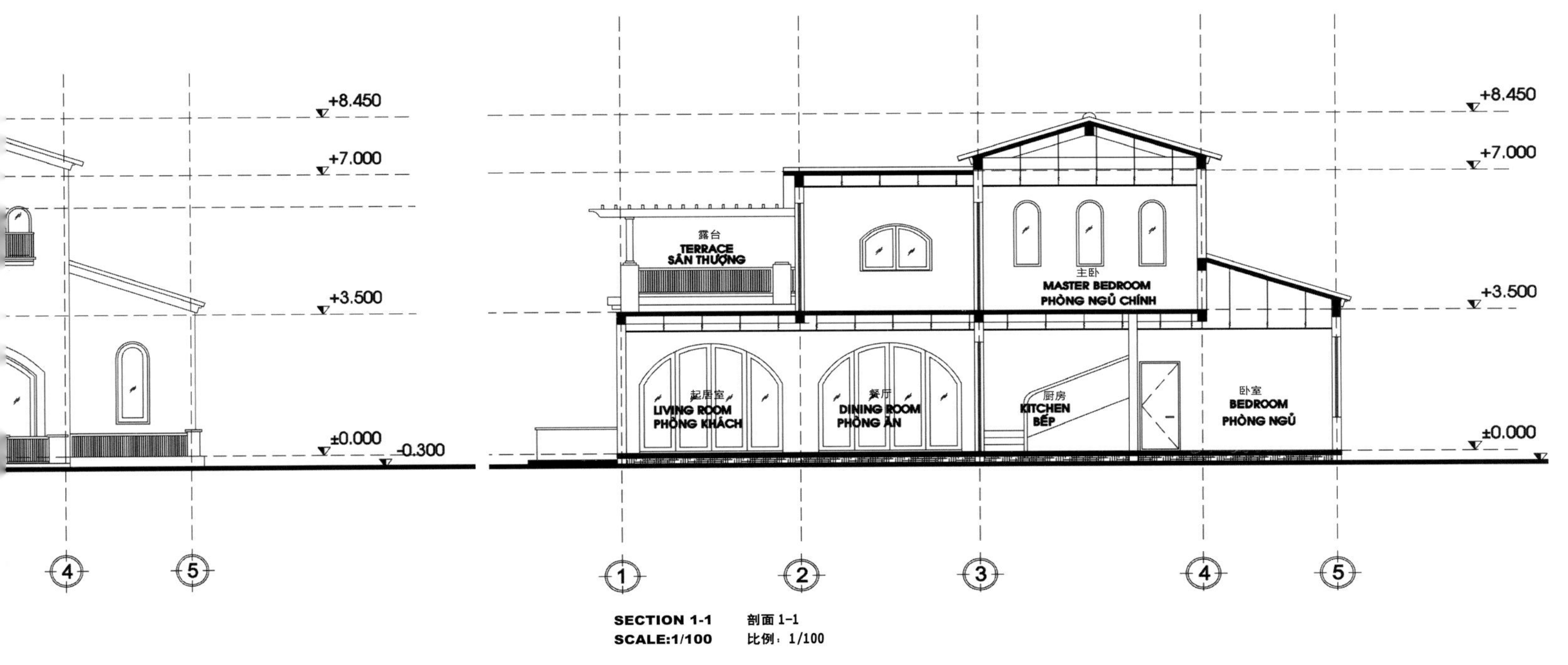

SECTION 1-1 剖面 1-1
SCALE:1/100 比例：1/100

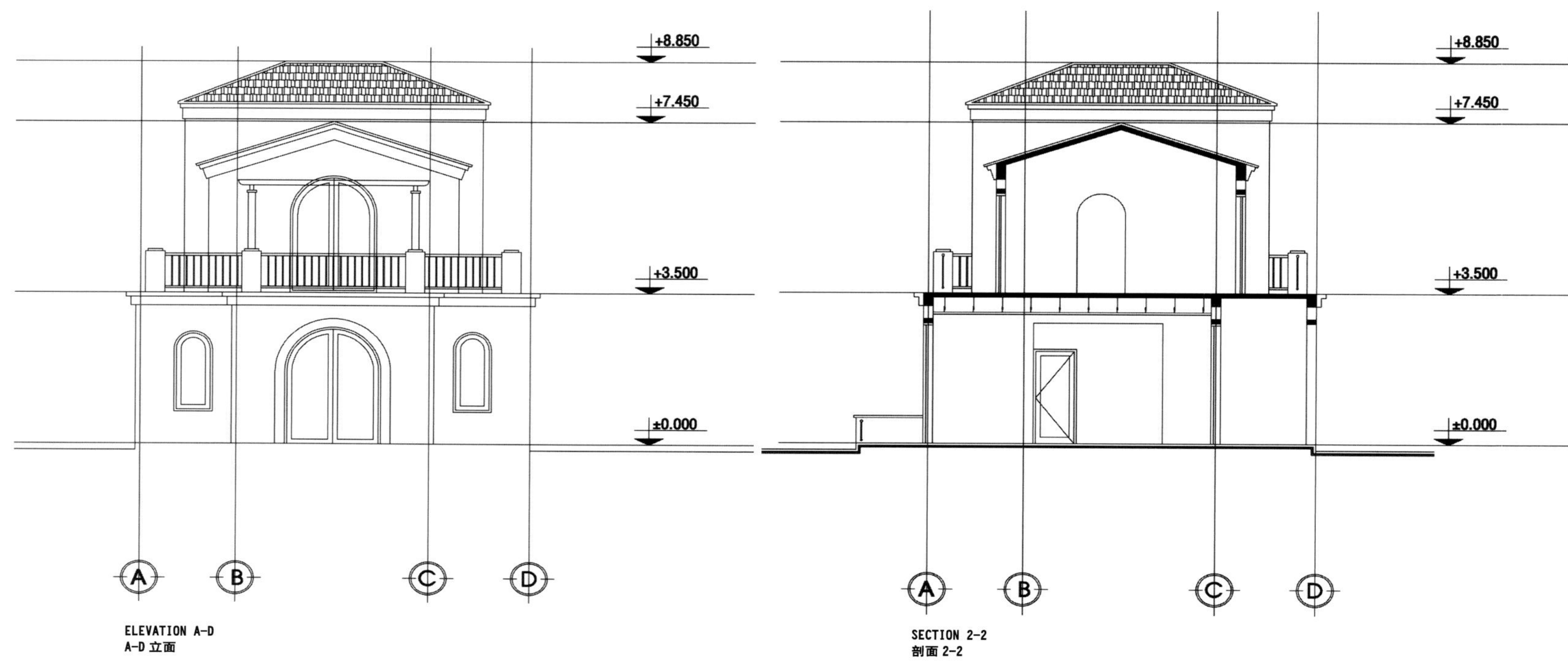

ELEVATION A-D
A-D 立面

SECTION 2-2
剖面 2-2

SECTION 1-1
剖面 1-1

SECTION 2-2
剖面 2-2

ELEVATION 1-5
1-5 立面

SECTION 2-2 剖面 2-2
SCALE:1/100 比例：1/100

ELEVATION A A 立面
SCALE:1/100 比例：1/100

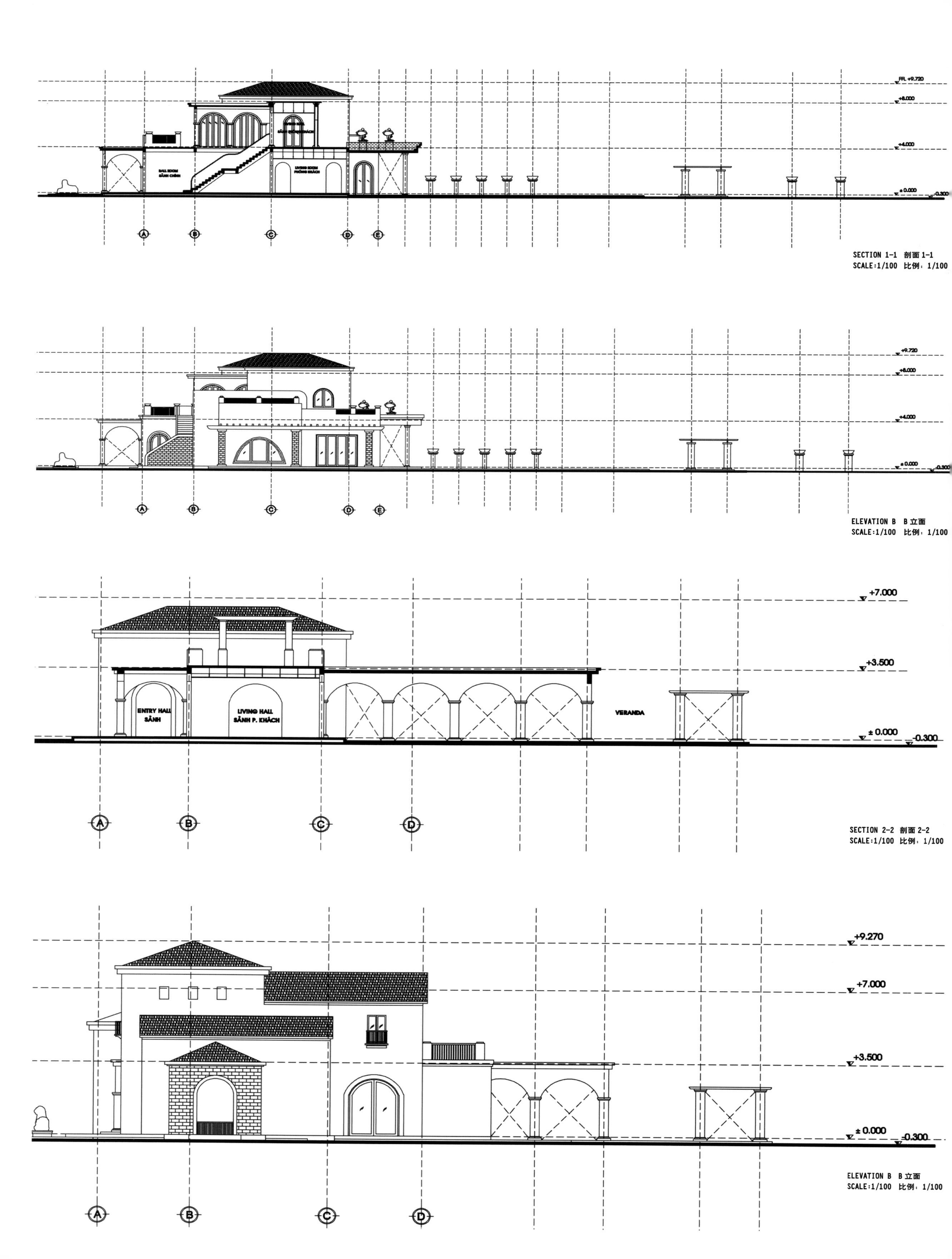

+9.720
+8.000
+4.000
± 0.000
-0.300
SECTION 1-1 剖面 1-1
SCALE:1/100 比例：1/100
ELEVATION B B立面
SCALE:1/100 比例：1/100
+7.000
+3.500
ENTRY HALL
SẢNH
LIVING HALL
SẢNH P. KHÁCH
VERANDA
SECTION 2-2 剖面 2-2
SCALE:1/100 比例：1/100
+9.270
ELEVATION B B立面
SCALE:1/100 比例：1/100
A
B
C
D
E

+10.400
+7.600
+3.600
±0.000

1 2 3 4 5 6 7 8 9

ELEVATION 1-9 1-9 立面
SCALE:1/100 比例：1/100

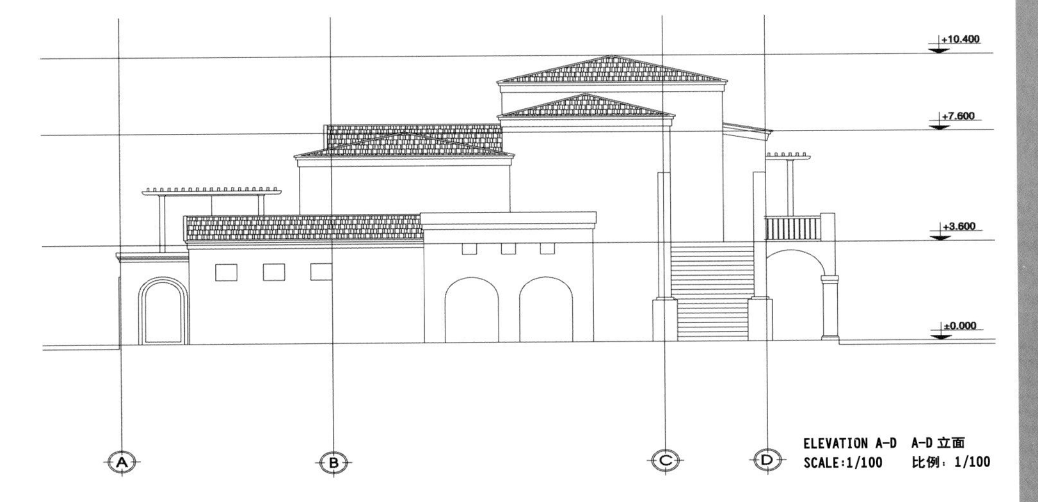

ELEVATION A-D A-D 立面
SCALE:1/100 比例：1/100

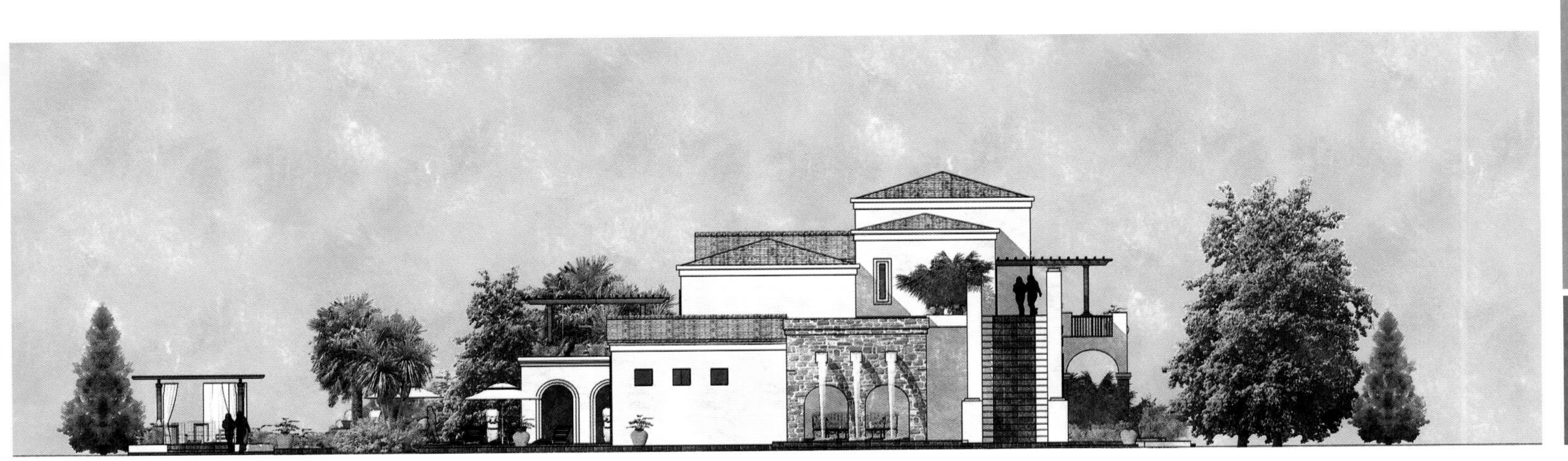

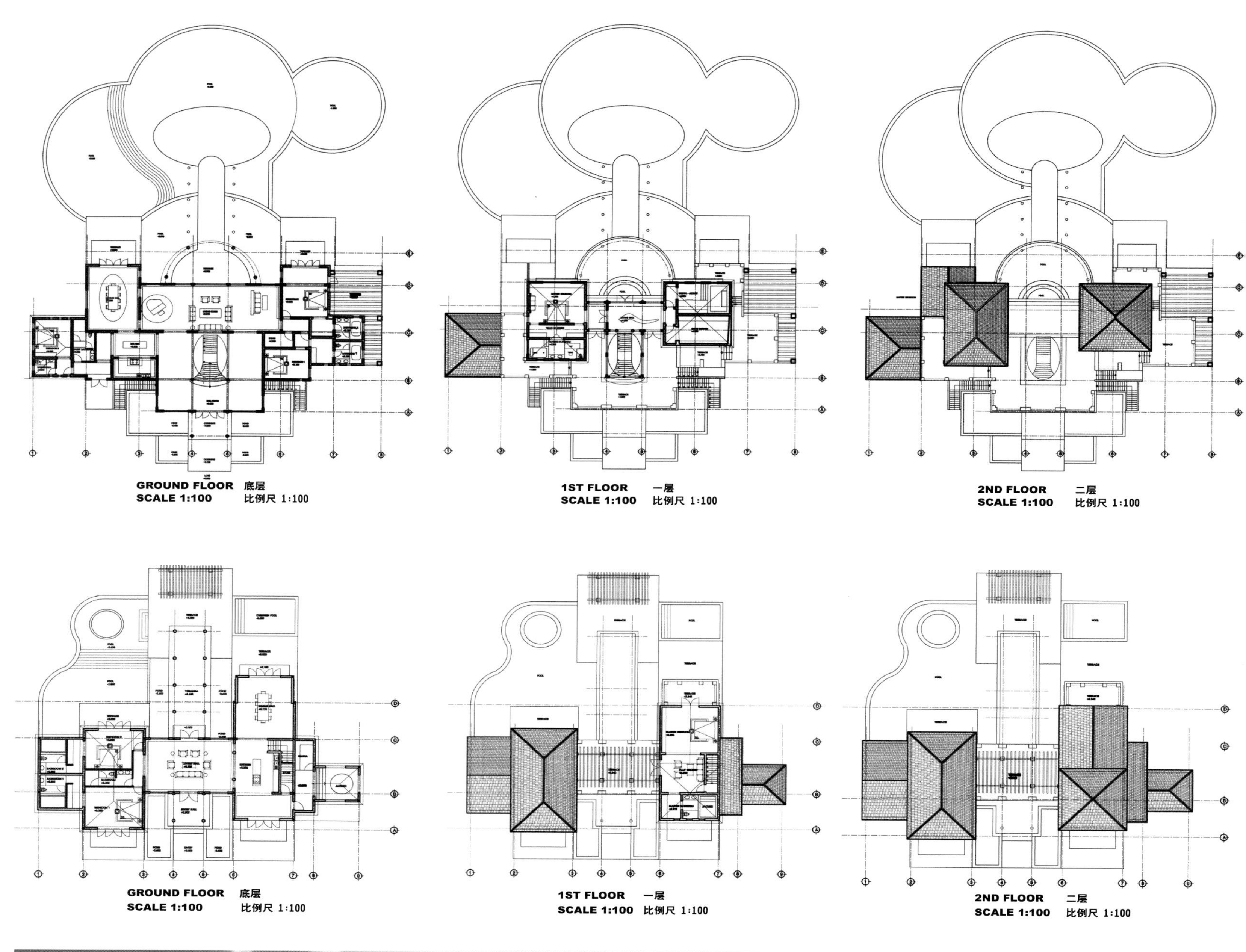
GROUND FLOOR 底层
SCALE 1:100 比例尺 1:100
1ST FLOOR 一层
SCALE 1:100 比例尺 1:100
2ND FLOOR 二层
SCALE 1:100 比例尺 1:100
GROUND FLOOR 底层
SCALE 1:100 比例尺 1:100
1ST FLOOR 一层
SCALE 1:100 比例尺 1:100
2ND FLOOR 二层
SCALE 1:100 比例尺 1:100

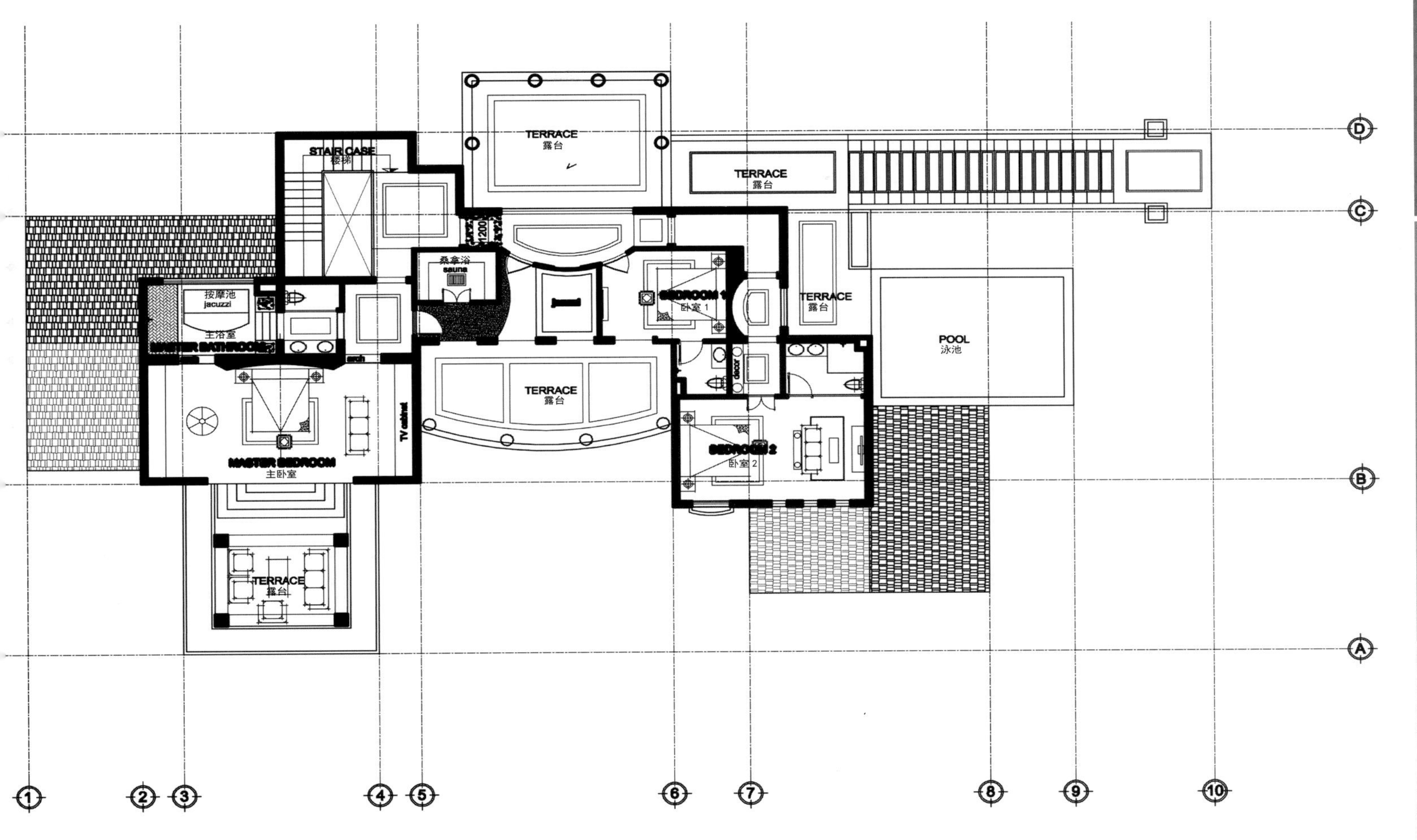
TERRACE
露台
STAIR CASE
楼梯
TERRACE
露台
桑拿浴
sauna
按摩池
jacuzzi
主浴室
MASTER BATHROOM
BEDROOM 1
卧室 1
TERRACE
露台
POOL
泳池
TERRACE
露台
TV cabinet
MASTER BEDROOM
主卧室
BEDROOM 2
卧室 2
TERRACE
露台
A
B
C
D
1
2
3
4
5
6
7
8
9
10

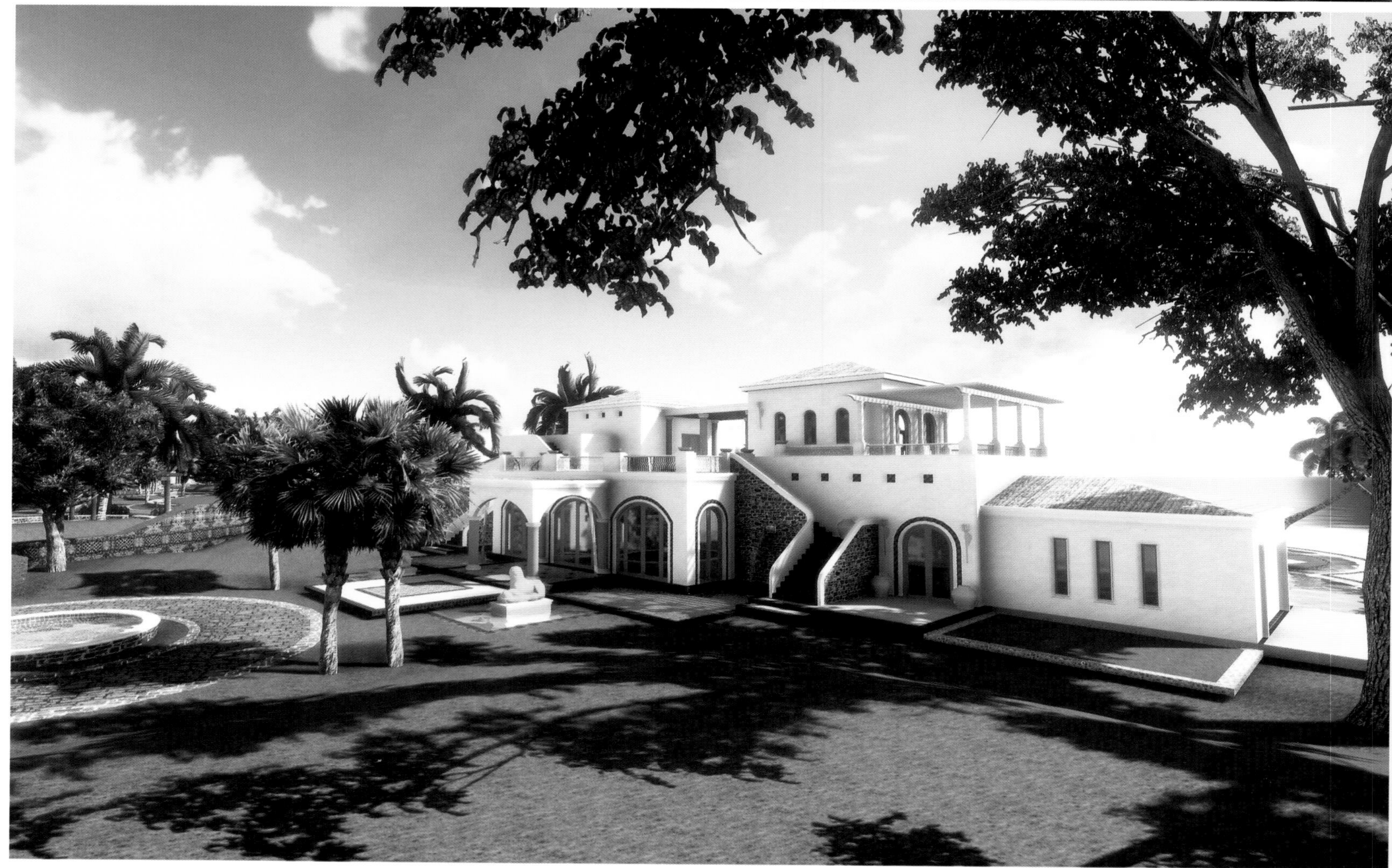

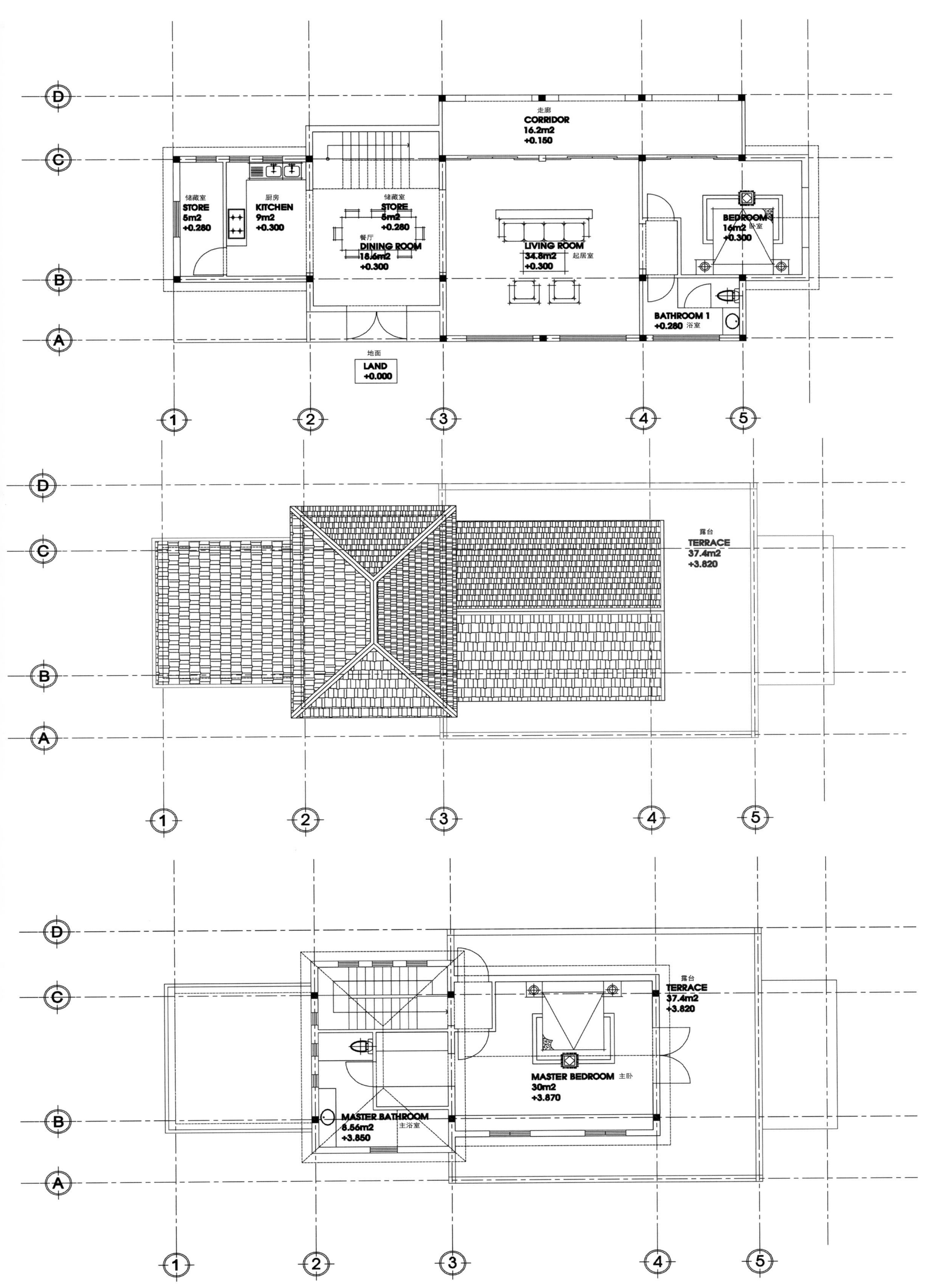
走廊
CORRIDOR
16.2m2
+0.150
储藏室
STORE
5m2
+0.280
厨房
KITCHEN
9m2
+0.300
储藏室
STORE
5m2
+0.280
餐厅
DINING ROOM
18.6m2
+0.300
LIVING ROOM
34.8m2
+0.300
起居室
BEDROOM
16m2
卧室
+0.300
BATHROOM 1
+0.280
浴室
地面
LAND
+0.000
露台
TERRACE
37.4m2
+3.820
露台
TERRACE
37.4m2
+3.820
MASTER BEDROOM
主卧
30m2
+3.870
MASTER BATHROOM
8.56m2
主浴室
+3.850

越南芽庄白沙度假村

White Sand Resort

设计单位： Campbell Shillinglaw and Partners
合作单位： UXC
开发商： White Sand Resort Investment and Development Company
项目地址： 越南芽庄
场地面积： 120 000 m²
建筑面积： 92 358.2 m²
建筑密度： 21.62%
容积率： 0.67

Designed by: Campbell Shillinglaw and Partners
Collaboration: UXC
Developer: White Sand Resort Investment and Development Company
Location: Nha Trang, Vietnam
Site Area: 120,000 m²
Construction Area: 92,358.2 m²
Building Density: 21.62%
Plot Ratio: 0.67

项目概况

项目位于越南东南部港口城市芽庄，是一个集酒店、别墅及高级住宅于一体的五星级度假村。

建筑设计

项目位于一个12公顷的地块上，总建筑面积达80 000平方米，包括46间海景别墅、18层高的五星级酒店以及高级度假住宅区。

将当今普遍重视的绿色设计与热带风景结合起来是别墅建筑设计的主要特色。所有的别墅都配备有面向海洋的私人游泳池，当地独特的地势特征也使别墅拥有优越的观海视野，给居者提供了一份独特的体验和感官享受。

设计采用了木材、石材和玻璃等多种天然材料和透明材料，并将两种材料结合起来，营造出一个舒适、轻松、自然的度假休闲环境。

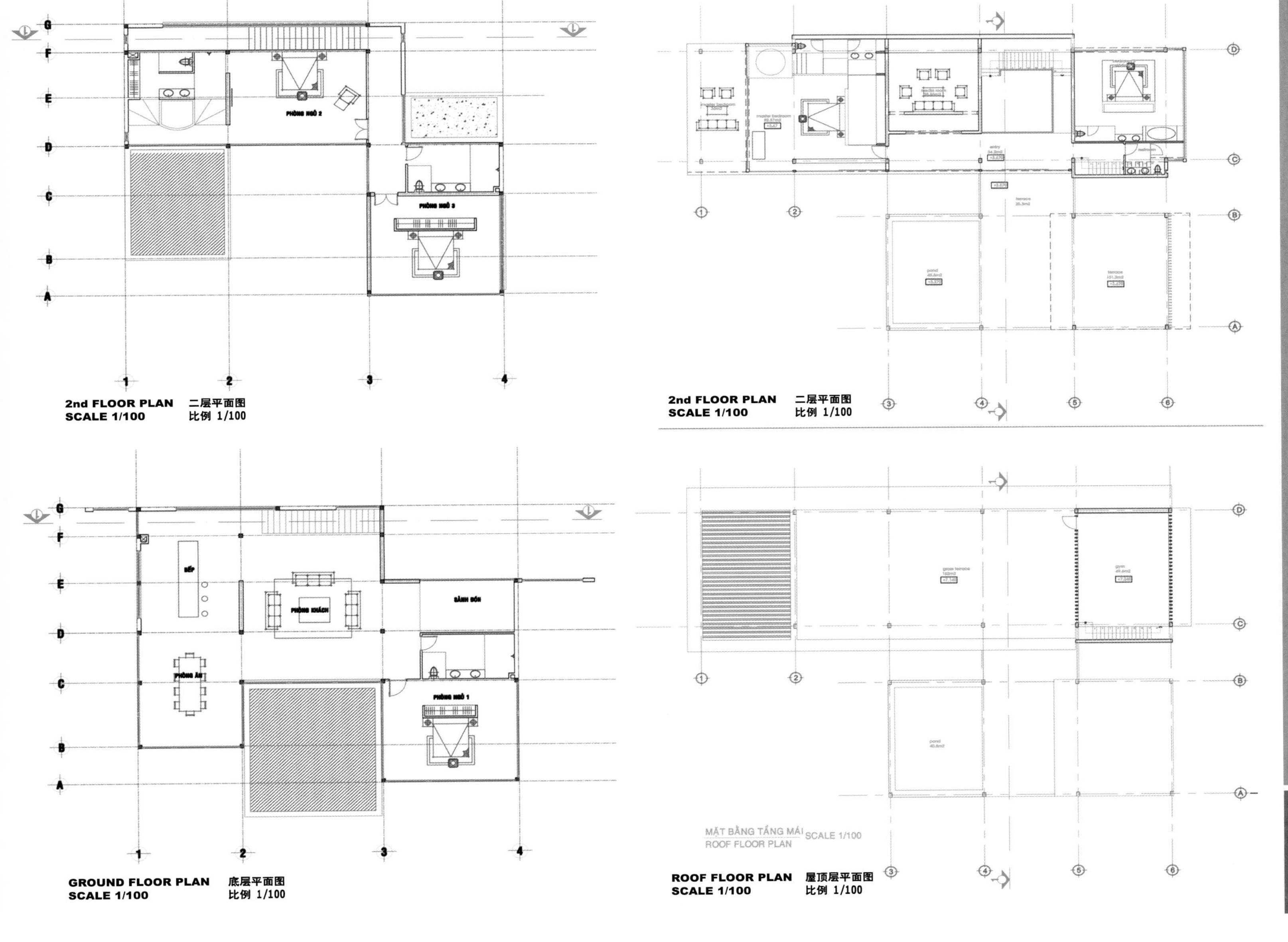
2nd FLOOR PLAN 二层平面图
SCALE 1/100 比例 1/100
2nd FLOOR PLAN 二层平面图
SCALE 1/100 比例 1/100
GROUND FLOOR PLAN 底层平面图
SCALE 1/100 比例 1/100
MẶT BẰNG TẦNG MÁI SCALE 1/100
ROOF FLOOR PLAN
ROOF FLOOR PLAN 屋顶层平面图
SCALE 1/100 比例 1/100

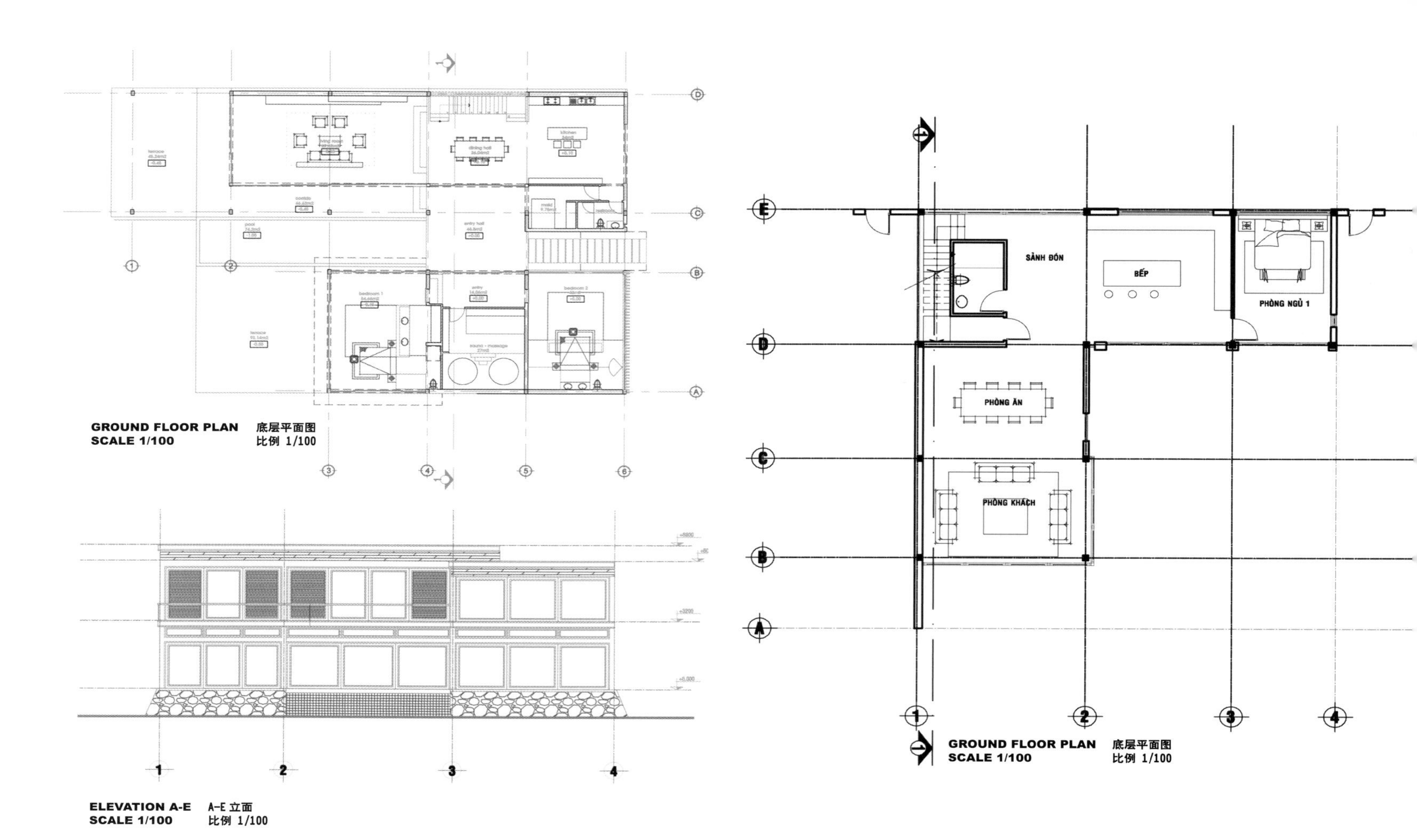

Profile

The project is located in Nha Trang, the southeast port city of Vietnam. It is a 5-star resort integrating hotels, villas and apartments.

Architectural Design

The project is located on a site of 12 hectares with a total construction area of 80,000 square meters. It includes 46 ocean views villas, a 5-star hotel on 18 floors, and luxury apartments.

The main architectural feature of the villas is the green and tropical contemporary design. All the villas are designed with private infinity pool to the ocean. Local unique terrain features give to every villa a full ocean view experience.

The combination of natural and transparent materials such as wood, stone and glass create a fully relax, leisure and natural experience and integration of each villas into the surrounding landscape.

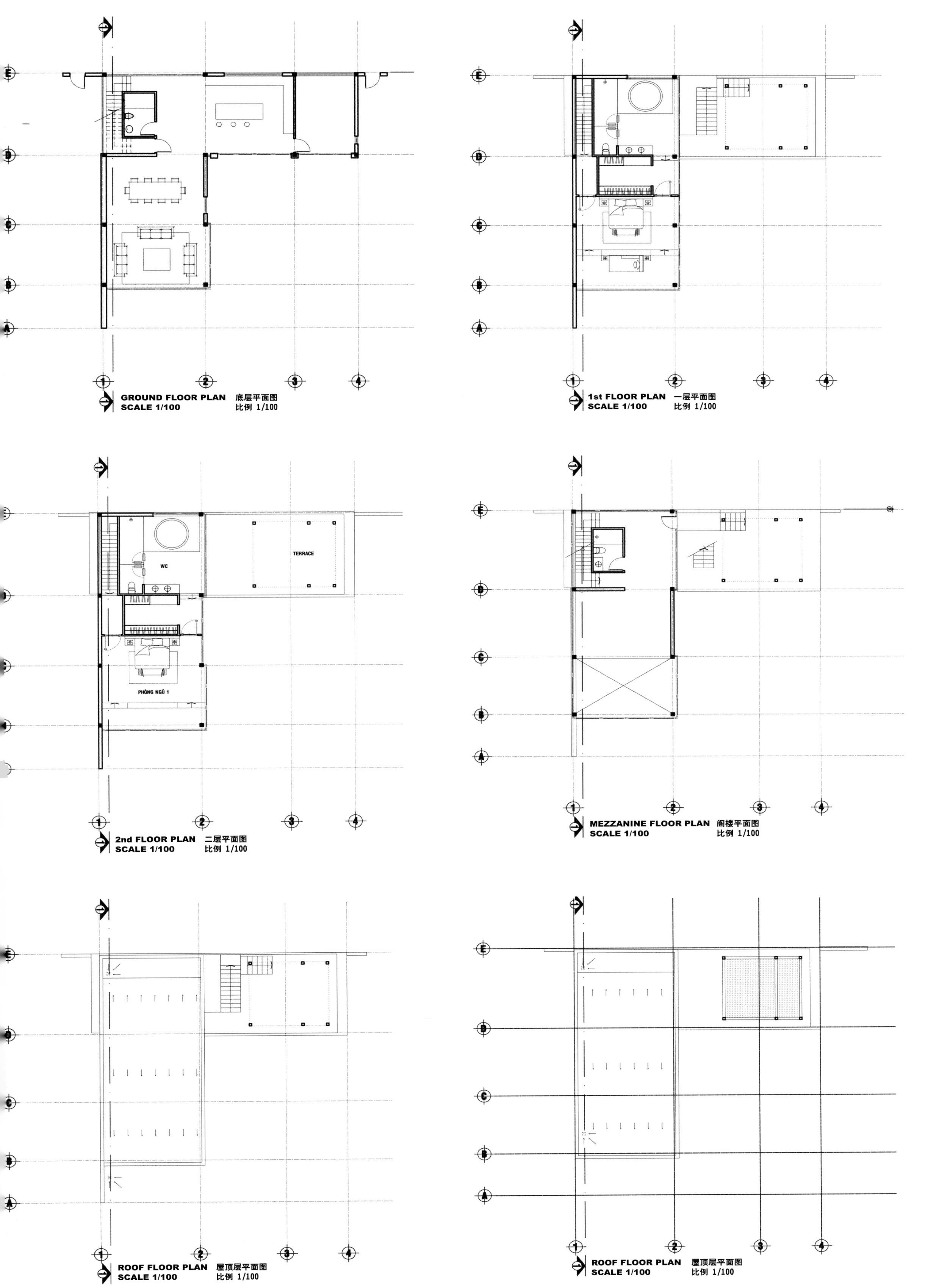

GROUND FLOOR PLAN 底层平面图
SCALE 1/100 比例 1/100

1st FLOOR PLAN 一层平面图
SCALE 1/100 比例 1/100

2nd FLOOR PLAN 二层平面图
SCALE 1/100 比例 1/100

MEZZANINE FLOOR PLAN 阁楼平面图
SCALE 1/100 比例 1/100

ROOF FLOOR PLAN 屋顶层平面图
SCALE 1/100 比例 1/100

ROOF FLOOR PLAN 屋顶层平面图
SCALE 1/100 比例 1/100

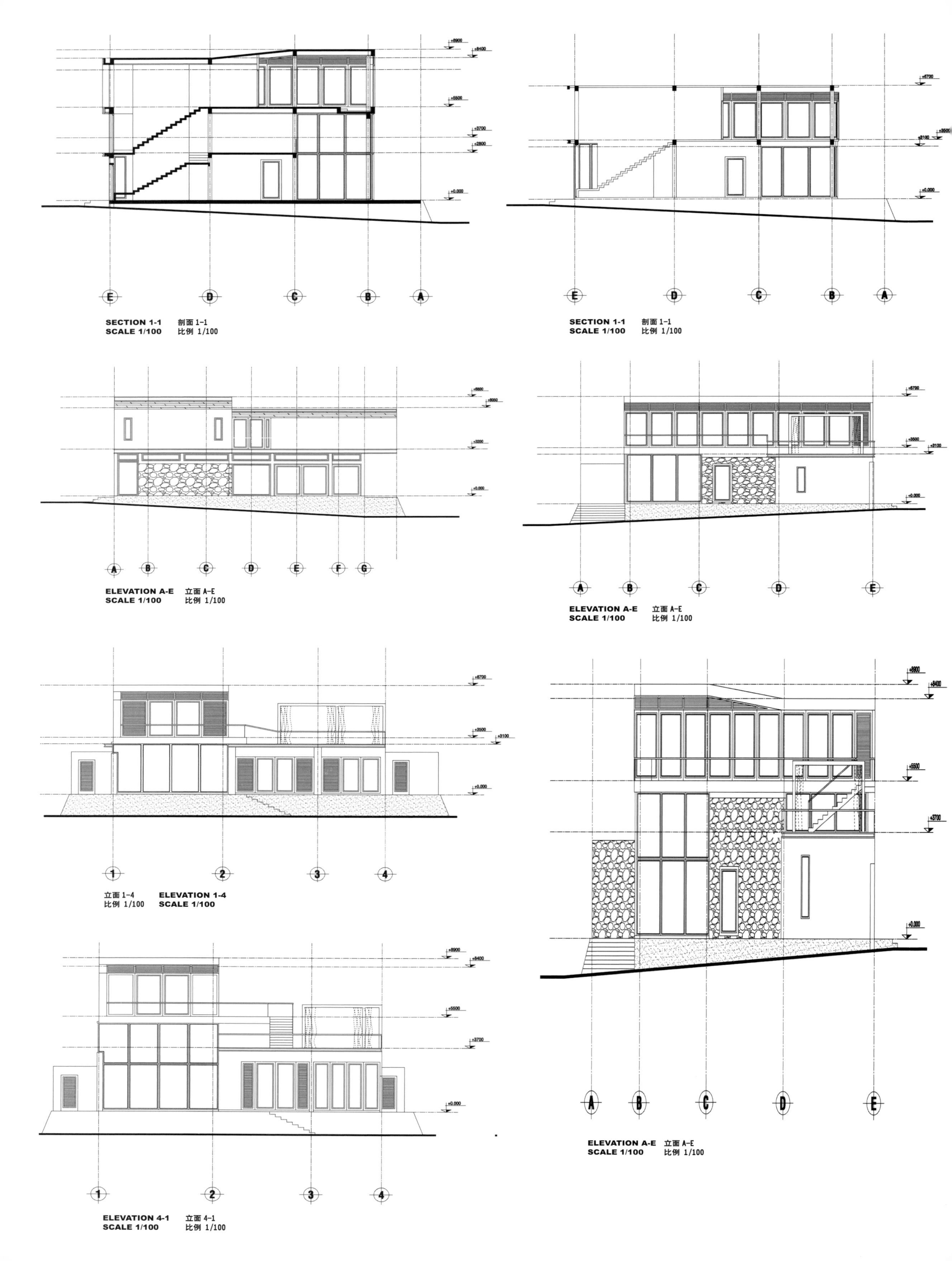
SECTION 1-1 剖面 1-1
SCALE 1/100 比例 1/100
SECTION 1-1 剖面 1-1
SCALE 1/100 比例 1/100
ELEVATION A-E 立面 A-E
SCALE 1/100 比例 1/100
ELEVATION A-E 立面 A-E
SCALE 1/100 比例 1/100
立面 1-4 ELEVATION 1-4
比例 1/100 SCALE 1/100
ELEVATION 4-1 立面 4-1
SCALE 1/100 比例 1/100
ELEVATION A-E 立面 A-E
SCALE 1/100 比例 1/100

SECTION 1-1 剖面 1-1
SCALE 1/100 比例 1/100

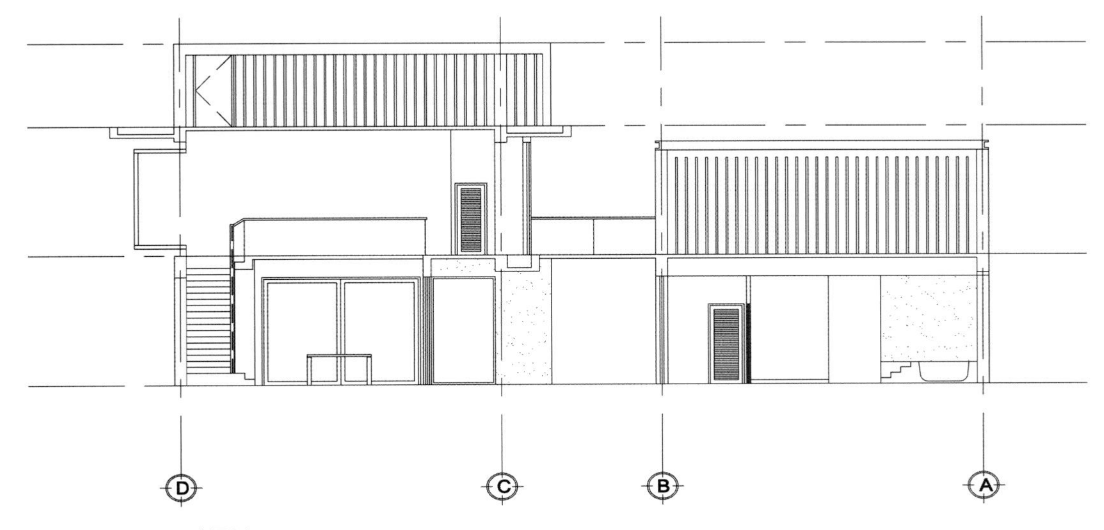

SECTION 1-1 剖面 1-1
SCALE 1/100 比例 1/100

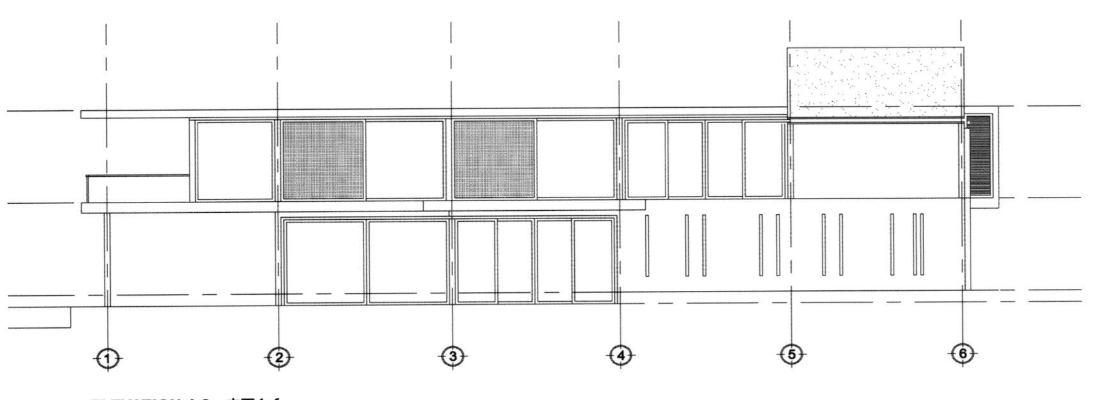

ELEVATION 1-6 立面 1-6
SCALE 1/100 比例 1/100

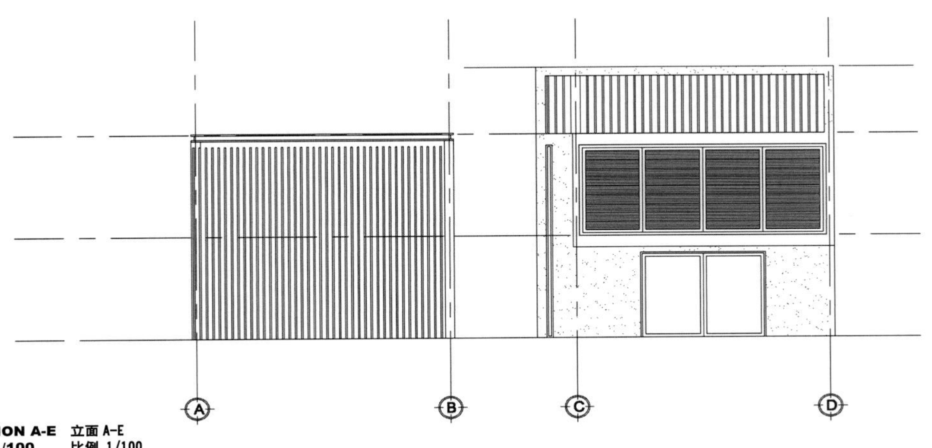

ELEVATION A-E 立面 A-E
SCALE 1/100 比例 1/100

海南三亚五号用地

Sanya Block 5

设计单位：NL Architects

开发商：海南万科地产开发有限公司

项目地址：中国海南省三亚市

Designed by: NL Architects

Client: Hainan Vanke Real Estate Development Co., Ltd.

Location: Sanya, Hainan, China

建筑设计

项目由 8 栋 6 层的度假式公寓组成，所有的餐厅、酒吧和零售空间都将集中在一层平台。这些大楼位于一个公共平台上，覆盖着容纳了地下车库和服务区的地下空间。每个大楼都包含 15 个独立的住宅单元。项目将会建造 4 个酒店，每个酒店由两座镜子形的大楼组成，其中一座朝向街道，另一座向周围景观开放。

设计特色

8 栋公寓建筑分成两列排布，相对的两栋建筑之间形成一个复杂的三维“迷宫”，“迷宫”里配备有走廊和电梯核心筒，从电梯核心筒向外延伸出的通道，代替了常规的走廊。这些交通设施产生了雕塑般的效果，蔚为壮观，同时也将两栋建筑联系起来。

酒店房间有着独特的空间结构，这些房间单元都是跃层的，形成了双高的空间，使空间更为宽敞舒适。双高的复式套房向外延伸，其外侧是一个宽敞的阳台，阳台可视为起居室向户外延伸，将配备有开放的厨房和浴室。

设计师巧妙地利用了阳台空间，在此设计了一个三角形的“大花盆”，使建筑立面更富有动感和节奏感。同时这个“大花盆”可以种植许多绿色植物，既提高了建筑的私密性，又可为建筑遮阳，创造出宜人的微气候。

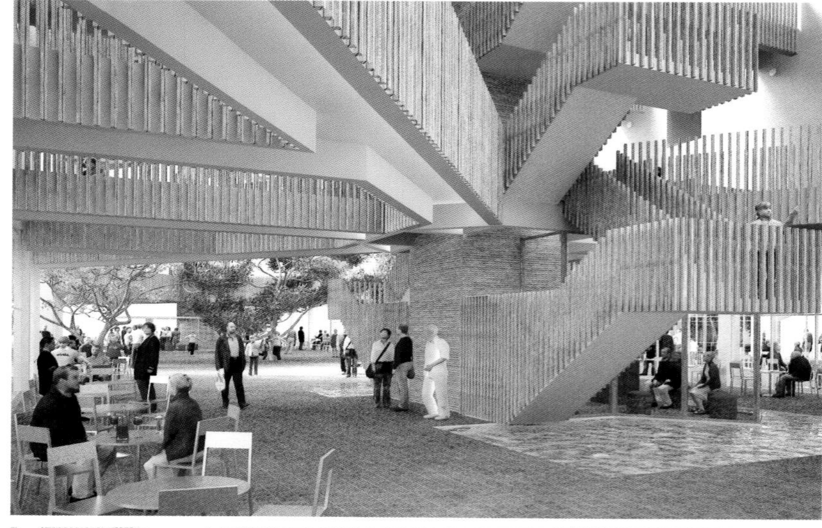

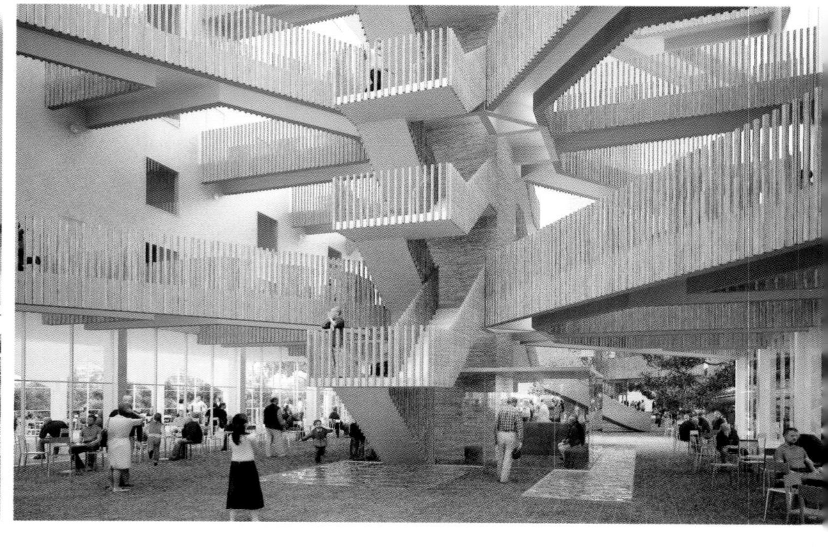

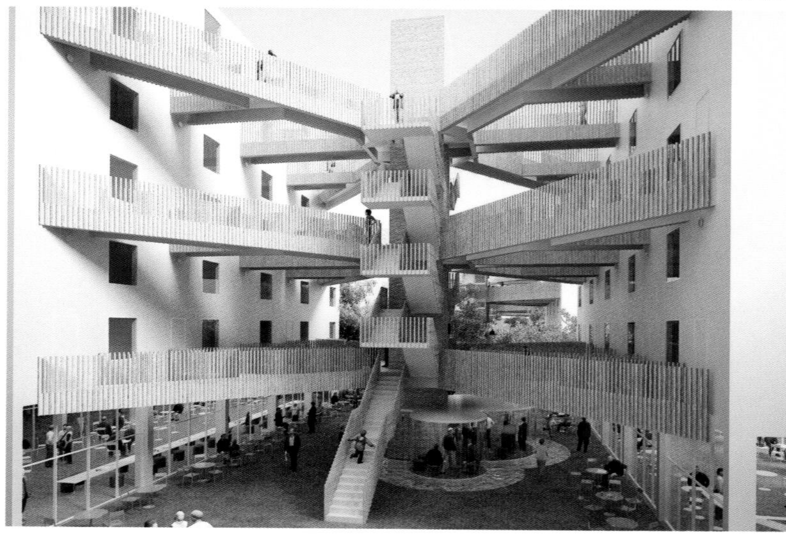

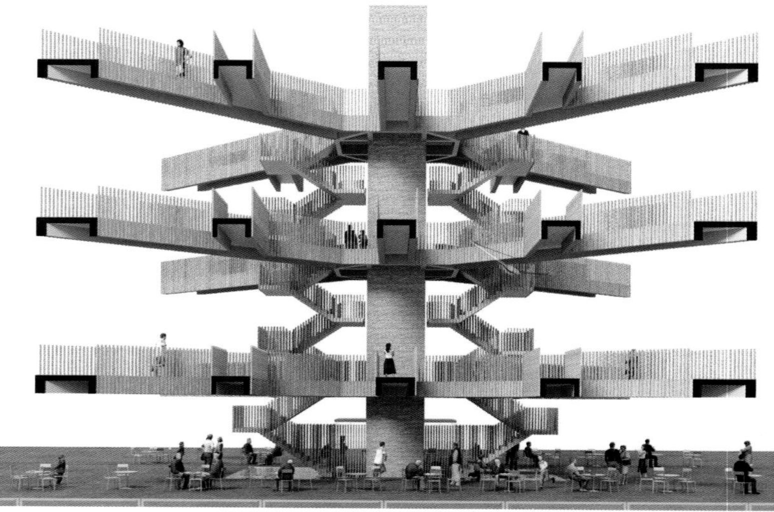

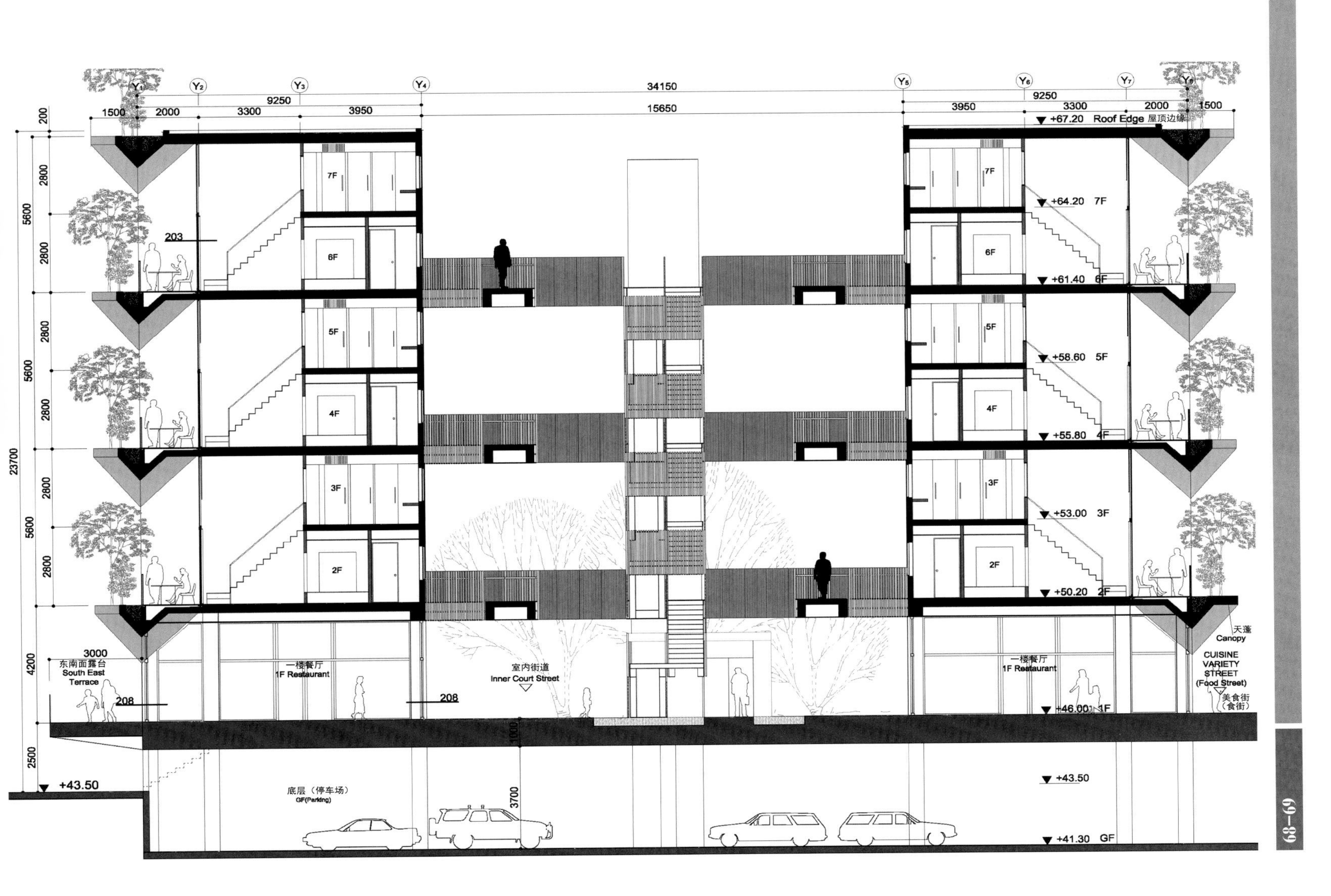
34150
9250
15650
1500
2000
3300
3950
200
2800
5600
23700
4200
2500
3700
1000
3000
+67.20 Roof Edge 屋顶边缘
+64.20 7F
+61.40 6F
+58.60 5F
+55.80 4F
+53.00 3F
+50.20 2F
+46.00 1F
+43.50
+41.30 GF
7F
6F
5F
4F
3F
2F
203
208
东南面露台
South East Terrace
一楼餐厅
1F Restaurant
室内街道
Inner Court Street
天蓬
Canopy
CUISINE VARIETY STREET (Food Street)
美食街
(食街)
底层（停车场）
GF(Parking)

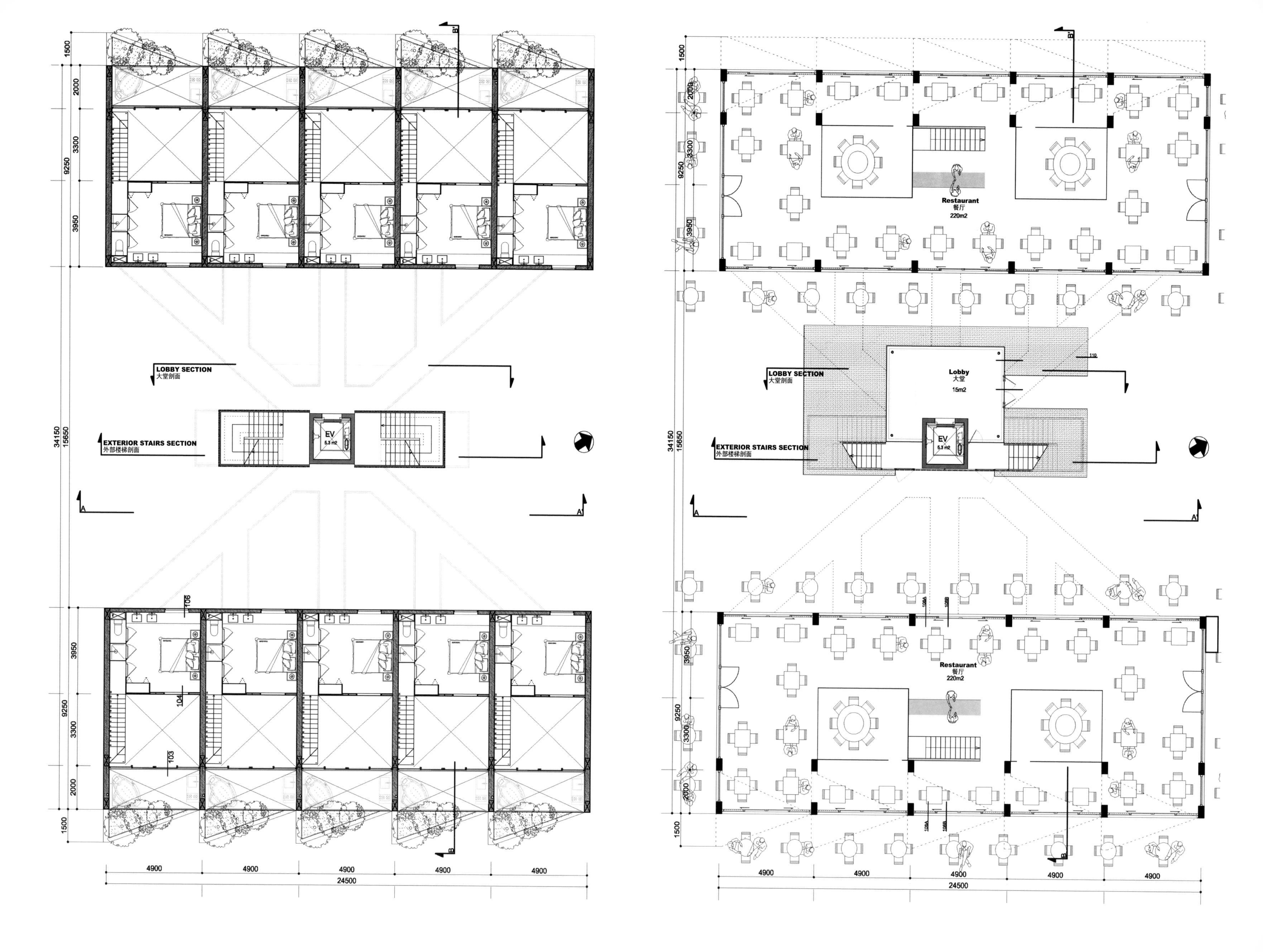

LOBBY SECTION
大堂剖面
EXTERIOR STAIRS SECTION
外部楼梯剖面
EV
5.3 m2
Restaurant
餐厅
220m2
Lobby
大堂
15m2
34150
15650
1500
9250
2000
3300
3950
4900
24500
103
104
106

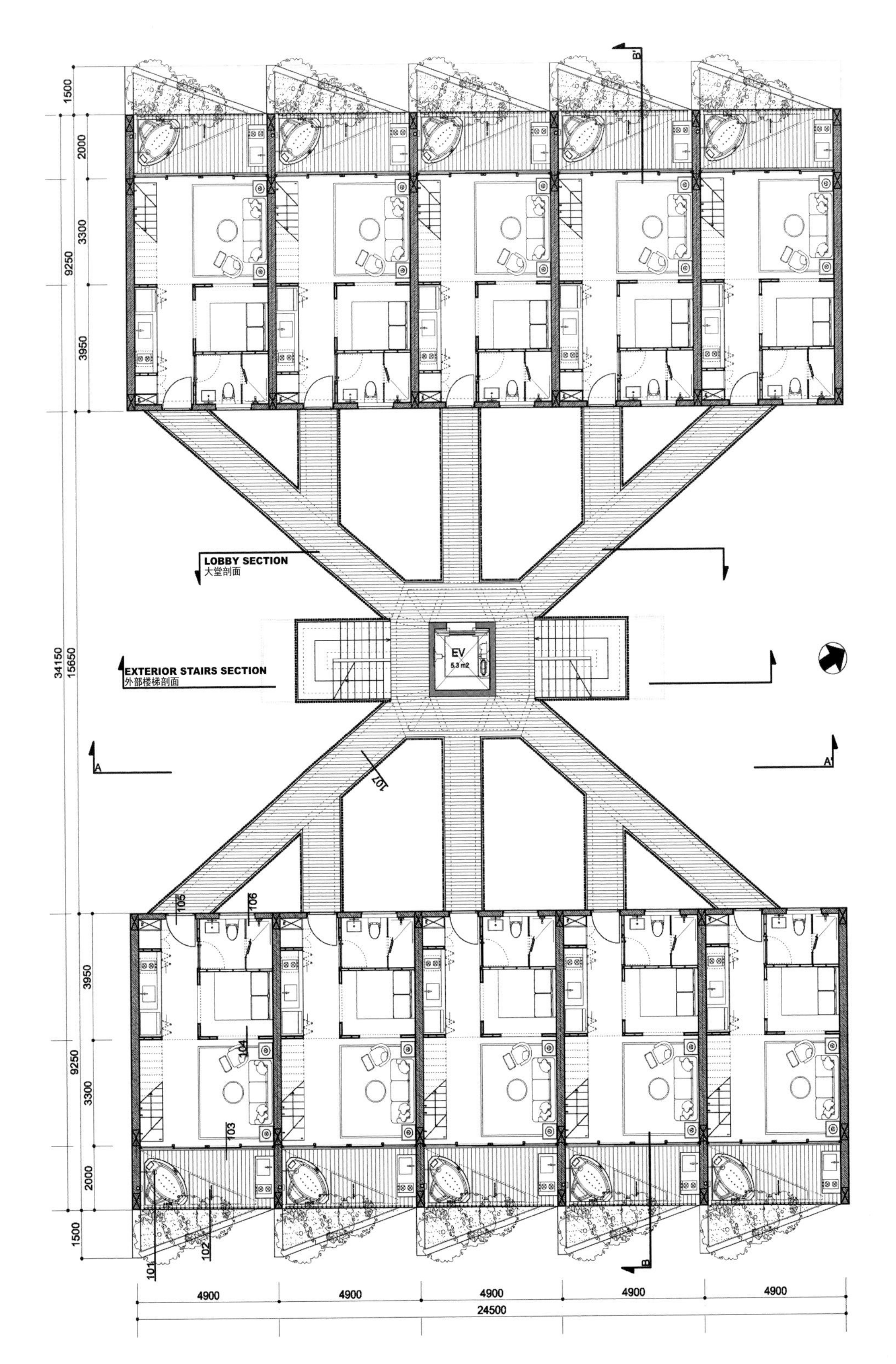
34150
15650
9250
1500
2000
3300
3950
4900
24500
EXTERIOR STAIRS SECTION
外部楼梯剖面
LOBBY SECTION
大堂剖面
EV
101
102
103
104
105
106
107
A
B
B'

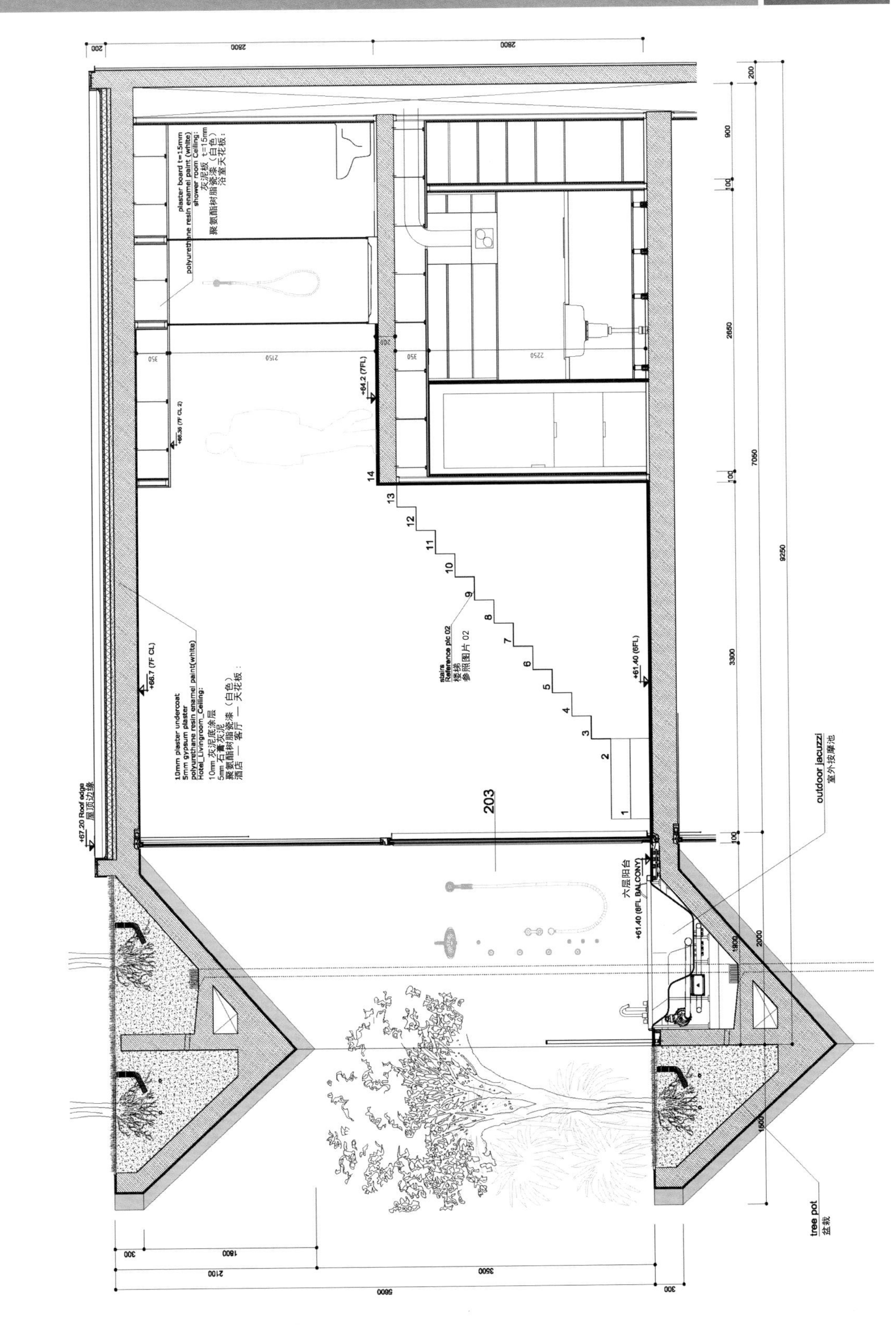
+67.20 Roof edge
屋顶边缘
+66.7 (7F CL)
+66.36 (7F CL 2)
10mm plaster undercoat
5mm gypsum plaster
polyurethane resin enamel paint(white)
Hotel_Livingroom_Ceiling:
10mm 灰泥底涂层
5mm 石膏灰泥
聚氨酯树脂瓷漆（白色）
酒店 — 客厅 — 天花板：
plaster board t=15mm
polyurethane resin enamel paint (white)
shower room Ceiling:
灰泥板 t=15mm
聚氨酯树脂瓷漆（白色）
浴室天花板：
+64.2 (7FL)
stairs
Reference pic 02
楼梯
参照图片 02
203
六层阳台
+61.40 (6FL BALCONY)
+61.40 (6FL)
tree pot
盆栽
outdoor jacuzzi
室外按摩池
3300
2650
900
7050
9250
2000
1500
1800

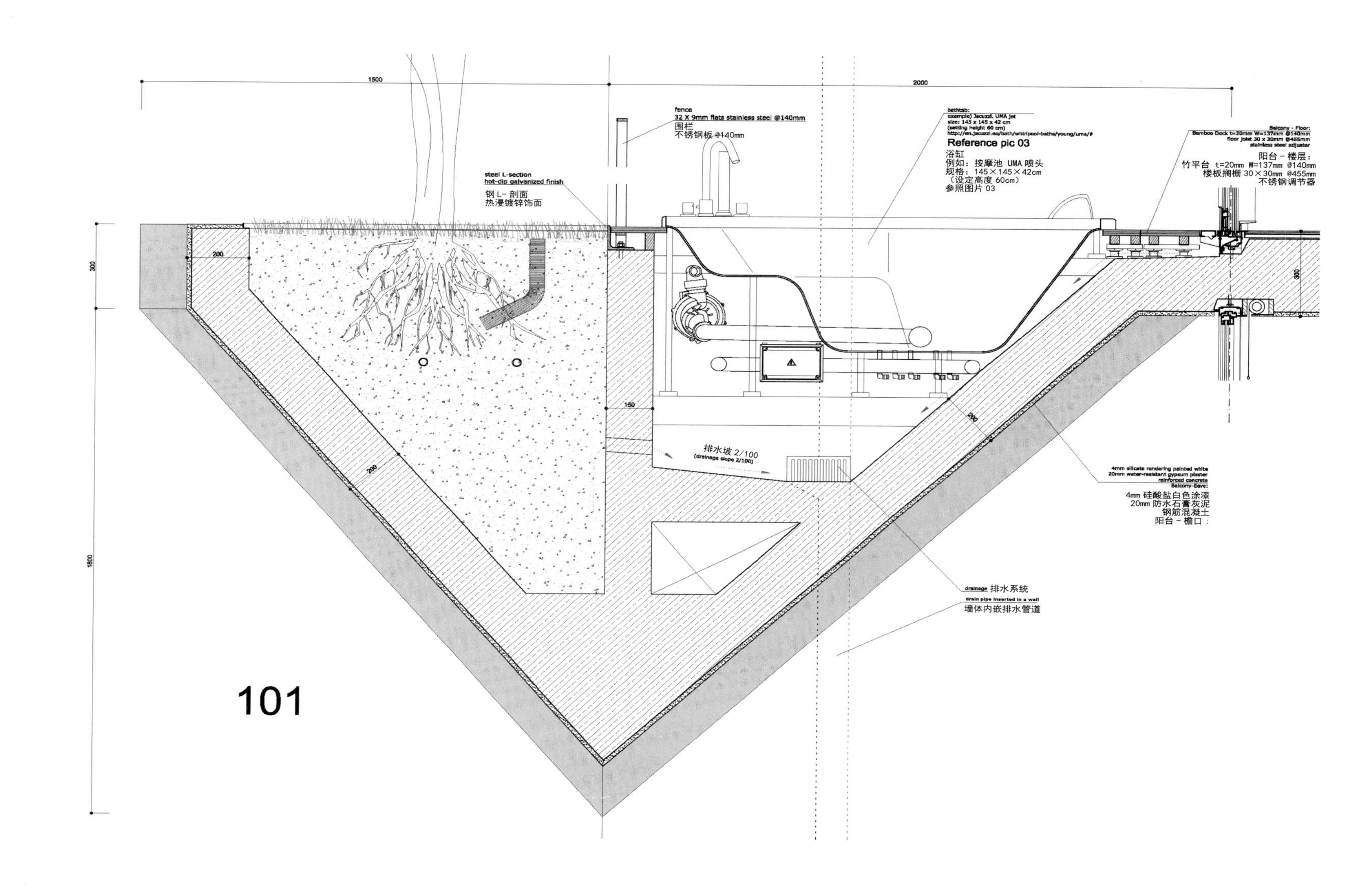

1500
2000
fence
32 X 9mm flats stainless steel @140mm
围栏
不锈钢板 @140mm
steel L-section
hot-dip galvanized finish
钢 L- 剖面
热浸镀锌饰面
bathtab:
example) Jacuzzi. UMA jet
size: 145 x 145 x 42 cm
(setting height 60 cm)
http://en.jacuzzi.eu/bath/whirlpool-baths/young/uma/#
Reference pic 03
浴缸
例如：按摩池 UMA 喷头
规格：145×145×42cm
（设定高度 60cm）
参照图片 03
Balcony - Floor:
Bamboo Deck t=20mm W=137mm @140mm
floor joist 30 x 30mm @455mm
stainless steel adjuster
阳台 – 楼层：
竹平台 t=20mm W=137mm @140mm
楼板搁栅 30×30mm @455mm
不锈钢调节器
排水坡 2/100
(drainage slope 2/100)
4mm silicate rendering painted white
20mm water-resistant gypsum plaster
reinforced concrete
Balcony-Eave:
4mm 硅酸盐白色涂漆
20mm 防水石膏灰泥
钢筋混凝土
阳台 – 檐口：
drainage 排水系统
drain pipe inserted in a wall
墙体内嵌排水管道
300
1800
200
150
101

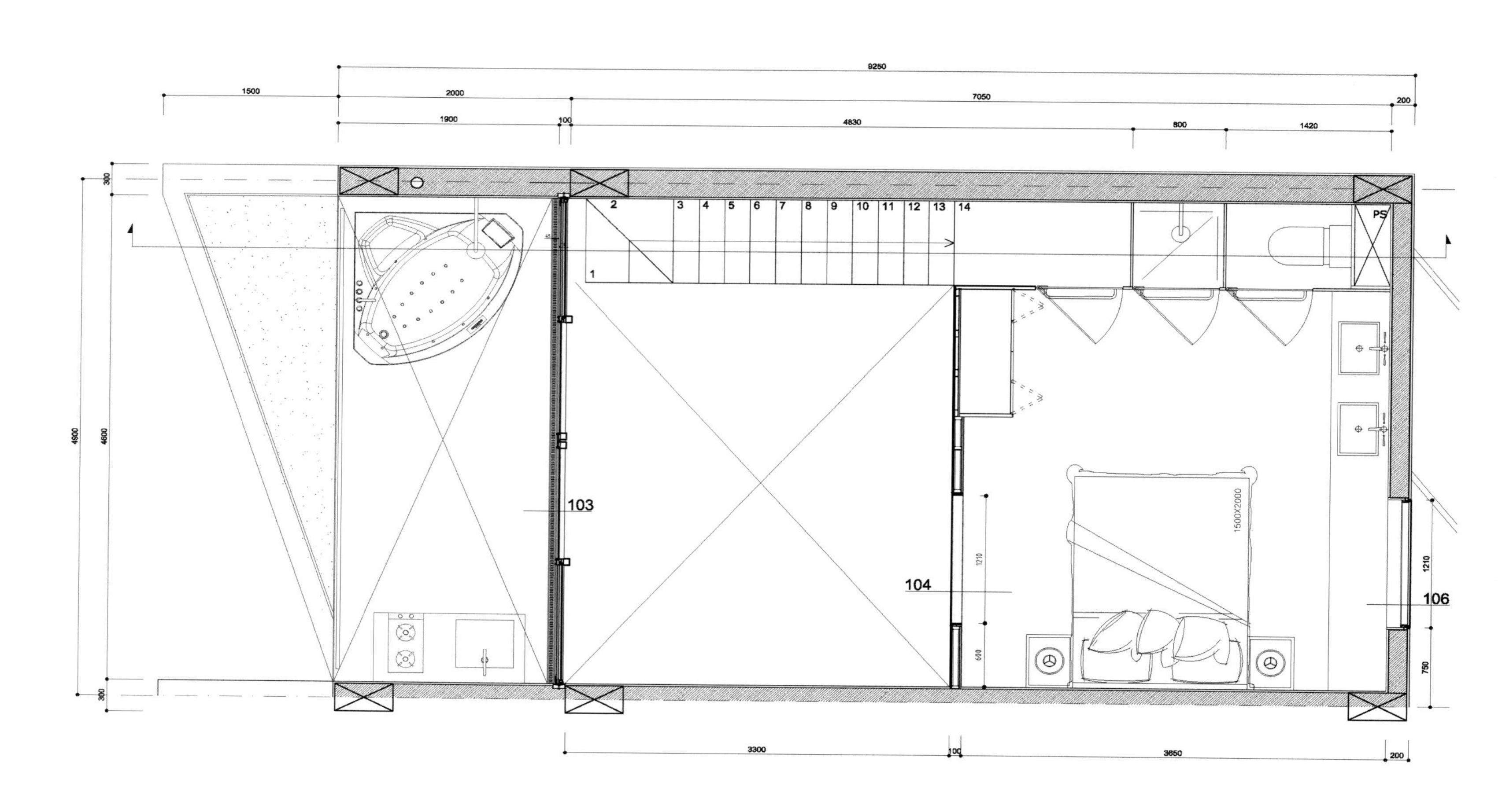

9250
1500
2000
7050
200
1900
100
4830
800
1420
1 2 3 4 5 6 7 8 9 10 11 12 13 14
PS
4900
4600
300
1500X2000
103
104
106
1210
600
750
3300
100
3650
200

fence
32 X 9mm flats stainless steel @140mm
围栏
32×9mm 不锈钢板 @140mm

steel L-section
hot-dip galvanized finish
钢 L- 剖面
热浸镀锌饰面

maintenance hatch
维修舱口

Balcony - Floor:
Bamboo Deck t=20mm W=137mm @140mm
floor joist 30 x 30mm @455mm
stainless steel adjuster
阳台 - 楼层:
竹平台 t=20mm W=137mm @140mm
楼板搁栅 30×30mm @455mm
不锈钢调节器

(drainage slope 2/100)
排水坡 2/100

4mm silicate rendering painted white
20mm water-resistant gypsum plaster
reinforced concrete
Balcony - Eave:
4mm 硅酸盐白色涂漆
20mm 防水石膏灰泥
钢筋混凝土
阳台 - 檐口

fence
32 X 9mm flats stainless steel @140mm
围栏
32×9mm 不锈钢板 @140mm

1500 2000 300 1500 200 150 200 200 1100 560 140 140 140 140

102

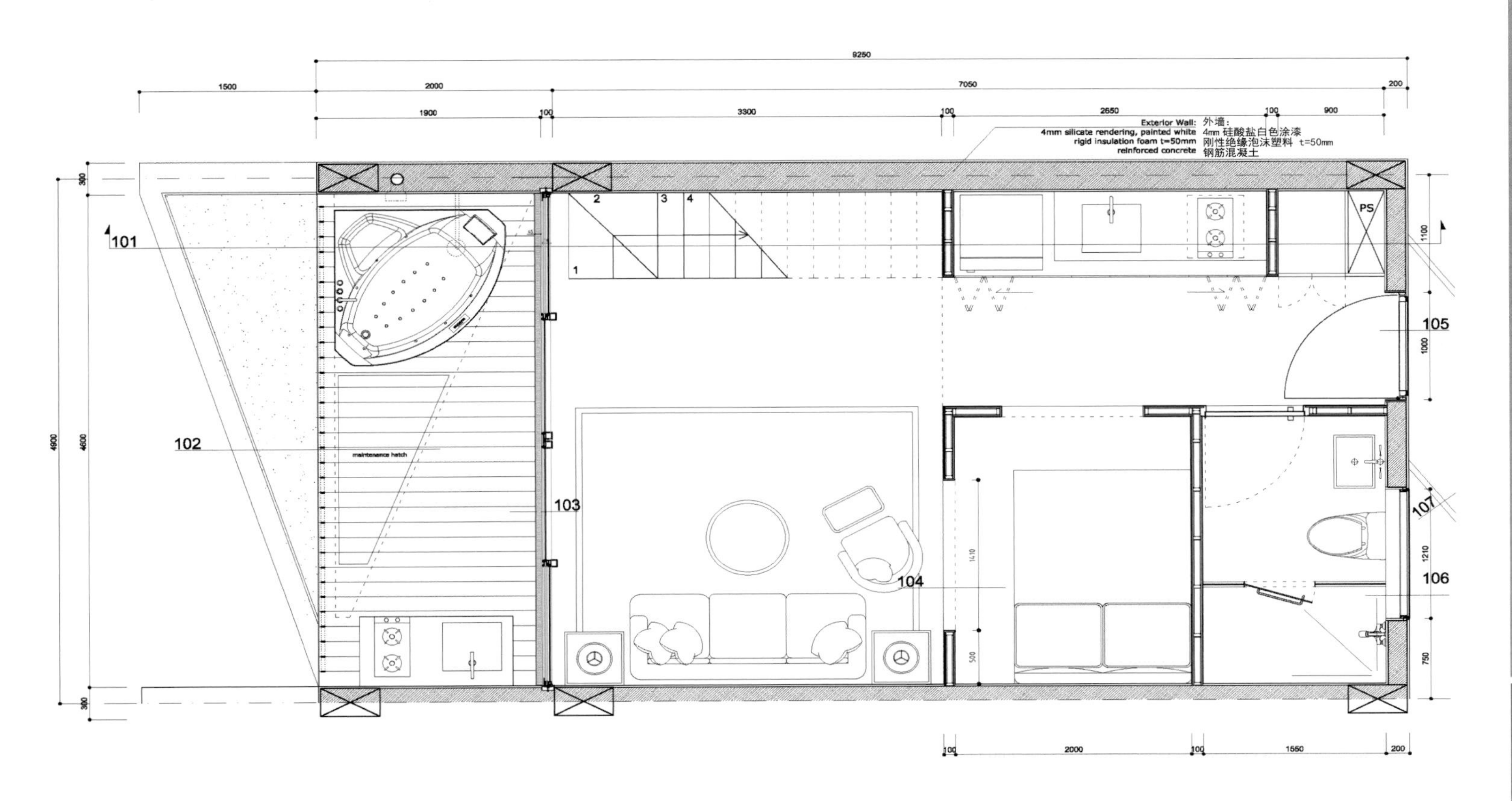

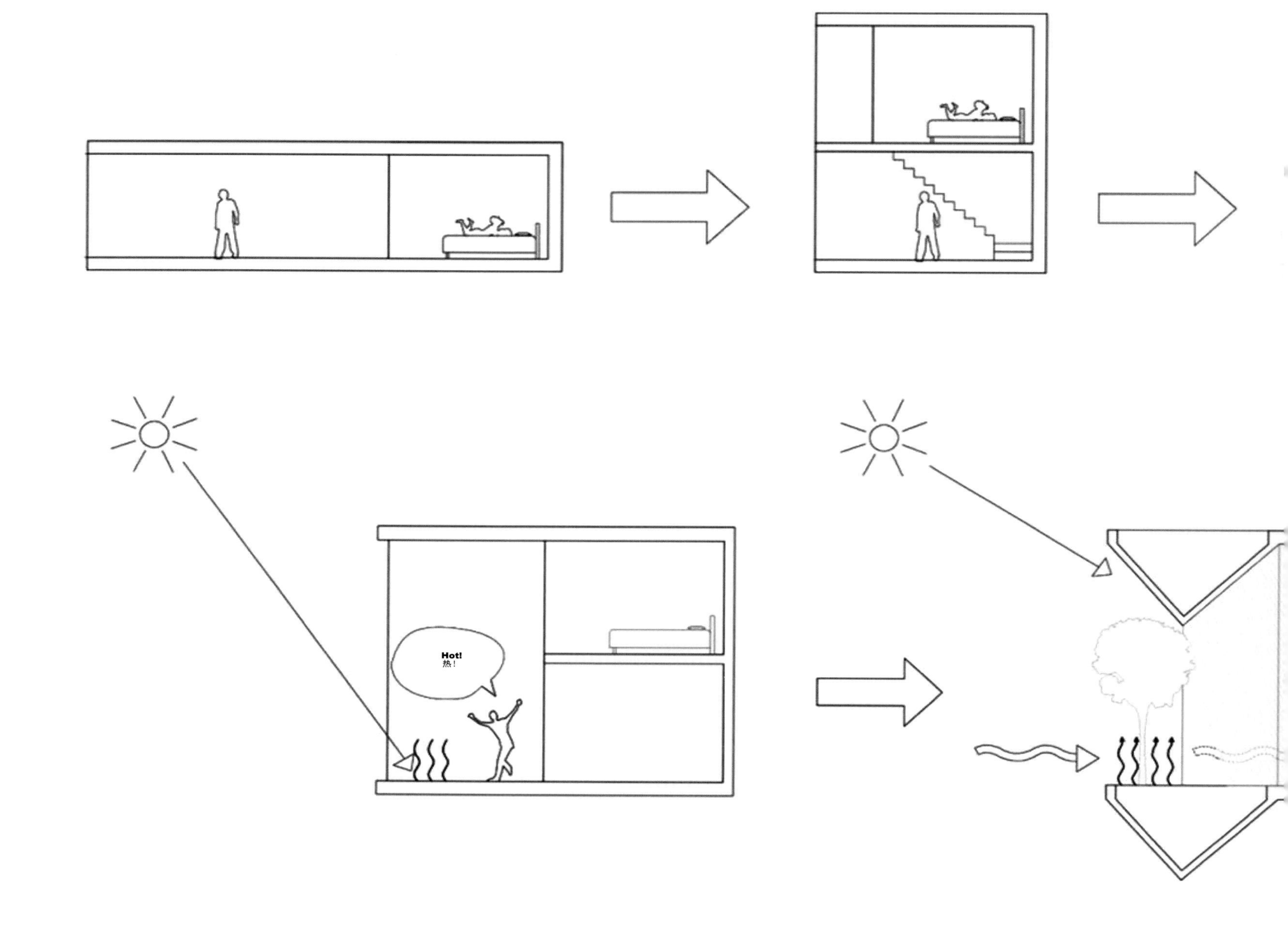
Hot!
热!

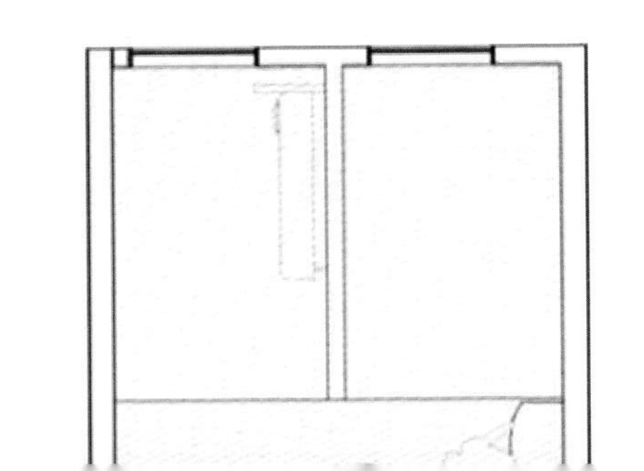

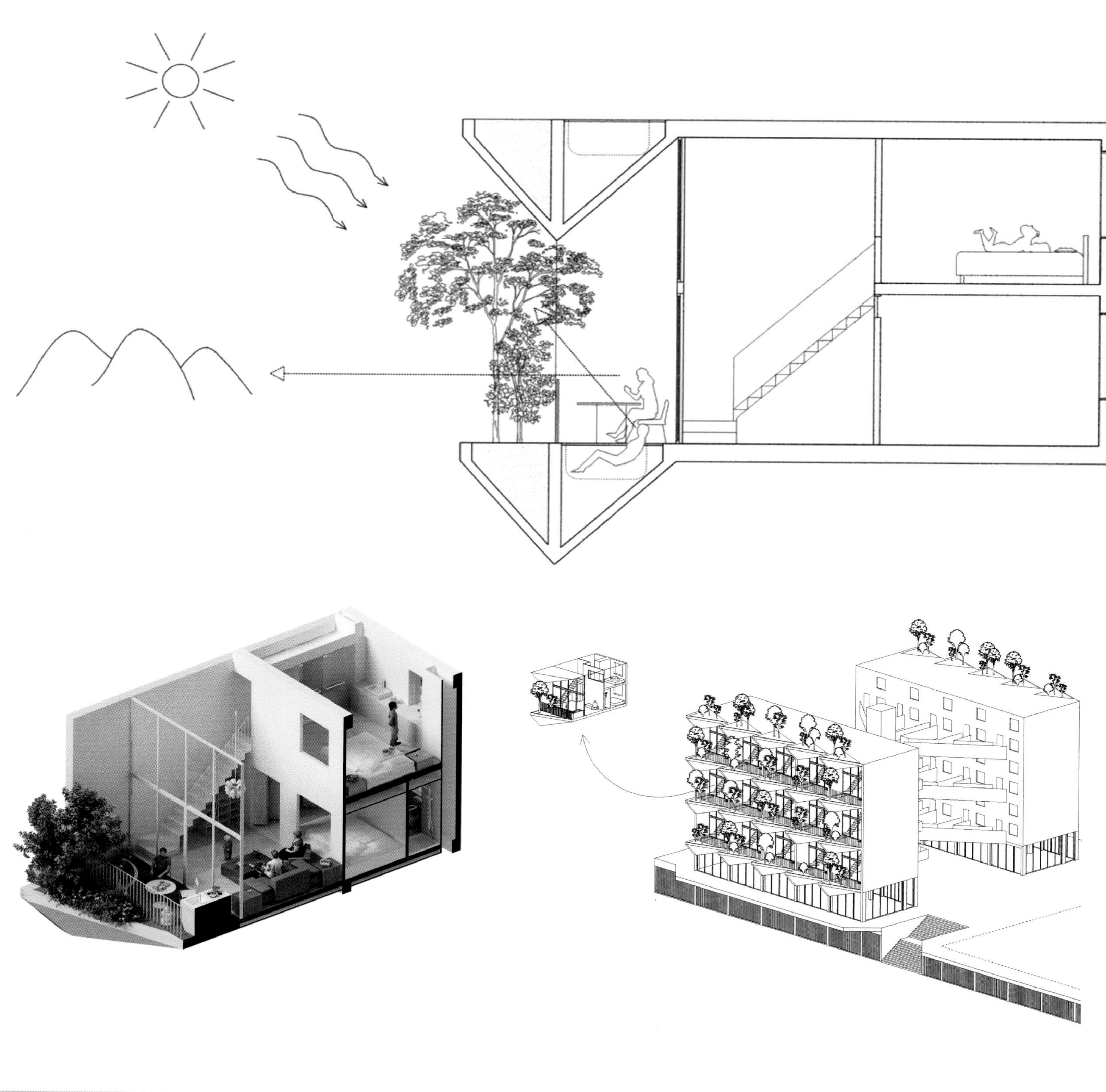

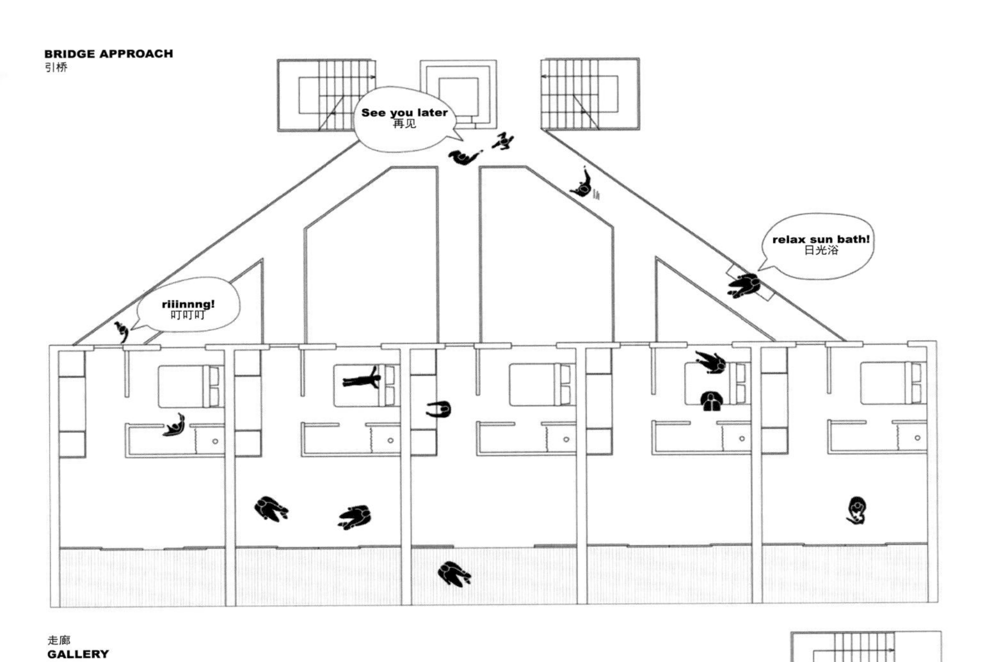

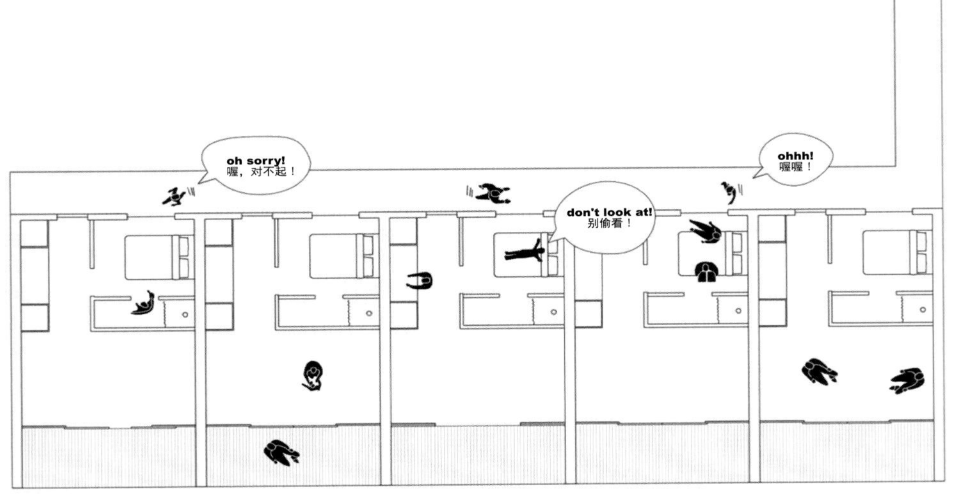

Architectural Design

The project consists of 8 blocks of 6 stories on top of a ground floor with restaurants, bars and retail. The blocks are placed on a public deck, a new ground level that covers the basement with parking spaces and service. Each block is built up from 15 individual units. Four hotels will be built consisting of two mirrored blocks each. One block is facing the street life and the other has a view over the surrounding landscape.

Design Feature

The two units are connected by an intricate 3D "Maze". From the elevator core a refined network branches out, replacing an obligatory gallery. This spectacular "infrastructure" forms a super-sized sculpture that "stitches" the blocks together.

The hotel rooms have a unique organization. The units are organized over two levels. A double height space can now be introduced, providing a spacious ambience. The double height of the duplex room is extended to the outside: a wonderful balcony comes into being, intimate but spacious. The balcony serves as an out-door living room. There will be an open-air kitchen and a hot-tub with shower.

The hotels feature a large triangular "Flowerpot" for each room. Together they create a rhythmic, dynamic pattern. The lush greenery aspires to increase privacy, provide shade and cooling and will create a natural atmosphere. Perhaps even a micro ecology.

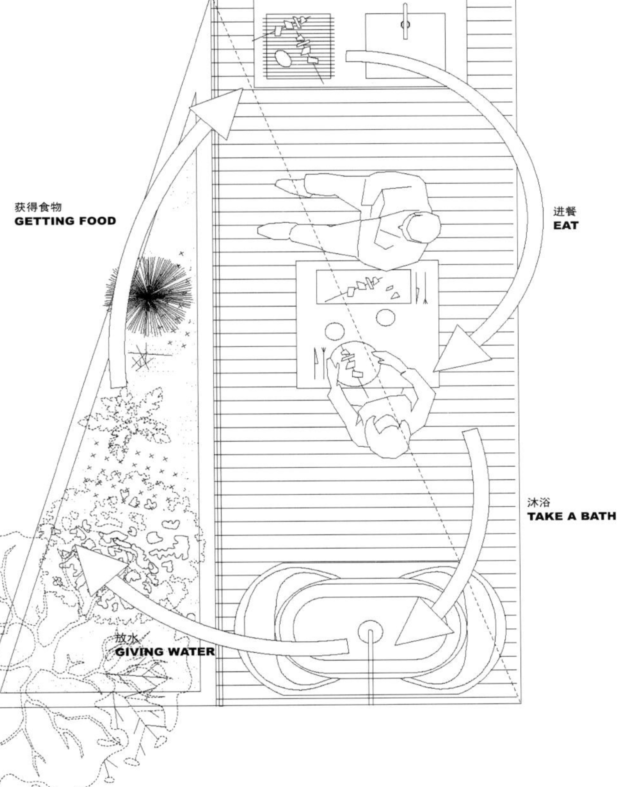

商业建筑
Commercial Building

土耳其伊斯坦布尔坎米利卡山电讯塔

Telecommunications Tower in Çamlica Hill

设计单位：RTA-Office 建筑事务所
合作单位：Santiago Parramón
Dome Partners
开发单位：伊斯坦布尔市政当局
项目地址：土耳其伊斯坦布尔
总用地面积：14 000 m²
总建筑面积：21 556 m²
摄影：RTA-Office 建筑事务所

Designed by: RTA-Office
Collaboration: Santiago Parramón; Dome Partners
Client: Istanbul Metropolitan Municipality
Location: Istanbul, Turkey
Total Land Area: 14,000 m²
Total Building Area: 21,556 m²
Photography: RTA-Office

项目概况

伊斯坦布尔的美丽和独特归功于周边环境的壮丽景观，独特的自然环境也使该地形成了“依地形而建”的建筑特色和传统。这个由 RTA-Office 建筑事务所设计的坎米利卡山电讯塔，重点在于对场地价值的挖掘以及对场地的整体塑造，建筑不是居于主导地位，而是成为对场地和绝佳的视野完美回应的载体。

设计理念

成功的设计理念往往取决于对场地的理解，设计师认为场地的景观不应该被建筑吞噬，而是应该建造一个展现自然元素的物体。设计虽简单，但是自然，它是一个不属于城市的建筑，却能自然地融于自然景观中，凸显当地的自然属性。同时，这也是一个形态自由的建筑，它可视为一个雕刻的尖塔，也可视为一座 21 世纪的灯塔，留给人们一个自由想象的空间。

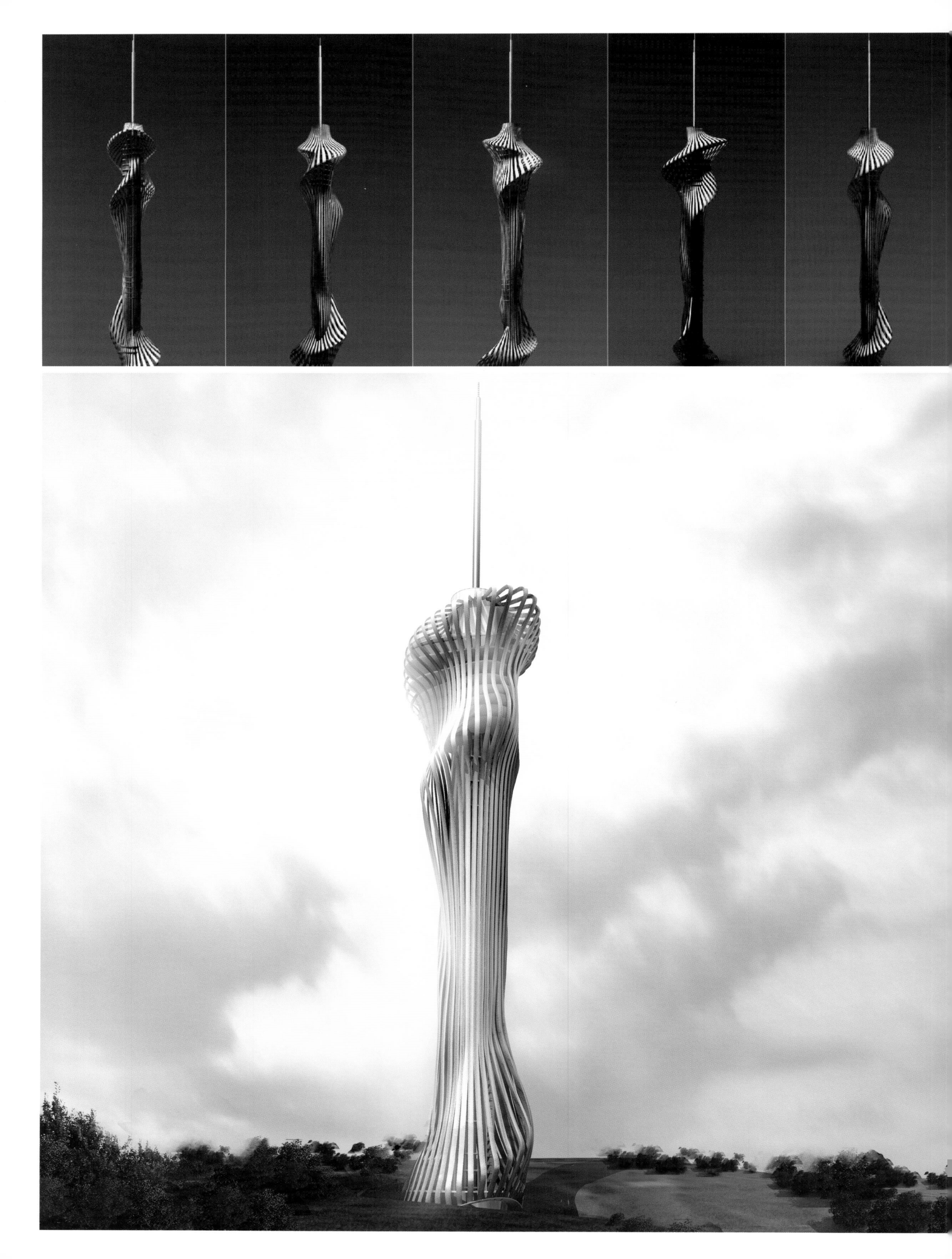

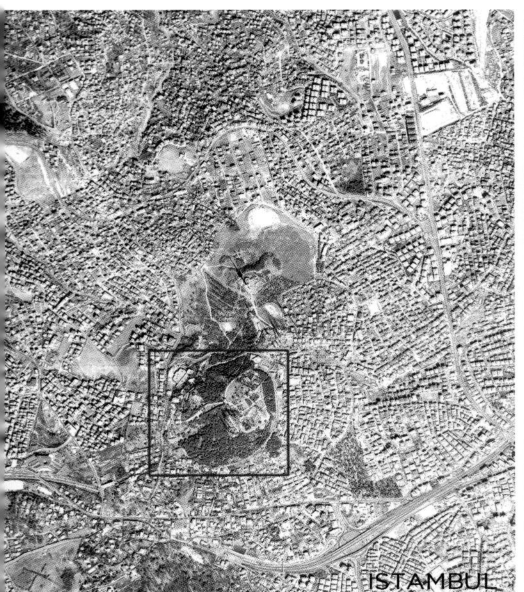

设计特色

这是一个自由形态的建筑，塔身是一个兼顾了结构性和技术性的核心体，由核心和其他功能组成，也包含了 MEP 和垂直交通空间。主体呈螺旋状盘旋上升，平面的大小也随之不断地变动，从而形成塔楼不同的功能区。

塔身的表皮覆盖着螺旋状的纹理，随着塔楼的延展，建筑表皮显得灵活而动感。塔身的纹理也会因观赏距离的远近而发生变化：从远处看，它是一个坚固的整体；走近后则会呈现出网面的造型。

为了突出建筑的形态，设计师在表皮中装置了 LED 灯，核心部分采用了逆向的 LED 灯，以突出核心。同时，设计包含了几种不同的光效，使光线能够穿透表皮。

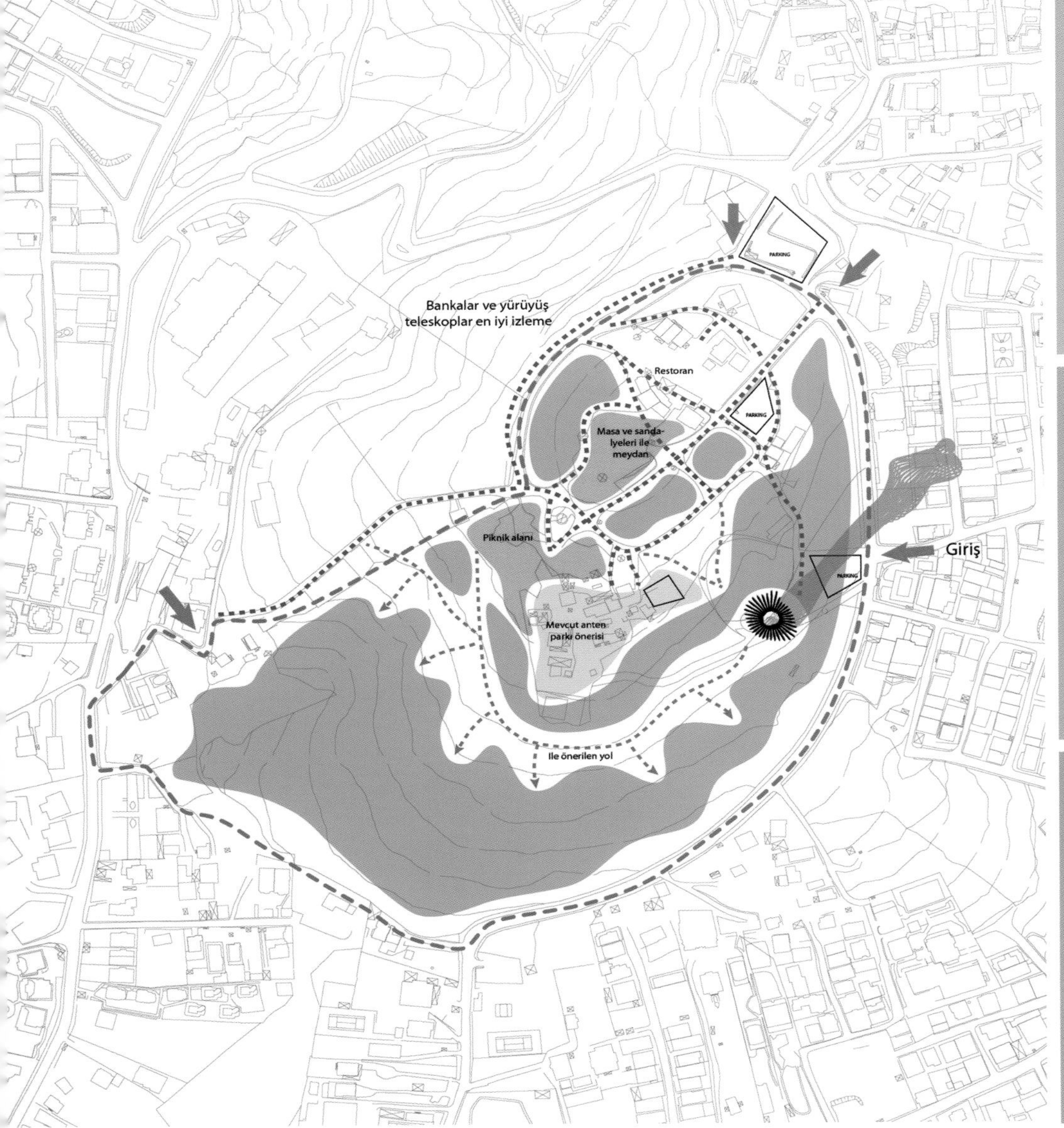

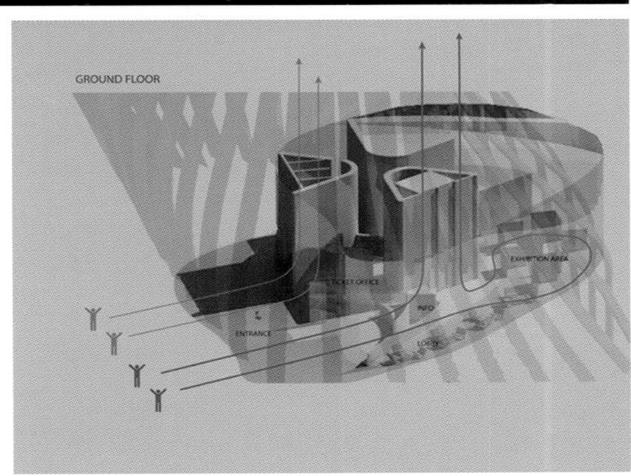

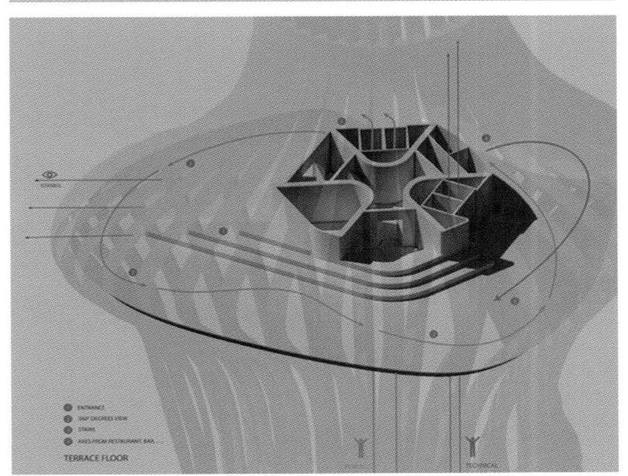

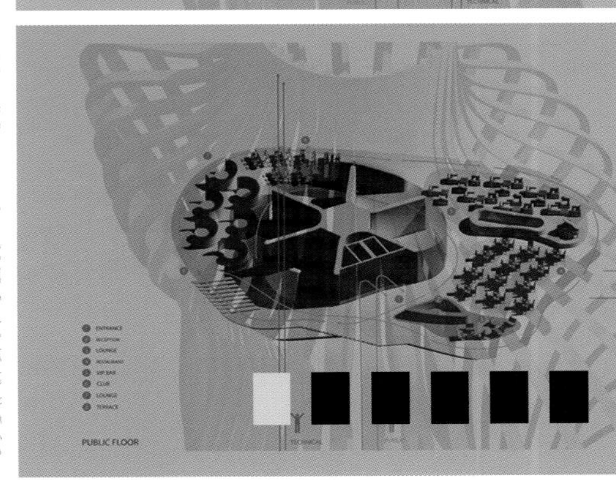

Seviye 40 (+210 m)
200 m2
Seviye 41 (+215 m)
200 m2
Seviye 36 (+190 m)
636 m2
Seviye 37 (+195 m)
723 m2
Seviye 31 (+165 m)
302 m2
Seviye 32 (+170 m)
621 m2
Seviye 33 (+175 m)
646 m2
874m2
Seviye 1 (+10 m)
877 m2
Seviye 15 (+85 m)
300 m2
Seviye 16 (+90 m)
302 m2

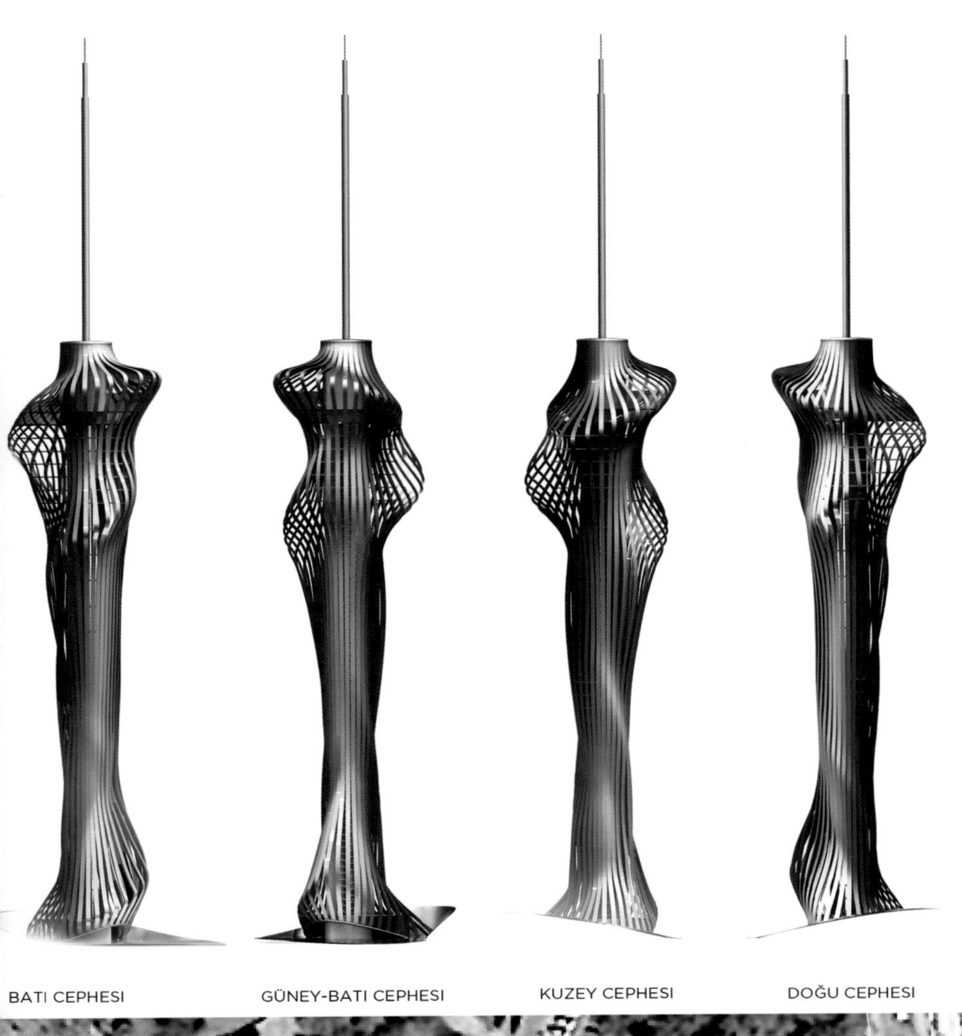
BATI CEPHESI
GÜNEY-BATI CEPHESI
KUZEY CEPHESI
DOĞU CEPHESI

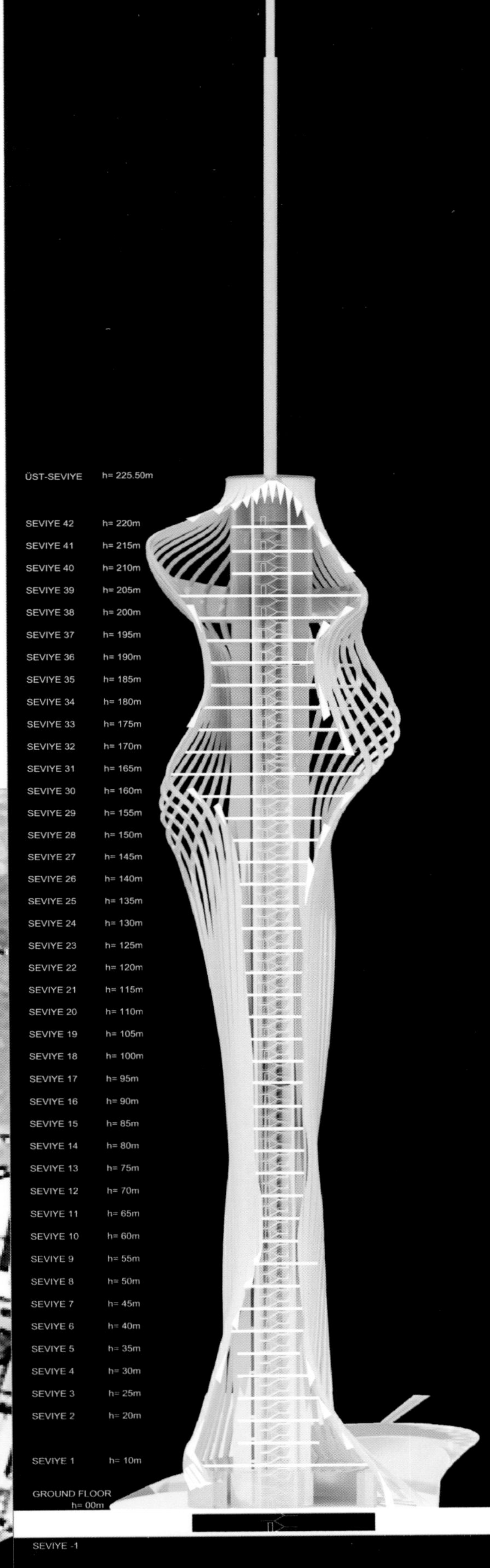
ÜST-SEVIYE h= 225.50m
SEVIYE 42 h= 220m
SEVIYE 41 h= 215m
SEVIYE 40 h= 210m
SEVIYE 39 h= 205m
SEVIYE 38 h= 200m
SEVIYE 37 h= 195m
SEVIYE 36 h= 190m
SEVIYE 35 h= 185m
SEVIYE 34 h= 180m
SEVIYE 33 h= 175m
SEVIYE 32 h= 170m
SEVIYE 31 h= 165m
SEVIYE 30 h= 160m
SEVIYE 29 h= 155m
SEVIYE 28 h= 150m
SEVIYE 27 h= 145m
SEVIYE 26 h= 140m
SEVIYE 25 h= 135m
SEVIYE 24 h= 130m
SEVIYE 23 h= 125m
SEVIYE 22 h= 120m
SEVIYE 21 h= 115m
SEVIYE 20 h= 110m
SEVIYE 19 h= 105m
SEVIYE 18 h= 100m
SEVIYE 17 h= 95m
SEVIYE 16 h= 90m
SEVIYE 15 h= 85m
SEVIYE 14 h= 80m
SEVIYE 13 h= 75m
SEVIYE 12 h= 70m
SEVIYE 11 h= 65m
SEVIYE 10 h= 60m
SEVIYE 9 h= 55m
SEVIYE 8 h= 50m
SEVIYE 7 h= 45m
SEVIYE 6 h= 40m
SEVIYE 5 h= 35m
SEVIYE 4 h= 30m
SEVIYE 3 h= 25m
SEVIYE 2 h= 20m
SEVIYE 1 h= 10m
GROUND FLOOR h= 00m
SEVIYE -1

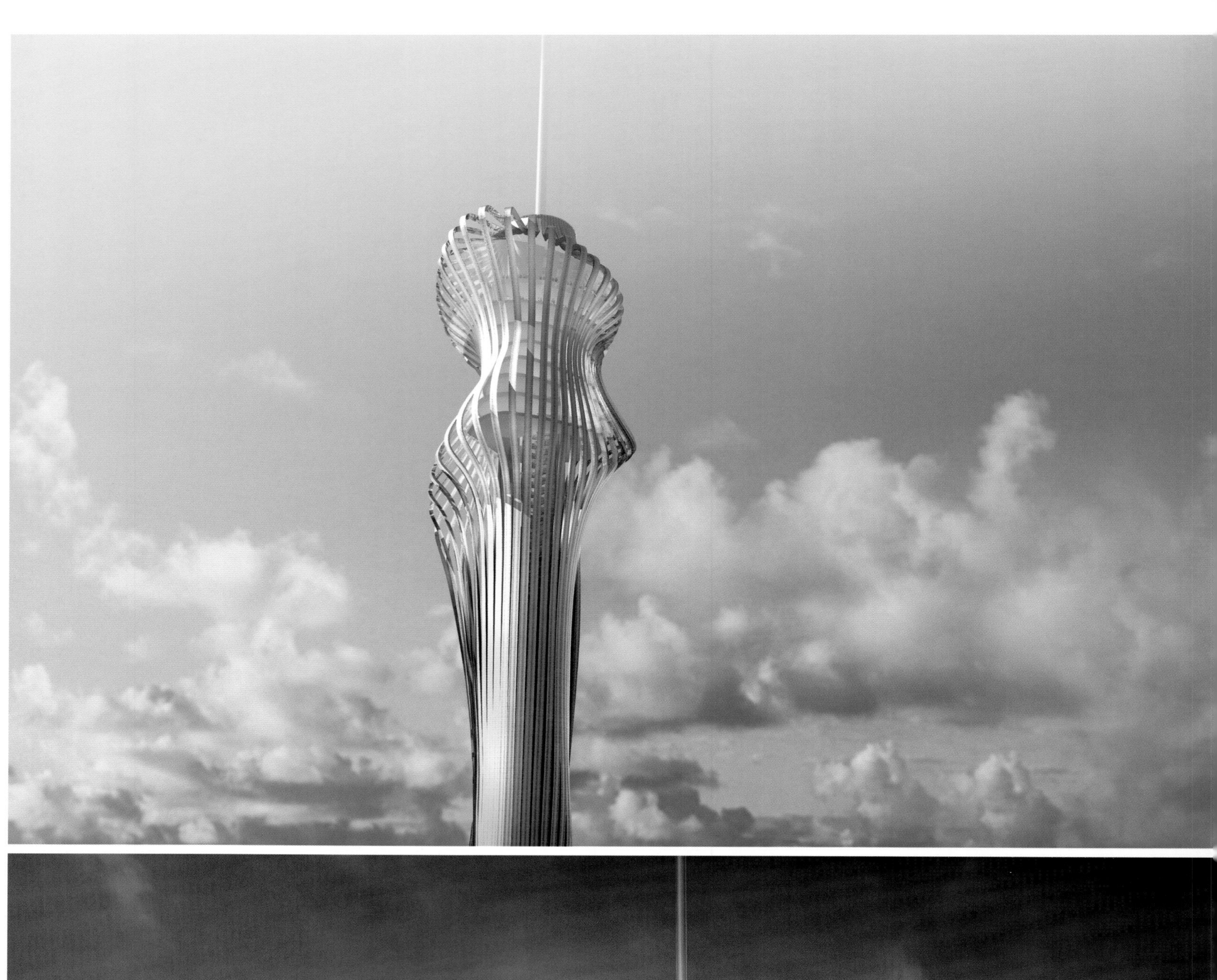

Profile

The beauty of Istanbul and this particular place is due, largely, to the environment, the landscape, the Bosporus, the Marmara Sea, the Black sea, the Golden Horn and the nearby mountains. The Telecommunications Tower designed by RTA-Office focuses on exploration of site value and overall shape of the site. The Tower no longer situates in a dominant position, but becomes a carrier perfectly in response to this site and wonderful views.

Design Concept

A successful design concept generally depends on the understanding of the site. Designers believe that natural landscape should not be swallowed by the architecture. The proposal offers an end to the conquest of the buildings and artifacts here. The design has not built any kind of building, but an object that clearly reveals the natural attributes of the place. In the meantime, it is a free-form building that can be regarded as a minaret sculpted by the wind, a twenty-first century lighthouse or a free imagination space.

Design Feature

It is a free-form tower whose body has a structural and technical role composed of the core, other functions, MEP and vertical transportation space. The thread is a flexible element. The sum of threads constitutes a flexible dress that adapts to the body and gives shape to the tower.

The texture of the tower changes depending on the distance in which the observer is located. It looks like a solid form from the distance and transforms into a mesh when the observer approaches to the tower.

LED lamps are embedded inside of the skin to highlight architectural form while the core using inverse LED lamps to heighten its effect. Several different kinds of lighting effects make light permeable.

上海芦潮港产业服务研发中心

Shanghai Luchao Port Industrial Service Research & Development Center

设计单位：加拿大 CPC 建筑设计顾问公司
开发商：临港芦潮港经济发展有限公司
项目地址：中国上海市
用地面积：15 204 m^2
建筑面积：53 000 m^2
建筑密度：35.1%
绿地率：30%
容积率：2.35
设计团队：邱 江　韩 强　Lisandro Ardusso　姚 颖

Designed by: Coast Palisade Consulting Group
Developer: Lingang Luchao Port Economic Development Co., Ltd.
Location: Shanghai, China
Site Area: 15,204 m^2
Floor Area: 53,000 m^2
Building Density: 35.1%
Greening Ratio: 30%
Plot Ratio: 2.35
Design Team: Qiu Jiang, Han Qiang, Lisandro Ardusso, Yao Ying

项目概况

项目位于浦东新区临港装备产业园区与临港新城主城区之间的芦潮港社区核心商贸区，位于港辉路以东，江山路以南，芦硕路以北，东侧紧邻芦潮港镇社区文化活动中心，地理位置优越。

建筑设计

设计延续了“在海上航行的两艘船只”这一设想，为两个建筑设计了起伏的屋顶。南面建筑屋顶的倒角可最大限度地为北块建筑引入自然光，也形成了由北到南、从高到低逐步递减的建筑形态。

建筑内部的立面设计以航海为主题，采用了远洋轮的元素，建有人行通道和凸窗。位于建筑内部的“空中街道”，演绎了当船舶停靠港口时，桥梁和舷梯从船只通向陆地的情景。

景观设计

在项目的景观设计中，设计师着重对建筑组群围合的内庭院和用地西北角的城市广场进行精心设计。建筑组群围合的内庭院被设计为完全步行，使其成为周边建筑在空间和功能上的延伸，同时将建筑群联结为一个整体。

与安静怡人的内庭院不同，城市广场紧邻城市干路，是该区的“脸面”和形象，也是汇集人流和人气的场所，同时考虑到小区绿地率的要求以及地下车库的实际情况，项目形成了以大片草坪为主、结合部分台地景观的立体绿化体系。

Profile

The project is located in Luchao Port Community core business district, between Pudong New District Lingang Equipment Industrial Park and Lingang New Town main urban district. It enjoys a superior location adjoining Ganghui Road to the east, Jiangshan Road to the south, Lushuo Road to the north, close to Luchao Port Community Cultural Center to the east.

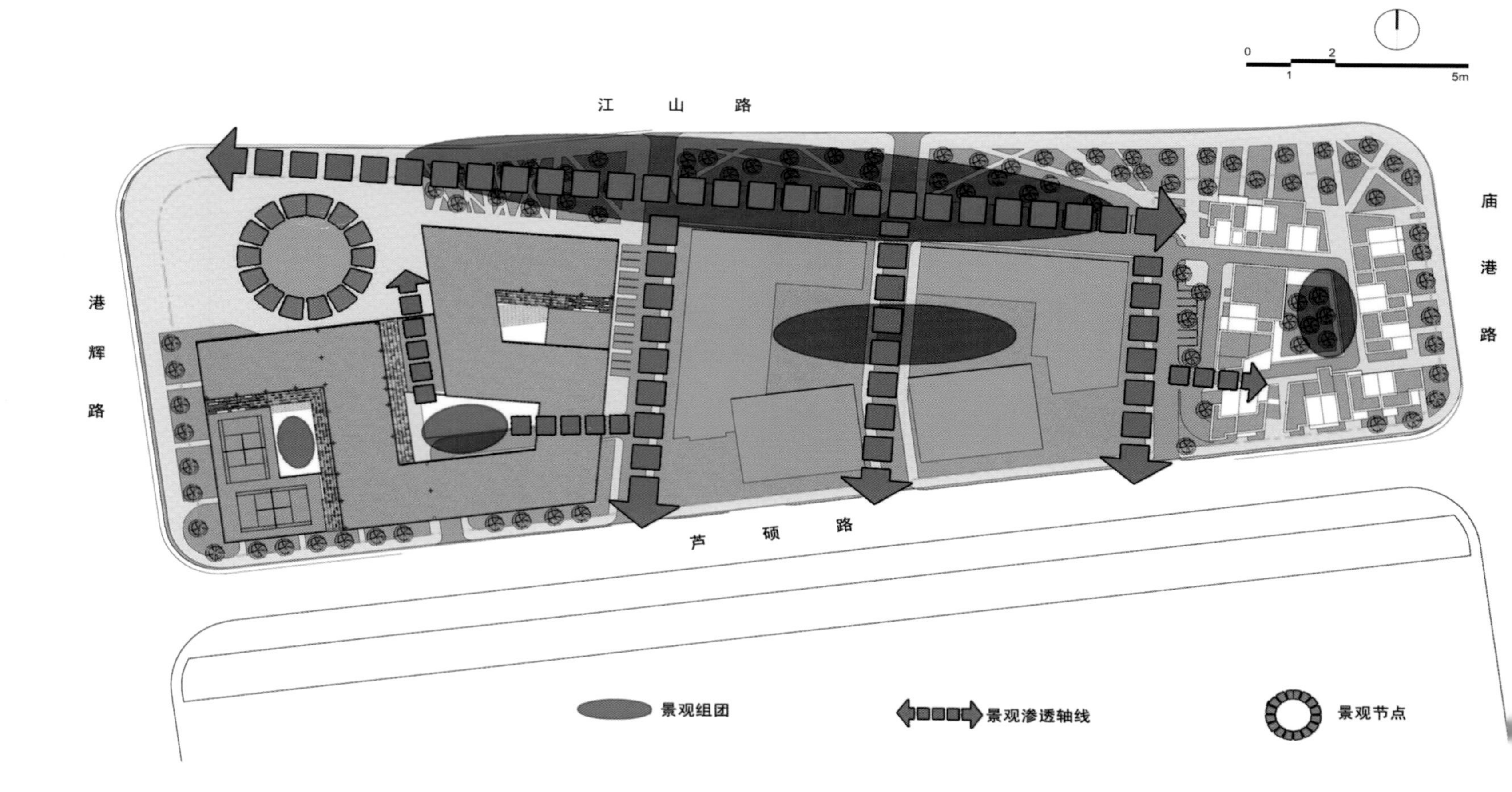

0
2
1
5m
江 山 路
港
辉
路
庙
港
路
芦 硕 路
景观组团
景观渗透轴线
景观节点

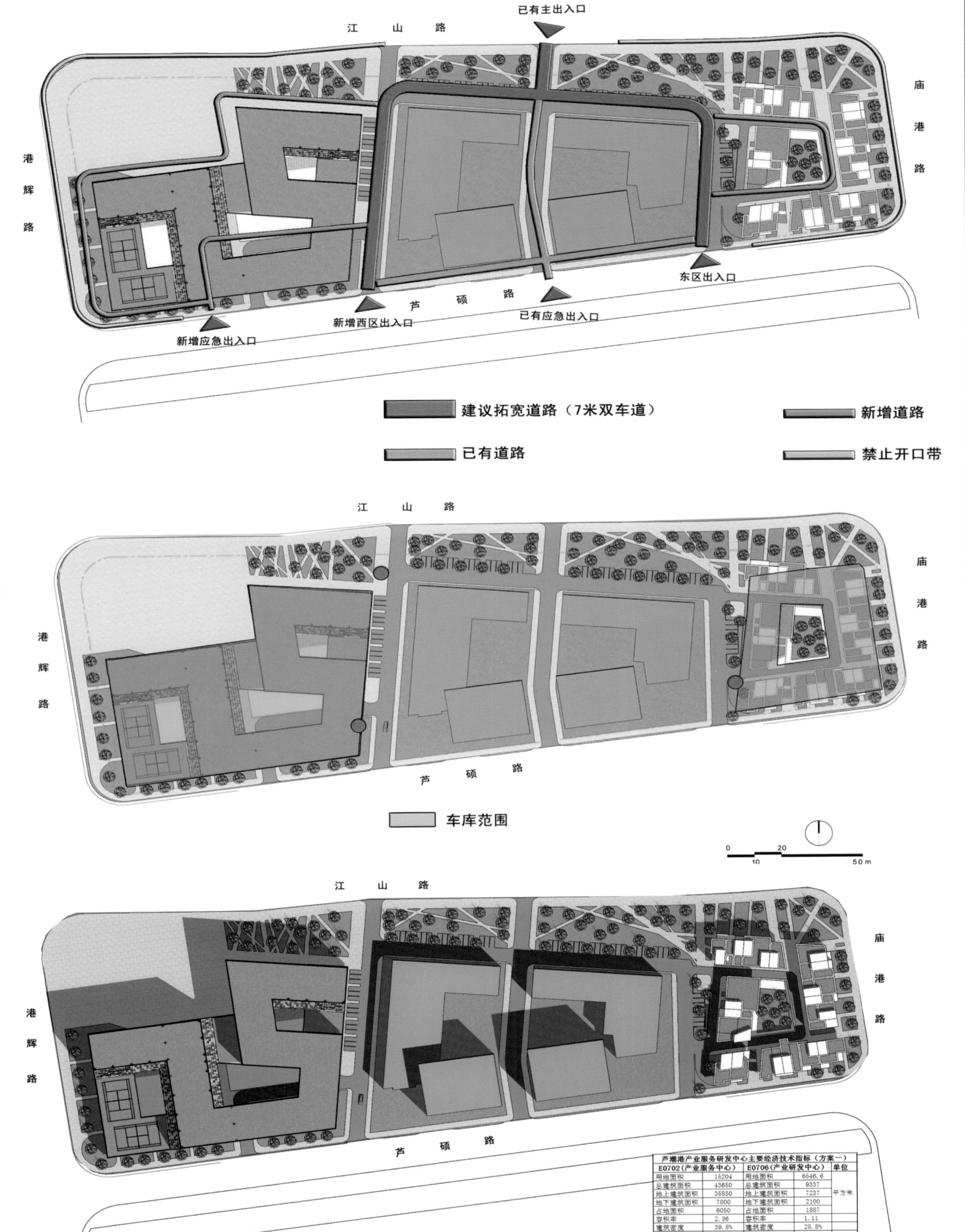

芦潮港产业服务研发中心主要经济技术指标（方案一）				
E0702（产业服务中心）		E0706（产业研发中心）		单位
用地面积	15204	用地面积	6546.6	平方米
总建筑面积	43650	总建筑面积	9337	
地上建筑面积	35850	地上建筑面积	7237	
地下建筑面积	7800	地下建筑面积	2100	
占地面积	6050	占地面积	1887	
容积率	2.36	容积率	1.11	
建筑密度	39.8%	建筑密度	28.8%	
绿化面积	4765	绿化面积	1996	
绿地率	31.3%	绿地率	30.5%	
单元套数	589	单元套数	25	套
机动车位	197	机动车位	55	辆
其中，地面停车	20	其中，地面停车	6	
地下停车	177	地下停车	49	
建筑限高	45	建筑限高	20	米

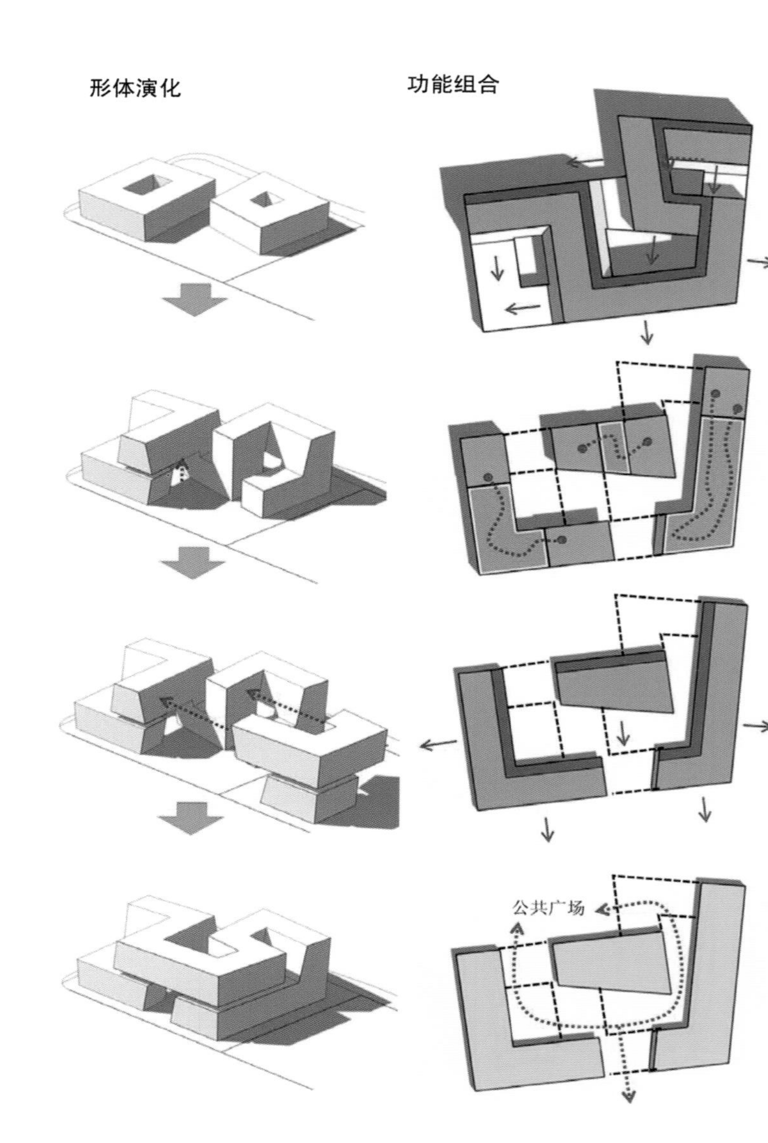

6-9层
公寓（蓝色）被布置在尽可能有利的朝向。走廊（灰色）则利用相对不利的朝向。

5层
这一层是低区和高区的转换层。基本上所有的可以用做休闲游乐区的空中花园都布置在这里。

2-4层
公寓（蓝色）被布置在尽可能有利的朝向。走廊（灰色）则利用相对不利的朝向。

1层和B1层
这里是布置商业空间和大堂的地方。利用建筑之间的空间自然形成中庭。同时也有利于外部公共空间与内部中庭和公共广场的连接。

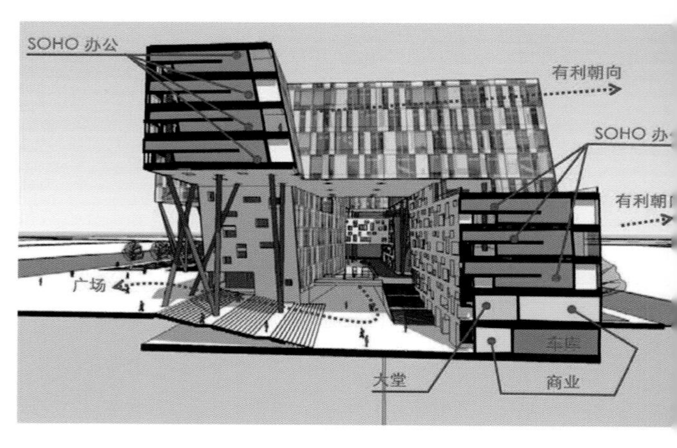

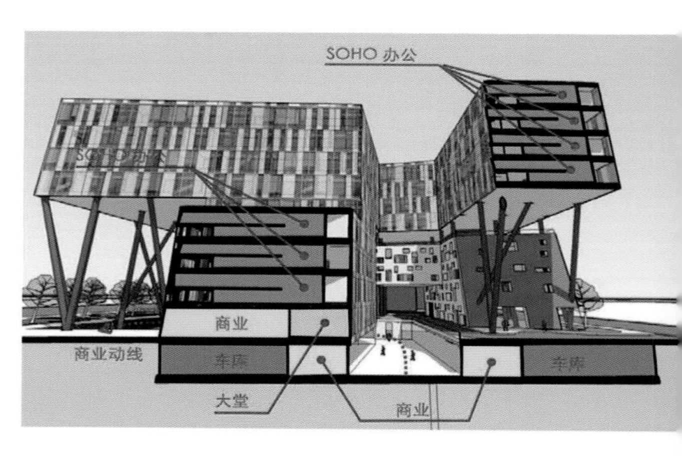

E0706就没有E0702地块的不利因素。相对方正的空间以及办公功能，都非常有利于中庭空间的布置。

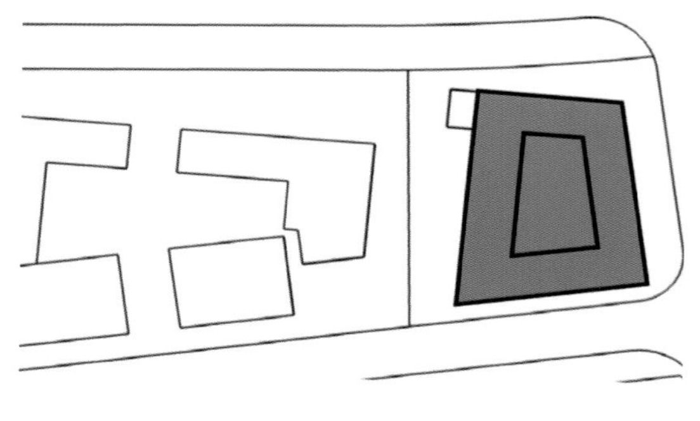

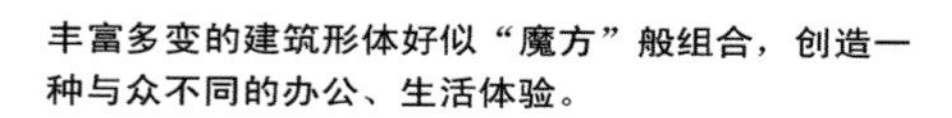

丰富多变的建筑形体好似“魔方”般组合，创造一种与众不同的办公、生活体验。

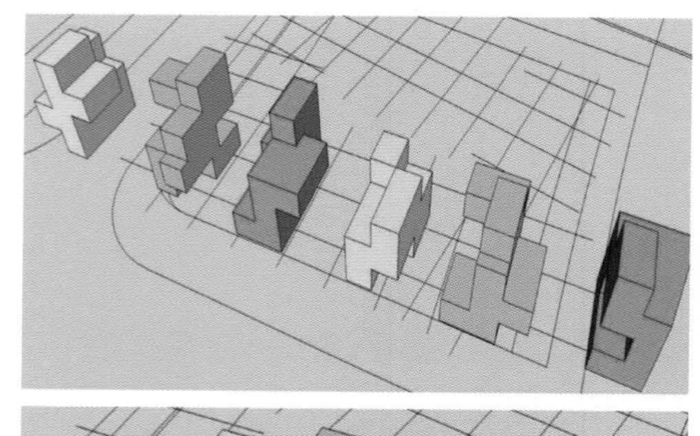

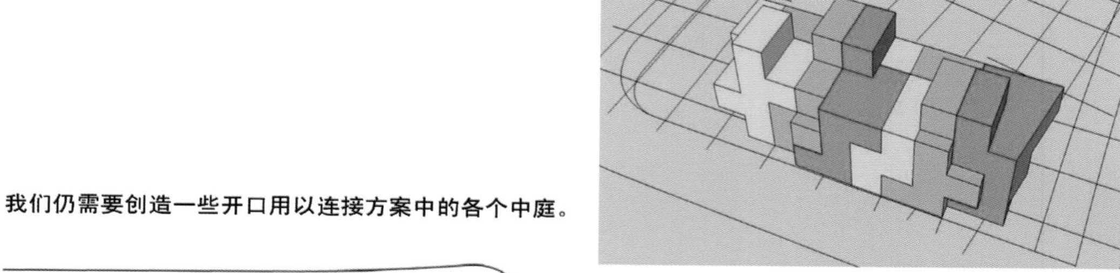

我们仍需要创造一些开口用以连接方案中的各个中庭。

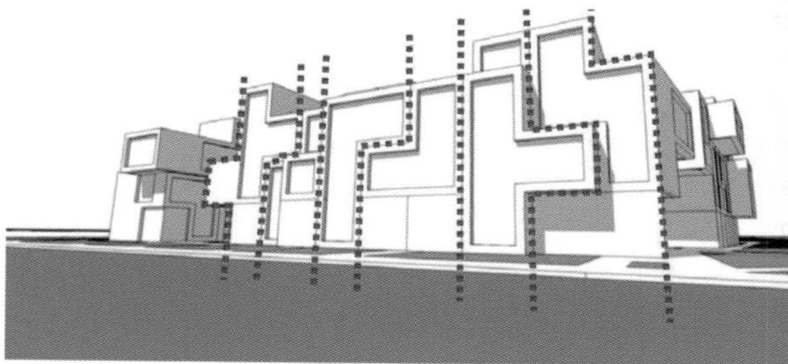

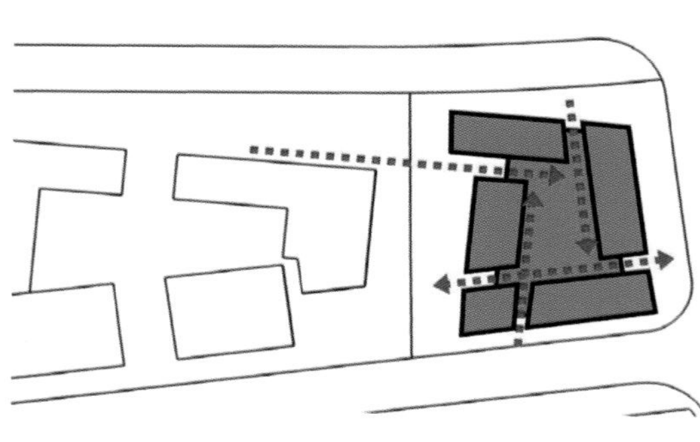

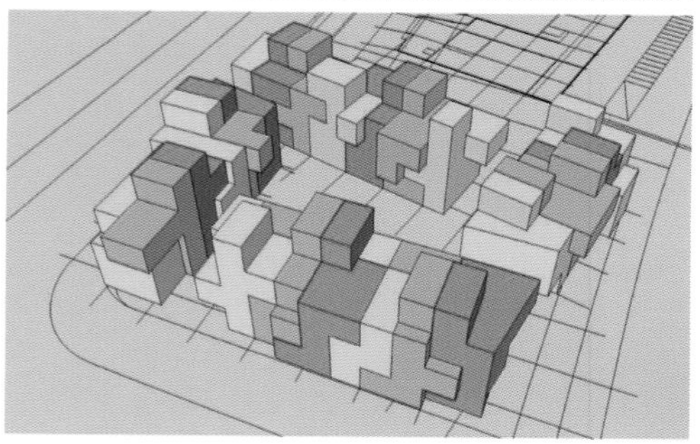

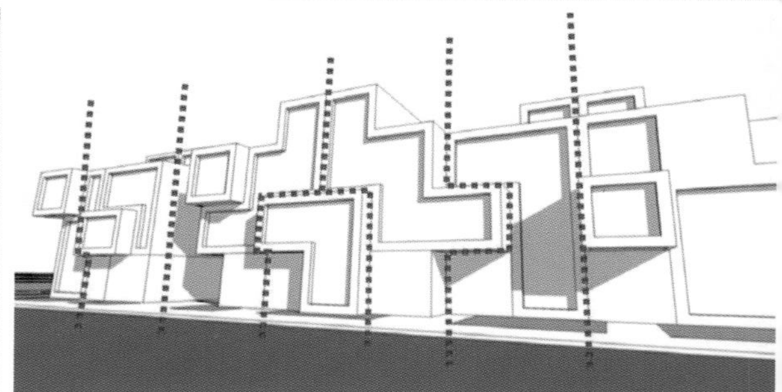

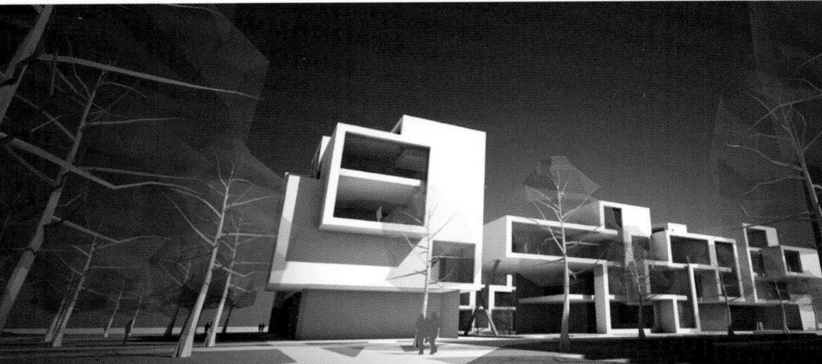

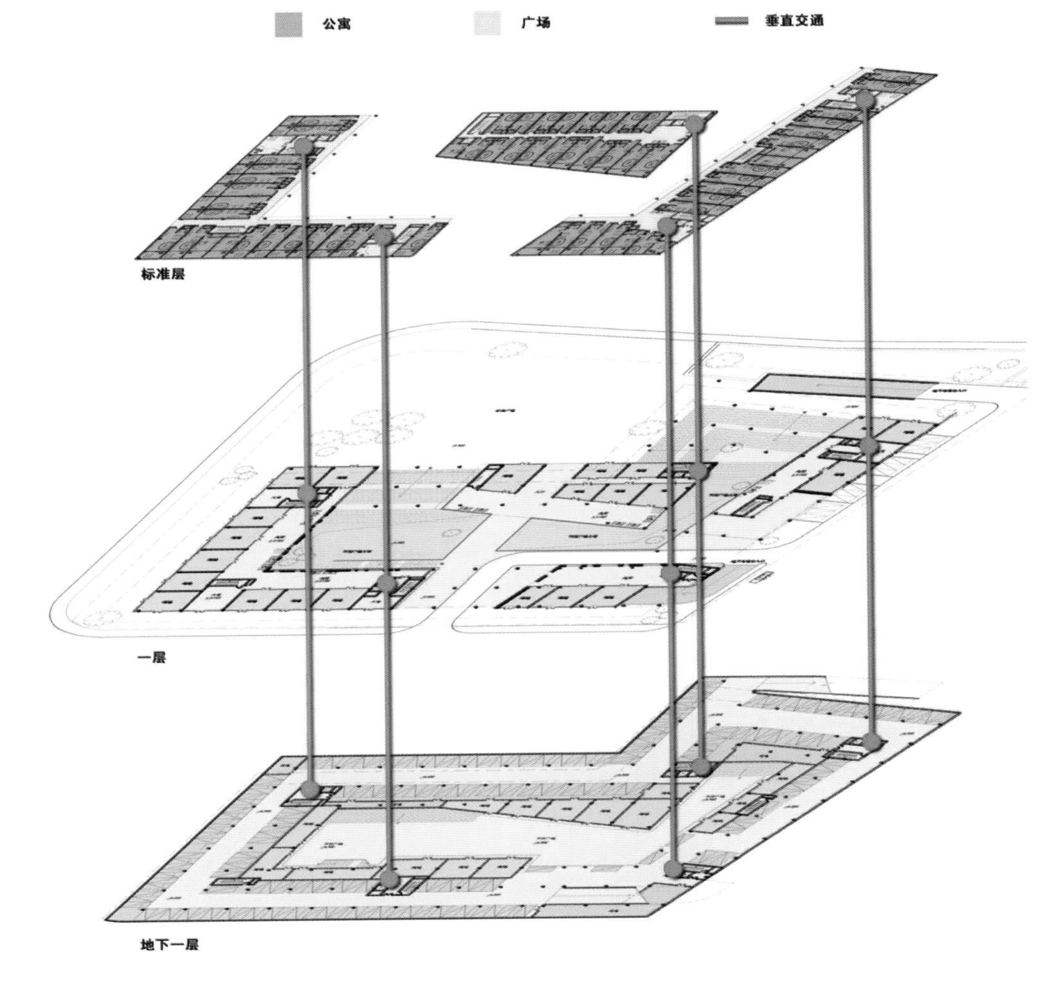

Architectural Design

Project design has continued the vision of "Two Ships Sailing on the Sea", constructing undulating roofs for two buildings. The chamfer of the building in the south maximizes natural light for the building in the north, creating a progressively decreasing building height from north to south.

The internal elevation design of the buildings is themed with navigation, characterized by the element of ocean liner as well as decorated by passages and bay-windows. The "Air Street" inside the buildings presents the scene of bridge and ladder stretching to the land when ships are anchoring.

Landscape Design

In the landscape design of the project, designers have exquisitely designed the internal courtyard enclosed by building cluster and the northwest urban plaza. The internal courtyard enclosed by building cluster is a courtyard catering for pedestrians. It is supposed to be a continuity of surrounding buildings on space and function as well as to integrate the building cluster into a whole.

Different from the tranquil pleasant internal courtyard, the urban plaza closely adjoins urban main road, which is regarded as the "Face" and image of this district accumulating pedestrians and popularity. For the consideration of greening ratio and actual condition of underground garage, the project has established stereoscopic green system dominated by large-scale lawns and local terrace landscape.

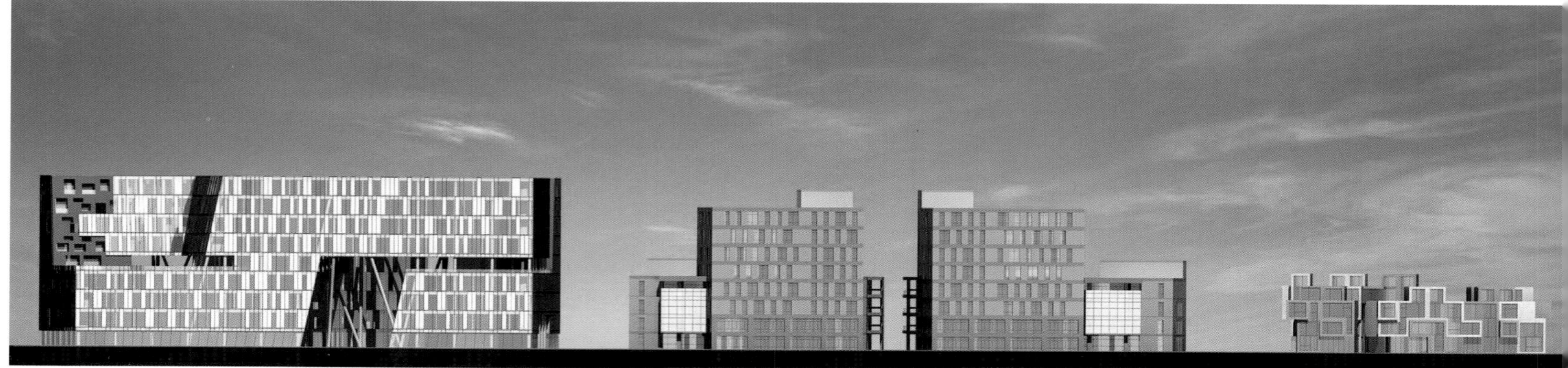
沿街南立面

沿街北立面

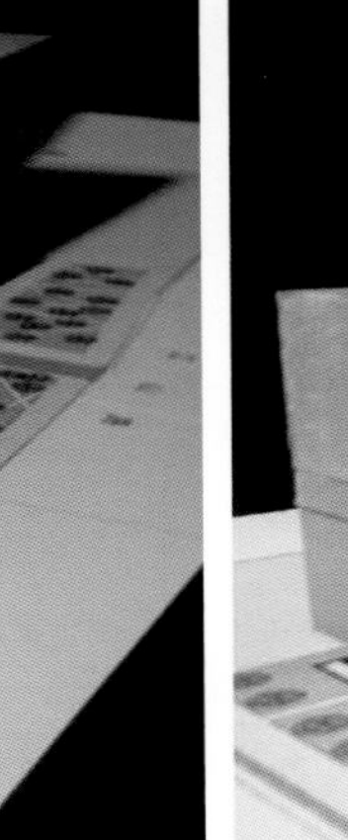

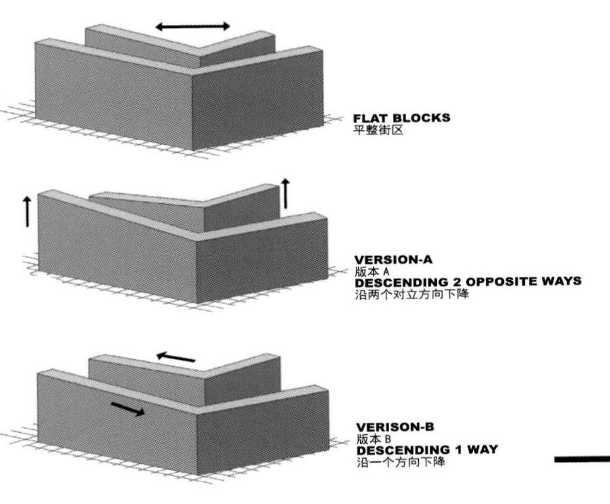
FLAT BLOCKS
平整街区
VERSION-A
版本 A
DESCENDING 2 OPPOSITE WAYS
沿两个对立方向下降
VERISON-B
版本 B
DESCENDING 1 WAY
沿一个方向下降

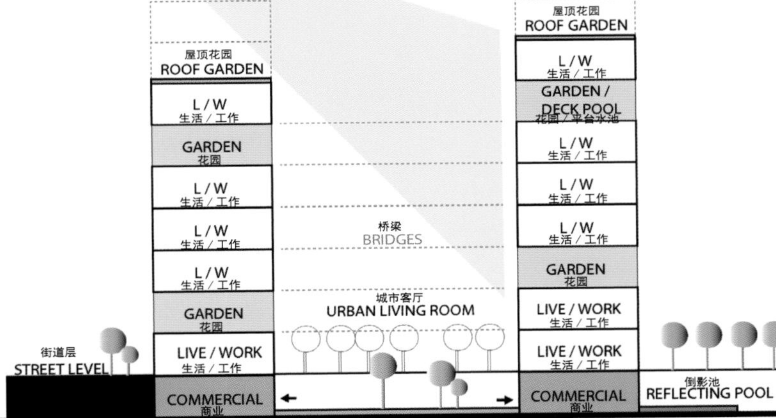
屋顶花园
ROOF GARDEN
L / W
生活 / 工作
GARDEN
花园
L / W
生活 / 工作
L / W
生活 / 工作
L / W
生活 / 工作
GARDEN
花园
LIVE / WORK
生活 / 工作
COMMERCIAL
商业
街道层
STREET LEVEL
桥梁
BRIDGES
城市客厅
URBAN LIVING ROOM
屋顶花园
ROOF GARDEN
L / W
生活 / 工作
GARDEN / DECK POOL
花园 / 平台水池
L / W
生活 / 工作
L / W
生活 / 工作
L / W
生活 / 工作
GARDEN
花园
LIVE / WORK
生活 / 工作
LIVE / WORK
生活 / 工作
COMMERCIAL
商业
倒影池
REFLECTING POOL
街道层
STREET LEVEL
NORTH
北面

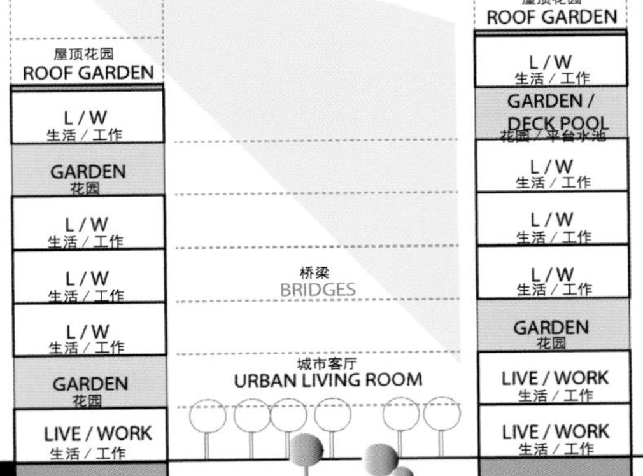

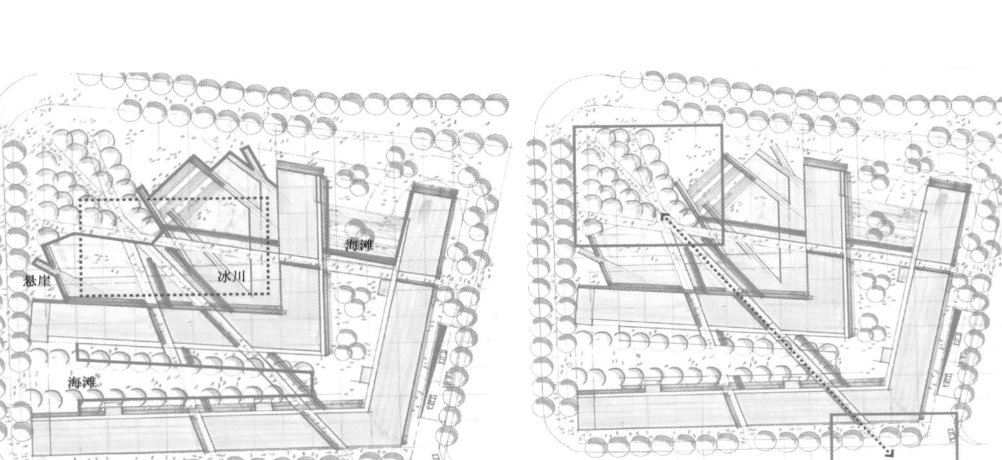

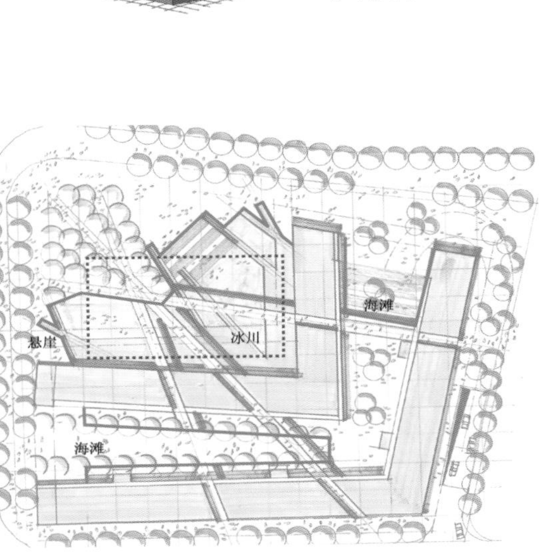
海滩
悬崖
冰川
海滩

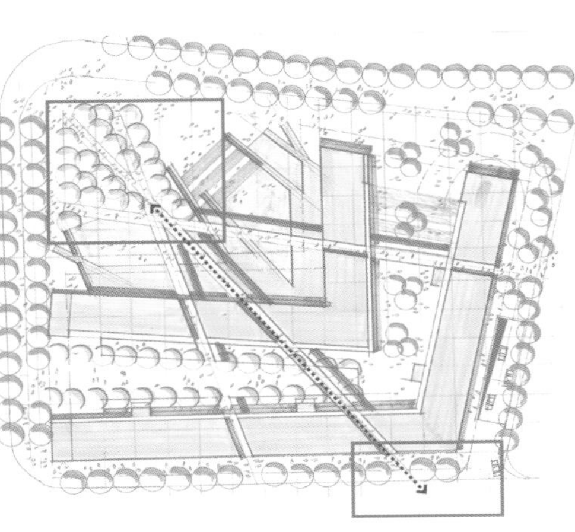

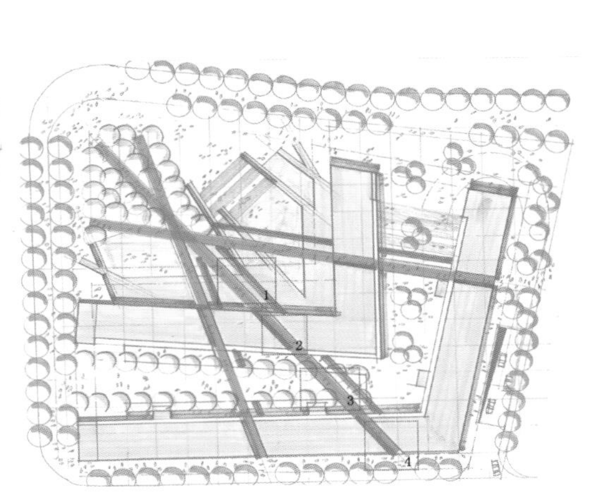

克罗地亚萨格勒布 123 商业大厦

Business Tower 123

设计单位：3LHD
开发商：Donji Grad d.o.o.
项目地址：克罗地亚萨格勒布
占地面积：3 689 m²
建筑面积：44 347 m²
设计团队：Sasa Begovic　Marko Dabrovic
Silvije Novak　Nives Krsnik Rister
Margareta Spajic　Krunoslav Szoersen
Irena Mazer　Ljerka Vucic
Tatjana Grozdanic Begovic
摄影：Studio HRG　Domagoj Blazevic

Designed by: 3LHD
Client: Donji Grad d.o.o
Location: Zagreb, Croatia
Site Area: 3,689 m²
Floor Area: 44,347 m²
Project Team: Sasa Begovic, Marko Dabrovic, Silvije Novak, Nives Krsnik Rister, Margareta Spajic, Krunoslav Szoersen, Irena Mazer, Ljerka Vucic, Tatjana Grozdanic Begovic
Photography: Studio HRG, Domagoj Blazevic

项目概况

123 商业大厦高 123 米，是克罗地亚及萨格勒布地区最高的建筑，为解决城市公共空间匮乏而建，同时，该项目也是萨格勒布第一座采用“DIAGRID”结构概念的建筑。

建筑设计

项目场址位于萨格勒布的入口，该地区同时存在现代都市和乡村风格这两种迥异的城市形态，无法为城市的发展提供充足的公共空间。项目定义为一个主要的商务办公区，以为这座城市带来全新的面貌。

设计采用一流的 DIAGRID 结构，这是一种对角线的格栅结构，能最大限度地保障建筑的张力及安全性，同时达到最大化节约建设成本的效果。

建筑底座没有梁柱，空间安排的随意性比较大。大厦内部为商业和公共空间，底层设有多层次大厅，有多个通道可以通往各个方向和各个楼层。咖啡馆和餐厅位于大厦的中间楼层和顶部，设有 360 度观景台以方便人们欣赏周围的风景。屋顶区域预留做直升飞机场。

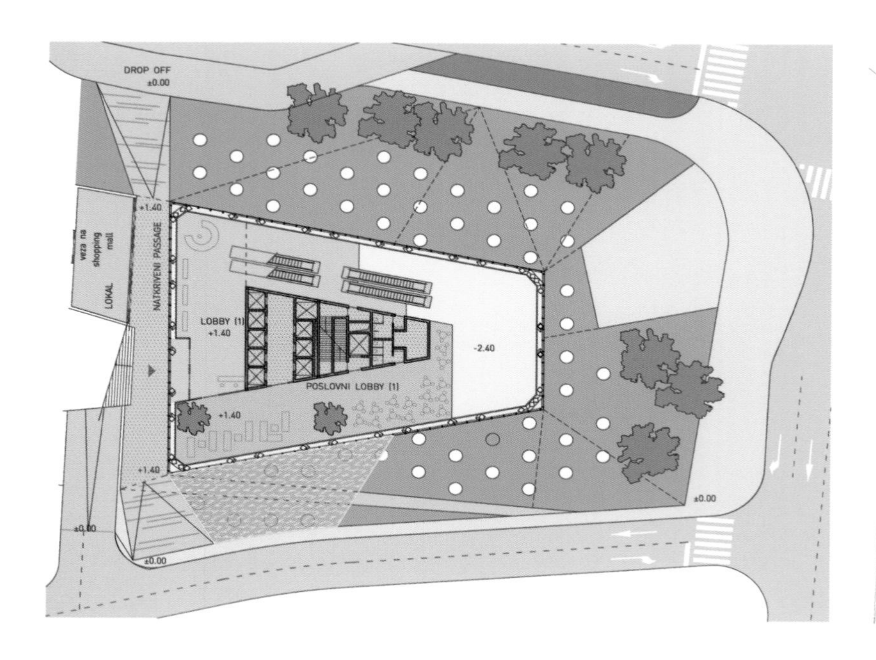
DROP OFF
±0.00
+1.40
veza na shopping mall
LOKAL
NATKRIVENI PASSAGE
LOBBY (1)
+1.40
-2.40
POSLOVNI LOBBY (1)
+1.40
+1.40
±0.00
±0.00
±0.00

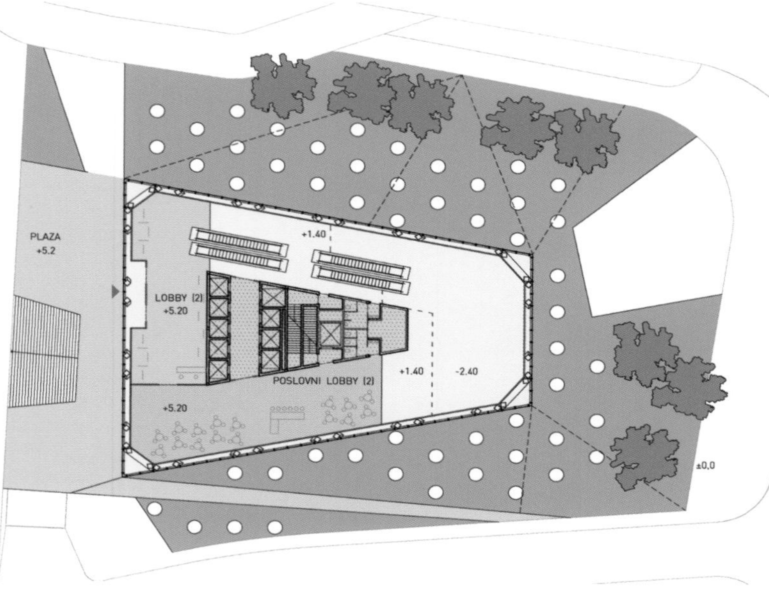
PLAZA
+5.2
+1.40
LOBBY (2)
+5.20
POSLOVNI LOBBY (2)
+1.40
-2.40
+5.20
±0,0

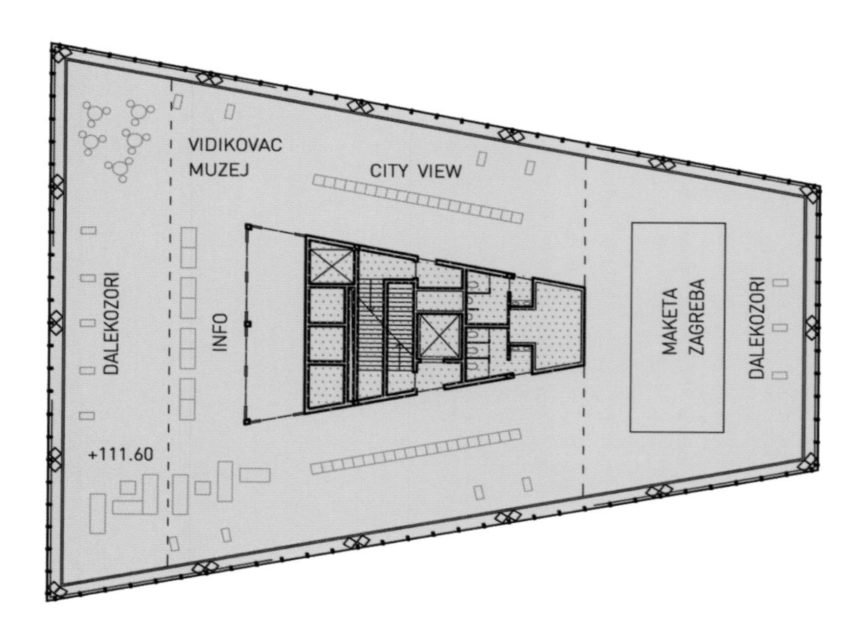
VIDIKOVAC
MUZEJ
CITY VIEW
DALEKOZORI
INFO
MAKETA
ZAGREBA
DALEKOZORI
+111.60

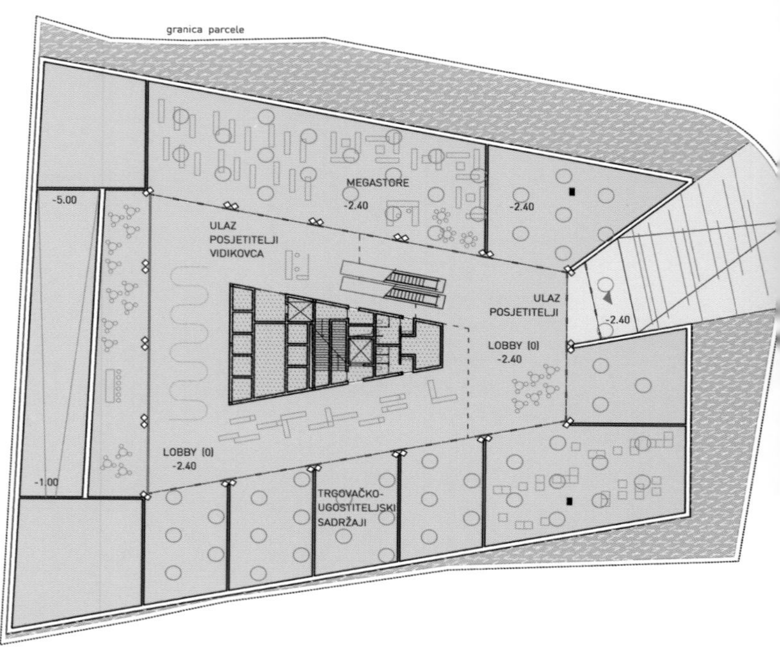
granica parcele
-5.00
MEGASTORE
-2.40
-2.40
ULAZ
POSJETITELJI
VIDIKOVCA
ULAZ
POSJETITELJI
-2.40
LOBBY (0)
-2.40
LOBBY (0)
-2.40
-1.00
TRGOVAČKO-
UGOSTITELJSKI
SADRŽAJI

rofile

he business Tower 123 is a design by invitation for the new nd tallest building in Zagreb and in Croatia. The solution or the design has been made after identifying the shortage f quality public space on the location of the Tower and the urrounding district. In the meantime, the project is also the first ower adopting "DIAGRID" structural concept in Zagreb.

rchitectural Design

he principal avenue at the entry to Zagreb has been onceived as the main business area, thus giving to the image f the city a thoroughly new appearance. The analysis revealed ne coexistence of two matrices-one modern and urbane, and ne other, the remaining rural architecture which has been bsorbed by the city space. It was evident that both matrices acked quality public space.

he Business Tower 123 is planned to become the key element f the whole area, with its height of 123 meters and the dvanced DIAGRID structure which is a strong, safe and cost-ffective diagonal structural concept.

trapezoidal ground plan is without columns which provides or various ground plan possibilities. The interior space of the ower is shared by business and public areas. The ground oor consists of a tall multi level lobby with access from all irections and different levels; the public facilities – a café and restaurant are located at mid height of the Tower and on he top floor, where a belvedere with a 360° sightseeing area nd a Museum of Zagreb Urban Development have also been lanned. The roof area is reserved for a heliport.

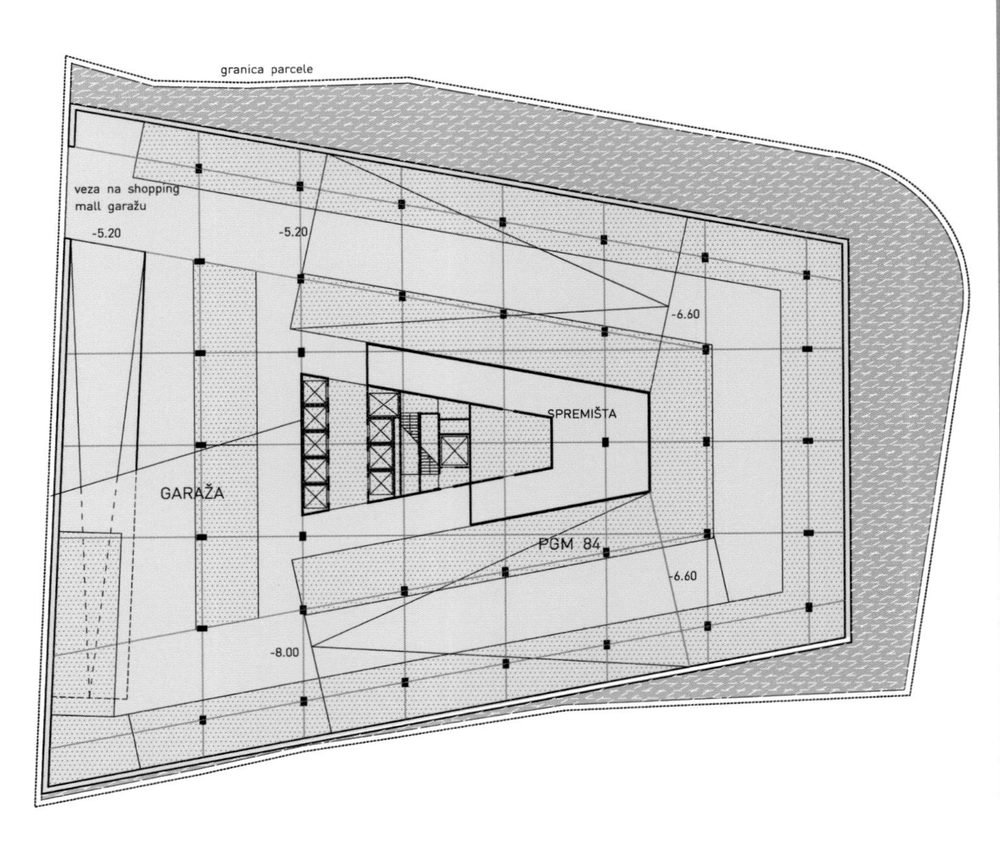

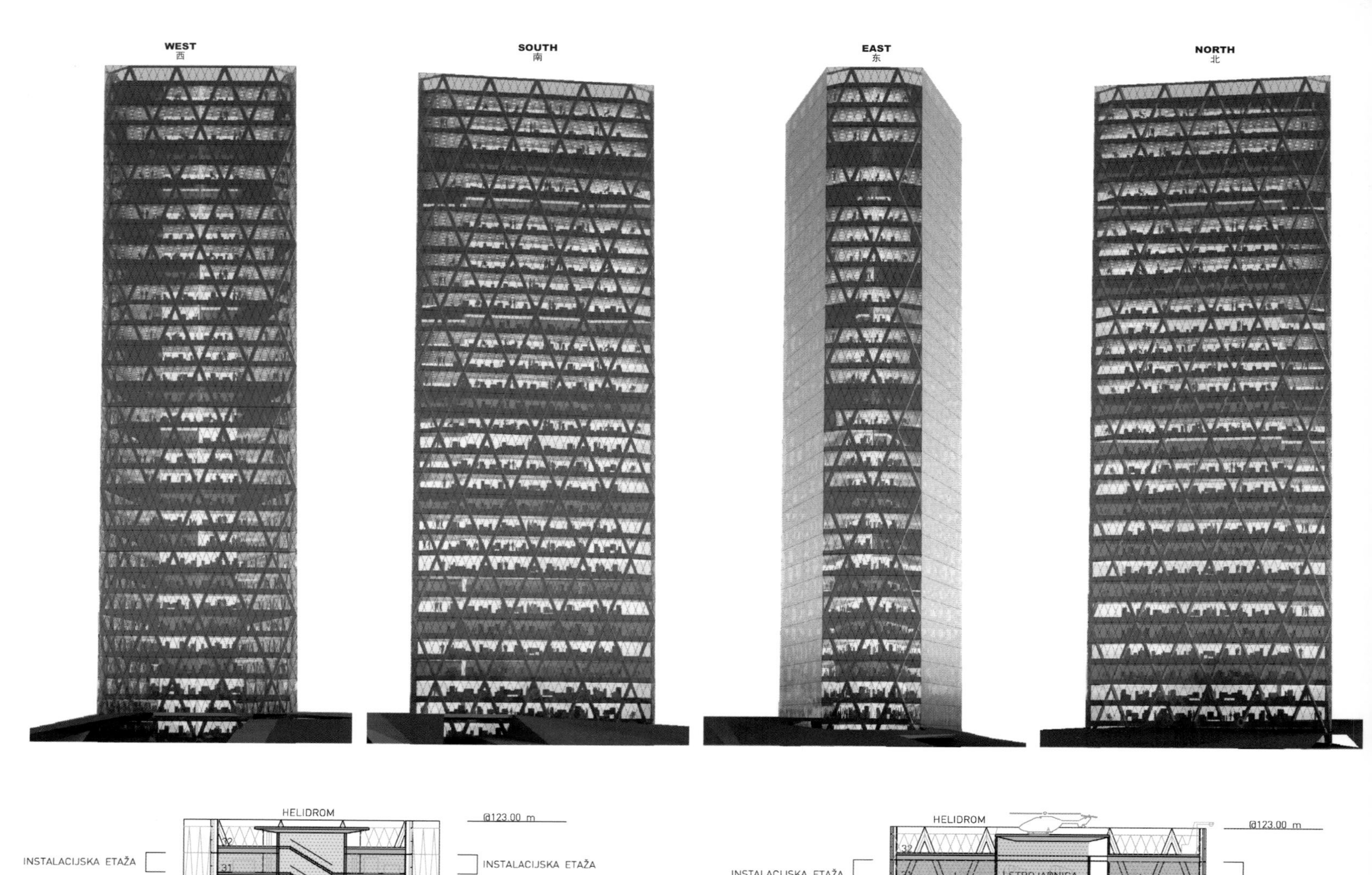
WEST
西
SOUTH
南
EAST
东
NORTH
北

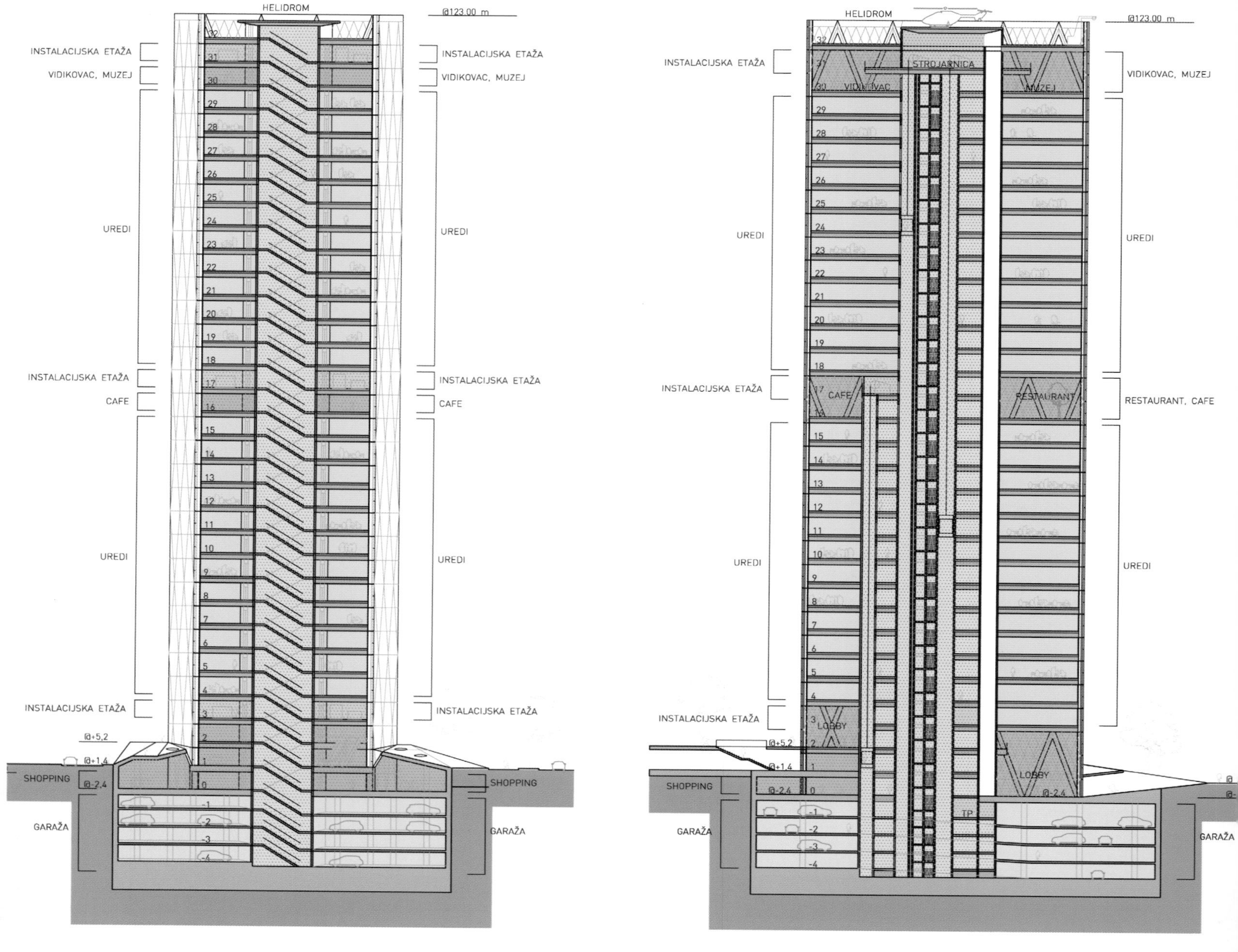
HELIDROM
ø123.00 m
INSTALACIJSKA ETAŽA
VIDIKOVAC, MUZEJ
UREDI
INSTALACIJSKA ETAŽA
CAFE
UREDI
INSTALACIJSKA ETAŽA
SHOPPING
GARAŽA
HELIDROM
ø123.00 m
STROJARNICA
VIDIKOVAC
MUZEJ
INSTALACIJSKA ETAŽA
VIDIKOVAC, MUZEJ
UREDI
INSTALACIJSKA ETAŽA
CAFE
RESTAURANT
RESTAURANT, CAFE
UREDI
INSTALACIJSKA ETAŽA
LOBBY
SHOPPING
GARAŽA
TP

江苏南京洪家园

Nanjing Hong's Garden

设计单位：日本 M.A.O. 一级建筑士事务所
项目地址：中国江苏省南京市
用地面积：18 195 ㎡
总建筑面积：139 413 ㎡
建筑密度：37.72%
容积率：4.1

Designed by: M.A.O.
Location: Nanjing, Jiangsu, China
Land Area: 18,195 m^2
Gross Floor Area: 139,413 m^2
Building Density: 37.72%
Plot Ratio: 4.1

项目概况

项目基地位于南京市秦淮区，距离中国商业密度最大的新街口商圈南部约 5 千米。如何在人流极度局限的基地范围内构建多样化的商业空间是设计的重点。

建筑设计

设计将基地在平面上分为 3 部分，并构建了 3 栋建筑，在这 3 栋建筑所围合的中庭空间中，设置跨层长梯以及串通流线的连廊平台，形成 4 种空间。

在剖面设计上，设计师特别注重地下、地上、空中三个不同空间的构成形式。地下层通过缓坡与地铁及西侧的绿地连接；地上空间包括与高架同标高的商业区域以及从高架上可以直接看到的内部活动空间；空中层即为可以眺望南京市南部地区的上部 SOHO 区。

设计通过平面及立体的多样动线为构建多样化的商业空间提供了必要条件，同时，通过将几种中等规模的明快动线进行有机组合，使商业设施的动线设计更加便捷有效。

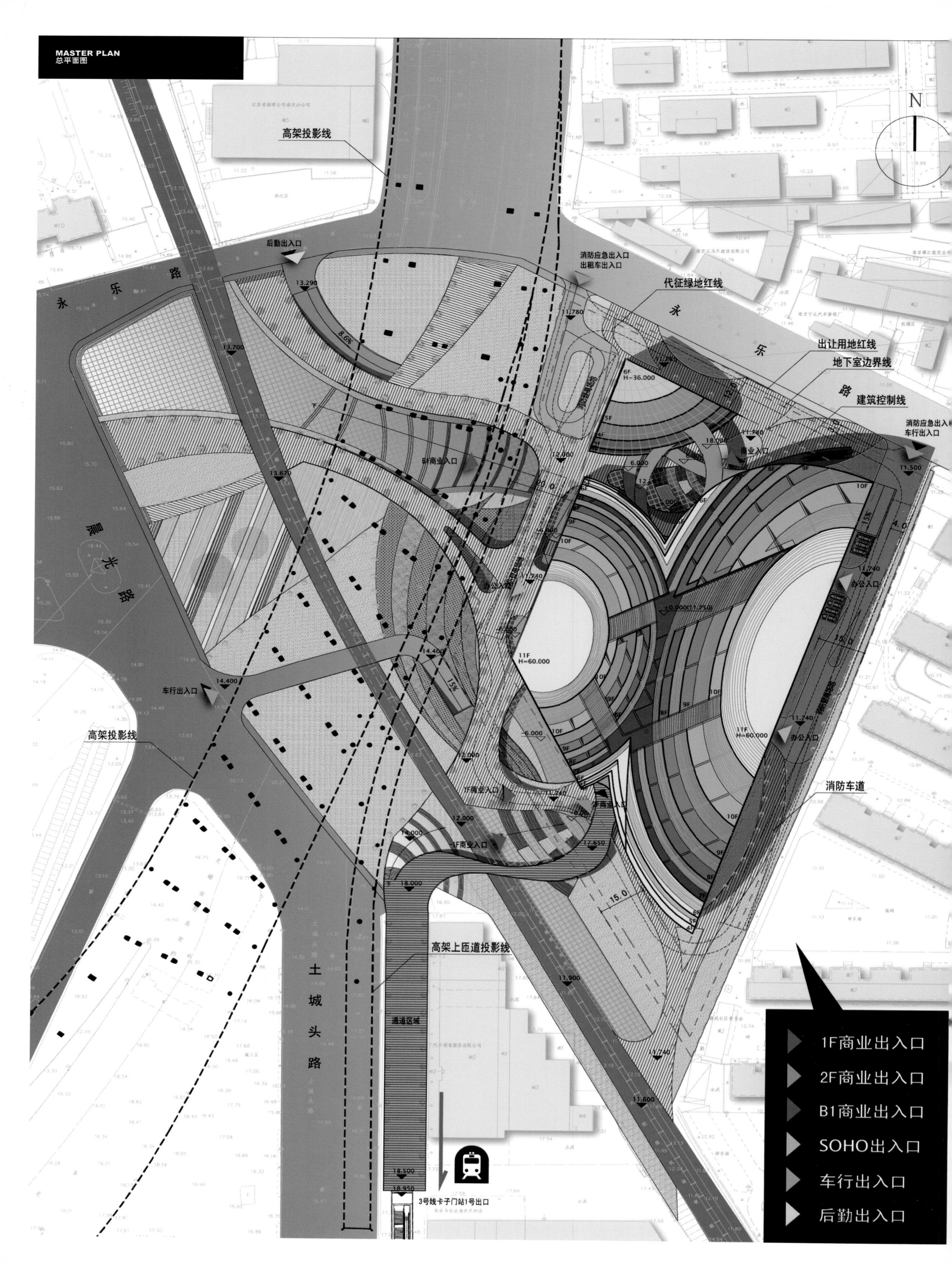
MASTER PLAN
总平面图
高架投影线
后勤出入口
消防应急出入口
出租车出入口
代征绿地红线
永乐路
出让用地红线
地下室边界线
建筑控制线
消防应急出入口
车行出入口
B1商业入口
晨光路
办公入口
车行出入口
高架投影线
1F商业入口
2F商业入口
-1F商业入口
消防车道
高架上匝道投影线
通道区域
土城头路
3号线卡子门站1号出口
1F商业出入口
2F商业出入口
B1商业出入口
SOHO出入口
车行出入口
后勤出入口

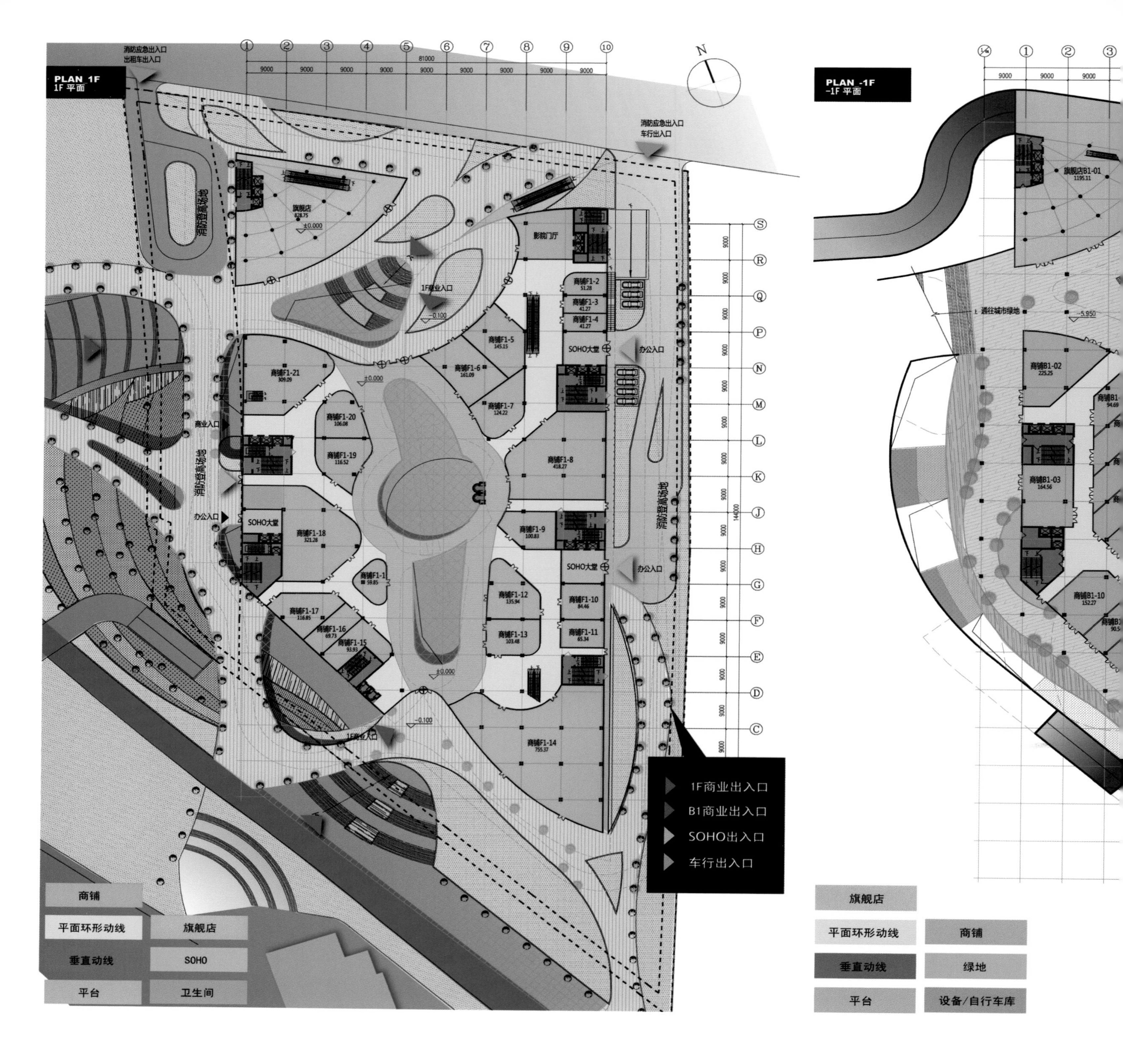

Profile

The site is located in Qinhuai District of Nanjing, about 5 kilometers from Xinjiekou Business Area with the highest commercial density. Constructing a diversified commercial space in a limited site crowded with pedestrians is a key to the design.

Architectural Design

The site plane is divided into three parts and has further constructed three buildings. The atrium enclosed by these three buildings has set cross-layer stairs and connecting corridors to create four different types of spaces.

On the section design, a special attention has been paid to the composition of underground, ground and above ground spaces. The underground floors connect with metro and the green space in the west through gentle slopes; ground space includes commercial areas with the same elevation of viaduct from which the internal activity space can be seen; space above ground is the upper SOHO District available for overlooking the southern district of Nanjing City.

Two-dimensional and three-dimensional dynamic lines of the project offer necessary conditions for constructing a diversified commercial space. Meanwhile, an organic combination of several medium-sized dynamic lines makes the design of commercial facilities more available and effective.

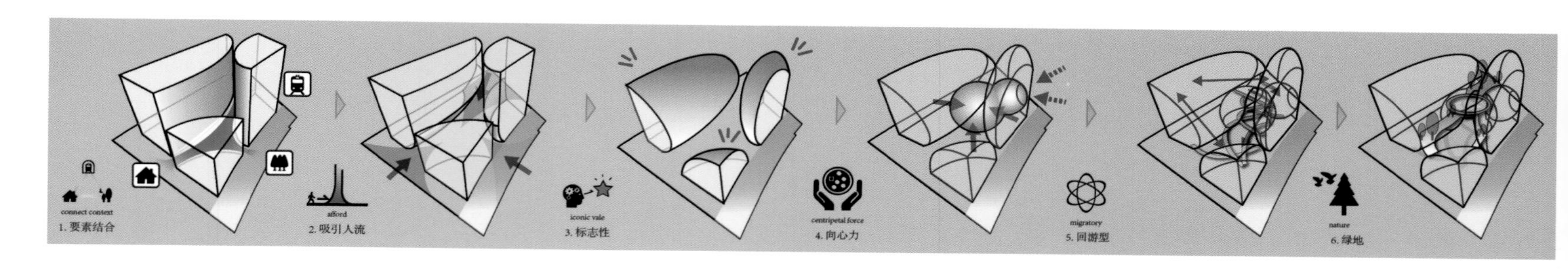

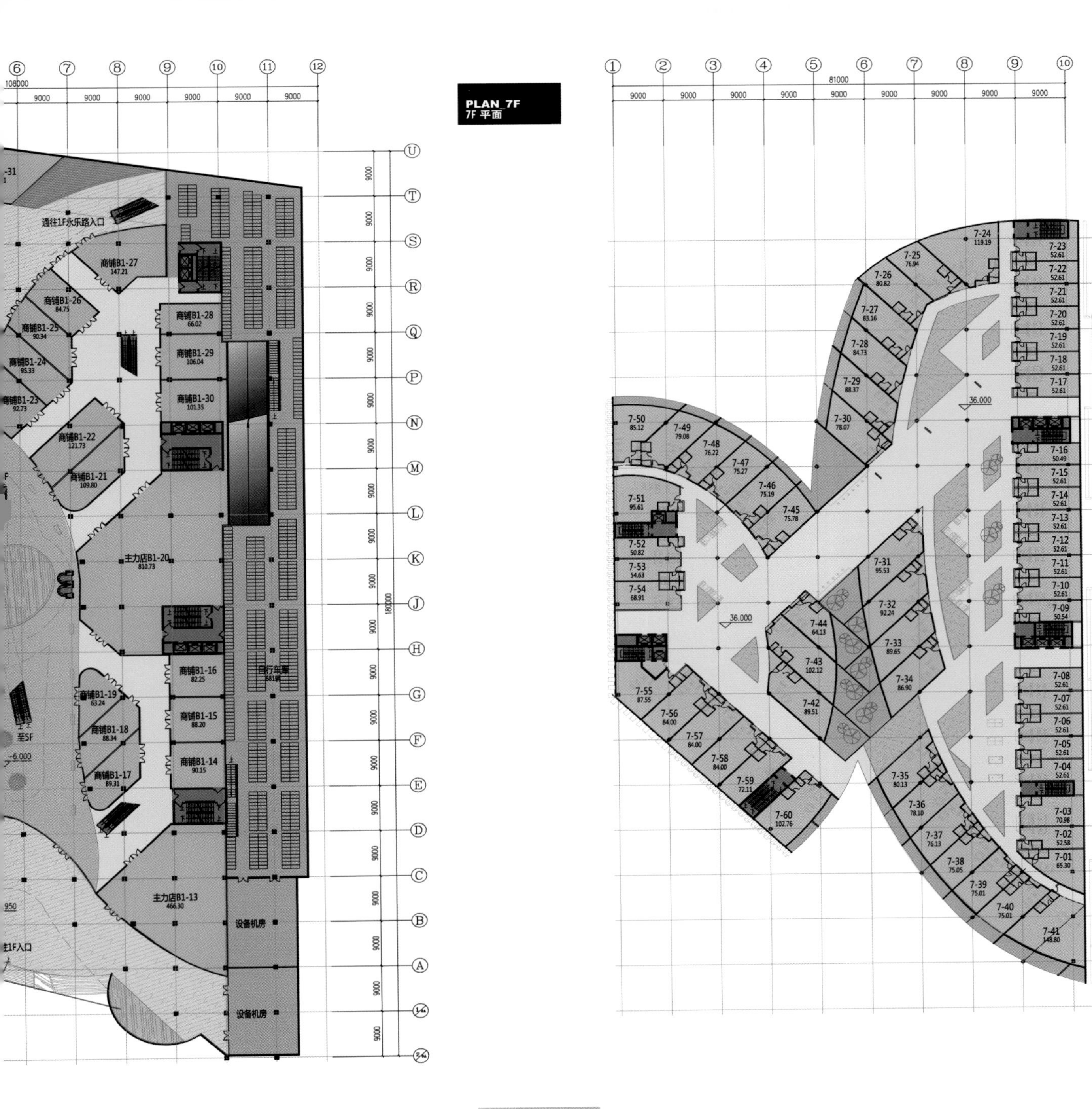

绿地	
垂直动线	平面环形动线
平台	SOHO

Chloé
CHANEL
FENDI

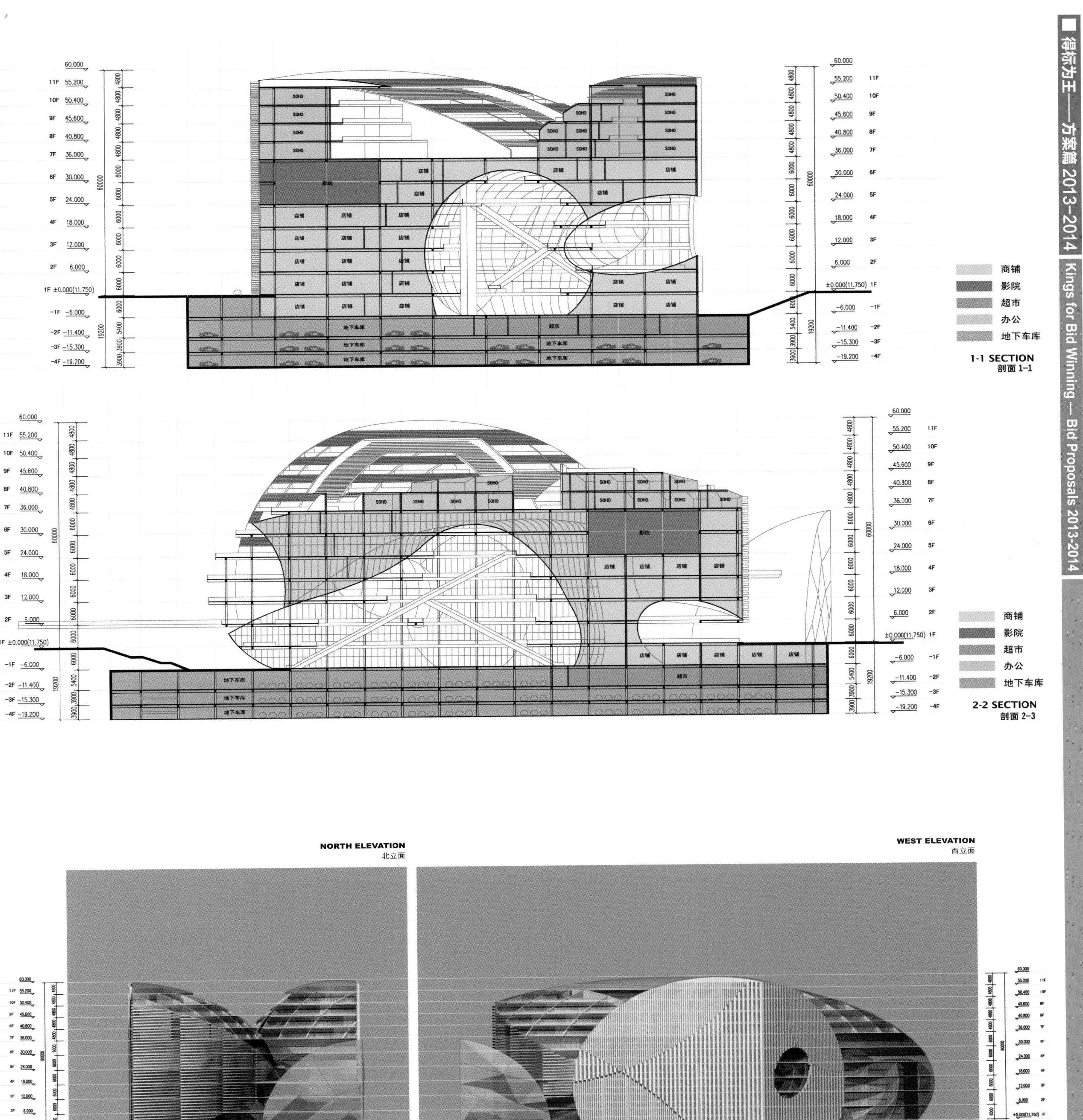

NORTH ELEVATION
北立面

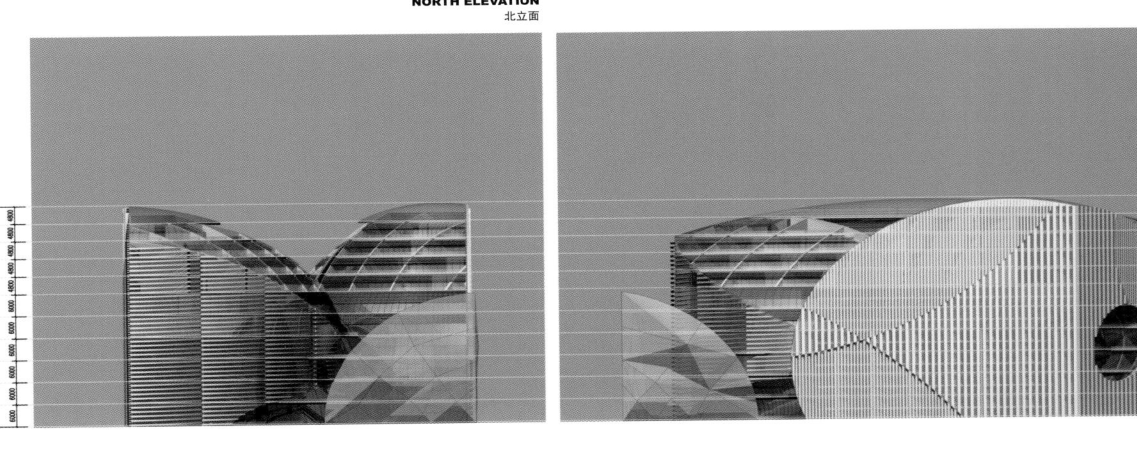

WEST ELEVATION
西立面

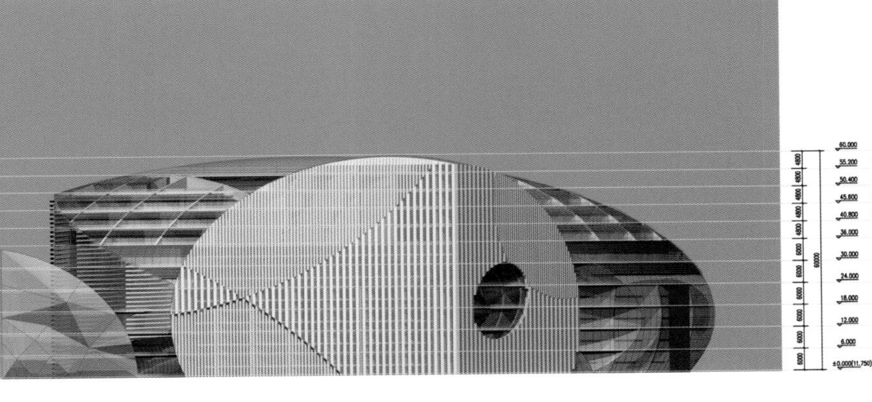

SOUHT ELEVATION
南立面

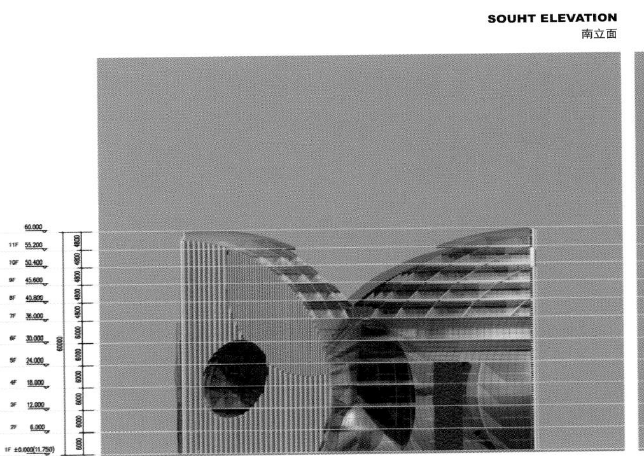

EAST ELEVATION
东立面

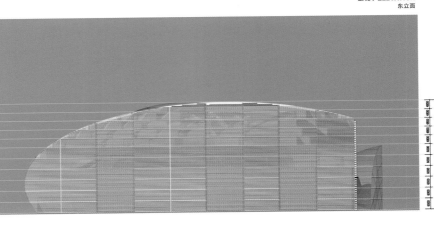

河南郑州瀚海北环路商业综合体

Hanhai Beihuan Road Commercial Complex, Zhengzhou, Henan

设计单位：amphibianArc
项目地址：中国河南省郑州市
项目面积：115 000 ㎡

Designed by: amphibianArc
Location: Zhengzhou, Henan, China
Area: 115,000 m²

项目概况

项目地块位于偏离城市中心的地带，不具备发展大型都市项目的背景文脉和空间条件。项目运用了极具雕塑感的设计，通过连续的流线型设计整合高层与裙房的关系，使之成为一个超越了传统设计模式的商业综合体。

设计特色

设计沿用了 amphibianArc 的“立体书法”设计理念，其雏形源自上海“空间书法”别墅项目。项目抛却了传统商务地产严肃呆板的形象，采用前卫大胆的流线型外观，鲜明的建筑风格突出了项目作为郑州北区商务地标的形象。

这是一个集高品质、时尚元素、高智能化于一体的区域性地标建筑，由两栋高 130 米的高楼组合而成。稳重的黑色辅以跳跃的白色，给人极强的视觉冲击感，打破了沉闷的建筑格局，为建筑增加了时尚元素。

高档金属板材在玻璃幕墙上呈现出流畅的线条，底层商业空间的柔美曲线，演绎了独特的立面风格，既彰显了建筑的品质感和时尚感，也使建筑流露出一丝艺术气息。

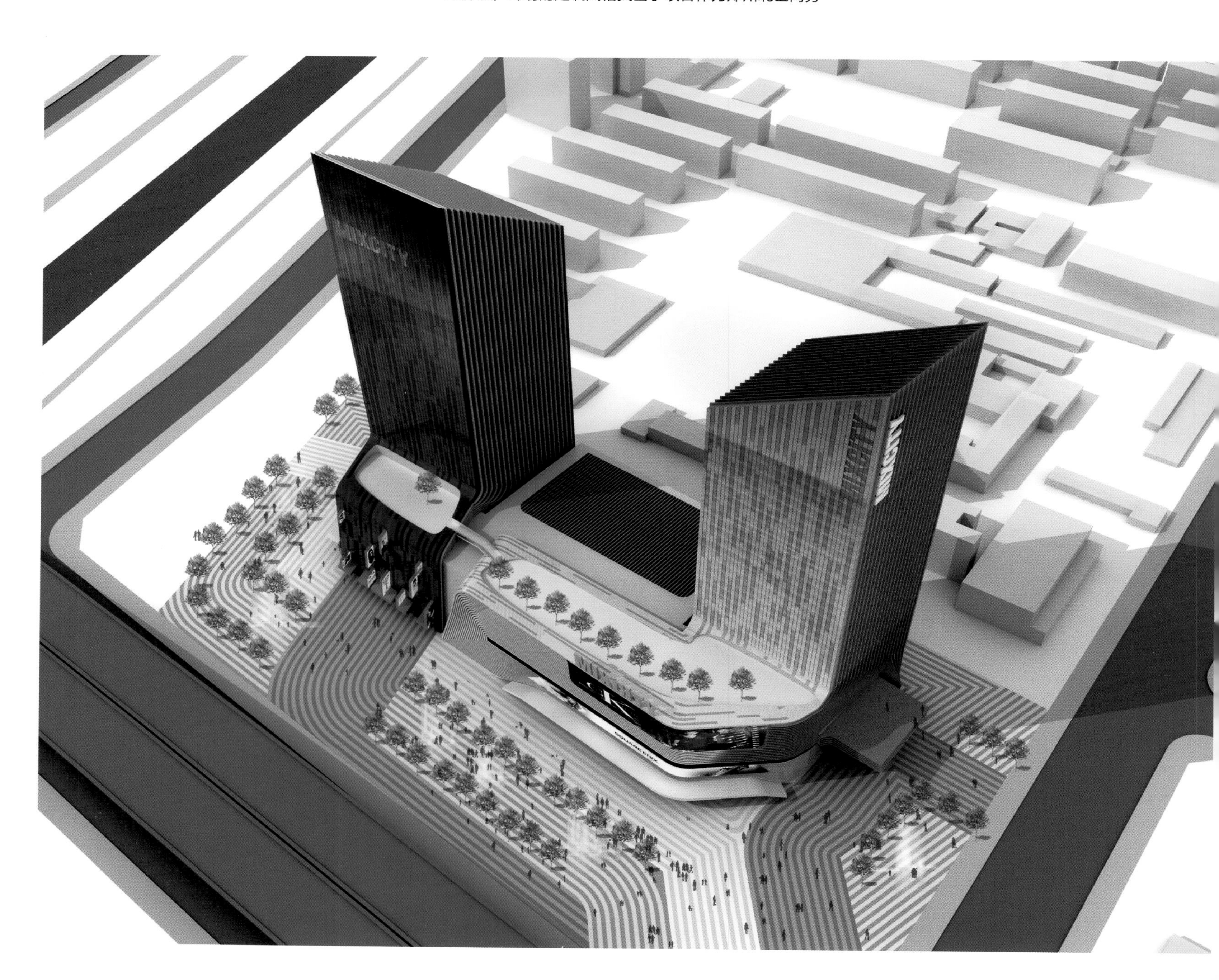

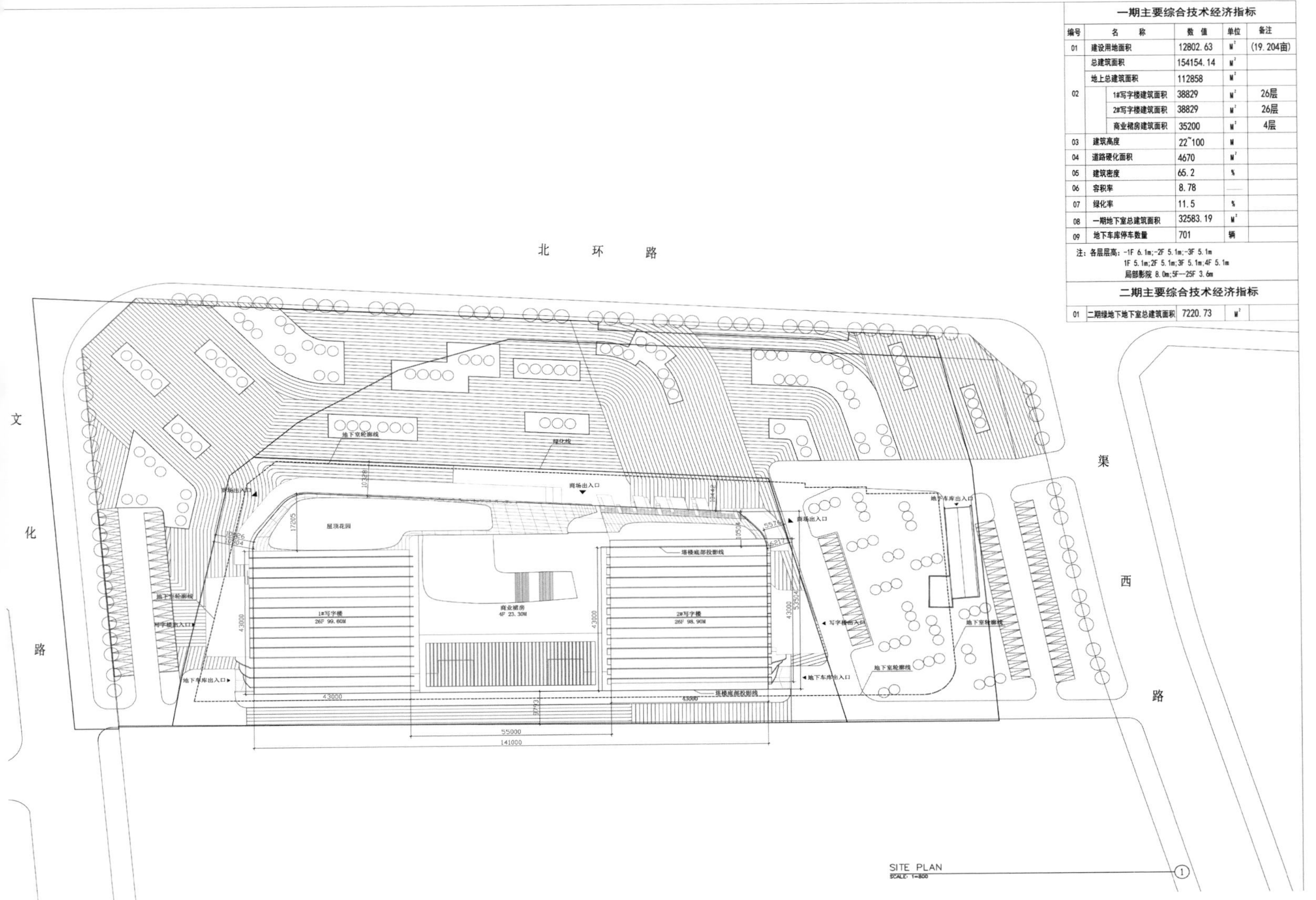

一期主要综合技术经济指标					
编号	名称		数值	单位	备注
01	建设用地面积		12802.63	M²	(19.204亩)
02	总建筑面积		154154.14	M²	
	地上总建筑面积		112858	M²	
		1#写字楼建筑面积	38829	M²	26层
		2#写字楼建筑面积	38829	M²	26层
		商业裙房建筑面积	35200	M²	4层
03	建筑高度		22~100	M	
04	道路硬化面积		4670	M²	
05	建筑密度		65.2	%	
06	容积率		8.78	——	
07	绿化率		11.5	%	
08	一期地下室总建筑面积		32583.19	M²	
09	地下车库停车数量		701	辆	

注：各层层高：-1F 6.1m;-2F 5.1m;-3F 5.1m
1F 5.1m;2F 5.1m;3F 5.1m;4F 5.1m
局部影院 8.0m;5F--25F 3.6m

二期主要综合技术经济指标				
01	二期绿地下地下室总建筑面积	7220.73	M²	

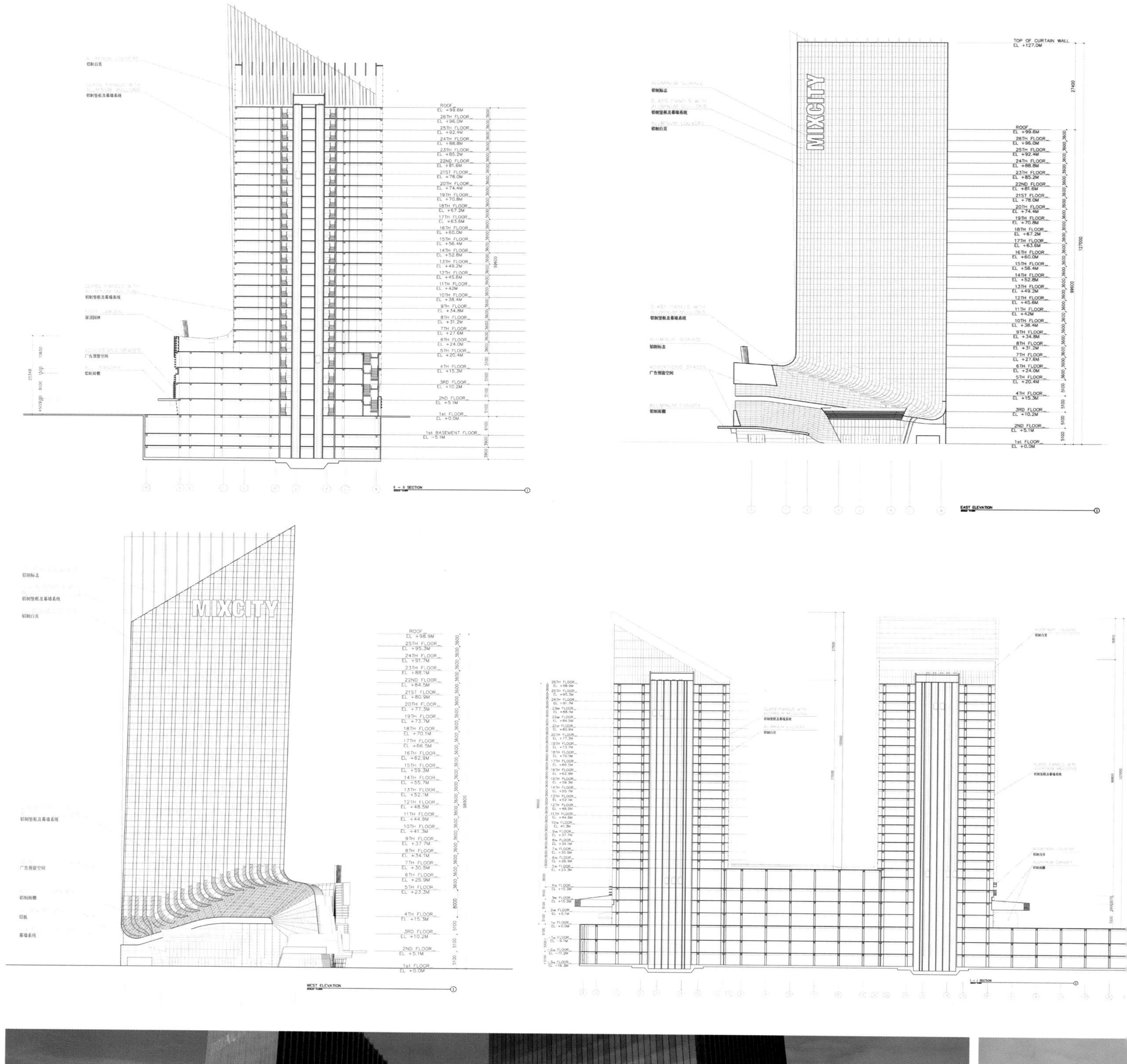

MIXCITY
SQUARE ENIX

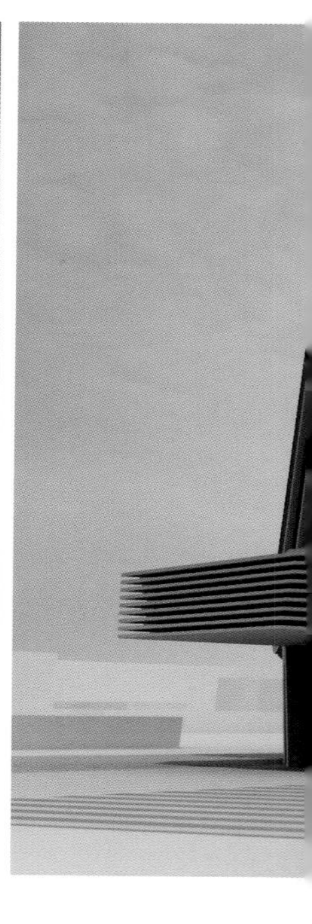

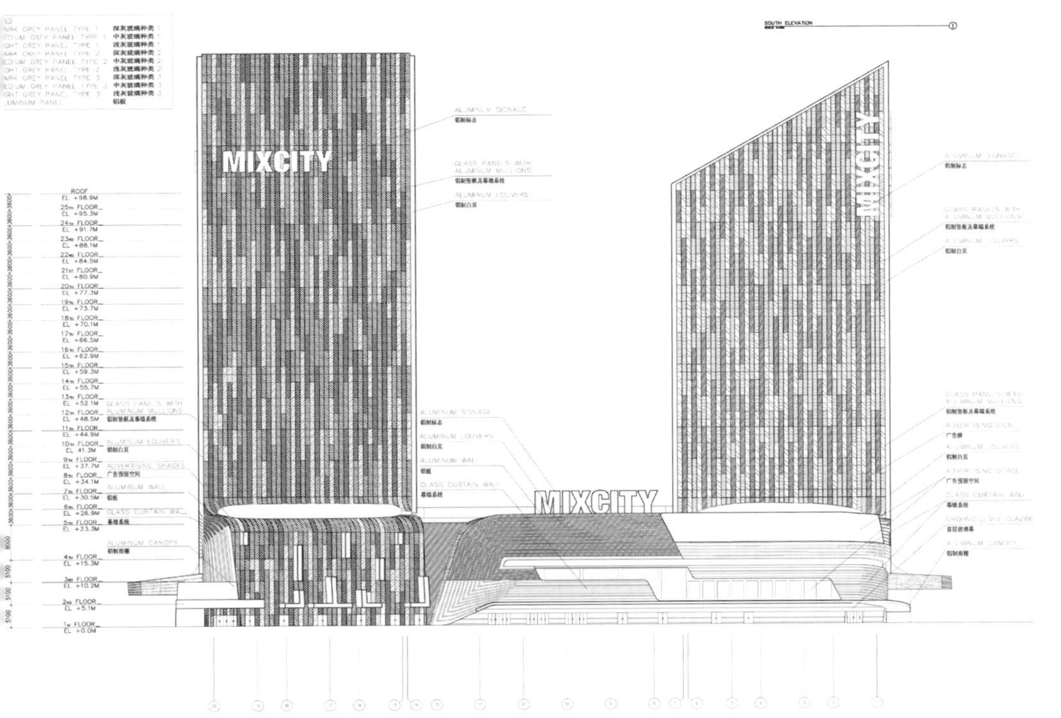

Profile

The site is located off the city center, with less cultural background compared to other city projects. The Hanhai Beihuan Road project is designed with strong sculptural element, integrating towers and podium with streamlines, making the site a commercial complex surpassing traditional design mode.

Design Feature

The design embodies amphibianArc's 3D calligraphy design concept, which was initially developed in a villa project Calligraphy in Space in Shanghai. Abandoned the serious rigid image of traditional commercial real estates, the project adopts bold streamlined facades and distinctive style to highlight this commercial landmark in the north district of Zhengzhou.

It is a regional landmark building integrating high quality, fashionable elements and high intelligence. The project is composed of two 130-meter towers. Solemn black accompanied with leaping white color present strong visual impact, breaking rigid architectural pattern and adding fashionable elements for the project.

Top-grade metal sheets used on glazing curtain wall reflect fluid lines and soft curves of ground floor commercial space, interpreting unique façade style, exposing sense of quality and fashion and revealing artistic temperament.

北京万豪世纪中心

Beijing Wanhao Century Center

设计单位：Moore Ruble Yudell Architects & Planners
合作单位：Yang Architects
项目地址：中国北京市
摄影：Jim Simmons

Designed by: Moore Ruble Yudell Architects & Planners
Collaboration: Yang Architects
Location: Beijing, China
Photography: Jim Simmons

项目概况

项目位于三环的北京使馆区，地理位置十分优越。这个包含了办公楼和大型酒店的项目是对北京这个城市特征的再现，设计通过建筑元素表现了中国浓厚的历史文化。

设计理念

设计灵感来源于故宫花园中粗糙的圆柱形岩石，有些岩石甚至高达 6 米。建筑体好似中国南方让人眼花缭乱的小丘陵，这些形状奇特的结构反映了独特的中国文脉，也是对大型城市形态一个贴切的暗喻。

设计特色

随着这些极具立体感的办公楼的建造，设计将从塔楼上分离出来的极其匀称的小结构构成了一个独立的如山般的体量，其柔和的曲线形一侧界定了这个大型体量的特征。在这个被切割的塔楼的轴线交会处，一座三层的桥梁将办公楼与酒店连接起来。

酒店所带来的城市效益源自它所提供的公共空间。整个亚洲的酒店为公众及酒店的客人提供了极具吸引力的聚会空间，通常设有各色餐厅、娱乐空间和购物中心。考虑到北京的气候环境，设计将酒店设想为一个中高层的建筑，这个酒店的中庭三面被住宅楼层围绕，可视为一个空气清新的公共广场。酒店较低的体量考虑了阴阳风水等因素，加强了城市组织结构，另有两个补充形态设置在一个小型的城市公园内。

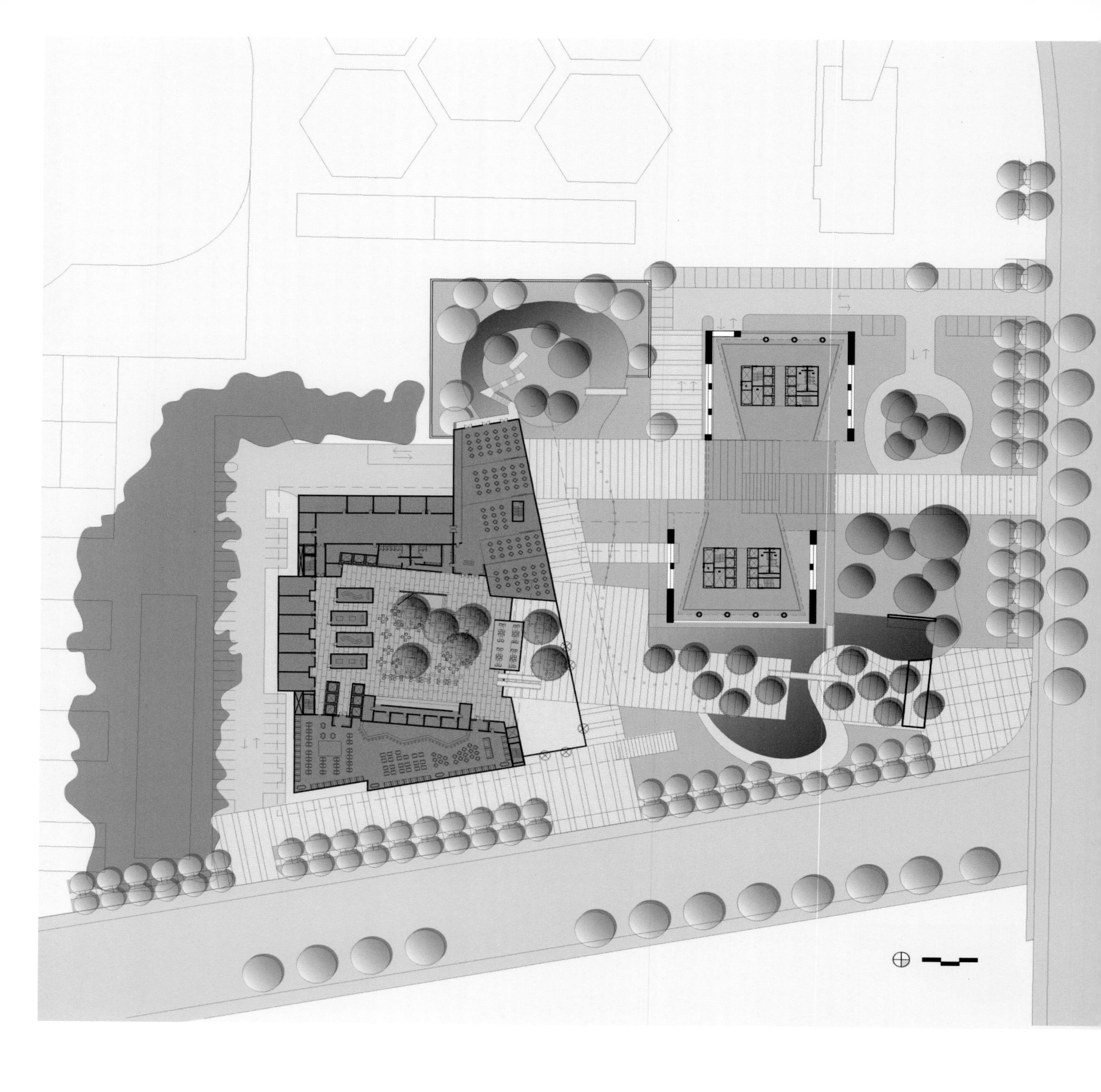

Profile

The project is located on a prominent Third Ring site in Beijing's diplomatic quarter. The project consisting of office towers and large hotel is a representation of urban characters. Architectural elements of this project reflect dense cultural history of China.

Design Concept

The project design is inspired by gnarled, columnar rocks in gardens in the Forbidden City, some nearly 6 meters tall. Like the vertiginous mini-mountains of hilly topography in southern China, these grotesque forms seem uniquely Chinese, and a fine metaphor for a large-scale urban form.

Design Feature

As the modeling of the office buildings proceeded, these shapes were abstracted in a nearly-symmetrical pairing of the towers into a single monumental form, whose gently bowed sides featured massive projecting volumes. At the cross-axis of the split tower a three-story bridge links the offices to the hotel.

The urban benefit of the hotel is the public space it offers. Hotels throughout Asia provide extremely popular gathering places, for the public as well as the guests, and typically have a rich program of restaurants, entertainment, and shops. Thinking of Beijing's climate, designers have organized the hotel as a mid-rise, with room floors wrapping three sides of a grand atrium—an acclimatized public square where the air is always clean. The relatively low volume of the hotel provides a yin-yang companion for the towers, strengthening the urban composition into two complementary forms set in a small urban park.

南立面图 1:500

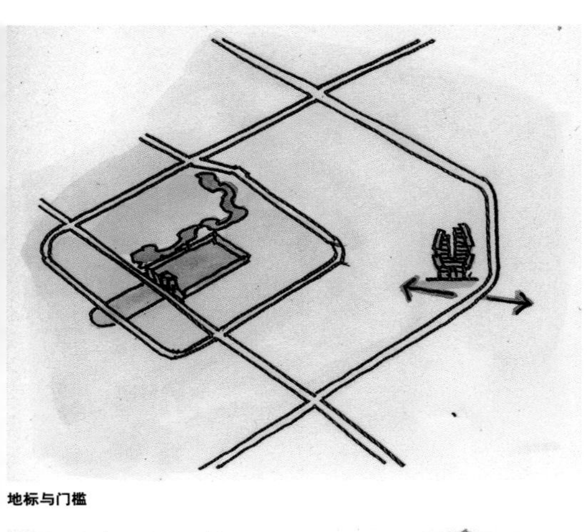
地标与门槛

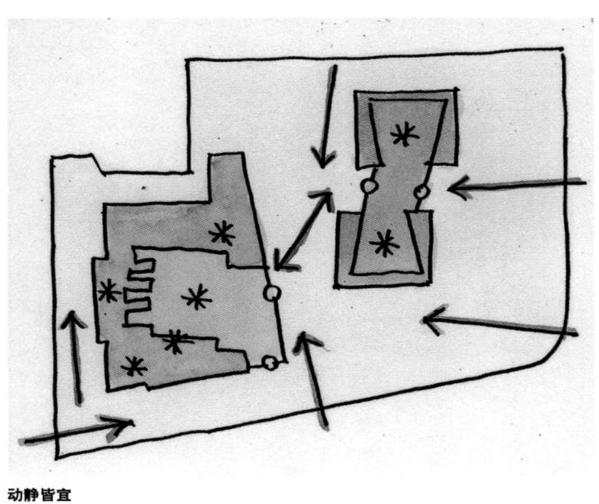
动静皆宜

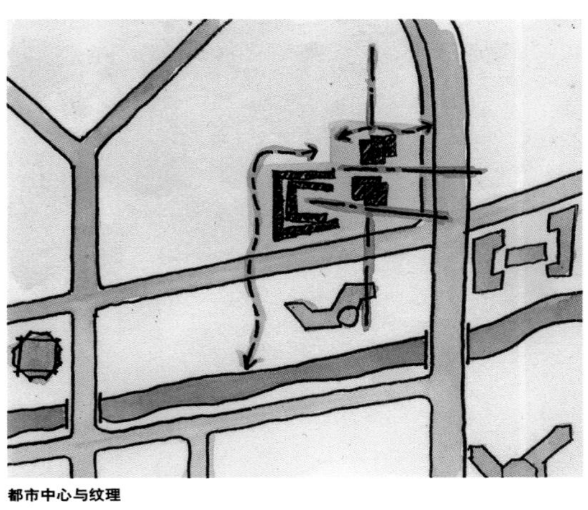
都市中心与纹理

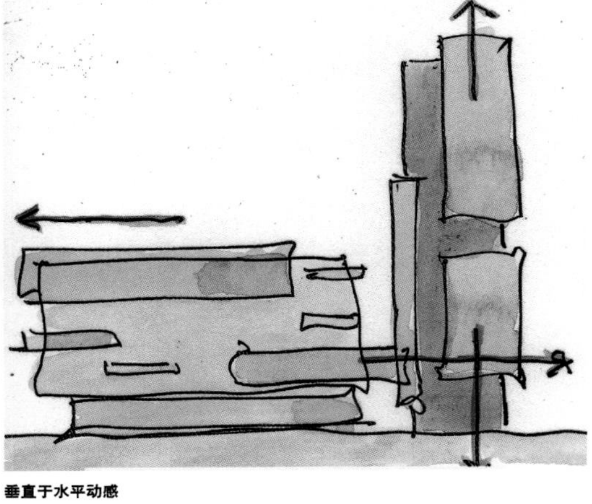
垂直于水平动感

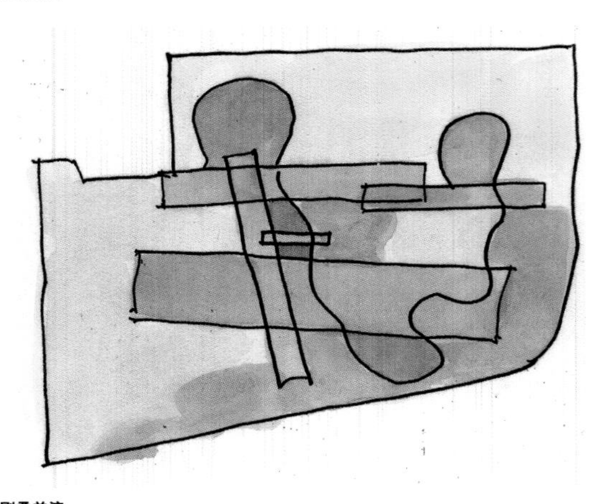
刚柔并济

虚实对比

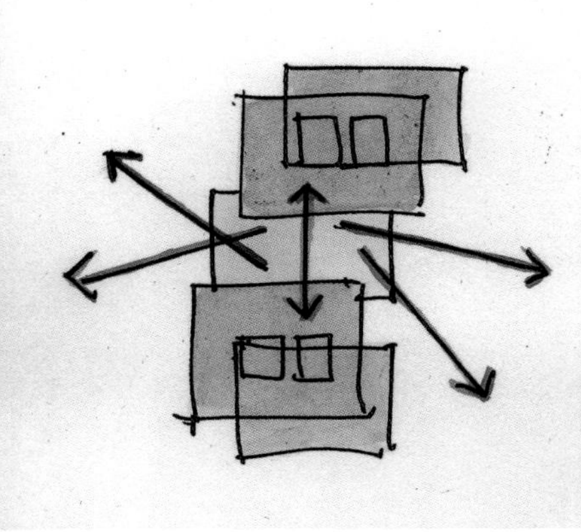
退缩 取景

外墙与中庭

广场与绿地

巴西圣保罗 Bandeirantes

Bandeirantes

设计单位： Oppenheim Architecture+Design
项目地址： 巴西圣保罗
项目面积： 232 257 ㎡

Designed by: Oppenheim Architecture + Design
Location: Sao Paulo, Brazil
Area: 232,257 m²

项目概况

项目位于圣保罗市郊的山侧地块上，如何将这个232 257 平方米的大型项目建筑在这个可用面积极少的场地上，是设计的一大挑战。

建筑设计

连续的裙房和零落布置的不同高度的塔楼，使整个项目呈现出蜿蜒的建筑形态，这也使每一栋建筑都可获得朝向城市和周围山体的良好视野。

更具立体感的裙房与通透的塔楼之间的界线打破了建筑的整体格局，为处于屋顶平台结构之上的大型康乐设施层提供了一个有遮蔽的、通风性能良好的空间。为住宅单位和零售空间服务的停车场全部位于地下层，为在地上空间设置更多的休闲娱乐空间创造了条件。

设计不是从地面上建造一栋商场型的体量，也不是将屋顶机械设备暴露出来，而是将所有的商业空间下沉至地下空间，以下沉庭院为中心分布。

被动的可持续性设计运用于整个设计的始终，灰水处理、住宅单元的自然通风、当地材料的运用以及屋顶太阳能热水器的配置，都提升了项目的经济和环保效益。

Profile

The Bandeirantes project is situated on a hill-side plot on the outskirts of Sao Paulo. To distribute the 2.5 million square feet of units among just a few sparse acres of land is a great challenge of this project design.

Architectural Design

The project has a winding scheme consisting of a continuous podium and intermittent towers of varying heights; all, of course, with magnificent views of the city and surrounding mountains.

The scale of the project was broken down with a clear separation between the more solid podium and the glazed towers, creating a covered and well-ventilated space for a generous amenities level on the roof-top of the podium structures. Parking for condominium units and the retail component are completely underground, allowing the use of valuable ground floor space for recreational areas.

Instead of building a mall-type structure from the ground up, exposing uninviting roof top mechanical equipment, all commercial spaces were sunken below ground with a major sunken courtyard as a center.

Passive sustainable design was used throughout the process, enhancing the value and quality of the project with features such as grey water treatment, cross ventilation on all units, application of local materials, and roof mounted solar hot water panels.

A 类单元角落 TYPE A CORNER
地毯 -93m²　Carpet - 93 m²
露台 -13m²　Terrace - 13 m²
总面积 -106m²　Gross Area - 106 m²

A 类单元中部 TYPE A MIDDLE
地毯 -88m²　Carpet - 88 m²
露台 -17m²　Terrace - 17 m²
总面积 -105m²　Gross Area - 105 m²

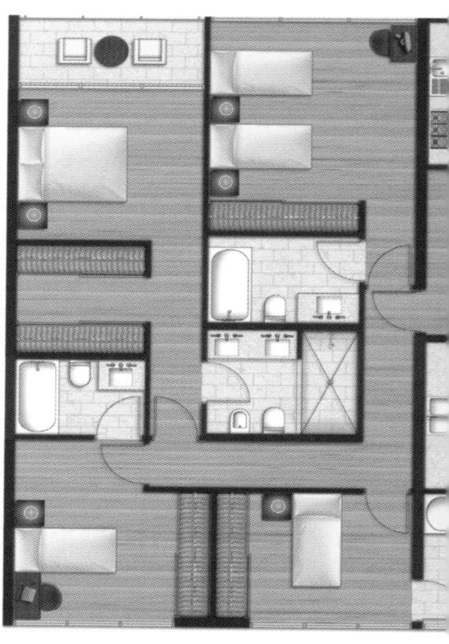

D1 类单元地面
地毯 -175
露台 -26
总面积 -201

B 类单元 TYPE B
地毯 -120m²　Carpet - 120 m²
露台 -21m²　Terrace - 21 m²
总面积 -141m²　Gross Area - 141 m²

B 类单元地面 TYPE B GROUND
地毯 -121m²　Carpet - 121 m²
露台 -27m²　Terrace - 27 m²
总面积 -148m²　Gross Area - 148 m²

TYPE D1 GROUND
Carpet - 175 m^2
Terrace - 26 m^2
Gross Area - 201 m^2

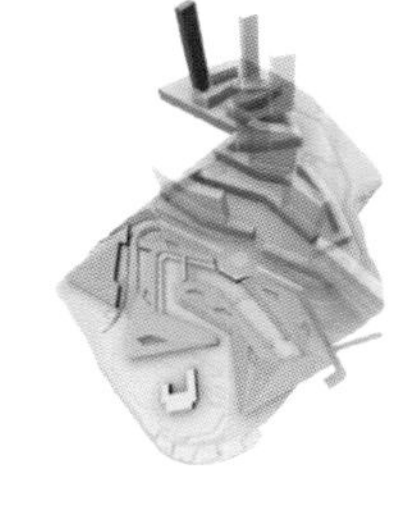

D1 类单元
地毯 -185m^2
露台 -28m^2
总面积 -213m^2

TYPE D
Carpet - 185 m^2
Terrace - 28 m^2
Gross Area - 213 m^2

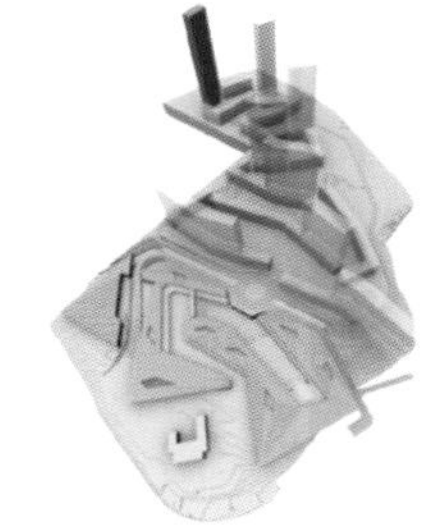

D2 类单元地面
地毯 -166m^2
露台 -36m^2
总面积 -202m^2

TYPE D2 GROUND
Carpet - 166 m^2
Terrace - 36 m^2
Gross Area - 202 m^2

SITE AMENITIES
场地便利设施

GROCERY 杂货店
Total 8000 m^2 总面积：8 000 ㎡

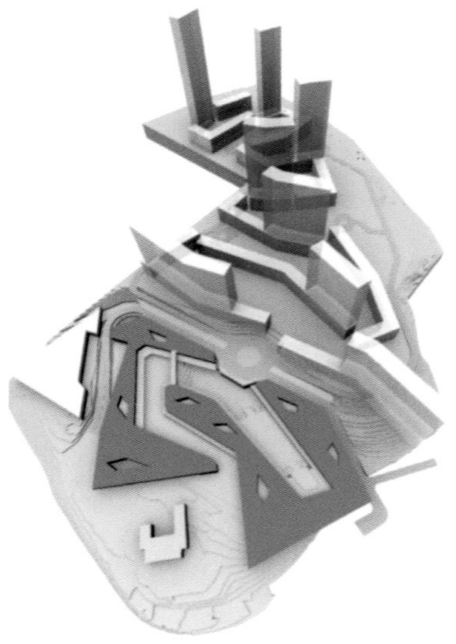

RETAIL 零售区
Total 60000m^2 总面积：6 000 ㎡
Anchor Stores 主力店
Boutique Shops 精品店

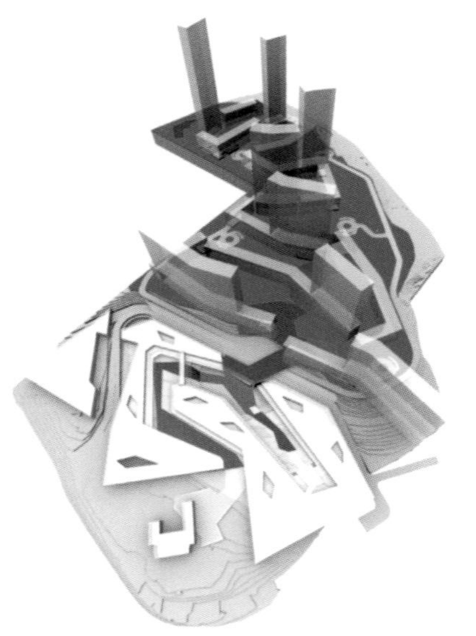

AEMENITIES GROUND LEVEL 底层便利设施
Total 28680m^2 总面积：28 680 ㎡
Lobby 大厅
Cafeteria/Bar 自助餐厅 / 酒吧
Spa Spa
Fitness 健身中心

AMENITIES ROOF LEVEL 屋顶层便利设施
Total 28680m^2 总面积：28 680 ㎡
Screening Rooms 放映室
Pools 水池
Business Centers 商业中心
Party Rooms 宴会厅

ALL AMENITIES DIAGRAM
所有便利设施图表

- **GREEN ROOFS** 绿色屋顶
- **GARDENS** 花园
- **RIVER** 河流
- **FITNESS** 健身中心
- **LOBBY** 大厅
- **PARTY ROOMS/SOCIAL HALLS** 宴会厅 / 交谊厅
- **SCREENINGS ROOMS** 放映室

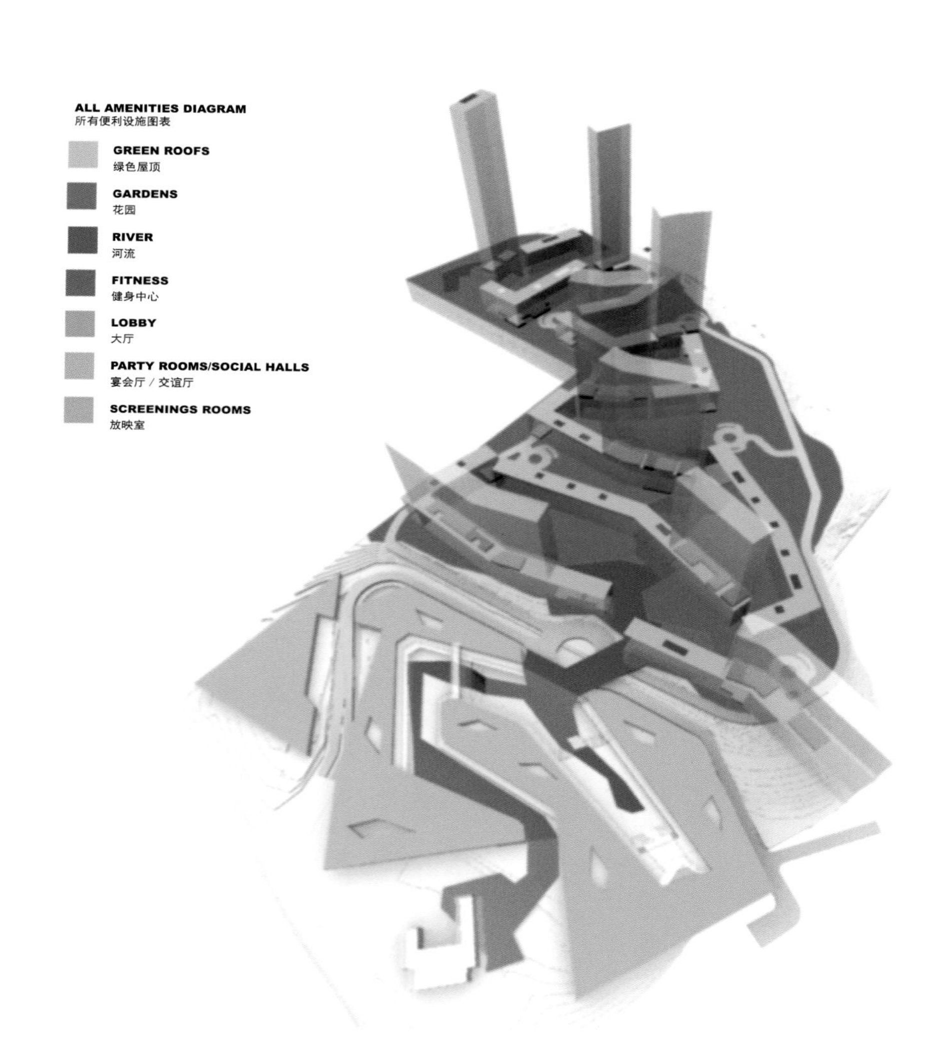

ALL UNITS DIAGRAM
所有单元简图

GREENSPACE DIAGRAM
绿色空间图表

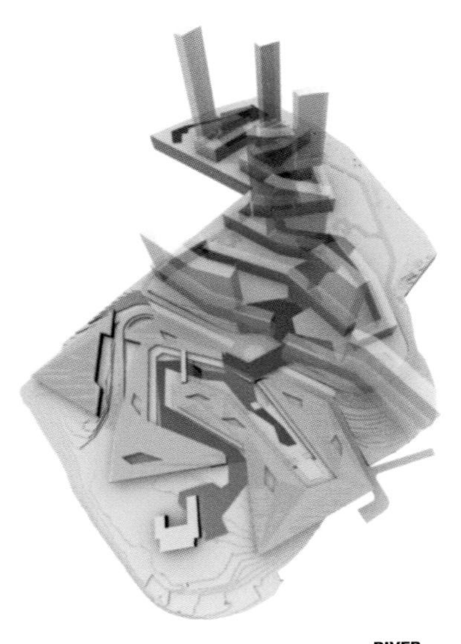

GREENSPACE
绿色空间

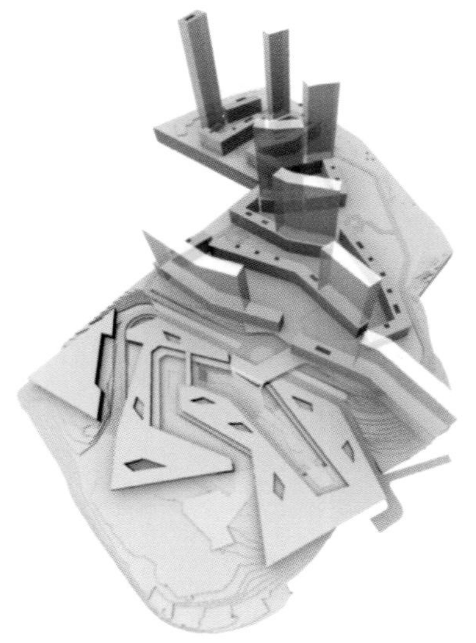

RIVER
河流

POOLS
水池

ALL
总图

olar Orientation: The building mass is stributed in such a way that it allows each uilding to have access to sun light. The ositioning and mix of low and tall buildings llows light to penetrate the site, providing olar exposure to the many landscape atures.
阳方位：建筑体量的分布确保每一个建筑可以享受自光照。高低错落的建筑布局和组合使太阳光线能渗透一场址，为多个景观元素提供太阳光照射。

olar Hot Water Panels: The roofs of the owers are proposed to be covered with olar hot water panels, providing hot water o residents without the consumption of lectricity. The panels will be mounted on he roof top at an angle and concealed by a arapet.
阳能热水面板：塔楼屋顶覆盖太阳能热水面板，在不费电能的情况下为居民提供热水。面板将安装在建筑顶，呈一定角度并且经护墙隐藏。

vaporative Cooling: The site-long water eature will provide a cooling effect through vaporation. Evaporative cooling is a hysical phenomenon in which evaporation f a liquid, typically into surrounding air, ools an object or a liquid in contact with it.
发冷却：场地的水景蒸发过程将引起制冷的效果。蒸制冷是一种物理现象，某种液体经蒸发进入周围空气中，在这一过程中，将使某一物体或是与之接触的某液体温度降低。

ain Water Harvesting and Storage: Rain vater falling on hard surfaces including roof ops will be collected and stored in the site vater feature. There, this natural source of vater will compensate for the evaporated iquid when temperatures rise.
水收集和储藏：落在硬质表面（包括屋顶）的雨水将集和储藏在场地的水景当中。这一天然水源在气温上的时候补充被蒸发的水体。

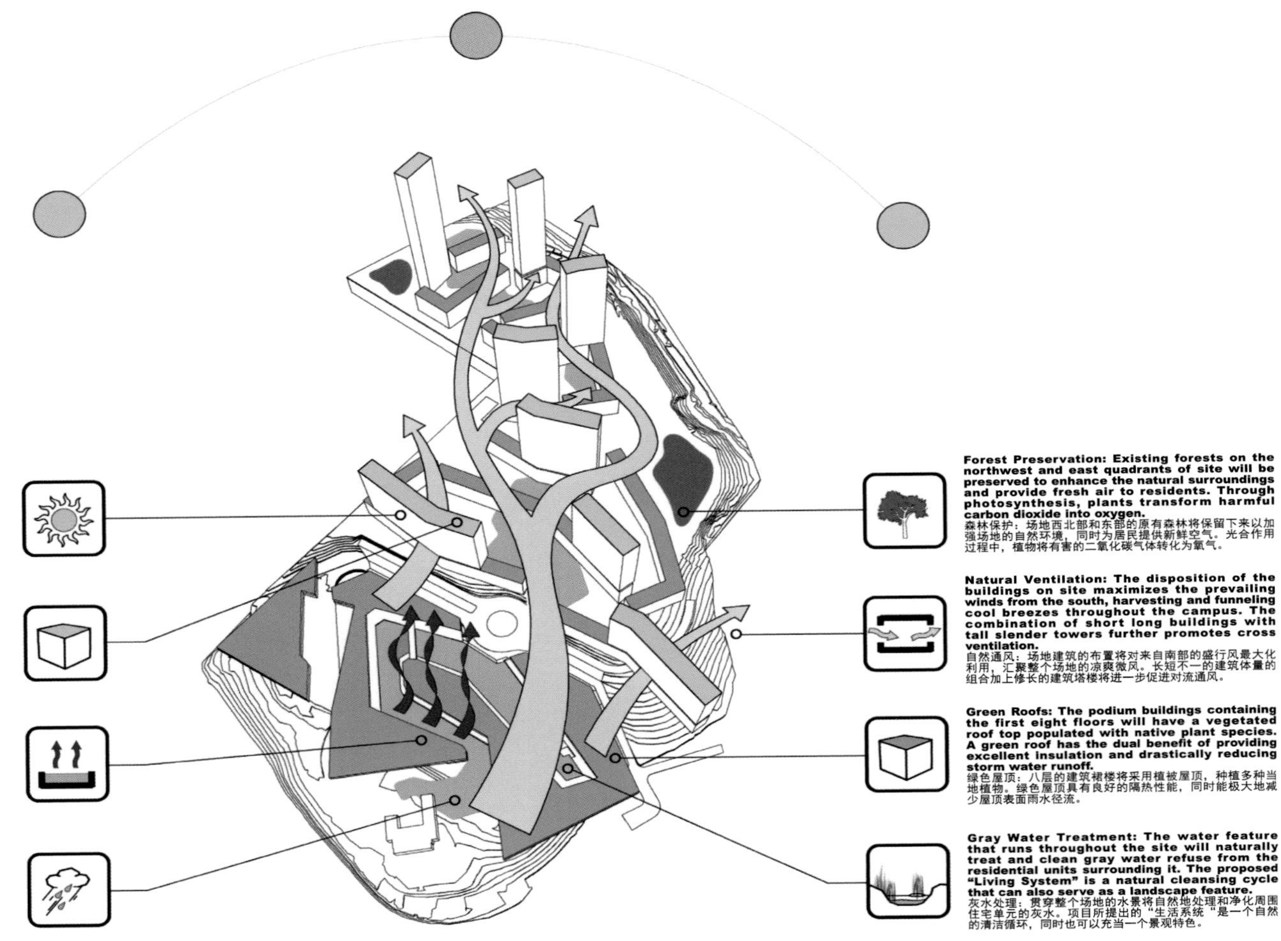

Forest Preservation: Existing forests on the northwest and east quadrants of site will be preserved to enhance the natural surroundings and provide fresh air to residents. Through photosynthesis, plants transform harmful carbon dioxide into oxygen.
森林保护：场地西北部和东部的原有森林将保留下来以加强场地的自然环境，同时为居民提供新鲜空气。光合作用过程中，植物将有害的二氧化碳气体转化为氧气。

Natural Ventilation: The disposition of the buildings on site maximizes the prevailing winds from the south, harvesting and funneling cool breezes throughout the campus. The combination of short long buildings with tall slender towers further promotes cross ventilation.
自然通风：场地建筑的布置将对来自南部的盛行风最大化利用，汇聚整个场地的凉爽微风。长短不一的建筑体量的组合加上修长的建筑塔楼将进一步促进对流通风。

Green Roofs: The podium buildings containing the first eight floors will have a vegetated roof top populated with native plant species. A green roof has the dual benefit of providing excellent insulation and drastically reducing storm water runoff.
绿色屋顶：八层的建筑裙楼将采用植被屋顶，种植多种当地植物。绿色屋顶具有良好的隔热性能，同时能极大地减少屋顶表面雨水径流。

Gray Water Treatment: The water feature that runs throughout the site will naturally treat and clean gray water refuse from the residential units surrounding it. The proposed "Living System" is a natural cleansing cycle that can also serve as a landscape feature.
灰水处理：贯穿整个场地的水景将自然地处理和净化周围住宅单元的灰水。项目所提出的"生活系统"是一个自然的清洁循环，同时也可以充当一个景观特色。

CIRCULATION DIAGRAM
循环图

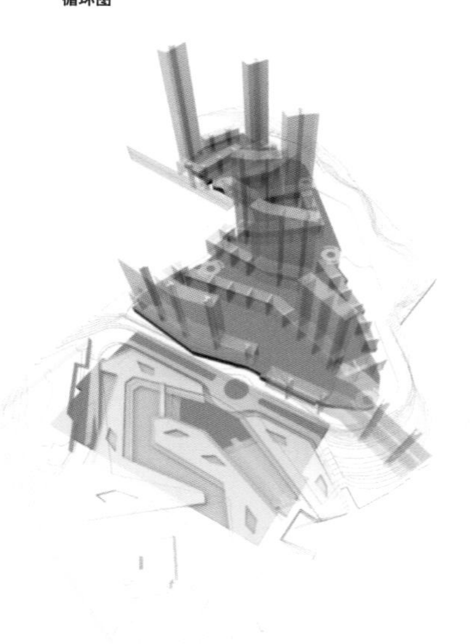

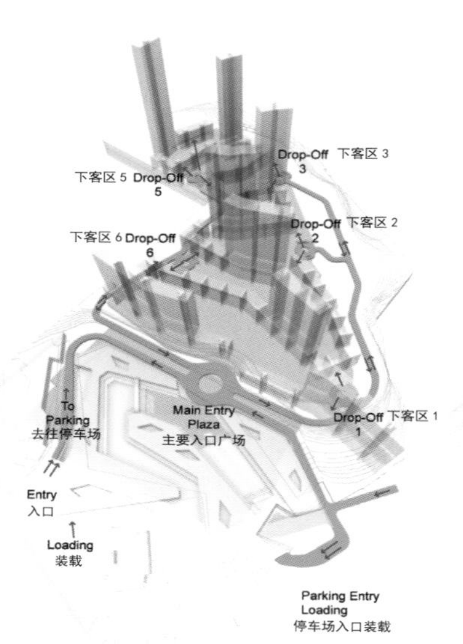

PARKING
停车场
Residential 3 level/6000 parking spaces 住宅三层 /6 000 停车位
Retail 3 Levels/4500 parking spaces 零售三层 /4 500 停车位
Total: 10500 parking spaces 共 10 500 停车位

ROADS
道路

CORES
核心
Type A: 1 per 4 units A 类单元：1/4 单元
Type B: 1 per 2 units B 类单元：1/2 单元
Type C: 1 per 2 units C 类单元：1/2 单元
Type D: 1 per 1 units D 类单元：1/1 单元

CORES/RAODS
核心 / 道路

SITE CIRCULATION DIAGRAM
场地循环表

SITE ROADS
场地道路

PROPOSED INFRASTRUCTURE
拟建基础设施

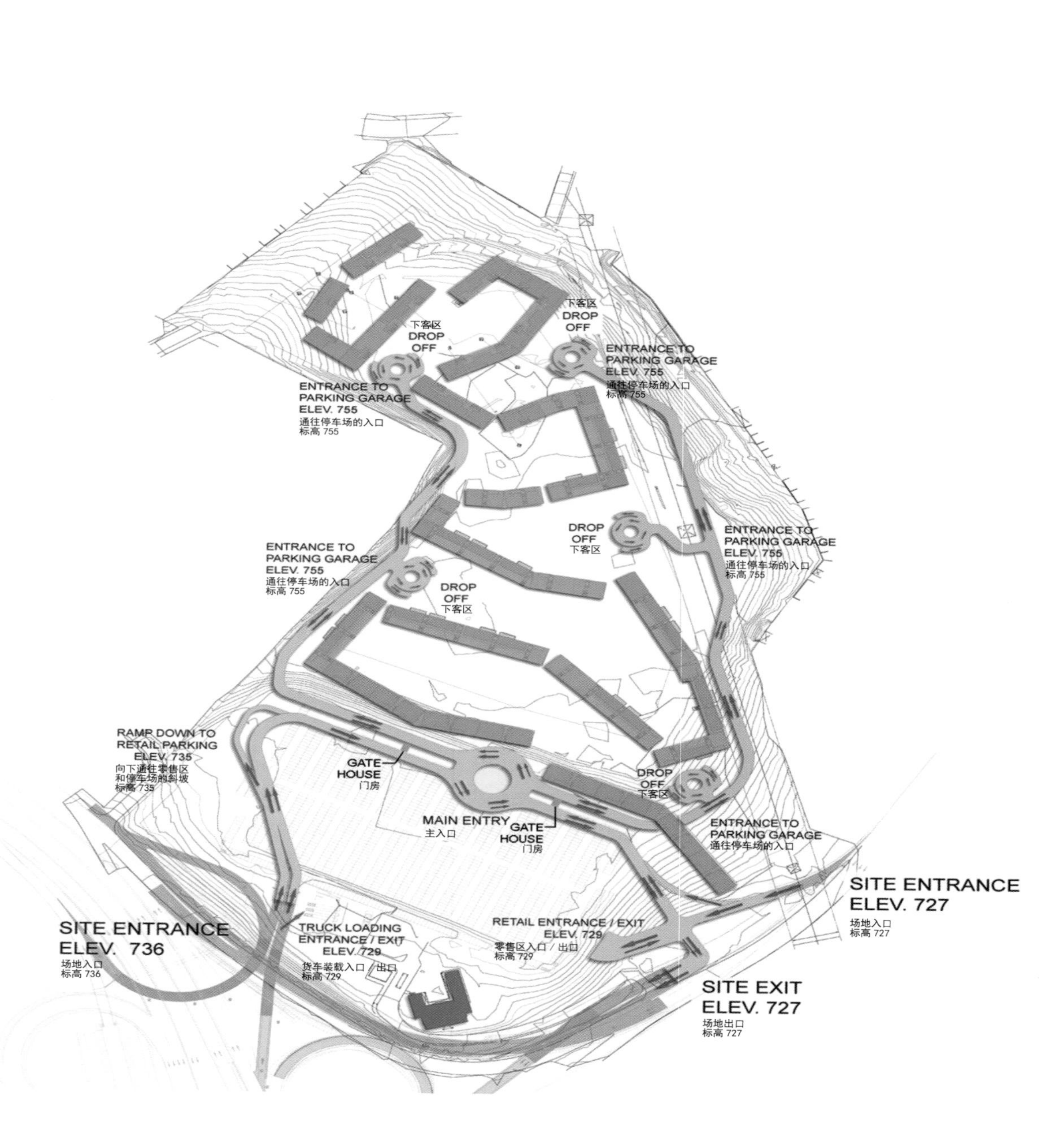

UNIT DIAGRAM
单元图表

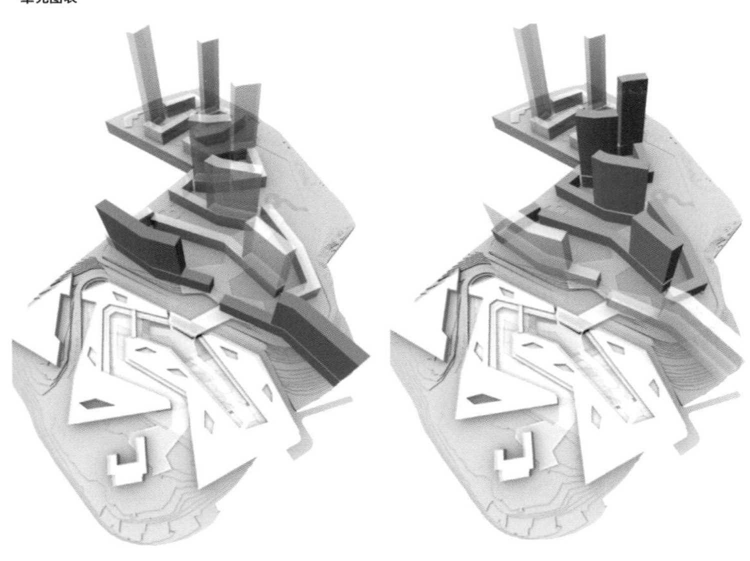

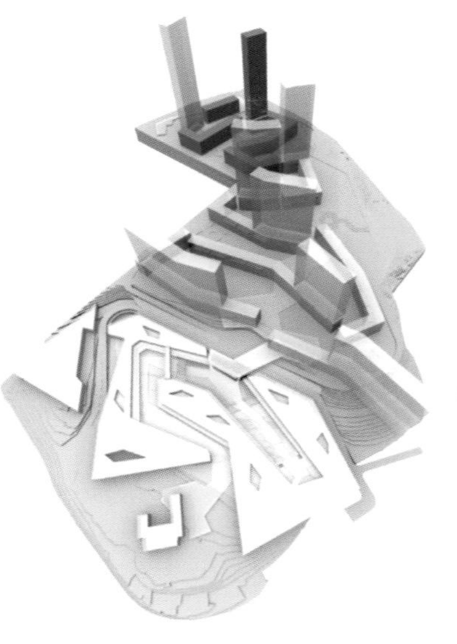

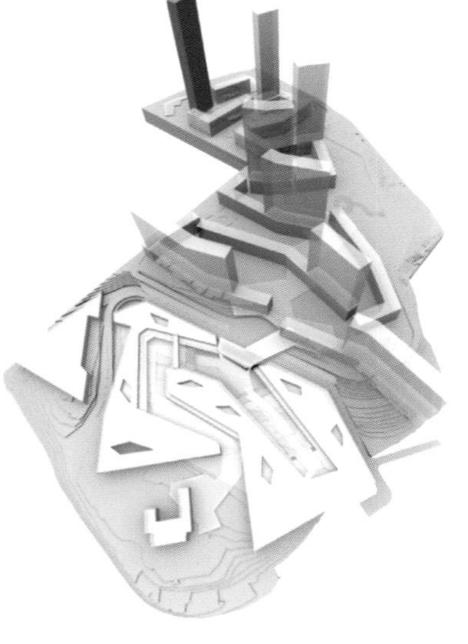

TYPE A A 类单元
3 BEDROOM 3 室
95 m² 95 m²
700 Units 700 单元

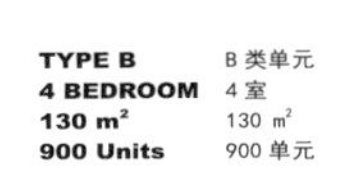

TYPE B B 类单元
4 BEDROOM 4 室
130 m² 130 m²
900 Units 900 单元

TYPE C C 类单元
4 BEDROOM 4 室
160 m² 160 m²
300 Units 300 单元

TYPE D D 类单元
4 BEDROOM 4 室
200 m² 200 m²
100 Units 100 单元

TOTAL UNITS=2000 总单元 =2 000
TOTAL M²=70.150 总面积：70 150 m²

SITE CONDITIONS
场地条件

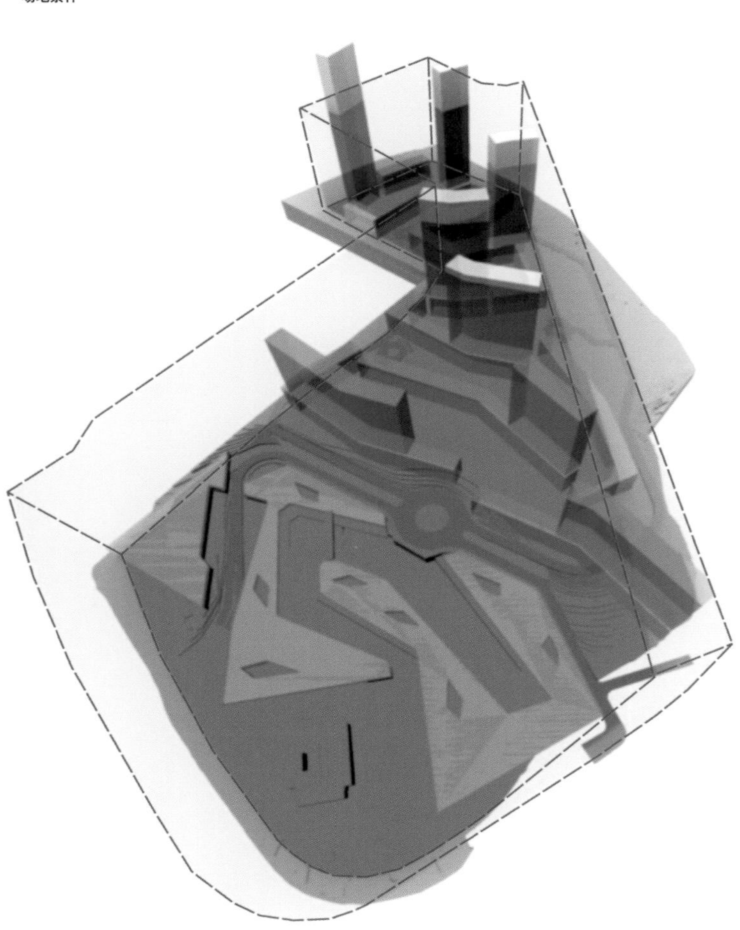

ZONING ENVELOPE
区域表层

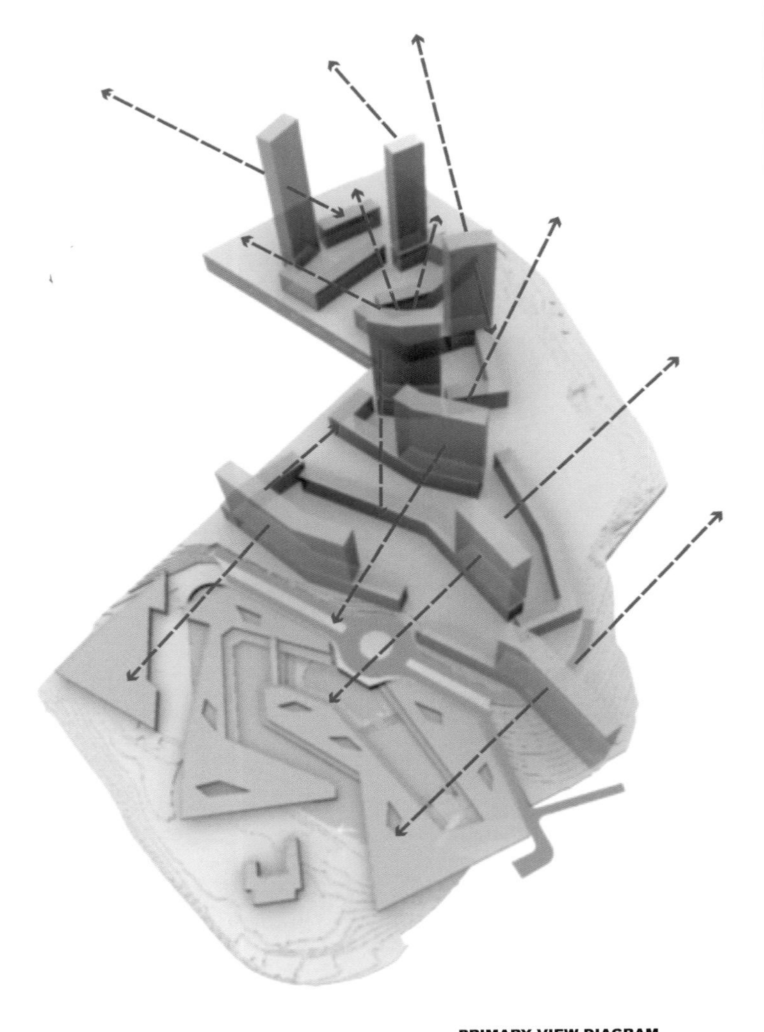

PRIMARY VIEW DIAGRAM
一级视图

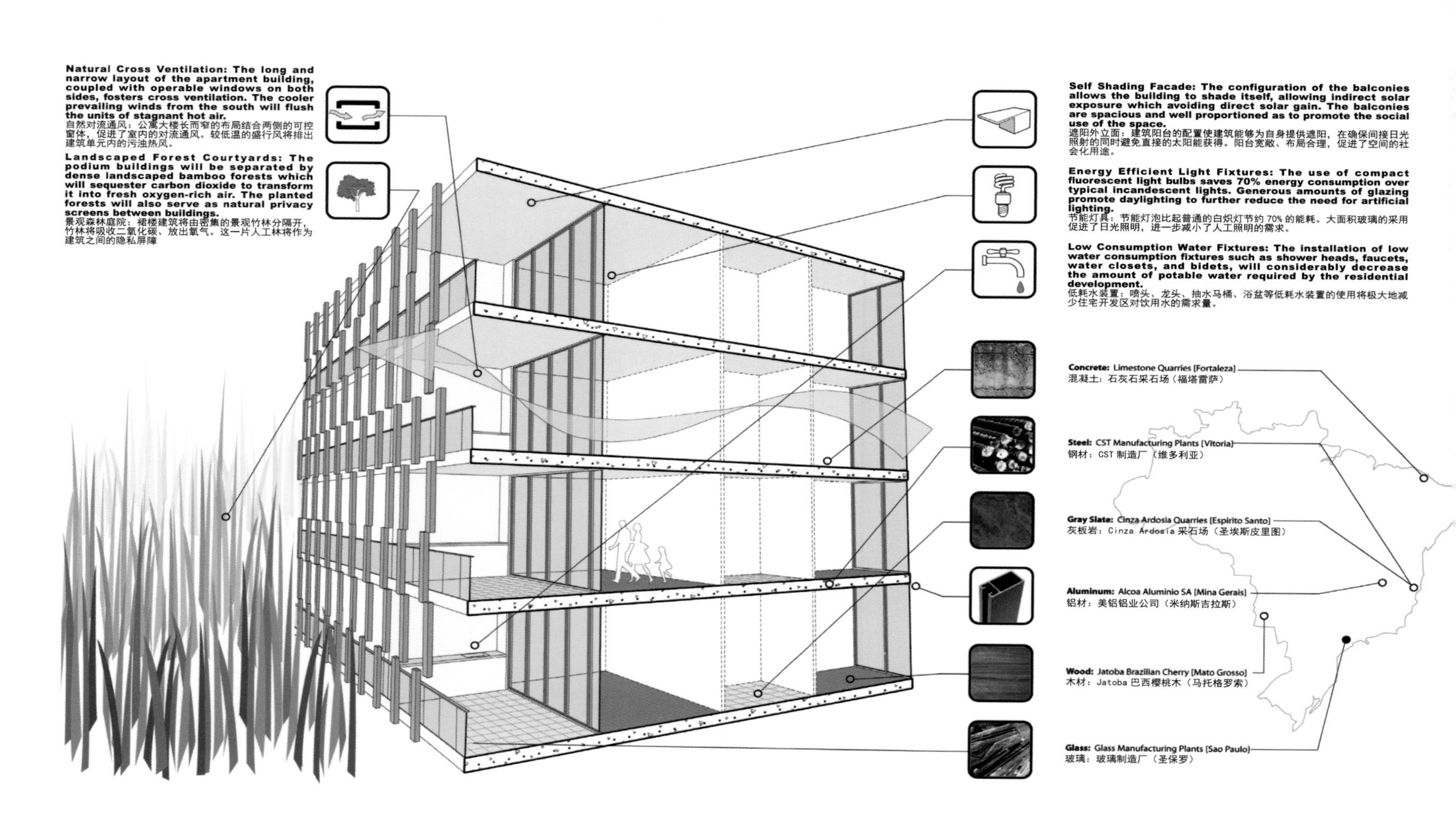
Natural Cross Ventilation: The long and narrow layout of the apartment building, coupled with operable windows on both sides, fosters cross ventilation. The cooler prevailing winds from the south will flush the units of stagnant hot air.
自然对流通风：公寓大楼长而窄的布局结合两侧的可控窗体，促进了室内的对流通风。较低温的盛行风将排出建筑单元内的污浊热风。
Landscaped Forest Courtyards: The podium buildings will be separated by dense landscaped bamboo forests which will sequester carbon dioxide to transform it into fresh oxygen-rich air. The planted forests will also serve as natural privacy screens between buildings.
景观森林庭院：裙楼建筑将由密集的景观竹林分隔开，竹林将吸收二氧化碳、放出氧气。这一片人工林将作为建筑之间的隐私屏障
Self Shading Facade: The configuration of the balconies allows the building to shade itself, allowing indirect solar exposure which avoiding direct solar gain. The balconies are spacious and well proportioned as to promote the social use of the space.
遮阳外立面：建筑阳台的配置使建筑能够为自身提供遮阳，在确保间接日光照射的同时避免直接的太阳能获得。阳台宽敞、布局合理，促进了空间的社会化用途。
Energy Efficient Light Fixtures: The use of compact fluorescent light bulbs saves 70% energy consumption over typical incandescent lights. Generous amounts of glazing promote daylighting to further reduce the need for artificial lighting.
节能灯具：节能灯泡比起普通的白炽灯节约 70% 的能耗。大面积玻璃的采用促进了日光照明，进一步减小了人工照明的需求。
Low Consumption Water Fixtures: The installation of low water consumption fixtures such as shower heads, faucets, water closets, and bidets, will considerably decrease the amount of potable water required by the residential development.
低耗水装置：喷头、龙头、抽水马桶、浴盆等低耗水装置的使用将极大地减少住宅开发区对饮用水的需求量。
Concrete: Limestone Quarries [Fortaleza]
混凝土：石灰石采石场（福塔雷萨）
Steel: CST Manufacturing Plants [Vitoria]
钢材：CST 制造厂（维多利亚）
Gray Slate: Cinza Ardosia Quarries [Espirito Santo]
灰板岩：Cinza Ardosia 采石场（圣埃斯皮里图）
Aluminum: Alcoa Aluminio SA [Mina Gerais]
铝材：美铝铝业公司（米纳斯吉拉斯）
Wood: Jatoba Brazilian Cherry [Mato Grosso]
木材：Jatoba 巴西樱桃木（马托格罗索）
Glass: Glass Manufacturing Plants [Sao Paulo]
玻璃：玻璃制造厂（圣保罗）

941 meters
903 meters
891 meters
885 meters
870 meters
858 meters
849 meters
843 meters
783 meters
755 meters

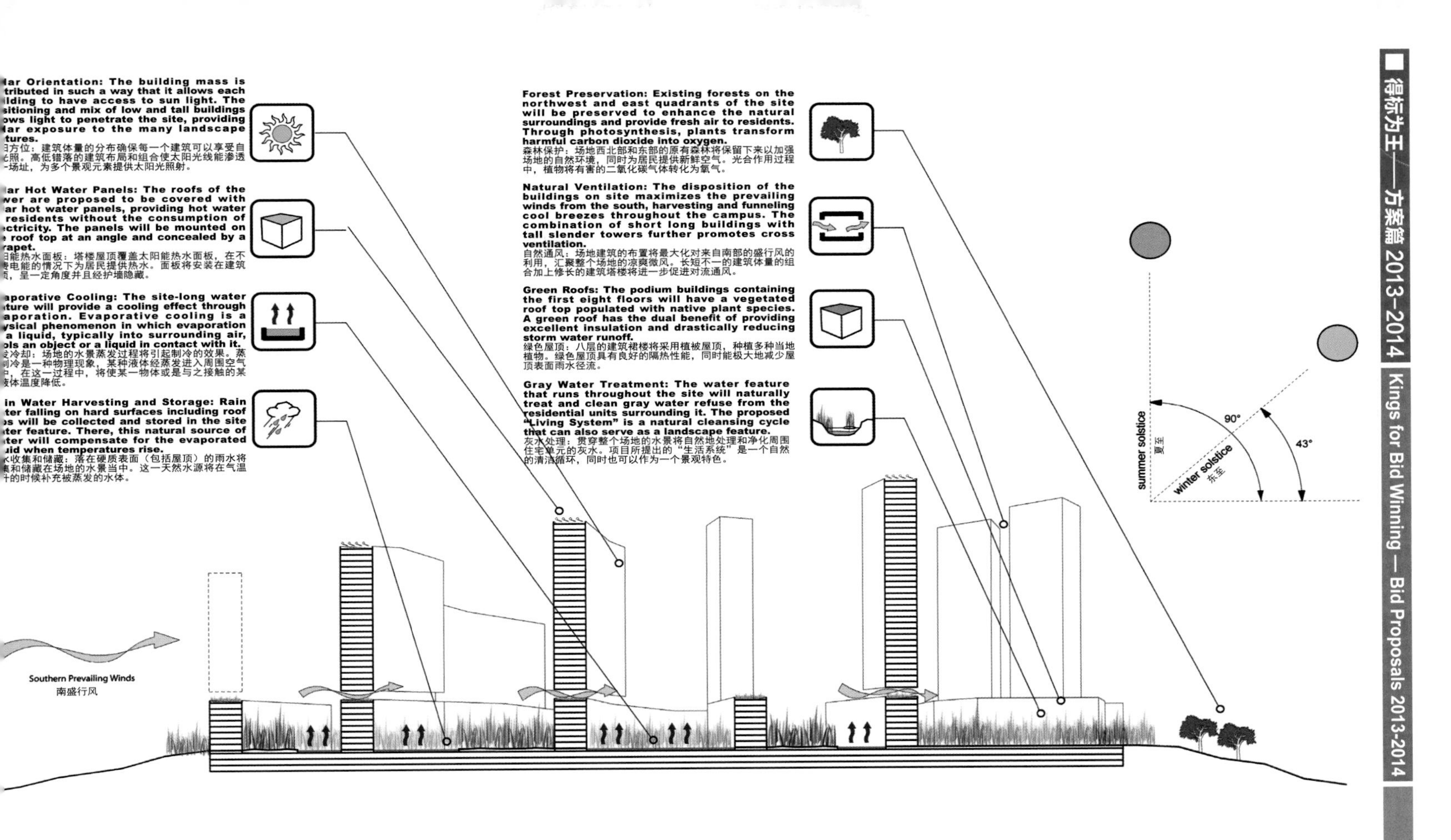
lar Orientation: The building mass is tributed in such a way that it allows each ilding to have access to sun light. The sitioning and mix of low and tall buildings ows light to penetrate the site, providing lar exposure to the many landscape tures.
日方位：建筑体量的分布确保每一个建筑可以享受自光照。高低错落的建筑布局和组合使太阳光线能渗透一场址，为多个景观元素提供太阳光照射。
lar Hot Water Panels: The roofs of the wer are proposed to be covered with ar hot water panels, providing hot water residents without the consumption of ctricity. The panels will be mounted on roof top at an angle and concealed by a rapet.
日能热水面板：塔楼屋顶覆盖太阳能热水面板，在不毛电能的情况下为居民提供热水。面板将安装在建筑顶，呈一定角度并且经护墙隐藏。
aporative Cooling: The site-long water ture will provide a cooling effect through aporation. Evaporative cooling is a ysical phenomenon in which evaporation a liquid, typically into surrounding air, ols an object or a liquid in contact with it.
发冷却：场地的水景蒸发过程将引起制冷的效果。蒸制冷是一种物理现象，某种液体经蒸发进入周围空气中，在这一过程中，将使某一物体或是与之接触的某液体温度降低。
in Water Harvesting and Storage: Rain ter falling on hard surfaces including roof ps will be collected and stored in the site ter feature. There, this natural source of ter will compensate for the evaporated uid when temperatures rise.
水收集和储藏：落在硬质表面（包括屋顶）的雨水将集和储藏在场地的水景当中。这一天然水源将在气温升的时候补充被蒸发的水体。
Forest Preservation: Existing forests on the northwest and east quadrants of the site will be preserved to enhance the natural surroundings and provide fresh air to residents. Through photosynthesis, plants transform harmful carbon dioxide into oxygen.
森林保护：场地西北部和东部的原有森林将保留下来以加强场地的自然环境，同时为居民提供新鲜空气。光合作用过程中，植物将有害的二氧化碳气体转化为氧气。
Natural Ventilation: The disposition of the buildings on site maximizes the prevailing winds from the south, harvesting and funneling cool breezes throughout the campus. The combination of short long buildings with tall slender towers further promotes cross ventilation.
自然通风：场地建筑的布置将最大化对来自南部的盛行风的利用，汇聚整个场地的凉爽微风。长短不一的建筑体量的组合加上修长的建筑塔楼将进一步促进对流通风。
Green Roofs: The podium buildings containing the first eight floors will have a vegetated roof top populated with native plant species. A green roof has the dual benefit of providing excellent insulation and drastically reducing storm water runoff.
绿色屋顶：八层的建筑裙楼将采用植被屋顶，种植多种当地植物。绿色屋顶具有良好的隔热性能，同时能极大地减少屋顶表面雨水径流。
Gray Water Treatment: The water feature that runs throughout the site will naturally treat and clean gray water refuse from the residential units surrounding it. The proposed "Living System" is a natural cleansing cycle that can also serve as a landscape feature.
灰水处理：贯穿整个场地的水景将自然地处理和净化周围住宅单元的灰水。项目所提出的"生活系统"是一个自然的清洁循环，同时也可以作为一个景观特色。
summer solstice
夏至
winter solstice
冬至
90°
43°
Southern Prevailing Winds
南盛行风

印度新德里古尔冈 Centra One

Centra One, Gurgaon

设计单位： Cervera & Pioz Arquitectos
开发商： 古尔冈 BPTP
项目地址： 印度新德里
项目面积： 37 161 m²

Designed by: Cervera & Pio
Client: BPTP, Gurgao
Location: New Delhi, Indi
Area: 37,161 m

项目概况

Centra One 是一个现代的高科技商业综合体，位于古尔冈地区发展最快的第 61 区。场地周边的大型开发项目 Ireo、Emmar MGF 等，为构建这个融合古尔冈城市肌理的完美地标式建筑创造了条件。

设计理念

设计师认为，除了建筑外观的表现力和建筑带给人的体验外，设计更多的是书写一段当地的历史和记忆。独特的建筑形式、精致的构造细节是诠释建筑内涵、精神的媒介。

建筑设计

该建筑包含了精致的办公和商业销售空间。商业销售区域面积为 10 869 平方米，占据该建筑地面的 1-3 层。办公空间覆盖 4-15 层，面积达 16 722 平方米。考虑到现代办公环境的需求，该综合体配备了世界一流的设备，以营造一个环境舒适、激发人创造力的工作场所。

项目的另一设计重点是打造一栋精致的绿色建筑。每一个楼层都有向两侧开启的通风设计，且每个功能空间都有一侧向外，一侧向中庭的自然采光。

Profile

Centra One is a contemporary and high-tech business complex in Gurgaon's fastest developed area of sector 61. One of largest housing complex by Ireo, Emmar MGF and many other developments are key factors which make Centra One project the perfect landmark merging with urban fabrics of Gurgaon.

Design Concept

The very essence of design is not simply the way of building's look and feel, but how the experience

GROUND FLOOR PLAN
底层平面图
Scale: 1/100
比例：1/100

CENTRA ONE, RETAIL&OFFICES BUILDING.
Centra One, 零售＆办公大楼

nprints a memory. A special form and fine detailing f each part of building elements are mediums to nterpret the spirit and soul of building.

rchitectural Design

he building contains exquisite office and retail paces. The retail space spreads over 10,869 square neters occupying the first three floors of the complex. he offices at Centra One are located across 12 oors above the retail sphere spreading over 16,722 quare meters. For the consideration of a modern work environment, the complex is equipped with world-class amenities to create an inspiring and comfortable work environment.

Another design focus of the project is to build an exquisite green building. Each floor opens up two sides for natural ventilation. Each function space enjoys natural lighting both from outside and from atrium.

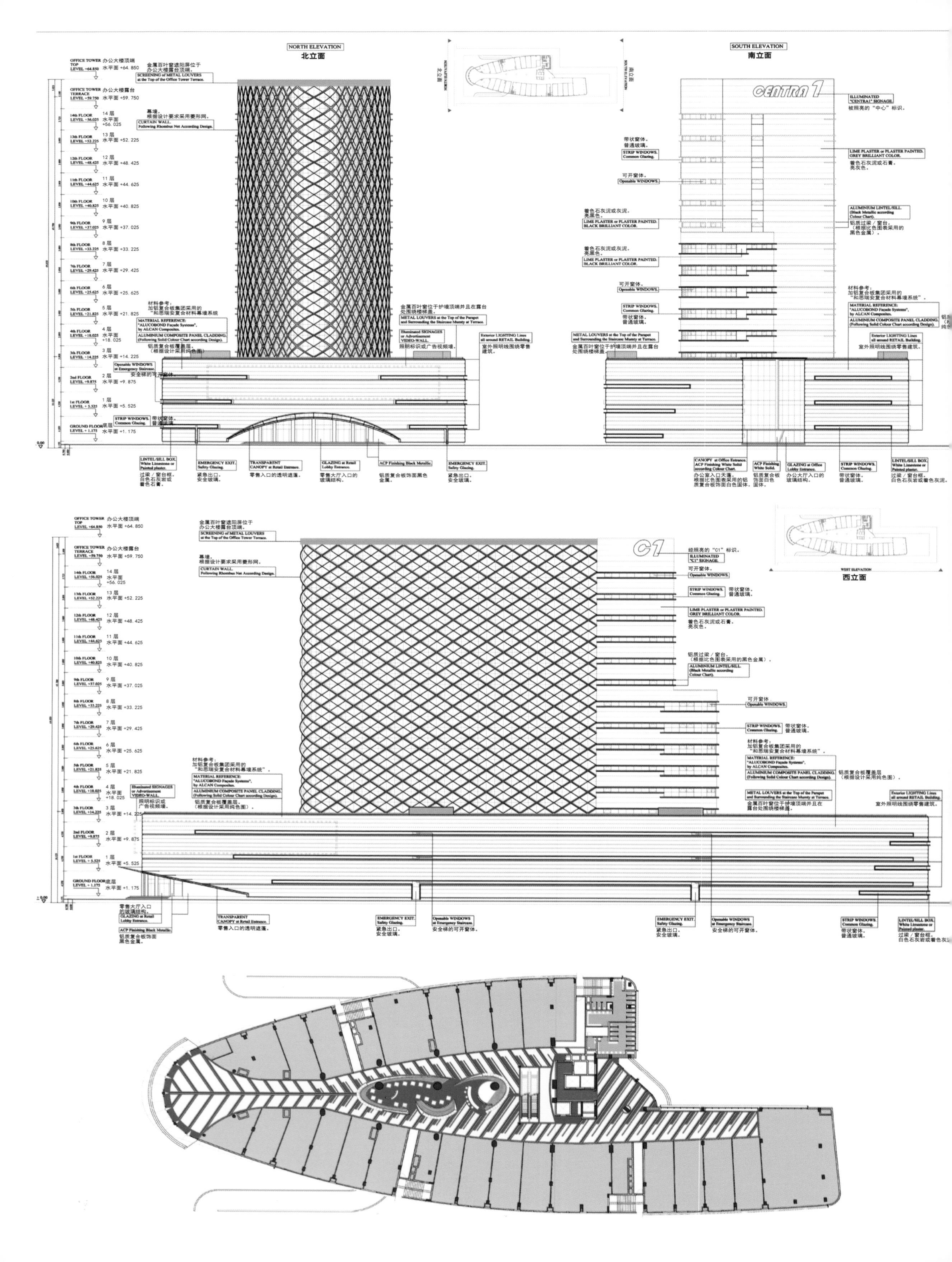

NORTH ELEVATION
北立面
SOUTH ELEVATION
南立面
WEST ELEVATION
西立面

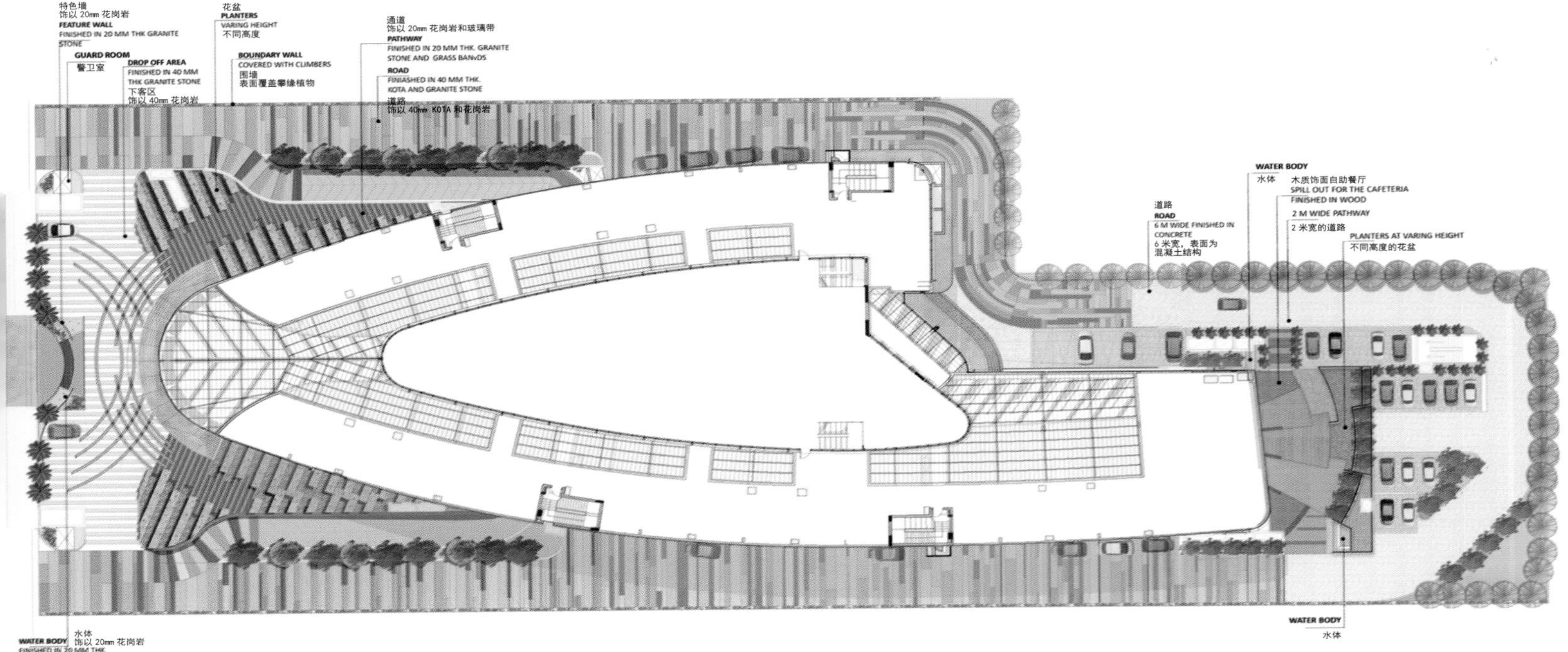
特色墙
饰以 20mm 花岗岩
FEATURE WALL
FINISHED IN 20 MM THK GRANITE STONE
GUARD ROOM
警卫室
DROP OFF AREA
FINISHED IN 40 MM THK GRANITE STONE
下客区
饰以 40mm 花岗岩
花盆
PLANTERS
VARING HEIGHT
不同高度
BOUNDARY WALL
COVERED WITH CLIMBERS
围墙
表面覆盖攀缘植物
通道
饰以 20mm 花岗岩和玻璃带
PATHWAY
FINISHED IN 20 MM THK. GRANITE STONE AND GRASS BANvDS
ROAD
FINIASHED IN 40 MM THK. KOTA AND GRANITE STONE
道路
饰以 40mm KOTA 和花岗岩
道路
ROAD
6 M WIDE FINISHED IN CONCRETE
6 米宽，表面为混凝土结构
WATER BODY
水体
木质饰面自助餐厅
SPILL OUT FOR THE CAFETERIA
FINISHED IN WOOD
2 M WIDE PATHWAY
2 米宽的道路
PLANTERS AT VARING HEIGHT
不同高度的花盆
WATER BODY
FINISHED IN 20 MM THK GRANITE STONE
水体
饰以 20mm 花岗岩
WATER BODY
水体

克罗地亚萨格勒布 Miramare 商务中心

Miramare Tower Center

设计单位：3LHD
项目地址：克罗地亚萨格勒布
基地面积：4 037 m²
设计团队：Tatjana Grozdanic Begovic
Saša Begovic
Marko Dabrovic
Silvije Novak
Tin Kavuric
Krunoslav Szorsen
Josko Kotula

Designed by: 3LHD
Location: Zagreb, Croatia
Site Area: 4,037 m²
Project Team: Tatjana Grozdanic Begovic, Saša Begovic, Marko Dabrovic, Silvije Novak, Tin Kavuric, Krunoslav Szorsen, Josko Kotula

项目概况

由 3LHD 设计的 Miramare 商业中心，位于克罗地亚萨格勒布市 Miramarska 街与 Bednjanska 街的交叉口处，这里也是两种城市景观的交界点：北半部分是萨格勒布市典型的贫民区，分布着萨格勒布城市铁路系统; 南半部分绝大部分是公共与社会建筑，现代特征鲜明显著。

建筑设计

设计方案旨在创造一栋统一的建筑物，既能协调周边的城市环境，又能满足项目的功能需求。该项目由两个相互关联的建筑体块构成，一个高 20 层，一个高 5 层，这样既能满足规划需求，又可将基地内两种截然不同的城市景观统一起来。简约但富有吸引力的立面设计是用伸进和拉出的玻璃模块构成的，形成了建筑的识别性和独特性。

在功能布局方面，建筑涵盖了 20 000 平方米的办公空间，一层主要用于商店、酒吧和餐厅，并规划了地下停车场。建筑的顶楼有一处独特的观景台，拥有良好的景观视角。一个半开放的中庭将两个建筑与周围的公用道路行人专用区连接起来。

Profile

The competition program is placed on a crossing of two streets, Miramarska and Bednjanska. It is a point of colliding two urban city concepts: north part with classical city blocks of the Zagreb's lower city enclosed by the railway and the south part primarily along the Vukovarska Street, outlined with modernistic dominant character of mostly public and social buildings.

Architectural Design

Concept of awarded project stems from forming a unified structure that will reconcile the surrounding urban matrix and meet the program requirements. Solution with two interacting volumes, the tower with 20 floors and the lower part of the 5 floors, united the planned goals and achieve high quality relations with the neighbouring buildings. Simple and at the same time very attractive effect of designing the facade by pulling the glass modules in and out, gave the building of business centre Miramare unique recognition and special value.

On functional layout, the volumes cover 20,000 square meters office space. The ground floor has been mainly used for stores, bar, restaurant as well as underground garage. Higher floors are reserved for business users, while the highest floor offers a unique view of the panorama of Zagreb.

Interspaces between two volumes - semi opened atrium, connect the inner area of the business center with the surrounding public space.

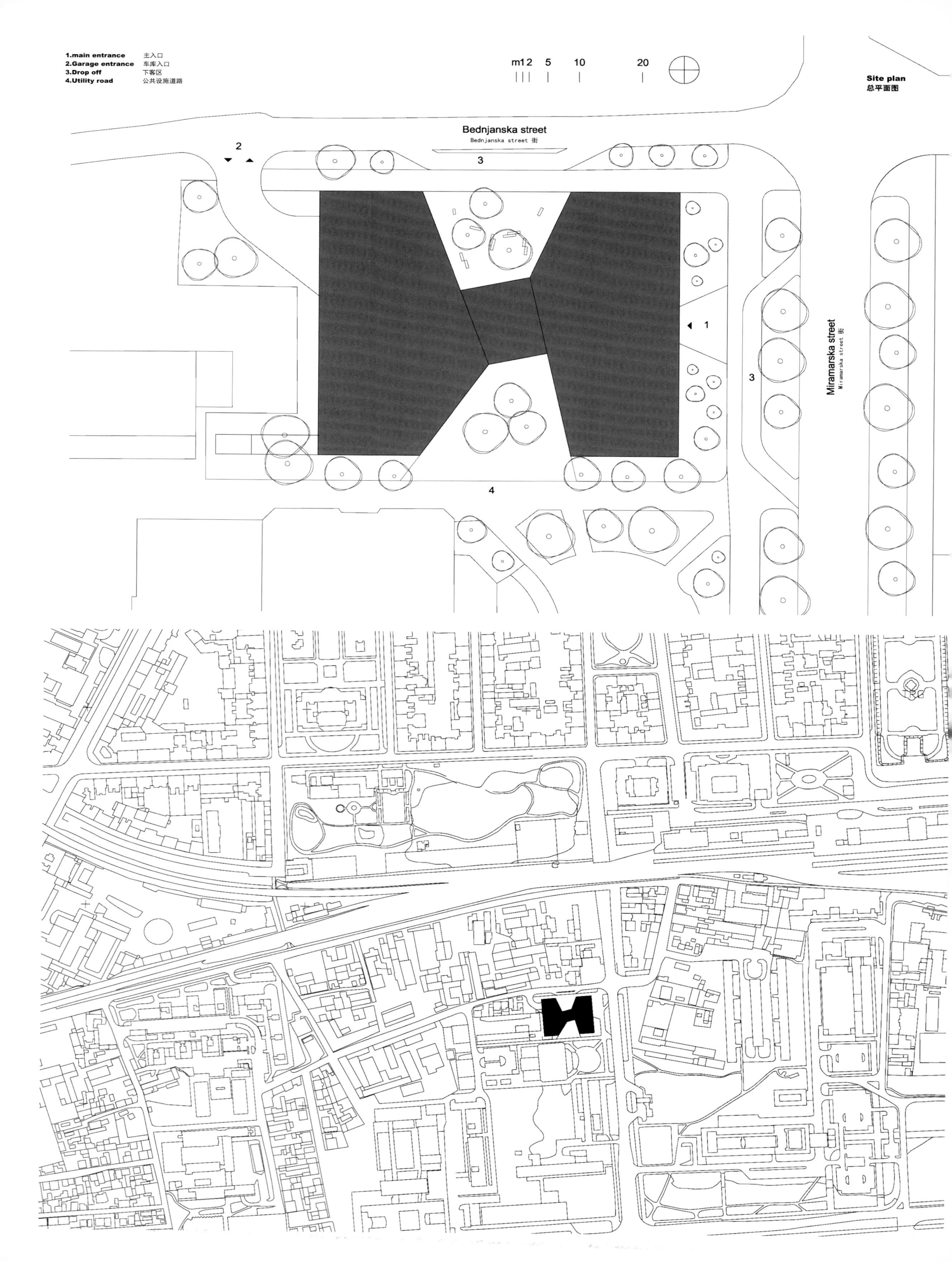
1.main entrance 主入口
2.Garage entrance 车库入口
3.Drop off 下客区
4.Utility road 公共设施道路
m1 2 5 10 20
Site plan
总平面图
Bednjanska street
Bednjanska street 街
Miramarska street
Miramarska street 街
1
2
3
3
4

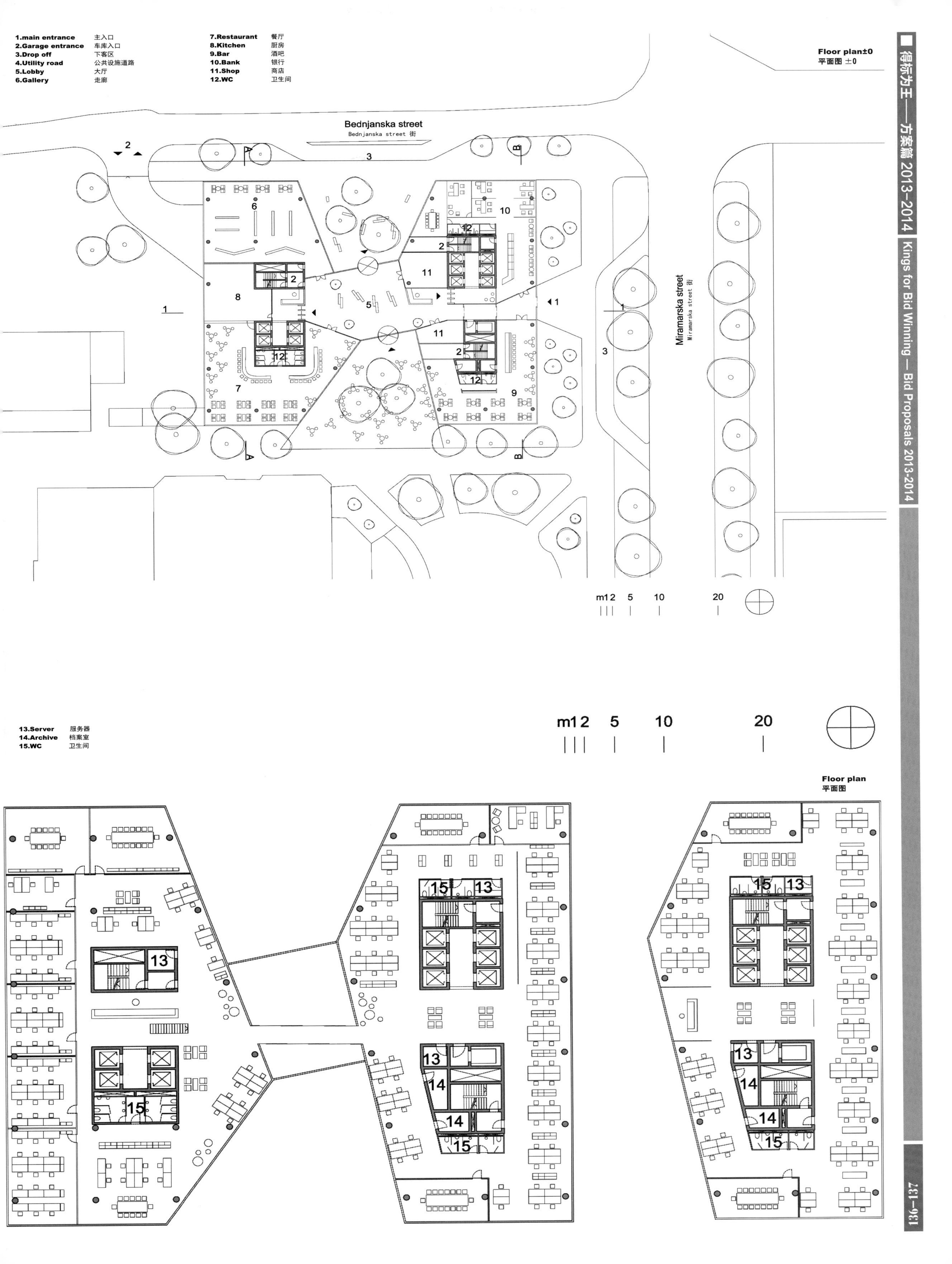
1.main entrance 主入口
2.Garage entrance 车库入口
3.Drop off 下客区
4.Utility road 公共设施道路
5.Lobby 大厅
6.Gallery 走廊
7.Restaurant 餐厅
8.Kitchen 厨房
9.Bar 酒吧
10.Bank 银行
11.Shop 商店
12.WC 卫生间
Floor plan±0
平面图 ±0
Bednjanska street
Bednjanska street 街
Miramarska street
Miramarska street 街
m1 2 5 10 20
13.Server 服务器
14.Archive 档案室
15.WC 卫生间
m1 2 5 10 20
Floor plan
平面图

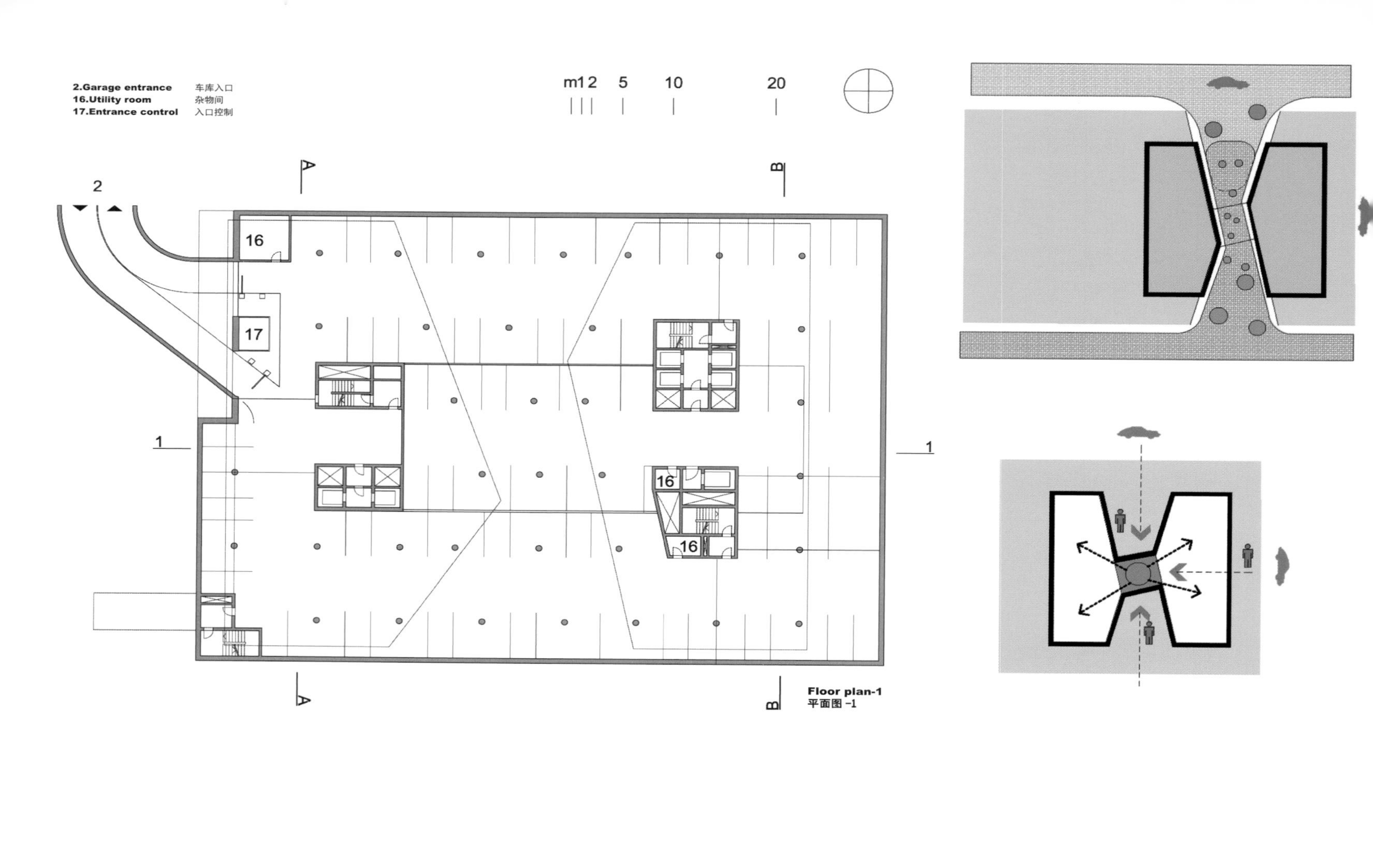

2.Garage entrance 车库入口
16.Utility room 杂物间
17.Entrance control 入口控制
m1 2 5 10 20
Floor plan-1
平面图 -1

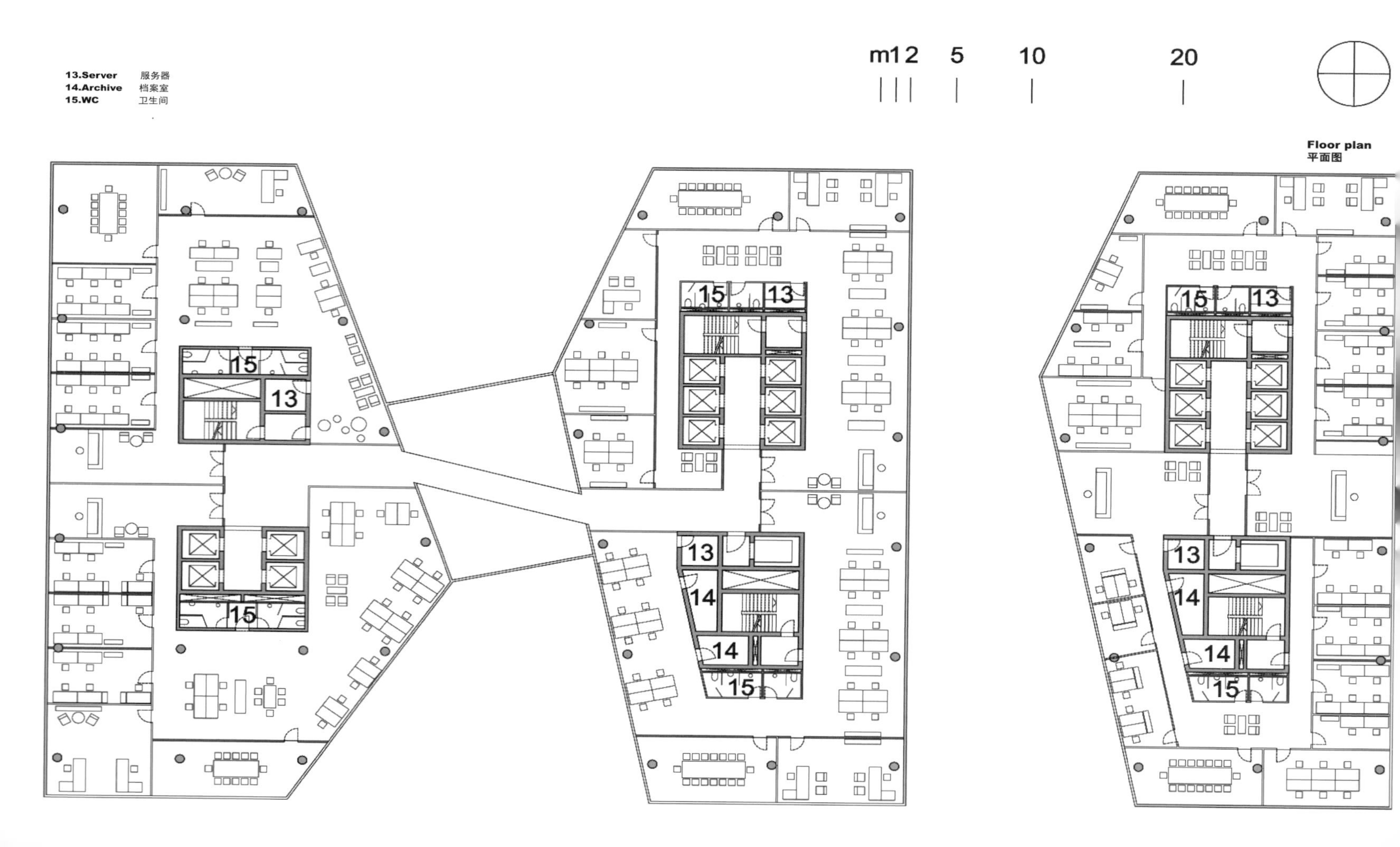

13.Server 服务器
14.Archive 档案室
15.WC 卫生间
m1 2 5 10 20
Floor plan
平面图

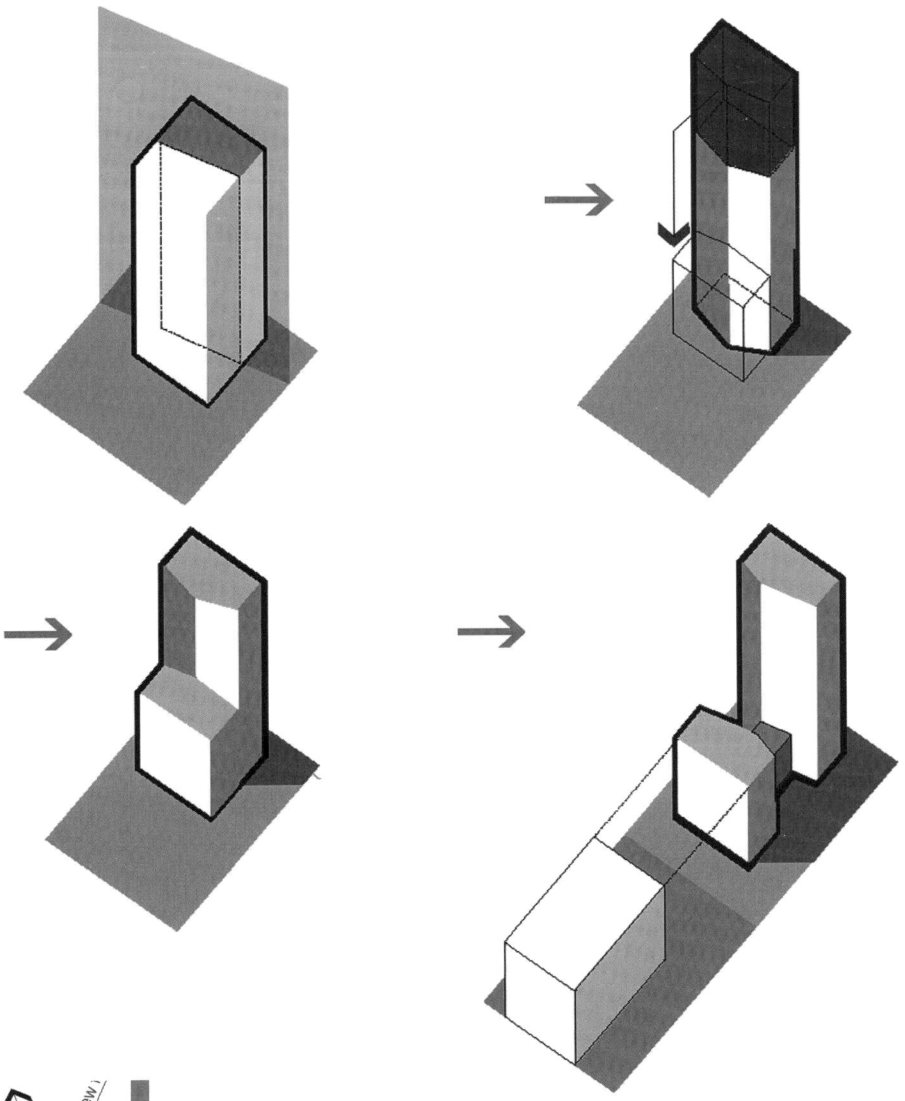

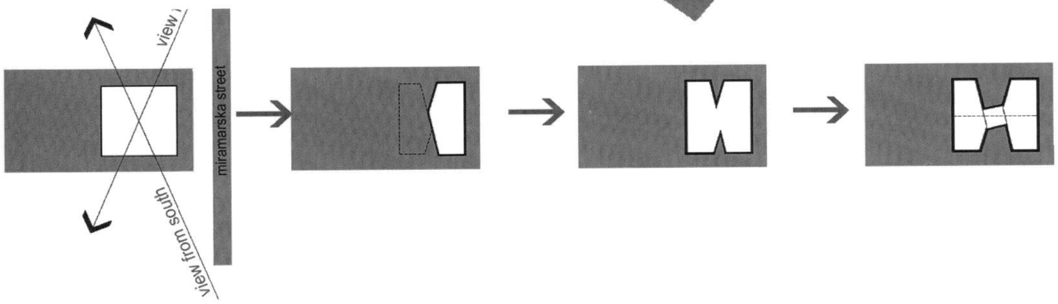

view from south

view

miramarska street

四川成都博瑞·最世界

Chengdu B-Ray Top World

设计单位：日本 M.A.O. 一级建筑士事务所

开发商：成都博瑞房地产开发有限公司

项目地址：中国四川省成都市

占地面积：107 759 m²

总建筑面积：233 456 m²

容积率：2.45

Designed by: M.A.O.

Developer: Chengdu B-Ray Real Estate Development Co., Ltd.

Location: Chengdu, Sichuan, China

Site Area: 107,759 m²

Gross Floor Area: 233,456 m²

Plot Ratio: 2.45

项目概况

项目位于成都市双流新城开发区的最北侧，西侧为白河防护绿化带，西北角为集商务、会议洽谈、休闲健身并结合国际比赛的四川国际网球中心，南侧为星慧集团投资的五星级酒店，优越的地理条件为这个城市综合型商业设施项目创造了条件。

设计理念

通过对这个综合型商业项目辐射范围内消费客群的综合分析，设计将“立体生态购物公园”作为主题元素融入其中，通过减少商业对环境的不利影响，赋予商业空间全新的购物体验，并将自然融入购物之中，从而创造出特殊的购物模式。

设计特色

将“平面变成山脉”是实现“立体生态购物公园”这一理念的主要手段。设计将立体化的公园向基地内部逐渐延伸，使精品商业在不同的标高上错落有致地分布，打破了常规的动线体系，有效地拉动了两层以上的商业价值，让人们在“一步一景一购物”的环境中享受休闲购物的乐趣。

为了体现商业与公园的关系，设计打破了传统的、大盒子形态的商业建筑模式，重新构建了一种开放的、有机联系的、多业态组合的“公园型”商业空间。它不像传统购物中心那样，将顾客引入封闭式的购物区，而是将商业区、餐饮区与开放空间完美地融合在一起。

“立体购物公园”这一理念的植入，使整个商业动线规划打破了传统的发展模式。设计师对标高和空间的巧妙设计及运用，赋予整个项目全新的立体化动线体验。

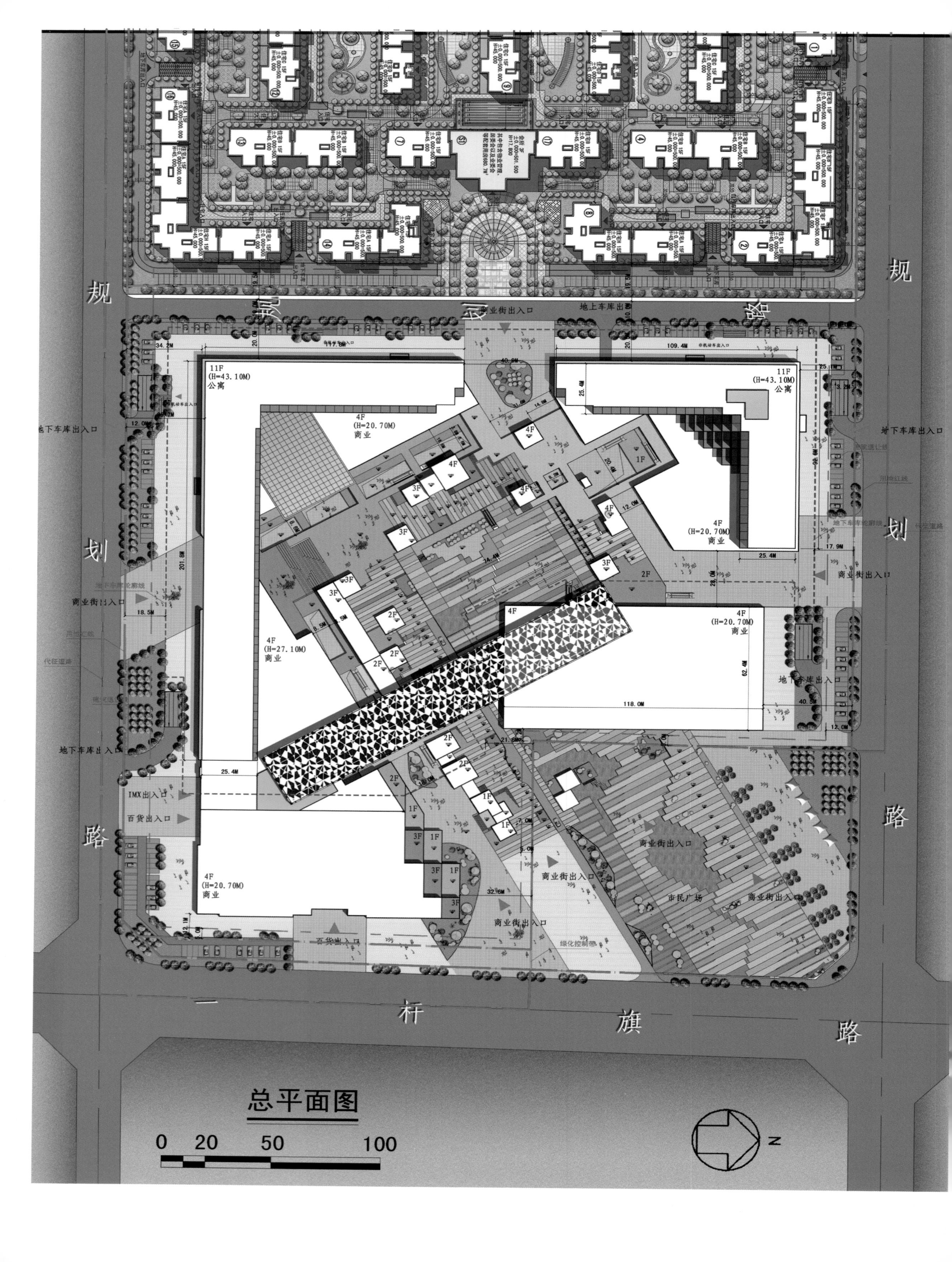

规划路
地上车库出入口
商业街出入口
11F
(H=43.10M)
公寓
4F
(H=20.70M)
商业
4F
(H=27.10M)
商业
地下车库出入口
商业街出入口
IMX出入口
百货出入口
市民广场
绿化控制带
一杆旗路
总平面图
0 20 50 100
N

Profile

The project is located in the north of Shuangliu New Town Development Zone of Chengdu City. It has Baihe River Protective Green Belt in the west. In the northwest corner locates the Sichuan International Tennis Center integrating business, conference, leisure fitness and international competition. Also, there is a five-star hotel in the south invested by Xinhui Group. Superior geographical location creates favorable conditions for this urban complex commercial facility.

Design Concept

For the overall consideration of customers in this comprehensive commercial project, the design shall blend a thematic element of "Stereoscopic Ecological Shopping Park" into the project. By decreasing the impact of commerce to environment and endowing commercial space with brand new shopping experience, the natural surroundings are brought to shopping experience, creating special shopping mode.

Design Feature

From "plane" to "mountain" is the main technique to realize a "Stereoscopic Ecological Shopping Park". The stereoscopic park gradually extending toward the internal of the site makes a staggered layout of exquisite commerce on different elevations, which breaks conventional system of dynamic lines, effectively promotes the commercial value above the 2nd floor as well as ensures the public to enjoy fun of shopping in a "landscaped shopping environment".

To reflect the relationship between business and the park, traditional box-shaped commercial building mode has been broken in this project, re-constructing open, organically connected, "Park Type" commercial space for combined multiple industries. Unlike traditional shopping center which leads customers to enclosed shopping districts, it strives to perfectly integrate commercial districts, dining areas and open spaces.

"Stereoscopic Shopping Park" has broken the traditional development mode of the whole planning of commercial lines. The intelligent design and application of elevations and spaces give this project fresh experience of stereoscopic dynamic lines.

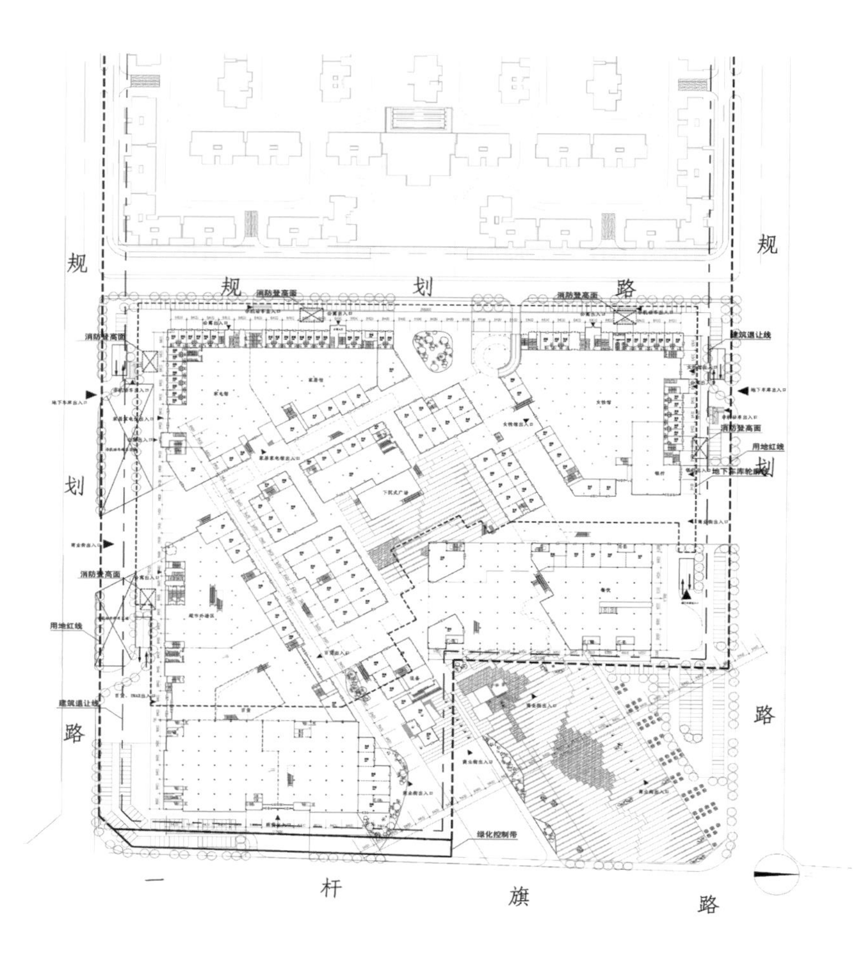

一层平面图 1:800

本层疏散楼梯宽度（M）= 本层营业厅建筑面积（M^2）× 面积折算系数 × 疏散人数换算系数（人/M^2）× 疏散宽度指标（M/百人）

	需要宽度	实际宽度
百货部分疏散楼梯宽度（M）=11308（M^2）×0.5×0.85（人/M^2）×1（M/百人）=48M	48×0.7=33.6M	34.1M
家居、家电部分疏散楼梯宽度（M）=4417（M^2）×0.5×0.85（人/M2）×1（M/百人）=18.7M	18.7×0.7=13.1M	17.1M
女人街部分疏散楼梯宽度（M）=2860（M^2）×0.5×0.85（人/M^2）×1（M/百人）=12.2M	12.2×0.7=8.5M	9.5M
银行部分疏散楼梯宽度（M）=3090（M^2）×0.5×0.85（人/M^2）×1（M/百人）=12.7M	12.7×0.7=8.8M	9.5M

二层平面图 1:800

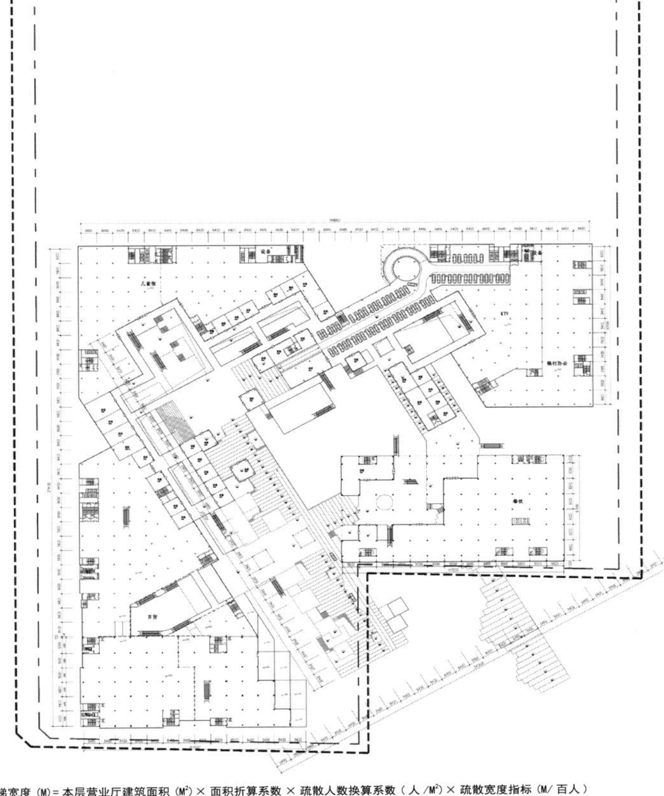

本层疏散楼梯宽度 (M)= 本层营业厅建筑面积 (M^2) × 面积折算系数 × 疏散人数换算系数（人 /M^2）× 疏散宽度指标 (M/ 百人）

	需要宽度	实际宽度
百货部分疏散楼梯宽度 (M)=10572 (M^2) ×0.5×0.77（人 /M^2）×1 (M/ 百人）=40.7M	40.7×0.7=28.5M	13.4M
儿童乐园部分疏散楼梯宽度 (M)=5373 (M^2) ×0.5×0.77（人 /M2）×1 (M/ 百人）=20.7M	20.7×0.7=14.5M	17.1M
KTV 部分疏散楼梯宽度 (M)=2612 (M^2) ×0.5×0.77（人 /M^2）×1 (M/ 百人）=10.5M	10.5×0.7=7.1M	9.5M
银行部分疏散楼梯宽度 (M)=3090 (M^2) ×0.5×0.77（人 /M^2）×1 (M/ 百人）=11.9M	11.9×0.7=6.4M	9.5M
餐饮部分疏散楼梯宽度 (M)=4795 (M^2) ×0.5×0.77（人 /M^2）×0.5 (M/ 百人）=9.3M	9.3×0.7=6.5M	8.4M

三层平面图 1:800

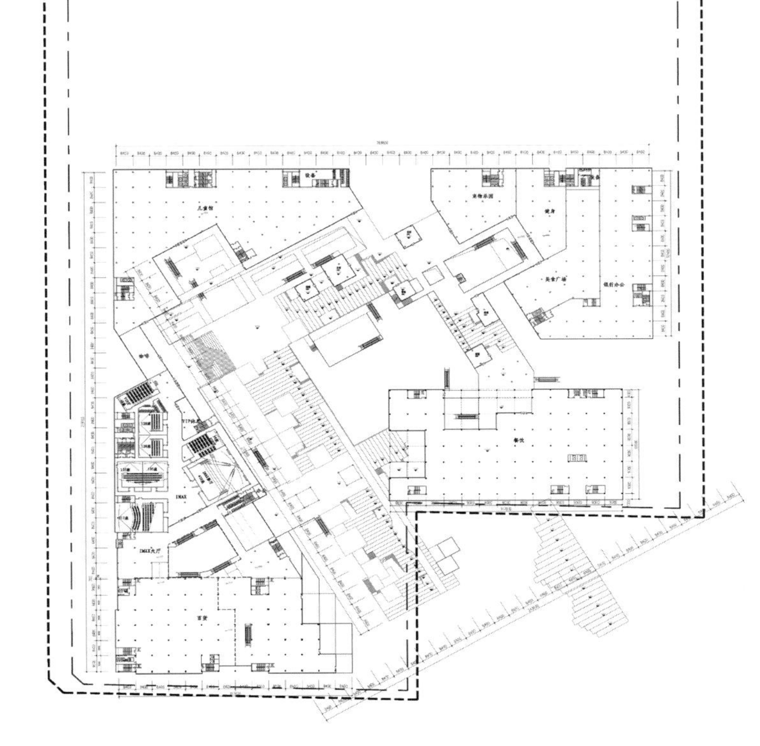

本层疏散楼梯宽度 (M)= 本层营业厅建筑面积 (M^2) × 面积折算系数 × 疏散人数换算系数（人 /M^2）× 疏散宽度指标 (M/ 百人）

	需要宽度	实际宽度
百货部分疏散楼梯宽度 (M)=5795 (M^2) ×0.5×0.6（人 /M^2）×1 (M/ 百人）=17.4M	17.4×0.7=12.2M	13.4M
影院部分疏散楼梯宽度 (M)=5285 (M^2) ×0.5×1（人 /M2）×1 (M/ 百人）=26.4M	26.4×0.7=18.5M	19M
儿童乐园部分疏散楼梯宽度 (M)=6228 (M^2) ×0.5×0.6（人 /M^2）×1 (M/ 百人）=18.7M	18.7×0.7=13.1M	17.1M
银行部分疏散楼梯宽度 (M)=3040 (M^2) ×0.5×0.6（人 /M^2）×1 (M/ 百人）=9.1M	9.1×0.7=6.4M	9.5M
餐饮部分疏散楼梯宽度 (M)=6077 (M^2) ×0.5×0.6（人 /M^2）×0.5 (M/ 百人）=9.1M	17.4×0.7=12.2M	10.5M

四层平面图 1:800

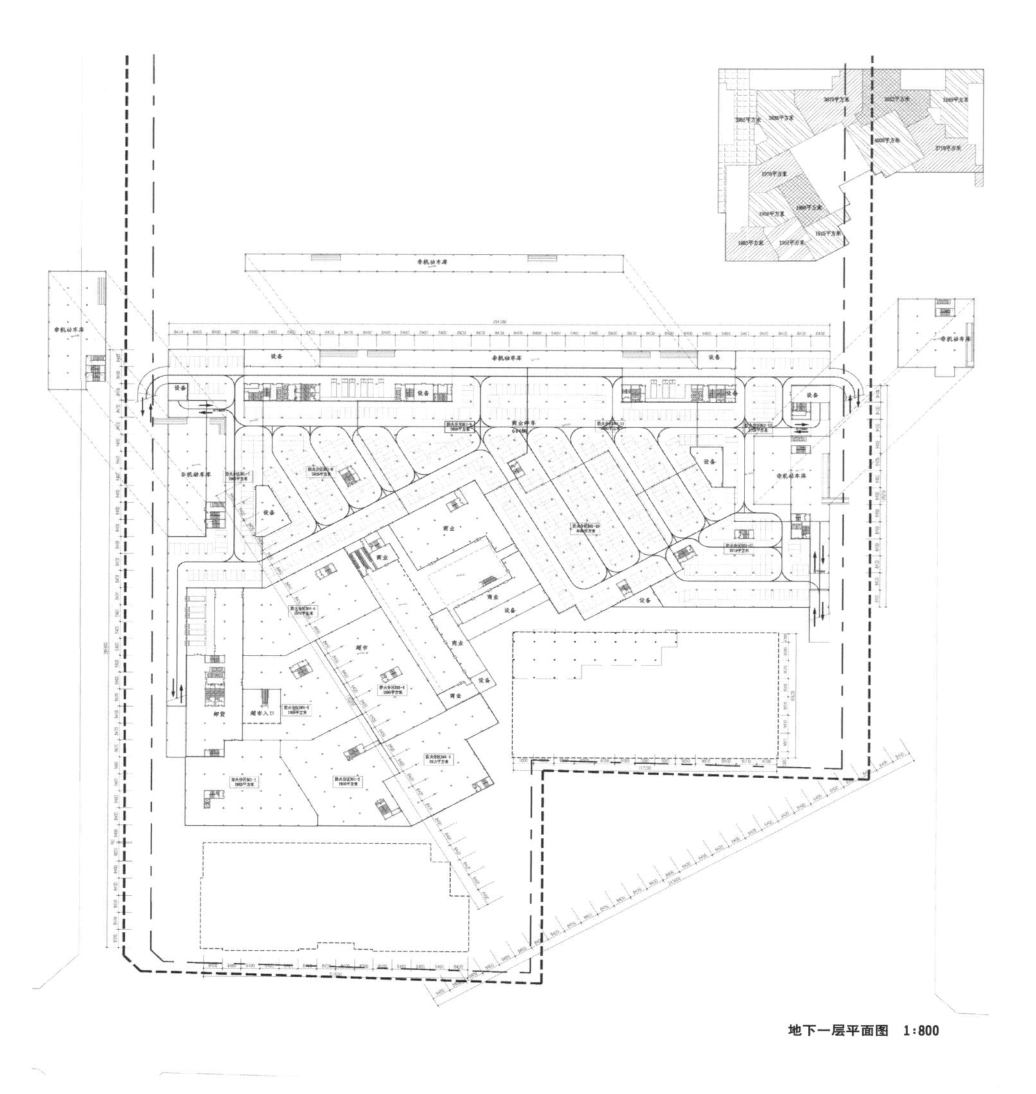

地下一层平面图　1:800

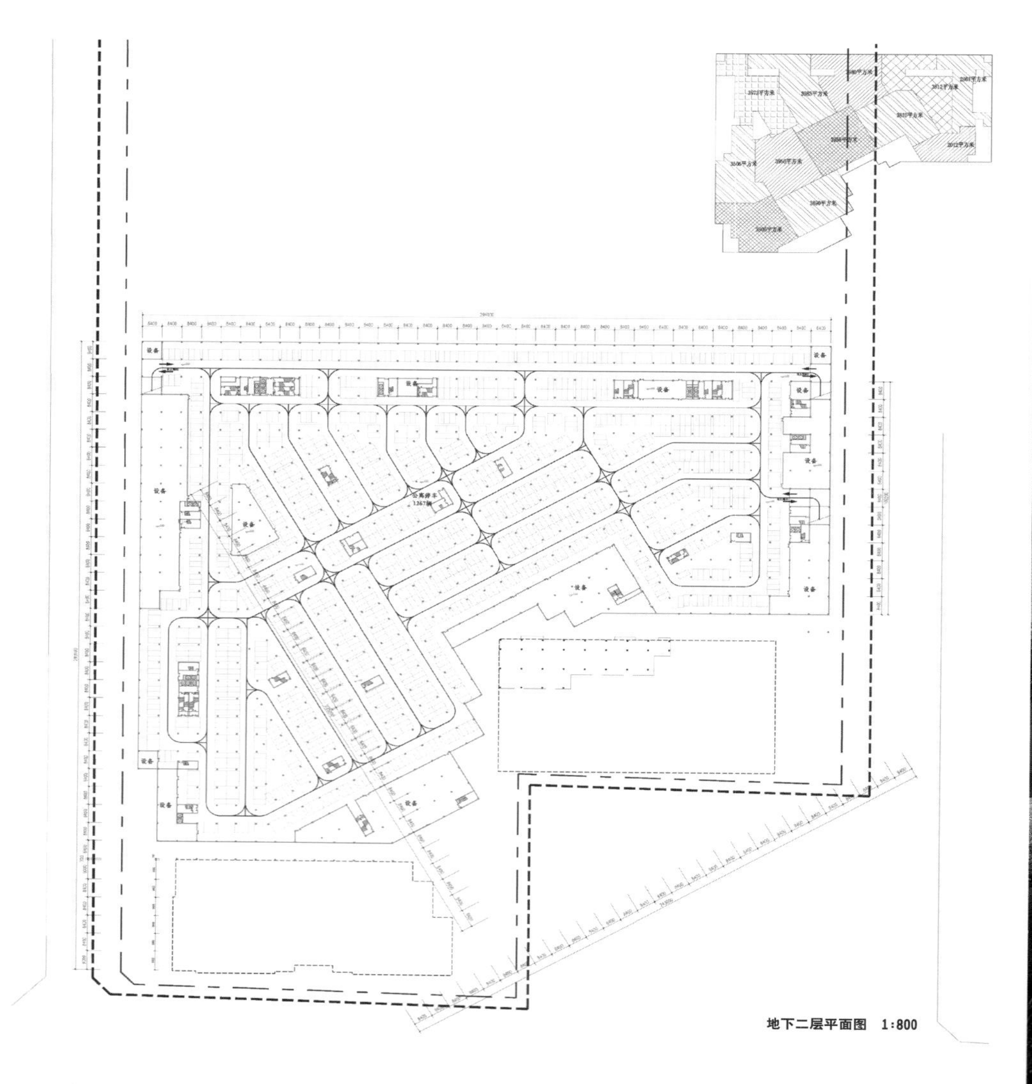

地下二层平面图　1:800

OXYGEN
MaxMara

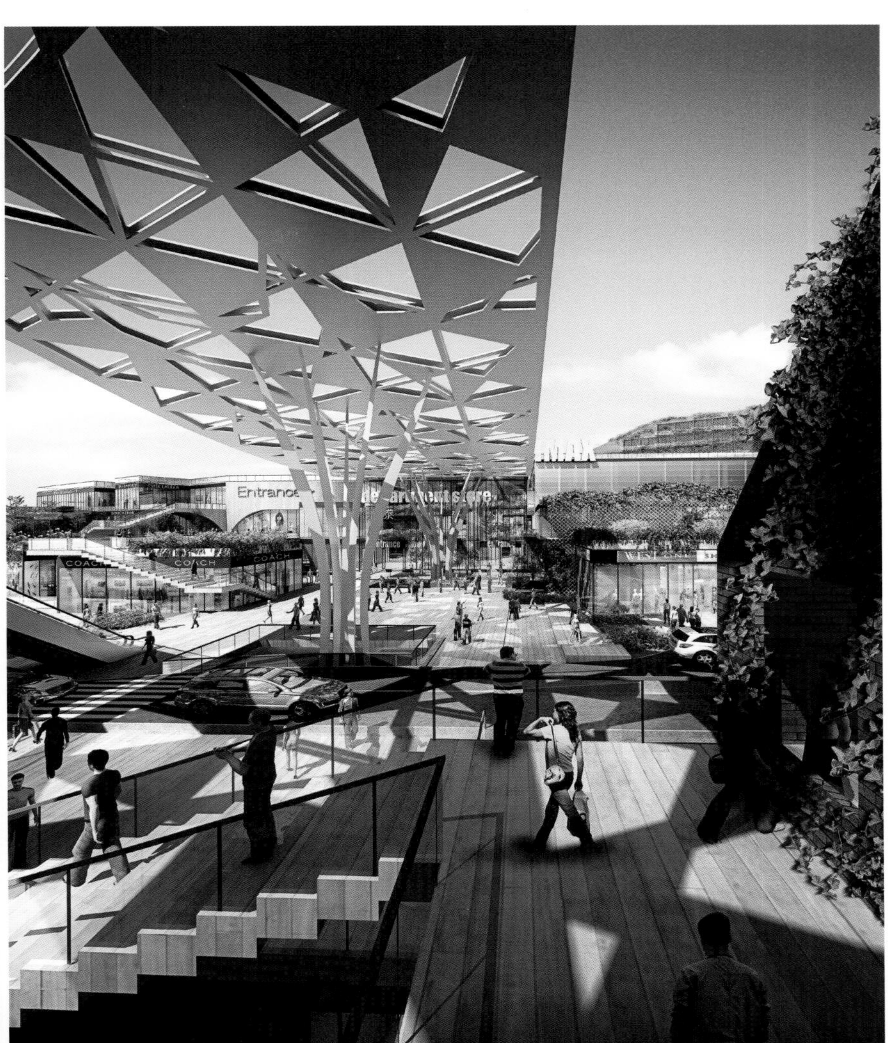
Entrance
COACH

Most

江西南昌玺悦城

Jiangxi Nanchang Xiyue Town

设计单位：日本 M.A.O. 一级建筑士事务所
开发商：南昌时代广场置业有限公司
项目地址：中国江西省南昌市
占地面积：41 593.54 ㎡
总建筑面积：124 780.62 ㎡
绿化率：30%
容积率：3.0

Designed by: M.A.O.
Developer: Nanchang Times Square Real Estate Co., Ltd.
Location: Nanchang, Jiangxi, China
Site Area: 41,593.54 m^2
Gross Floor Area: 124,780.62 m^2
Greening Ratio: 30%
Plot Ratio: 3.0

项目概况

项目处于南昌县核心区莲塘镇，紧邻莲塘镇的核心景点——澄碧湖。设计旨在打造一个集购物、餐饮、休闲娱乐、观光于一体的大型商业综合体。

设计理念

商业空间与主题公园都是需要具备强烈的风格与个性的城市空间，设计将两者统一起来，开创了一种非传统商业模式的“公园型商业”新模式。这一模式将澄碧湖的自然资源与项目主题相结合，以“水·游”文化为品牌，通过一系列的主题商品、游乐馆和餐厅的设计，整合主题资源，打造独一无二的“花园城购物中心”。

设计特色

在动线组织上，设计在传统的平面动线上引入一条“立体步行”动线，将“平面变成山脉”。这既有效地提升了两层以上空间的可售价值，模糊了“层”的概念，也将顾客从单调的购物模式中解脱出来，激发了他们对购物空间的探索。

项目采用了非日常性的商业模式，而不是传统的封闭式购物模式。在这里，人们可以享受到在公园中漫步、参观、购物、娱乐等多重乐趣；在这里，购物也成为了一种“经历”。

rofile

he project is located in Liantang Town of Nanchang ity, adjoining the core scenic spot of Liantang Town. he design aims to build a large commercial complex ntegrating shopping, F&B, leisure entertainment and ightseeing.

esign Concept

ommercial space and theme park are urban spaces hat need to have distinctive style and character. he design of this project unites these two aspects o initiate an unconventional commercial mode of Park-type Commerce". This mode shall combine he natural resources of Chengbi Lake with the roject theme. Taking "Water Tour" as cultural brand, he project strives to build a unique "Garden Town hopping Center" by designing of a series of theme roducts, recreation pavilions and restaurants.

esign Feature

"Stereoscopic Footpath" is introduced to traditional lanar lines, turning "plane" into "mountain". Thus shall effectively promote sales value of the space bove the 2nd floor, blur the impression of "Floor", s well as free customers from dull shopping mode, nspiring exploration of shopping space.

Uncommon commercial mode is adopted in this roject instead of traditional closed shopping node. Here, visitors can enjoy walking in the park, ightseeing, shopping and amusement — shopping ecomes a kind of "Experience".

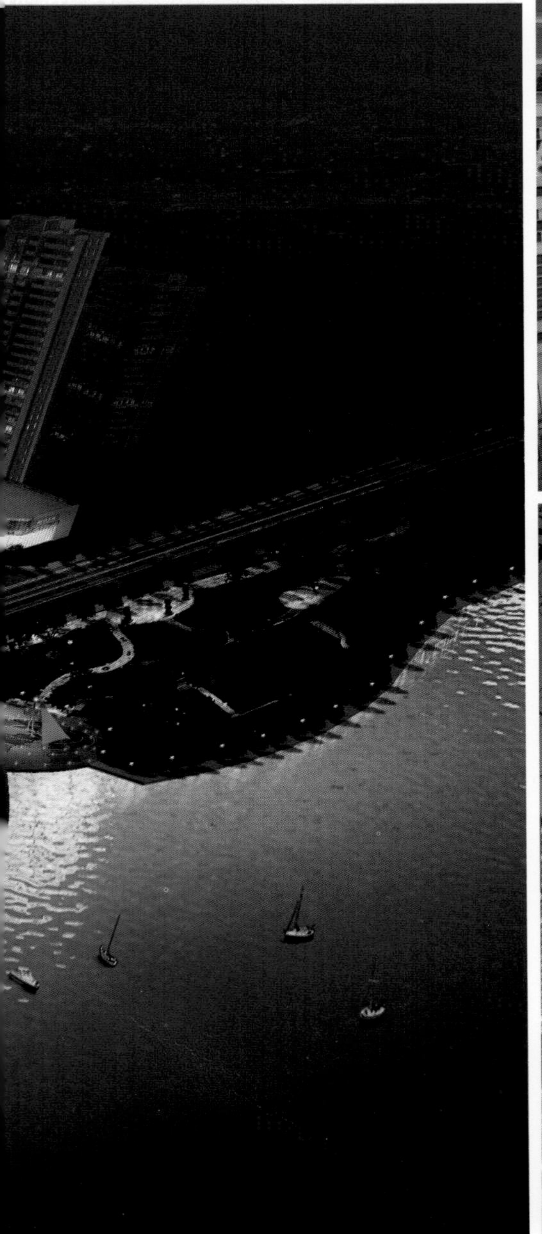

一层平面图

本层建筑面积 20781.6平方米

大型餐饮　外街店铺　临街店铺　百货　1+2店铺　内街店铺　IMAX入口　卫生间

三层平面图

本层建筑面积 16119.91平方米

外街店铺　百货　超市　内街店铺　交通空间

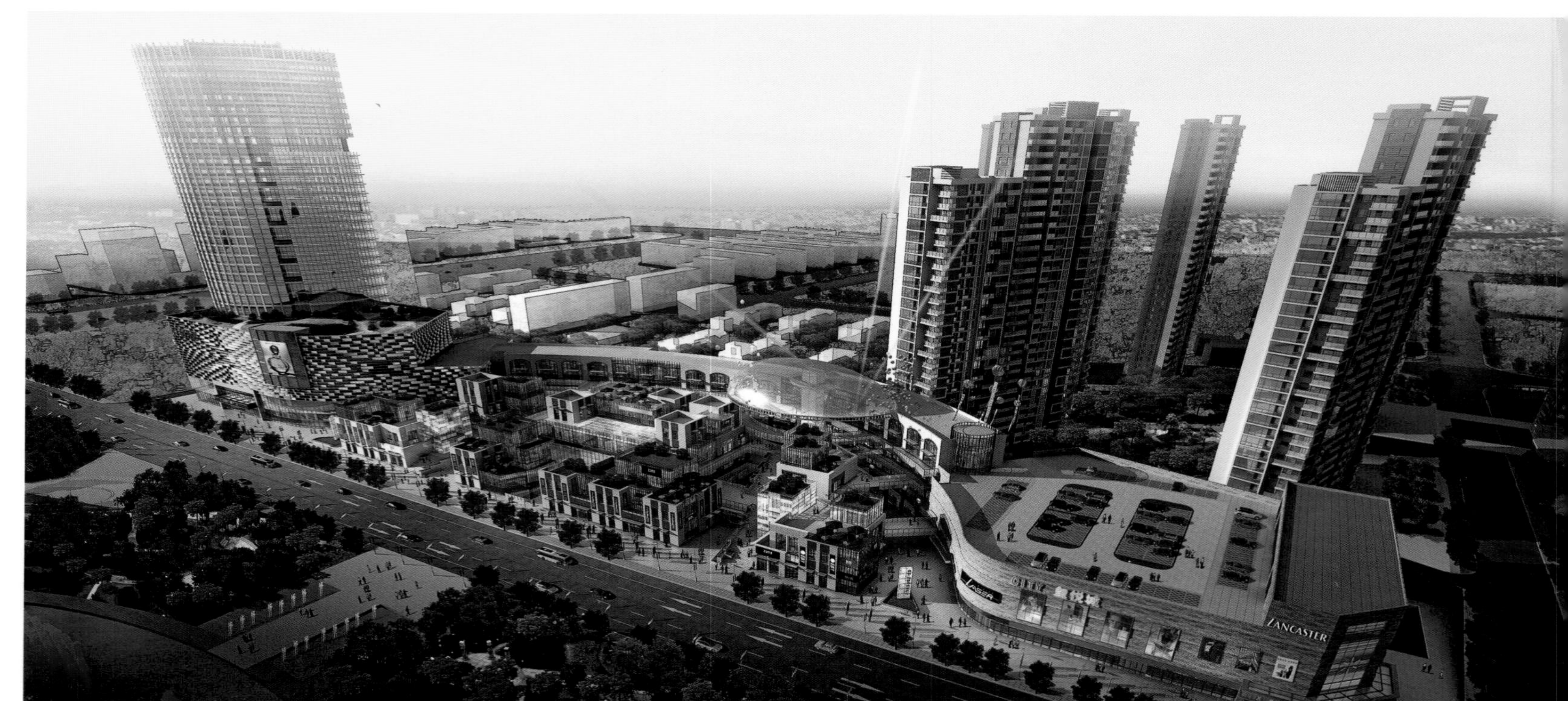

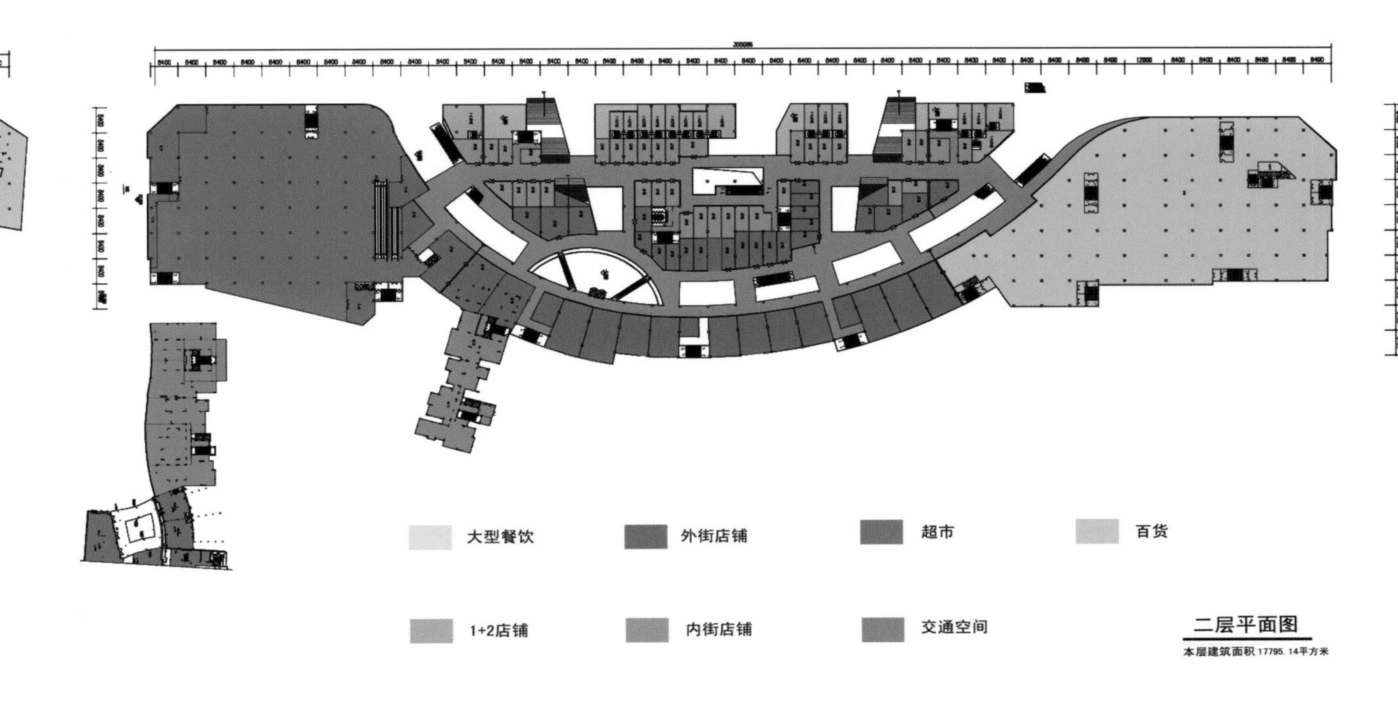

二层平面图

本层建筑面积 17795.14平方米

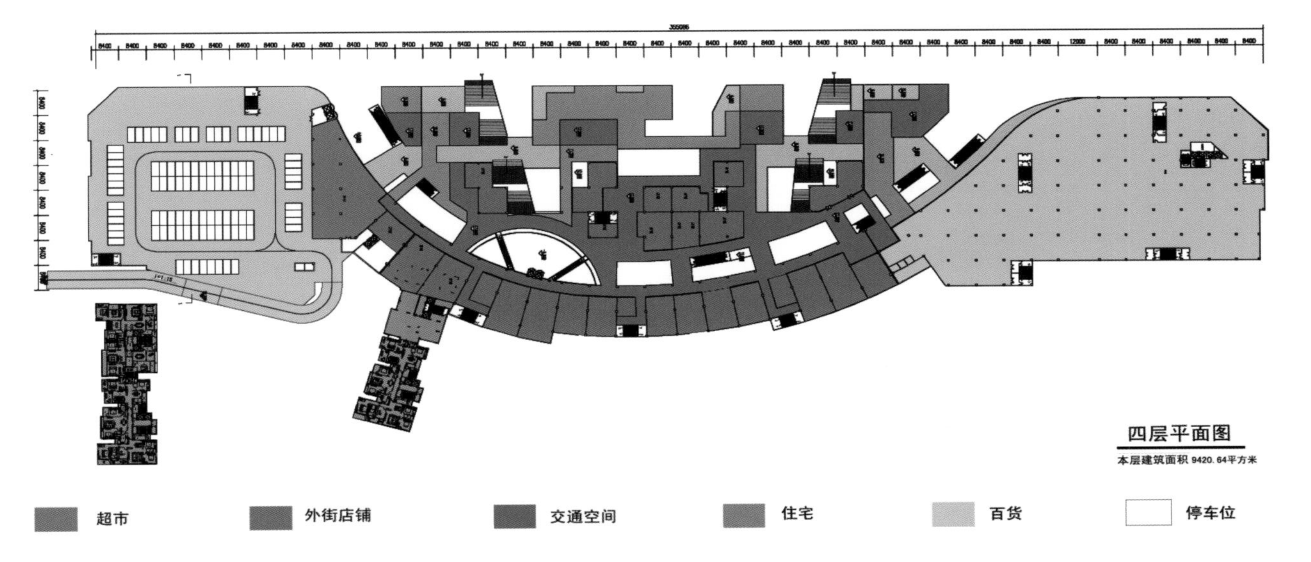

四层平面图

本层建筑面积 9420.64平方米

名称	房型	套型建筑面积	公摊面积	总建筑面积	得房率	赠送面积
A户型	四房两厅三卫	161.78㎡	27.91㎡	189.69㎡	0.85%	18.8㎡
B户型	三房两厅两卫	134.05㎡	22.09㎡	156.14㎡		22.37㎡
			楼层总建筑面积	345.83㎡		

■ 住宅平面图　1梯2户：方案一

名称	房型	套型建筑面积	公摊面积	总建筑面积	得房率	赠送面积
A户型	四房两厅三卫	161.86㎡	26.16㎡	188.02㎡	0.86%	15.97㎡
B户型	三房两厅两卫	135.96㎡	21.97㎡	157.93㎡		12.87㎡
			楼层总建筑面积	345.95㎡		

■ 住宅平面图　1梯2户：方案二

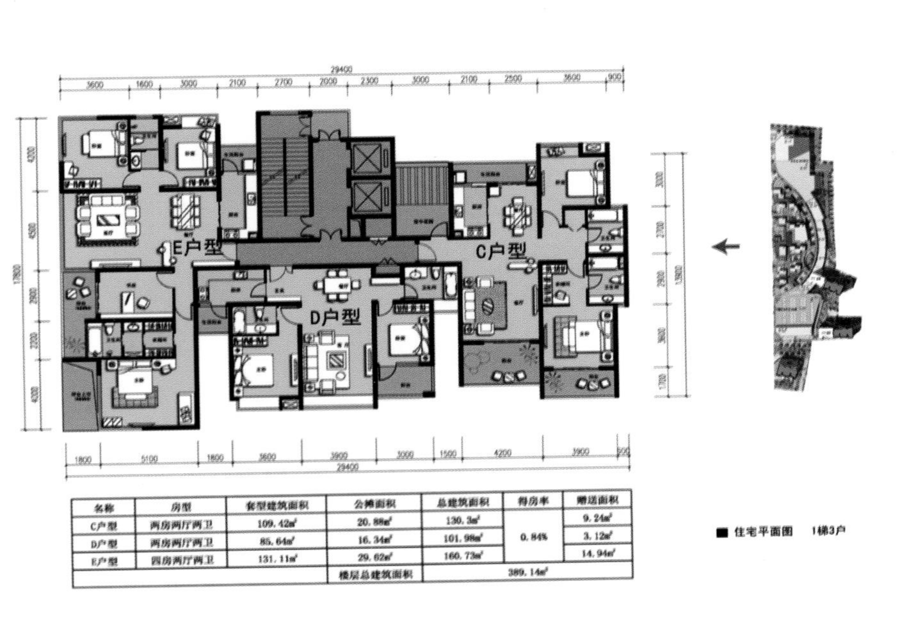

名称	房型	套型建筑面积	公摊面积	总建筑面积	得房率	赠送面积
C户型	两房两厅两卫	109.42㎡	20.88㎡	130.3㎡	0.84%	9.24㎡
D户型	两房两厅两卫	85.64㎡	16.34㎡	101.98㎡		3.12㎡
E户型	四房两厅两卫	131.11㎡	29.62㎡	160.73㎡		14.94㎡
			楼层总建筑面积	389.14㎡		

■ 住宅平面图　1梯3户

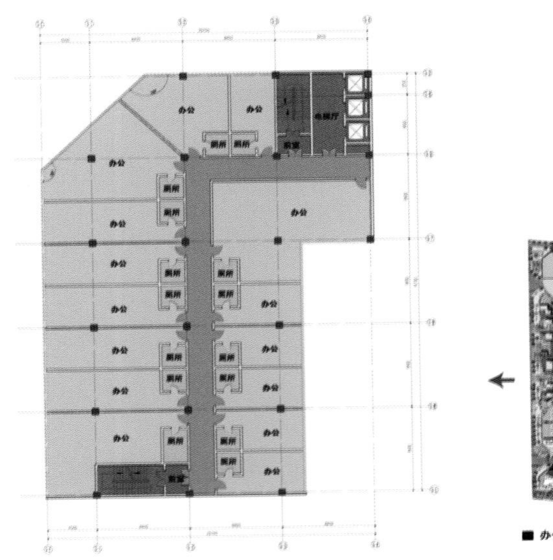

■ 办公平面图

重庆百年汇

Chongqing Parkland Complex

设计单位：日本 M.A.O. 一级建筑士事务所
主设计师：神山义浩
开发商：大连百年城
项目地址：中国重庆市
占地面积：20 360 m²
总建筑面积：265 722 m²
绿化率：20%
容积率：11.7

Designed by: M.A.O.
Chief Designer: Yoshihiro Kamiyama
Developer: Parkland Commercial Properties Co., Ltd.
Location: Chongqing, China
Site Area: 20,360 m²
Gross Floor Area: 265,722 m²
Greening Ratio: 20%
Plot Ratio: 11.7

项目概况

重庆百年汇是一个集商业及办公功能为一体的综合楼。项目基地位于重庆市渝中区朝天门地区，东临陕西路，西邻西华路，南侧隔曹家巷与朝天门商场相邻，东侧为金海洋商场。

商业设计

根据对当地居民日常行为模式的分析，设计在基地内设置了一条新的人行购物路线，不仅大大增加了沿街面店铺的数量，提高了沿街面的商业价值，同时也将建筑基地东西两侧的城市主要干道——陕西路和新华路连接起来。

这条人行购物路线设置在基地中央，即距离主要道路最远、商业价值最低的区域，这样的布局形成了一个看似封闭、实则具有丰富空间内涵的购物空间。

该方案平面设计的最大特色是充分利用了金海洋基地现有的商业条件和交通条件，既有效地解决了基地内商业人流的组织问题，又可以减少商业区所需的疏散楼梯数量。

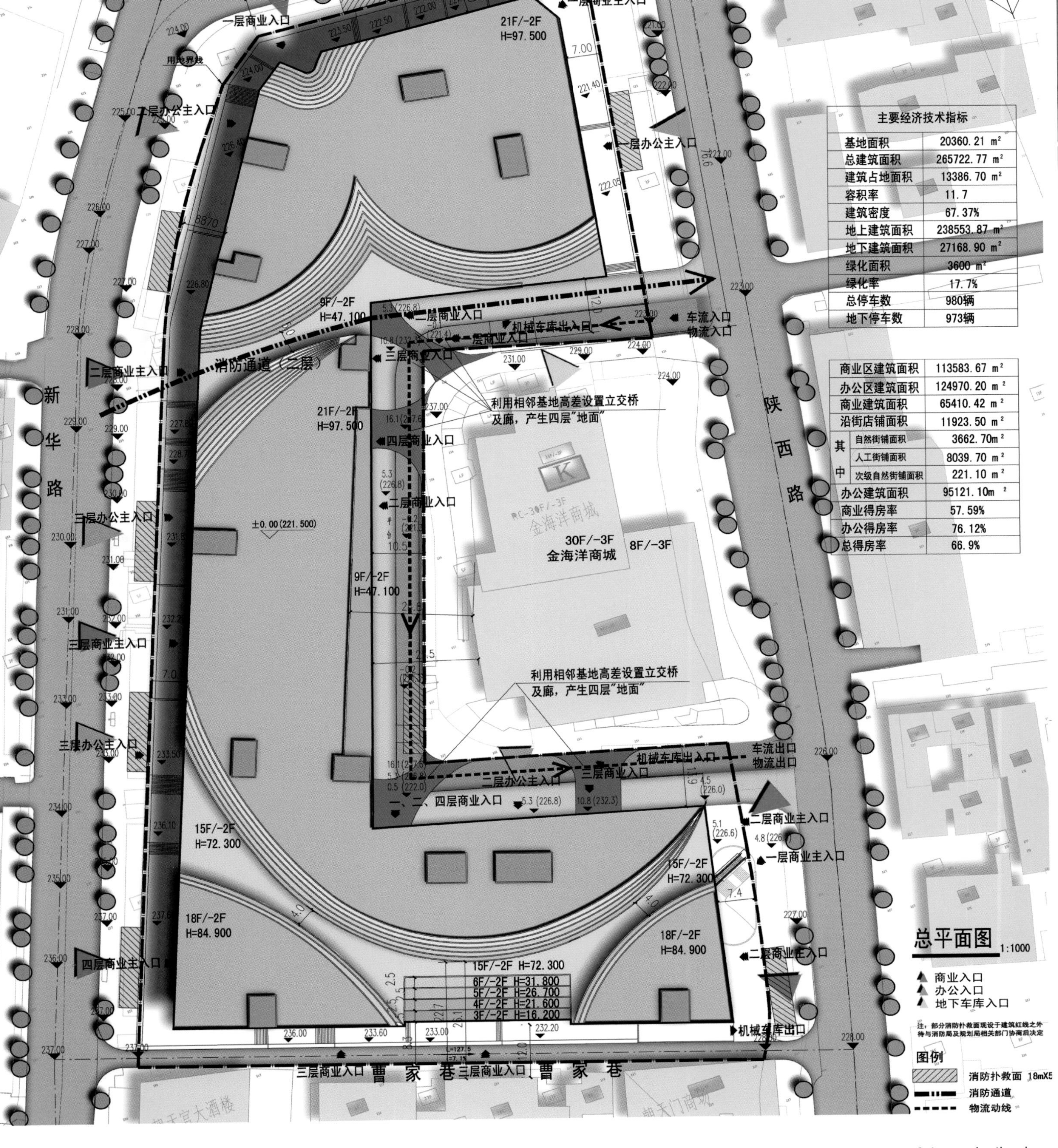

主要经济技术指标	
基地面积	20360.21 m²
总建筑面积	265722.77 m²
建筑占地面积	13386.70 m²
容积率	11.7
建筑密度	67.37%
地上建筑面积	238553.87 m²
地下建筑面积	27168.90 m²
绿化面积	3600 m²
绿化率	17.7%
总停车数	980辆
地下停车数	973辆

商业区建筑面积		113583.67 m²
办公区建筑面积		124970.20 m²
商业建筑面积		65410.42 m²
沿街店铺面积		11923.50 m²
其中	自然街铺面积	3662.70m²
	人工街铺面积	8039.70 m²
	次级自然街铺面积	221.10 m²
办公建筑面积		95121.10m²
商业得房率		57.59%
办公得房率		76.12%
总得房率		66.9%

Profile

Chongqing Parkland Complex is a multiple-use building integrating commerce and offices. The site is located in Chaotianmen of Yuzhong District, Chongqing City, facing Shaanxi Road to the east, Xihua Road to the west, adjoining Chaotianmen Mall across Caojia Lane to the south and close to Jinhaiyang Mall.

Commercial Design

Based on analysis of daily activities of local residents, the project has designed a new shopping route for pedestrians. It has not only greatly increased stores of the street front, elevated its commercial value, but also connected urban main roads (Shaanxi Road and Xinhua Road) on east and west sides of the site.

This shopping route for pedestrians is set in the middle of the site, that is to say, the district farthest from main roads and of lowest commercial value. The layout is seemingly enclosed while actually embraces rich space connotation.

The most distinctive feature of the project's plane design is the full use of existing business and traffic conditions of Jinhaiyang site. It has effectively organized pedestrian flows and reduced the number of necessary escape stairs in the commercial district.

■ 商业价值

□ 一般

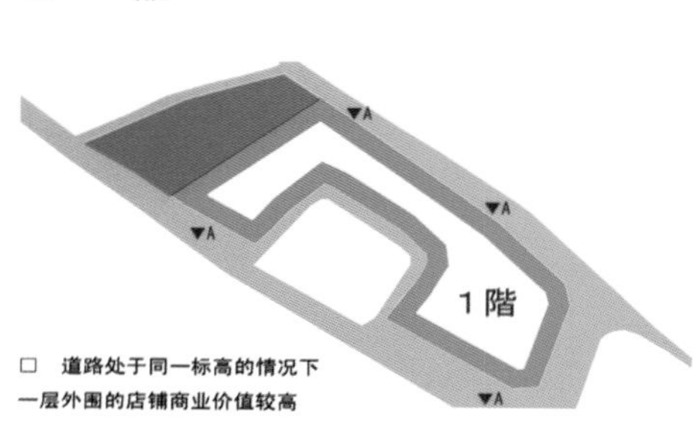

□ 道路处于同一标高的情况下
一层外围的店铺商业价值较高

■ 入口概念（四层分别设置入口）

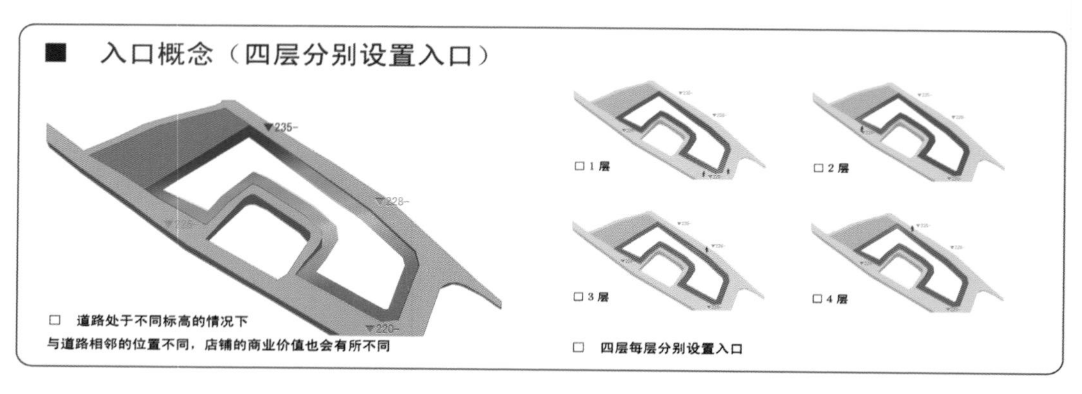

□ 道路处于不同标高的情况下
与道路相邻的位置不同，店铺的商业价值也会有所不同

□ 四层每层分别设置入口

□ 本案基地

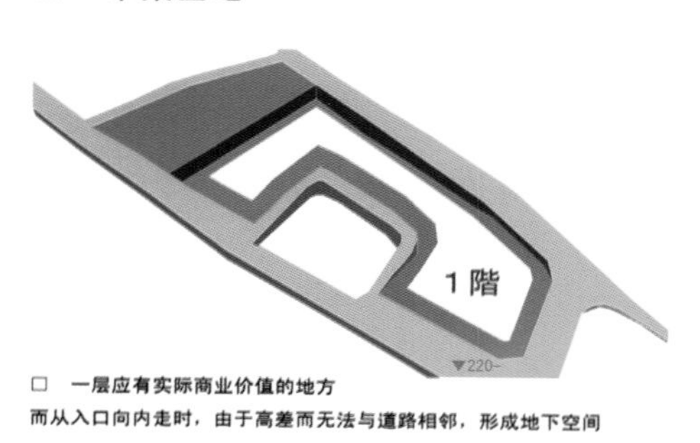

□ 一层应有实际商业价值的地方
而从入口向内走时，由于高差而无法与道路相邻，形成地下空间

■ 动线设计概念（商业价值存在于与道路相邻的部位，在基地内部设置道路）

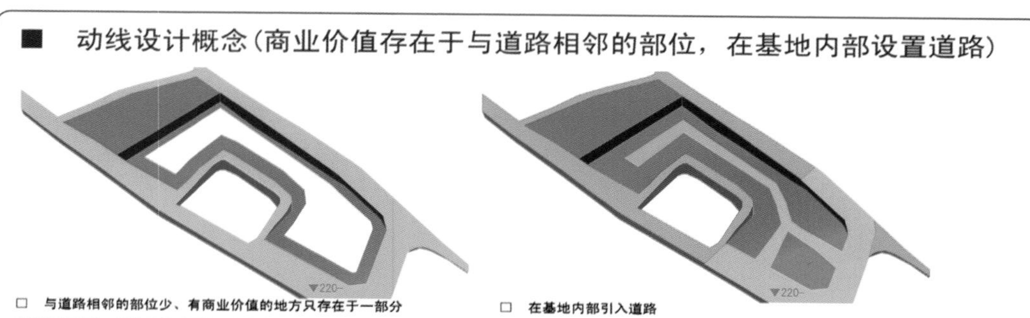

□ 与道路相邻的部位少、有商业价值的地方只存在于一部分
道路存在坡度，因此建筑外围并非全部存在商业价值

□ 在基地内部引入道路
在基地内部引入道路后，存在商业价值的部分增加

■ 商业价值2

■ 和动线相邻的店铺表面积增多

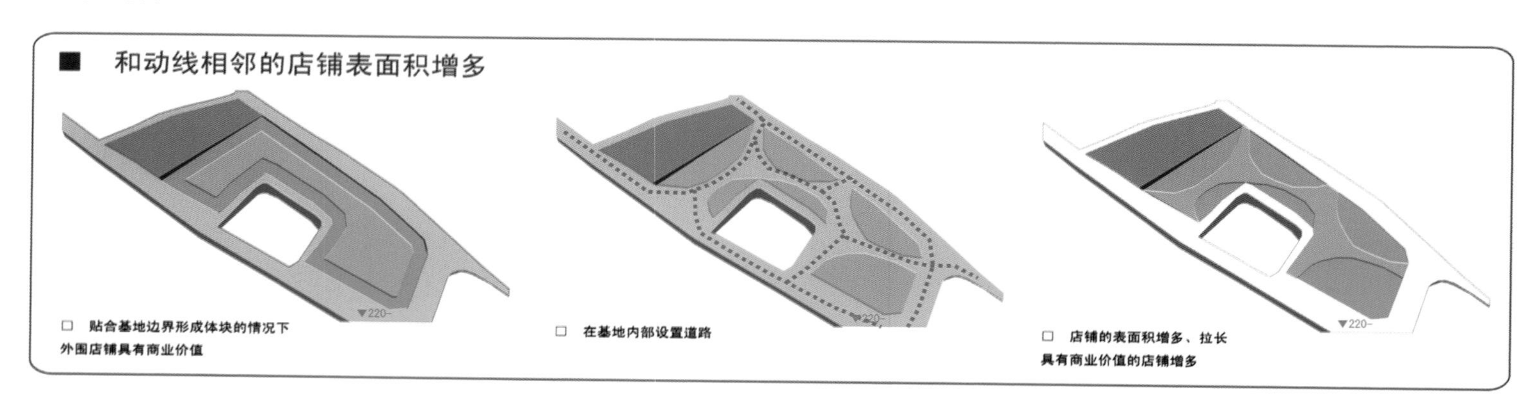

□ 贴合基地边界形成体块的情况下
外围店铺具有商业价值

□ 在基地内部设置道路

□ 店铺的表面积增多、拉长
具有商业价值的店铺增多

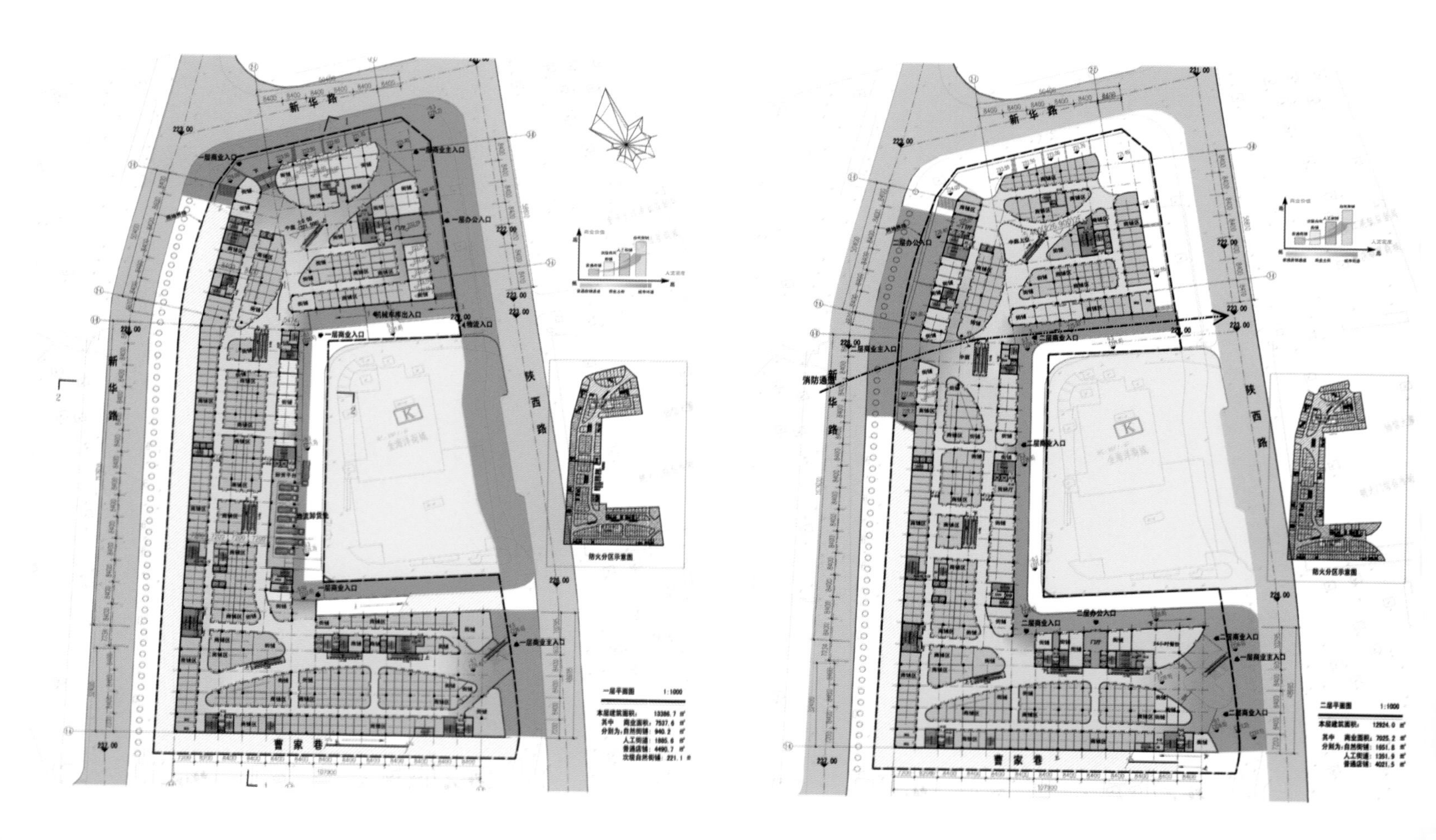

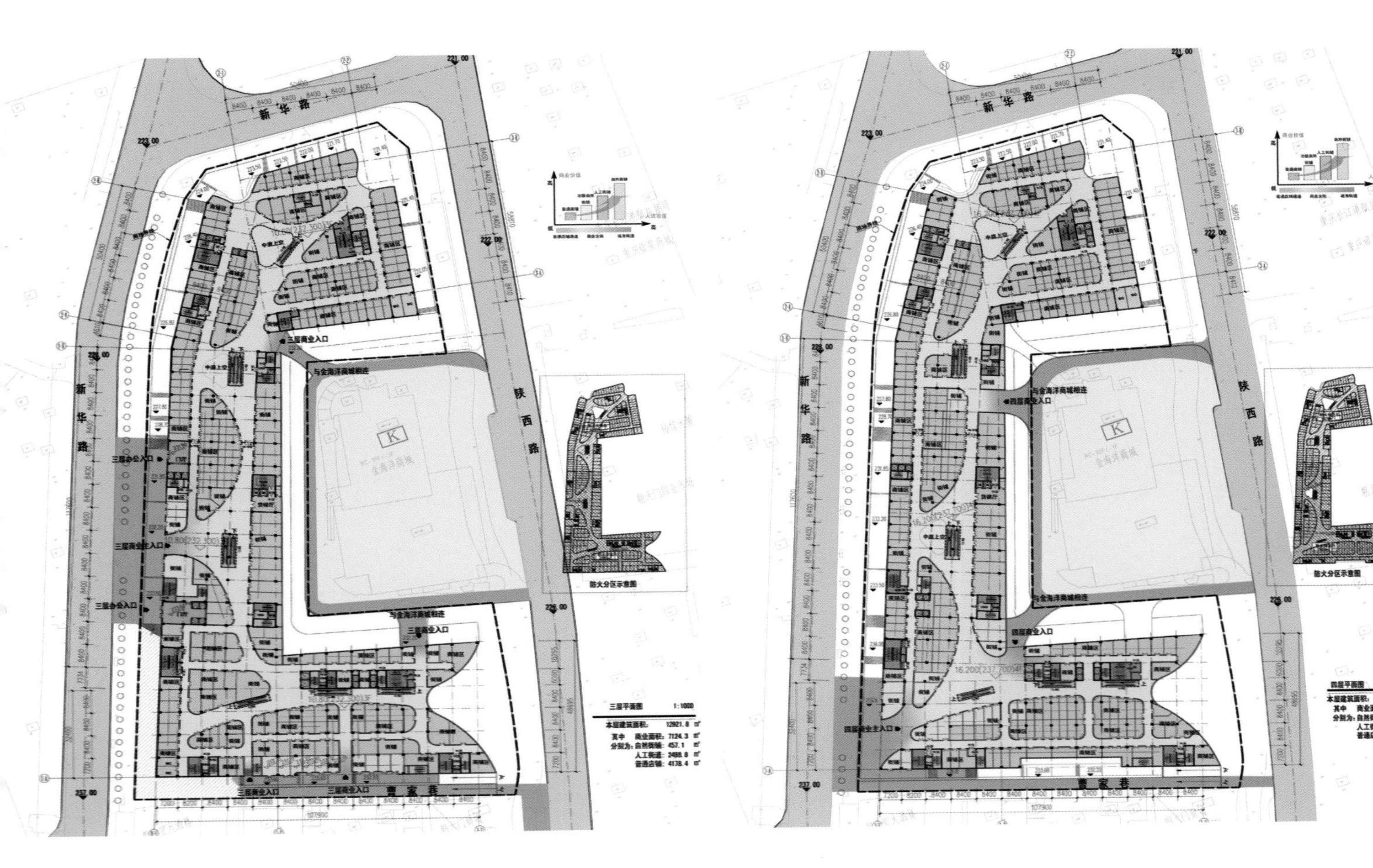

新华路
陕西路
防火分区示意图
三层平面图 1:1000
本层建筑面积: 12921.8 ㎡
其中 商业面积: 7124.3 ㎡
分别为: 自然街铺: 457.1 ㎡
人工街道: 2488.8 ㎡
普通店铺: 4178.4 ㎡
四层平面图 1:1000
本层建筑面积: 12607.9 ㎡
其中 商业面积: 7205.7 ㎡
分别为: 自然街铺: 613.6 ㎡
人工街道: 2313.4 ㎡
普通店铺: 4278.7 ㎡

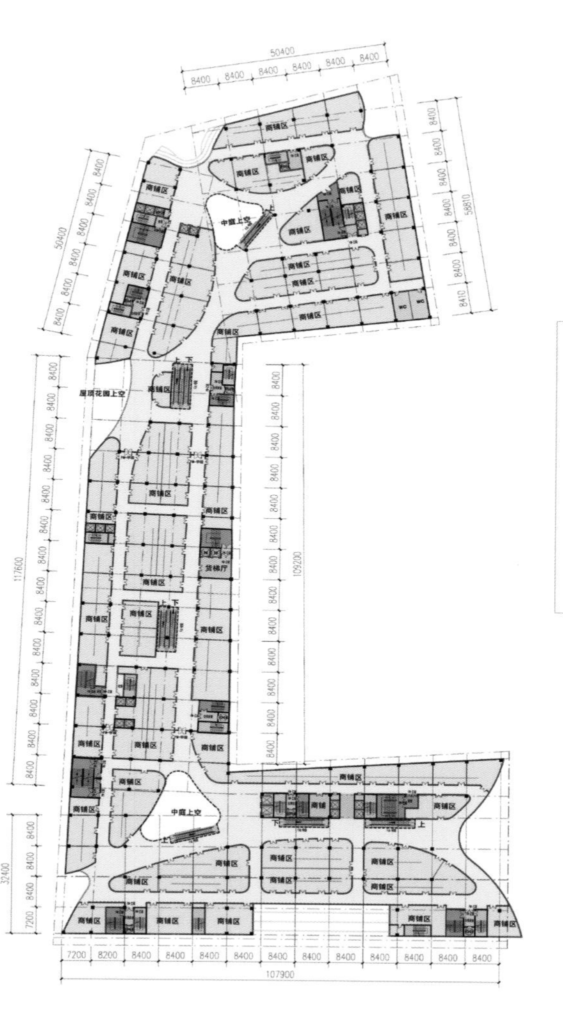

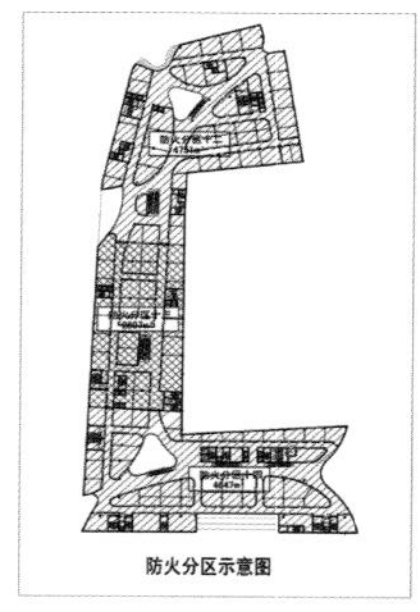

防火分区示意图

商铺(40-60平)：49个

商铺(60-80平)：60个

六层平面图　1:800

本层建筑面积：12205.28 m²

其中　商业面积：　7065.12 m²

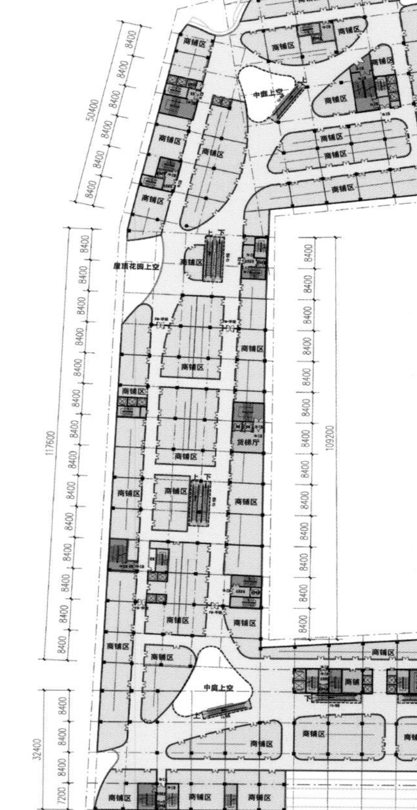

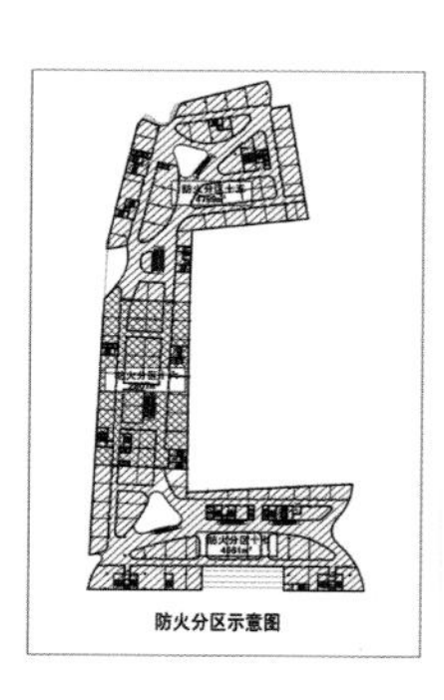

防火分区示意图

商铺(40-60平)：43个

商铺(60-80平)：60个

七层平面图　1:800

本层建筑面积：12267.78 m²

其中　商业面积：　7056.99 m²

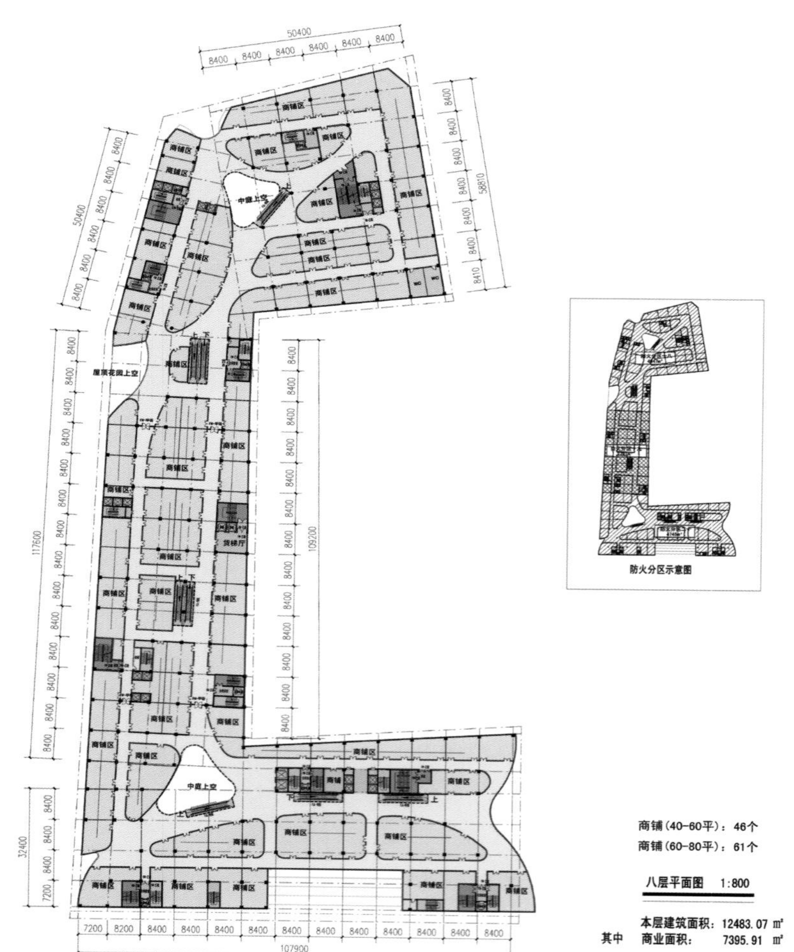

防火分区示意图

商铺(40-60平)：46个

商铺(60-80平)：61个

八层平面图　1:800

本层建筑面积：12483.07 m²

其中　商业面积：　7395.91 m²

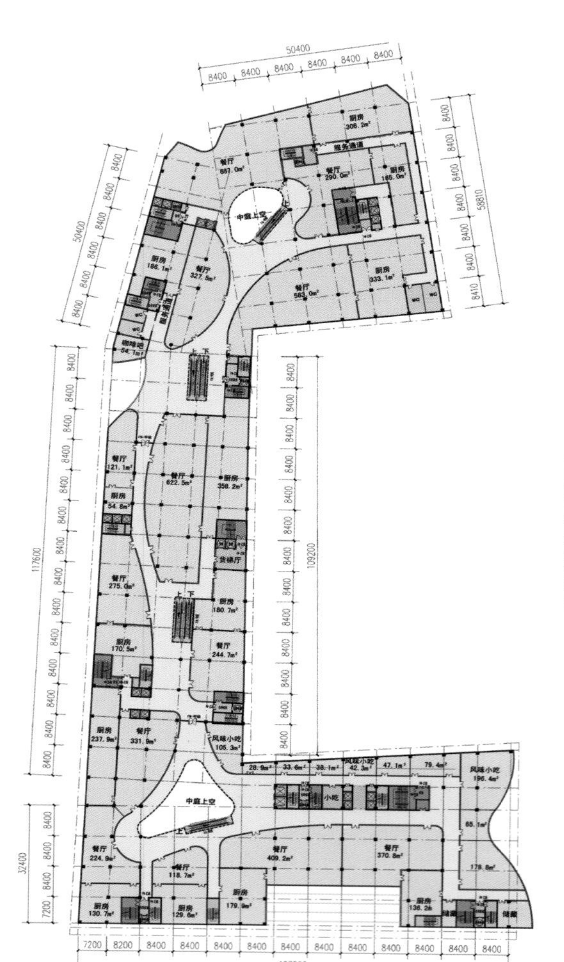

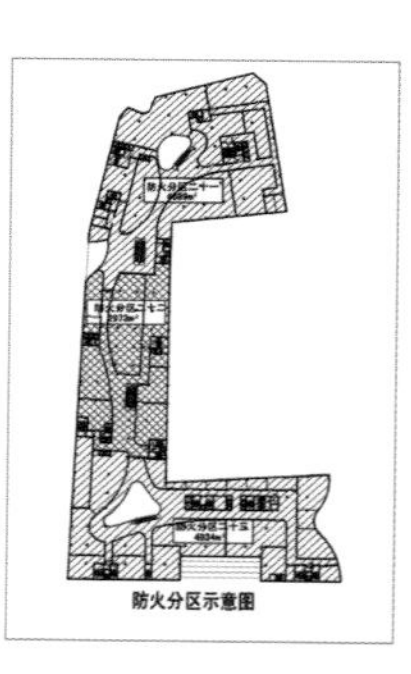

防火分区示意图

九层平面图　1:800

本层建筑面积：12496.81 m²

其中　商业面积：　7702.14 m²

消防垂直分析图

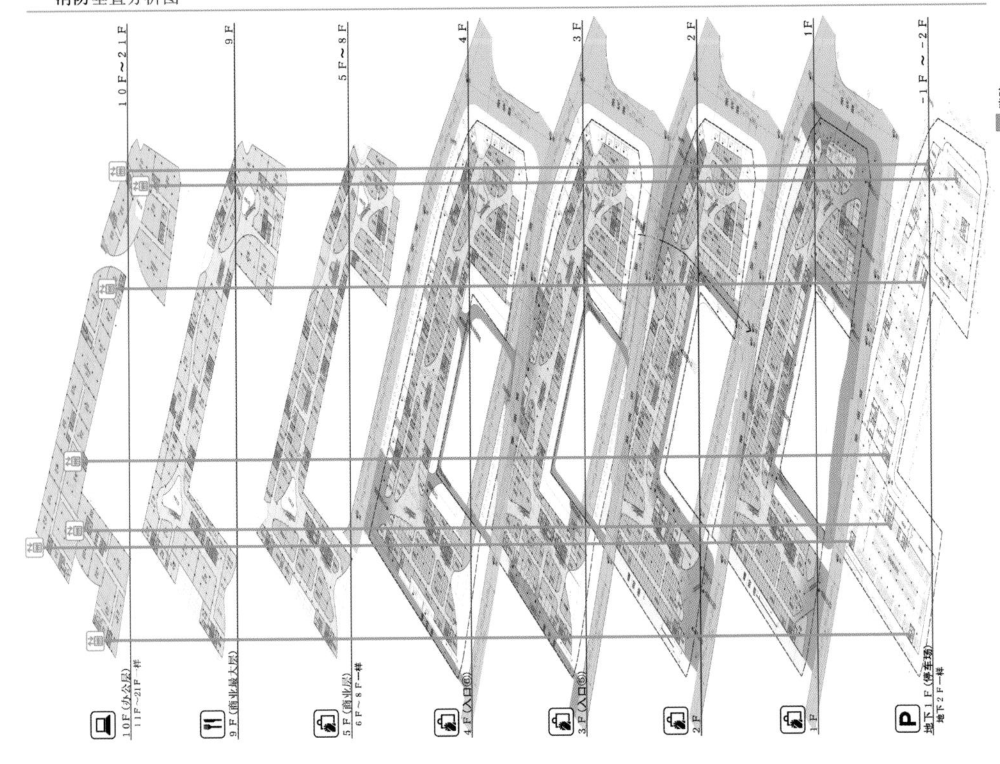

货流垂直分析图

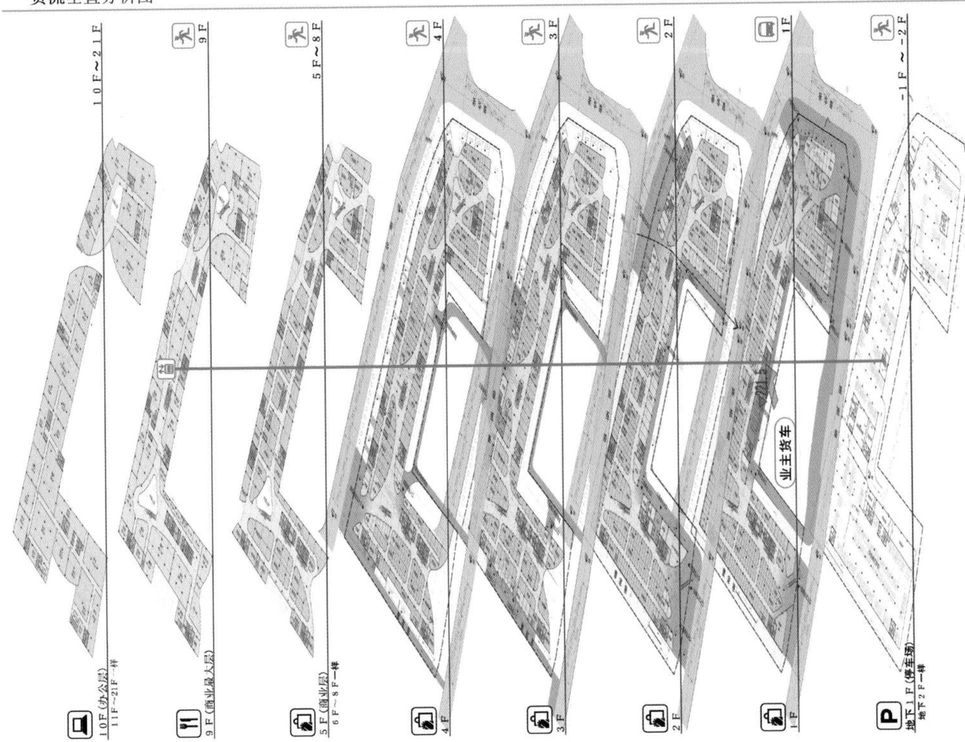

商业垂直分析图

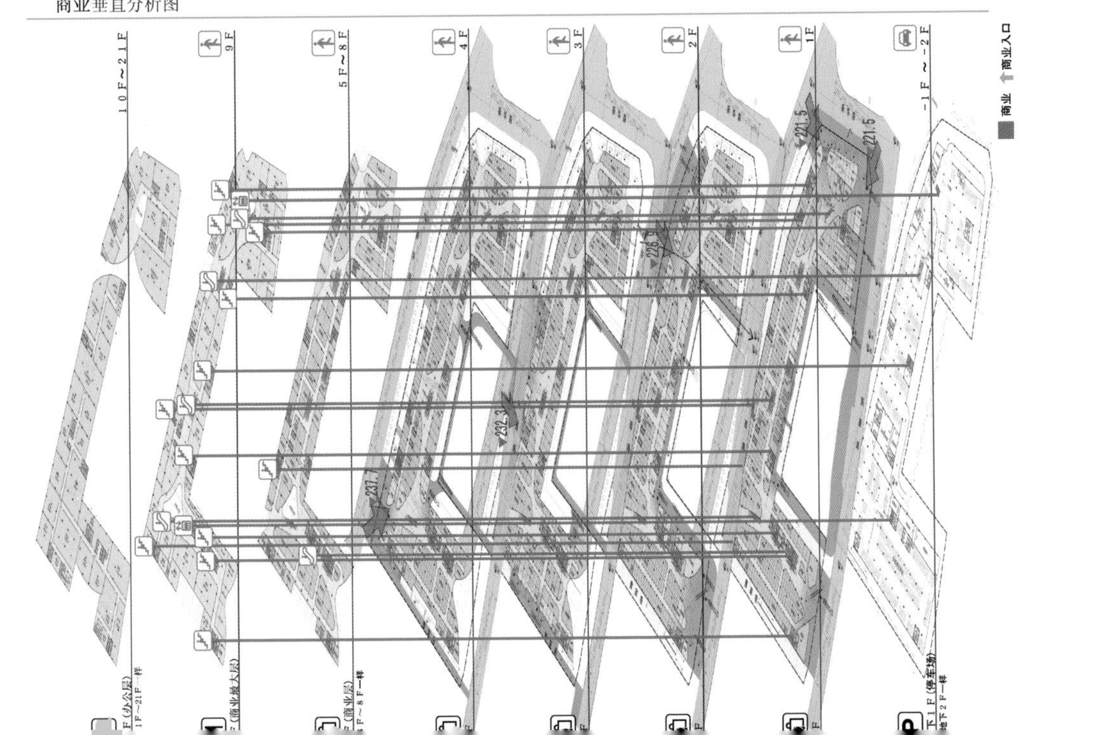

办公垂直分析图

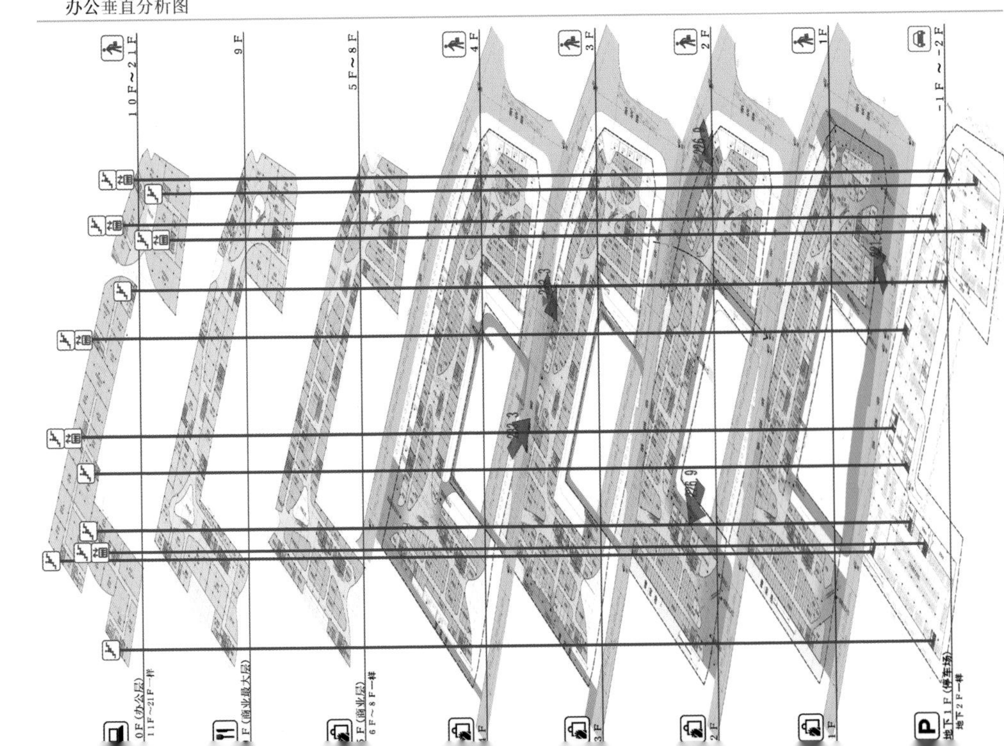

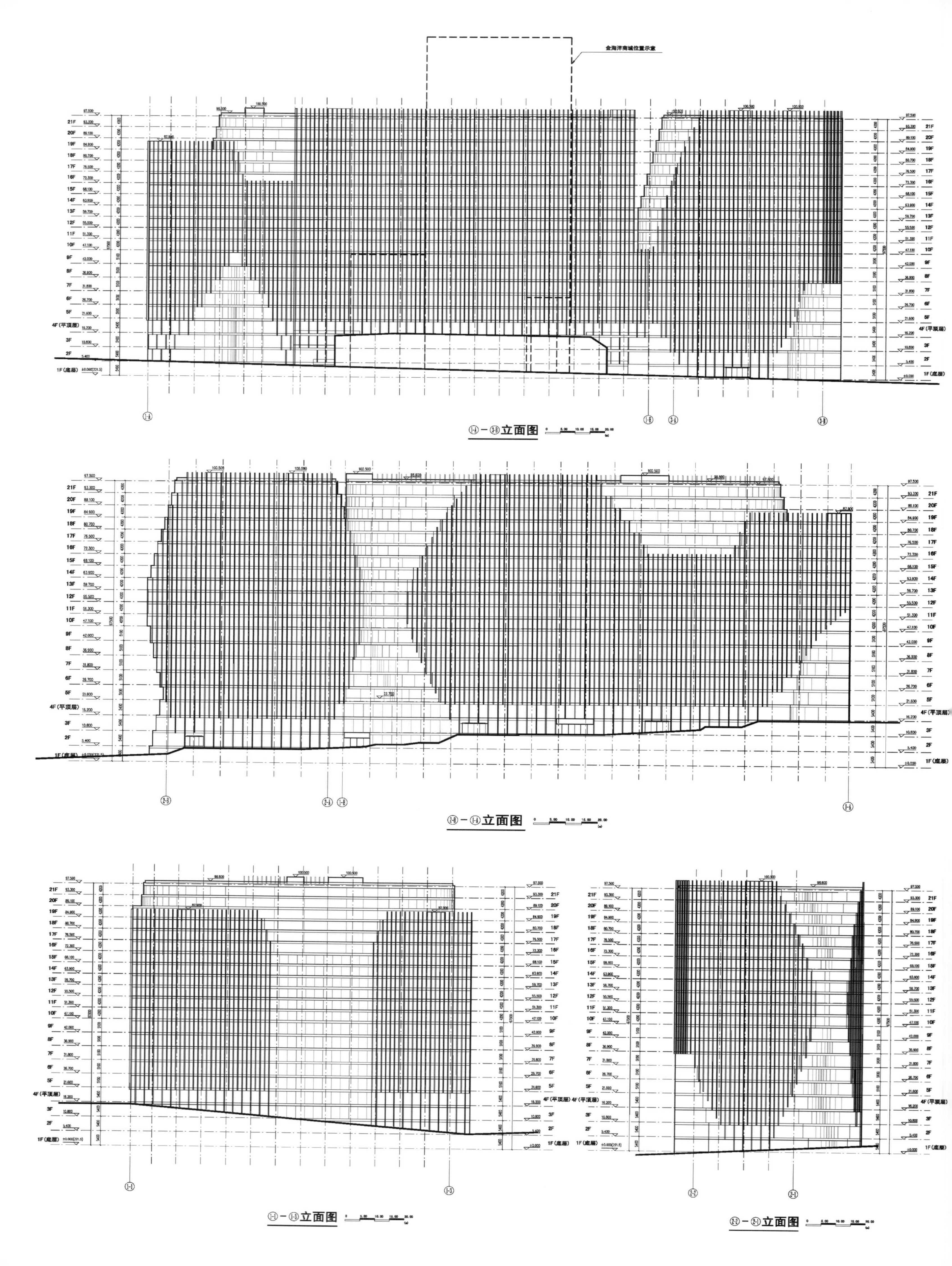

①-A — ③-B立面图

②-B — ①-A立面图

①-① — ①-③立面图

②-② — ②-①立面图

办公功能分区示意图

防火分区示意图

十层平面图 1:800

建筑面积：12260.65㎡
办公面积：9608.6㎡

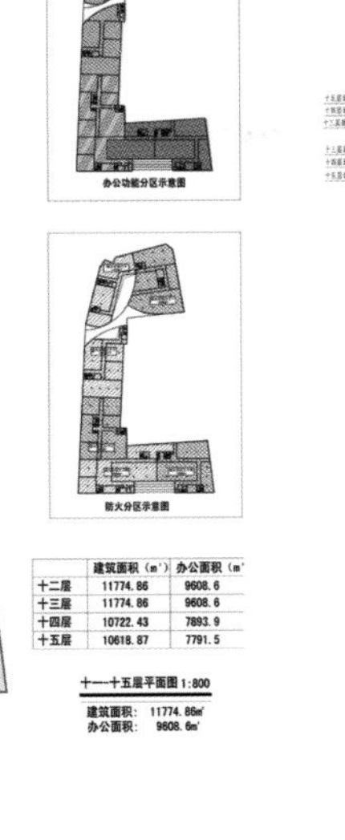

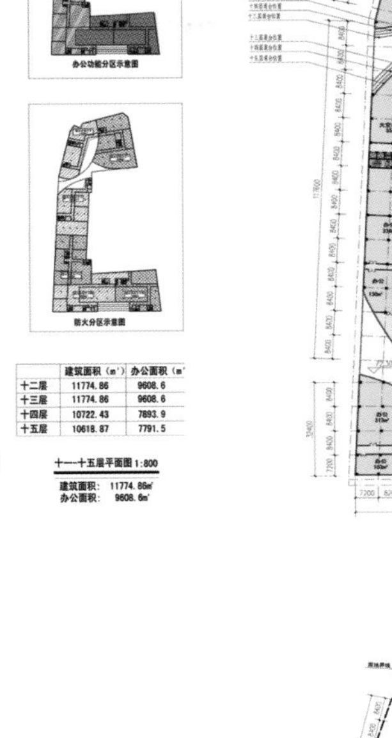

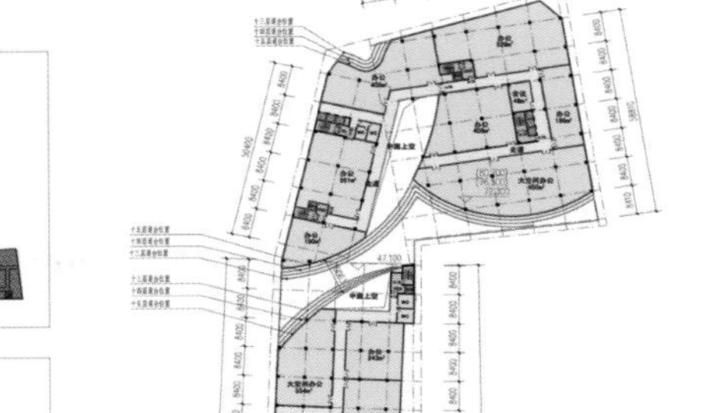

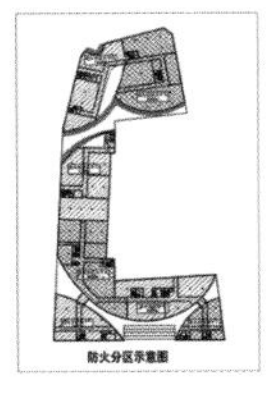

办公功能分区示意图

防火分区示意图

	建筑面积（㎡）	办公面积（㎡）
十二层	11774.86	9608.6
十三层	11774.86	9608.6
十四层	10722.43	7893.9
十五层	10618.87	7791.5

十一-十五层平面图 1:800

建筑面积：11774.86㎡
办公面积：9608.6㎡

办公功能分区示意图

防火分区示意图

	建筑面积（㎡）	办公面积（㎡）
十七层	10473.83	7558.5
十八层	10124.12	7318.7

十六-十八层平面图 1:800

建筑面积：10665.63㎡
办公面积：7837.1㎡

办公功能分区示意图

防火分区示意图

	建筑面积（㎡）	办公面积（㎡）
十九层	8313.06	6091.9
二十层	8069.61	5905.1

十九-二十一层平面图1:800

建筑面积：7809.44㎡
办公面积：5702.0㎡

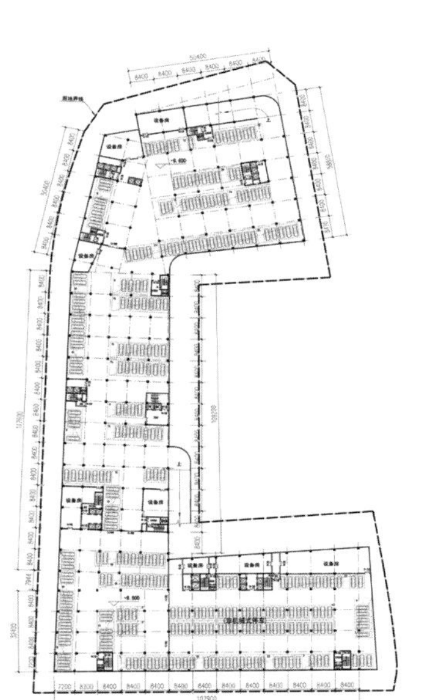

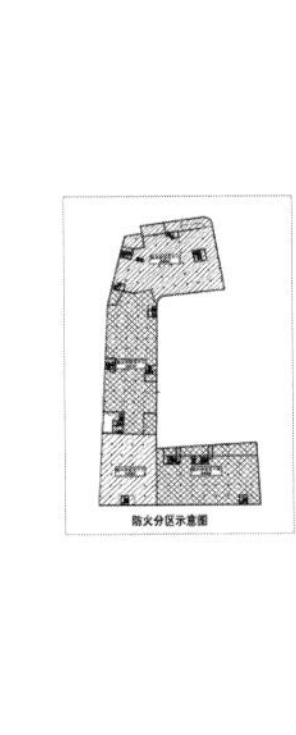

防火分区示意图

地下一层平面图 1:800

本层建筑面积：13626.93 ㎡
停车数量：460辆

防火分区示意图

地下二层平面图 1:800

本层建筑面积：13541.97 ㎡
停车数量：506辆

酒店建筑

Hotel

黑龙江哈尔滨冰雕酒店

Harbin Ice Hotel

设计单位：LAVA

项目地址：中国黑龙江省哈尔滨市

项目面积：530 000 ㎡

Designed by: LAVA

Location: Harbin, Heilongjiang, China

Area: 530,000 m^2

项目概况

这个新开发的冰雕酒店项目建筑面积达 530 000 平方米，除了一座全新的酒店，还包括中心、高档百货、写字楼等配套设施。哈尔滨冰雕酒店仿若一座四季长存的超大型冰雕，生动形象地表现了冰与火的冲突与交融。

设计理念

冰雕酒店的设计灵感来自于冰山以及石英晶体的外形和结构，暗合了哈尔滨这座国际闻名的冰雕城市。设计的另一个重要主题则来自优雅的天鹅，几个水晶形状的建筑围合在一起，突出了建筑的视觉效果，也为项目增添了一丝童话般的想象。

设计特色

项目由 10 栋不同高度的石英晶体建筑组成。灯光在冰雕一般的建筑表面上跳跃，从一栋塔楼跳跃至另一栋塔楼，建筑的外观也随之形成颜色的渐变，透明的外观也变得光鲜亮丽。

这个冰雕酒店冰冷透明的外观和温暖舒适的内部空间形成了鲜明的对比。参观者从看似冰冷透明的外立面和垂直的线性状冰山体量外面进入到酒店的内部，呈现在眼前的却是一个温暖、如洞穴状的空间。婀娜的栏杆、变形的楼层和天花板上钟乳石状的滴状物的设计灵感均源自熔浆，它们使室内空间充满了戏剧性和雕刻感。

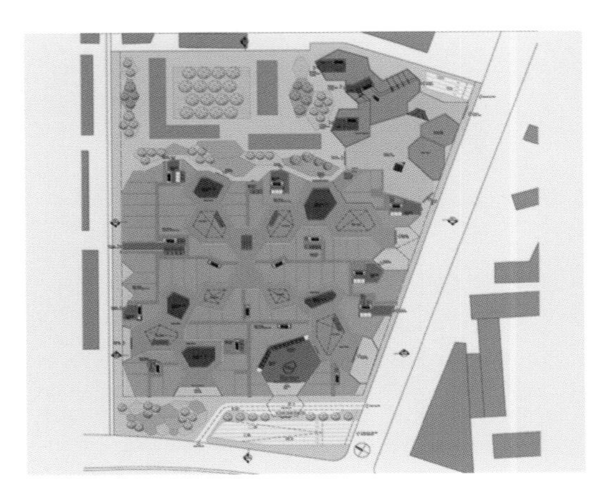

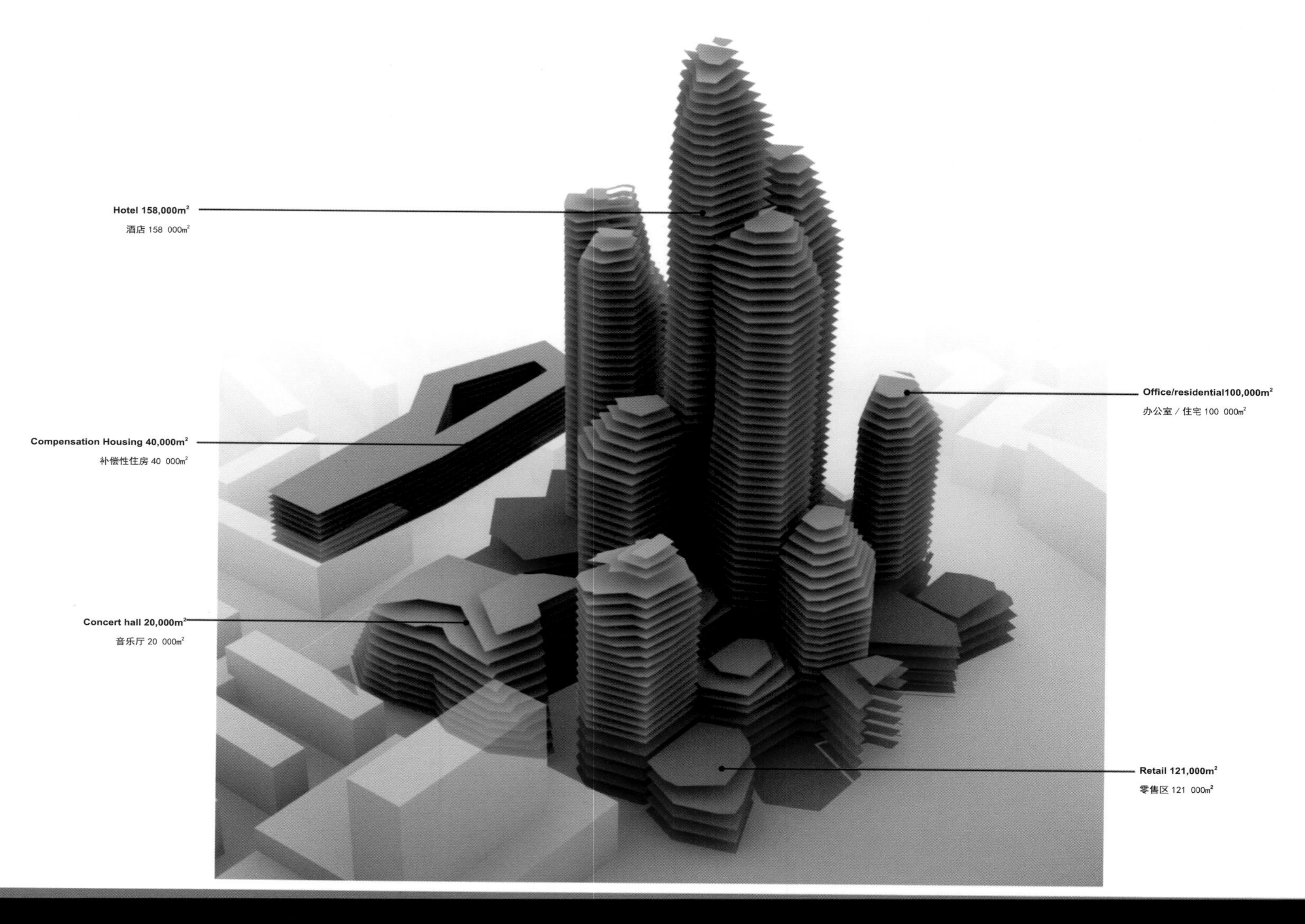
Hotel 158,000m²
酒店 158 000㎡
Compensation Housing 40,000m²
补偿性住房 40 000㎡
Concert hall 20,000m²
音乐厅 20 000㎡
Office/residential100,000m²
办公室 / 住宅 100 000㎡
Retail 121,000m²
零售区 121 000㎡

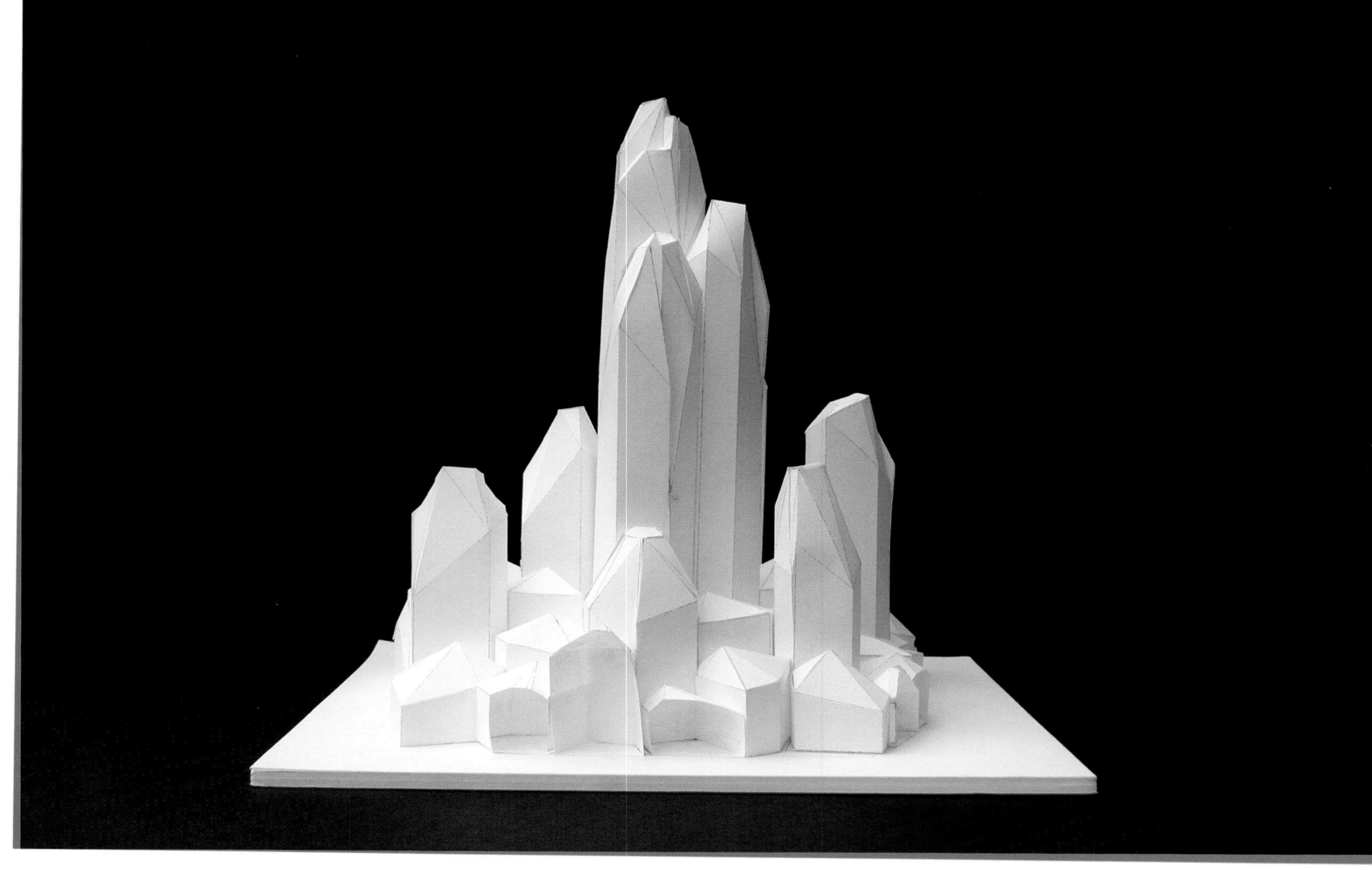

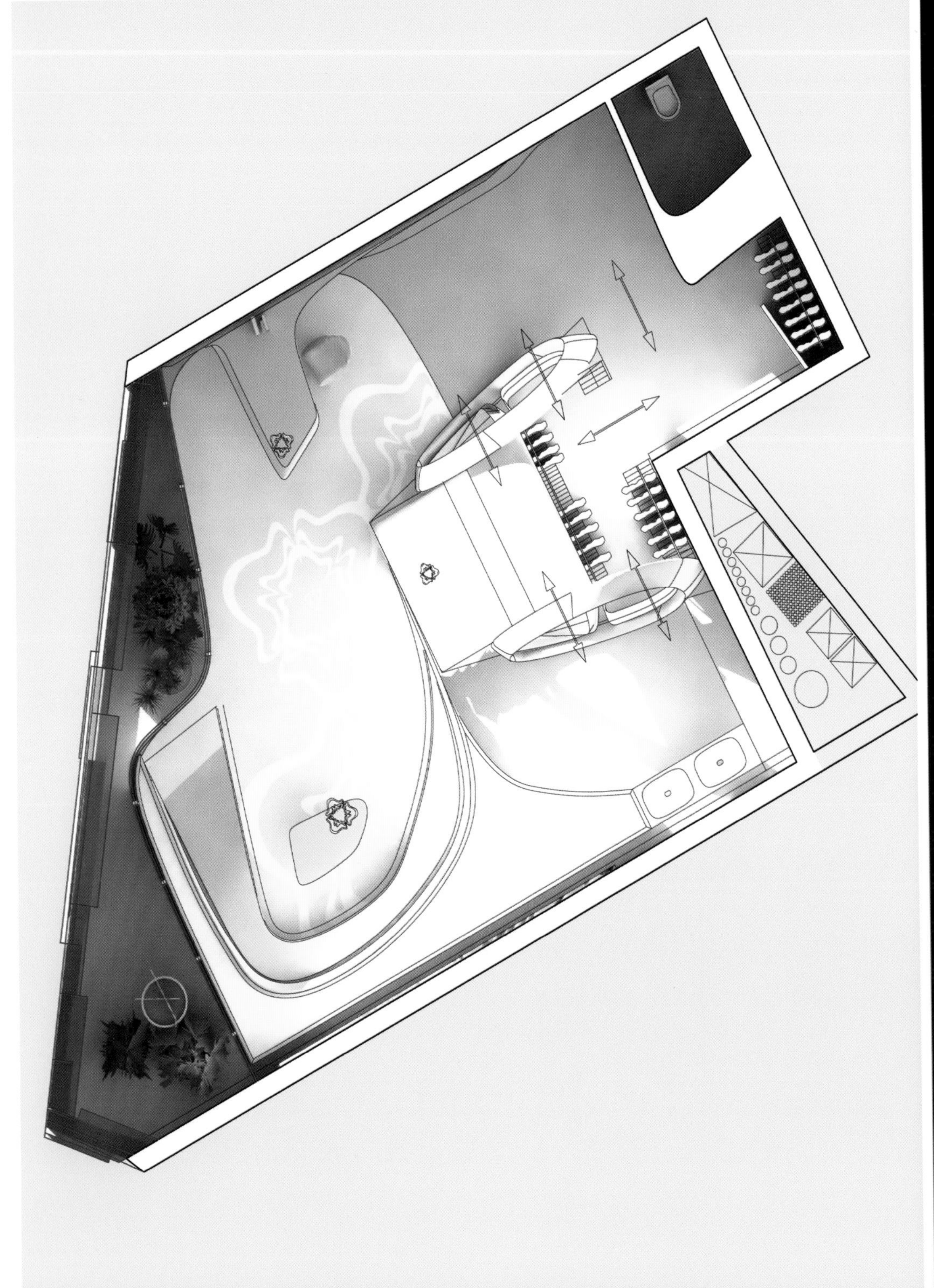

Profile

This newly developed Ice Hotel project has a building area of 530,000 square meters. As well as a brand new hotel, it also includes cultural center, high-end mall, office building and other supporting facilities. The Harbin Ice Hotel looks like an all-year super ice sculpture. It vividly reflects the clash and blend of fire and ice.

Design Concept

Icebergs, quartz crystal formations are the design inspiration referring to Harbin's famous Ice Festival. Another important theme of the design comes from the elegant swan. Several crystal buildings enclosed together highlight the visual effect and imagination as well.

Design Feature

The project is composed of ten quartz crystal towers of varying heights. Light leaps on the surface of this ice-like building, from one tower to another. The color of

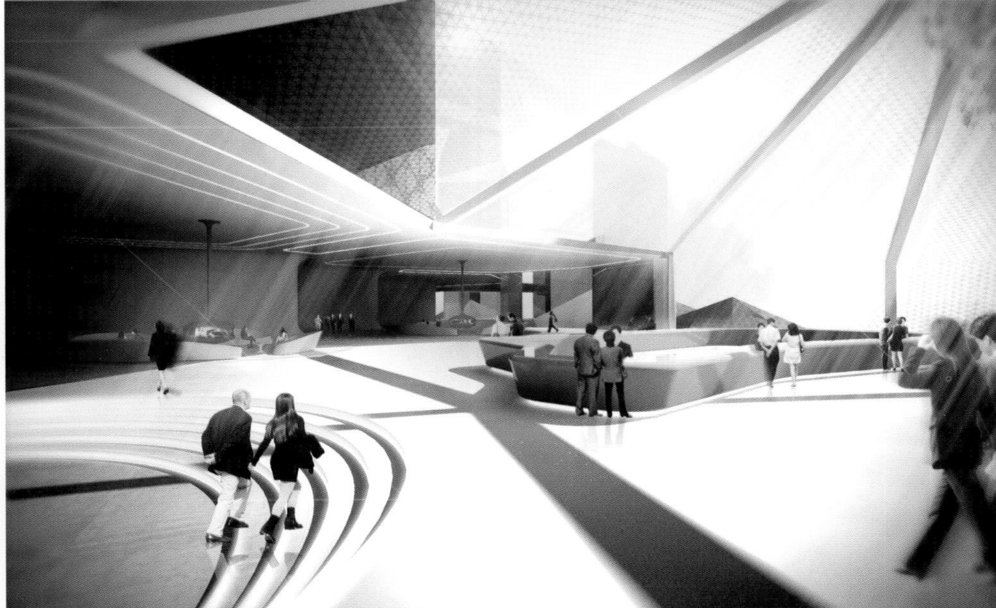

the building facades changes along with the light. The transparent appearance turns to be more bright and charming.

The Ice Hotel is a dynamic clash of fire and ice, the visitor moving from the perceptually "cold" exterior of a transparent facade to the dramatic crystalline hotel lobby with its warm, cave-like interior. Curvaceous balustrades, morphed floor levels and "stalagtite" drops are inspired by molten lava. They have jointly made the dramatic and sculptural interior space.

上海世茂新体验洲际酒店

Shanghai Shimao Wonderland Intercontinental Hotel

设计单位：阿特金斯 ATKINS
开发商：世茂集团
项目地址：中国上海市
建筑面积：58 494 m²

Designed by: Atkins
Developer: Shimao Group
Location: Shanghai, China
Floor Area: 58,494 m²

项目概况

在人们的通常概念当中，采石场并不是一个适合修建豪华酒店的理想场所。也许正是因为这样，这个位于松江的“深坑”酒店的设计概念才能显得如此与众不同。自从 2006 年阿特金斯赢得这个设计竞赛之后，在中国开发商世茂集团的支持之下，这个奇思妙想正逐渐成为现实。

设计理念

松江城因其独特的景观和自然资源成为当地以及国家重要的旅游休闲胜地，设计师的创作灵感来自于采石场坚硬的岩石峭壁、倾流而下的瀑布以及周边绵延的群山，设计方案很好地保留了采石场的自然景观，通过对周边生态环境的保护，构建有史以来最“绿色”的酒店。

设计特色

项目定位为五星级酒店，建设在一个深 90 米、底部注满水的已经废弃的采石场的侧壁上。在规划构思过程中，设计师考虑了多方面的因素，除了绿色覆土屋顶，地热以及太阳能的利用都被纳入设计之中。绿色覆土屋顶具有双重作用：它既使建筑完美地融入周边环境，成为“自然”地形的一部分，又起到环保和节能的作用。

除了令人惊叹的地貌，松江深坑酒店还拥有其他的特点：从崖顶倾流而下直至坑底湖面的瀑布、建筑主立面上的玻璃“瀑布”等设计都是项目的亮点。玻璃“瀑布”位于建筑的中心，设计师将建筑外形与室内公共空间结合，创造出形似瀑布的屋顶采光带，意在描绘瀑布从山上飞泻而下、直落崖底的壮观景象。

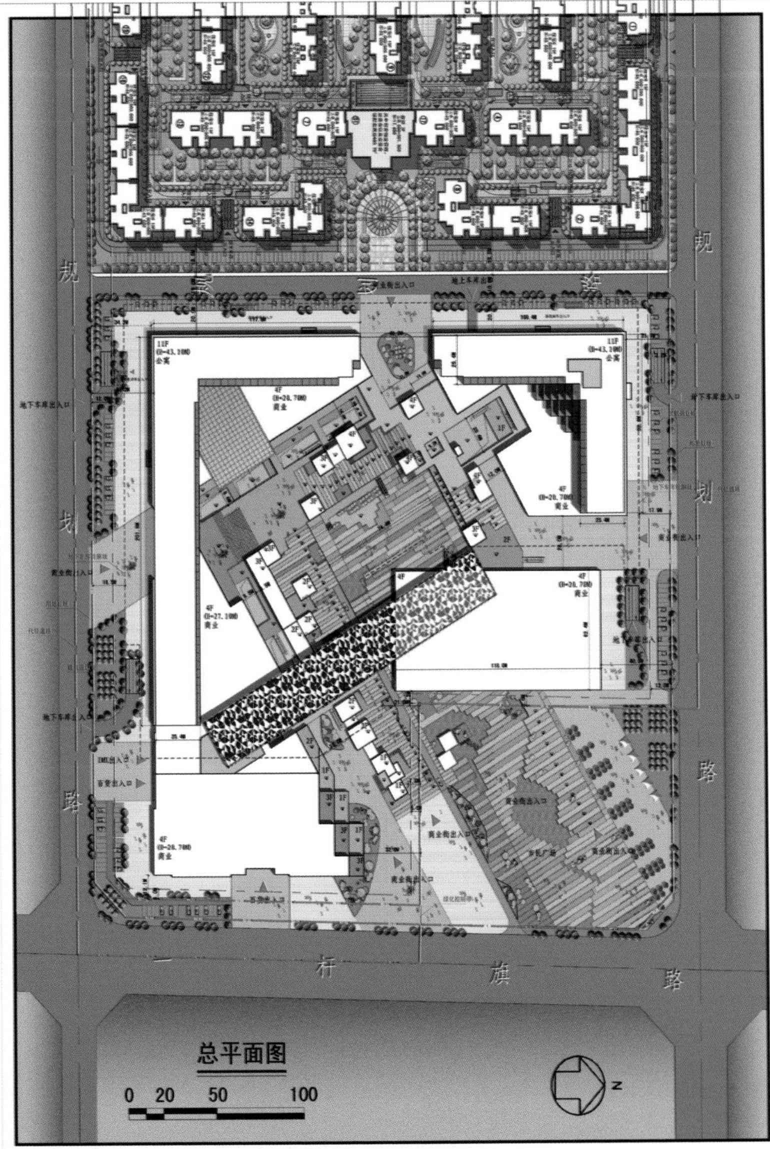
总平面图
0 20 50 100
N

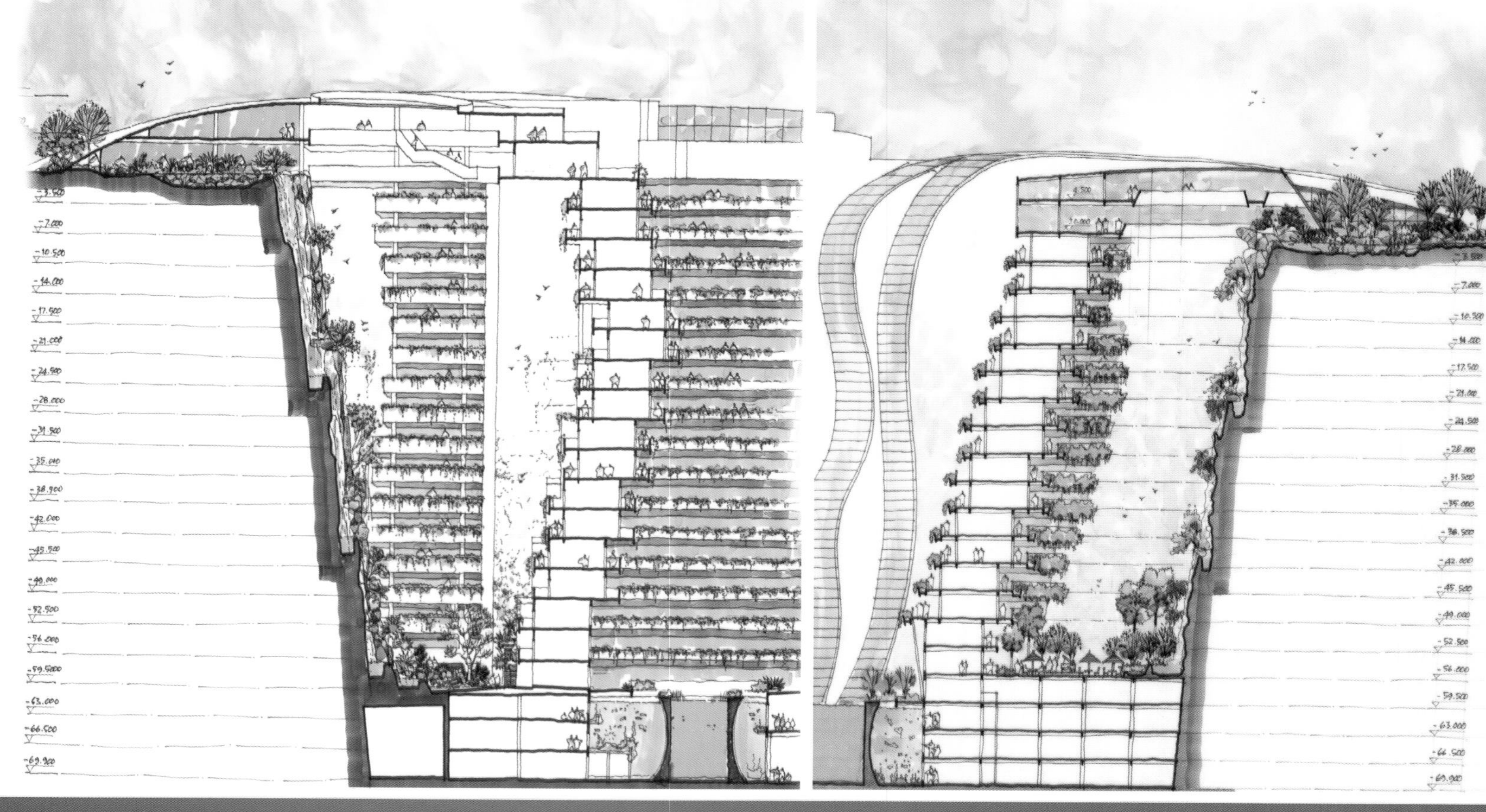

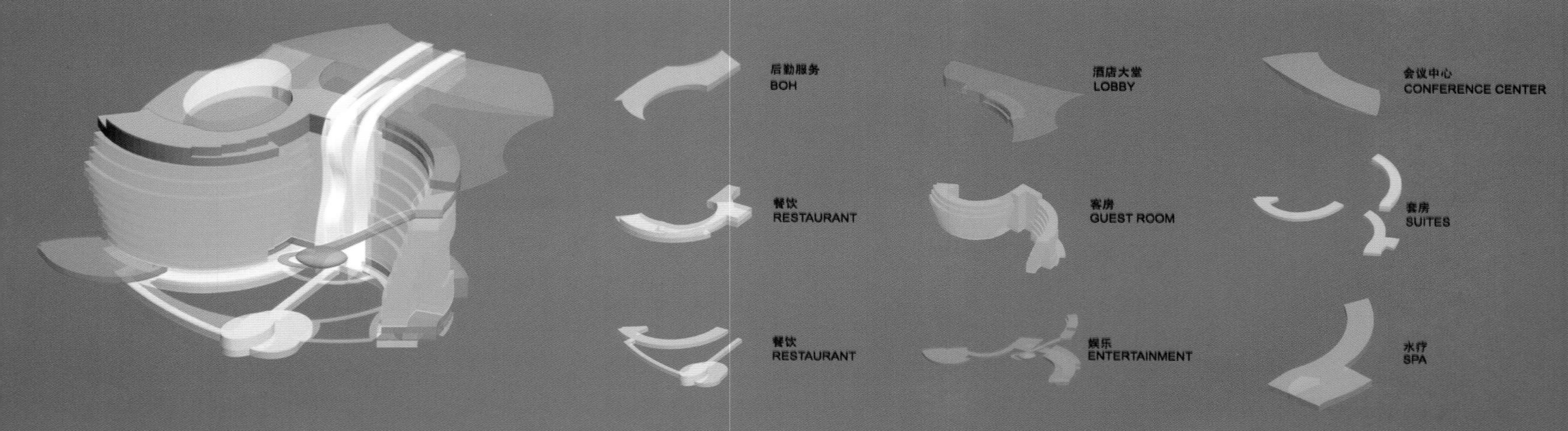
后勤服务
BOH
酒店大堂
LOBBY
会议中心
CONFERENCE CENTER
餐饮
RESTAURANT
客房
GUEST ROOM
套房
SUITES
餐饮
RESTAURANT
娱乐
ENTERTAINMENT
水疗
SPA

rofile

quarry is a rather unlikely destination for a swank hotel. Perhaps that's vhat makes the concept of Songjiang Quarry Hotel so awe-inspiring. ndeed, design and engineering consultancy Atkins won an international esign competition for the five-star hotel back in August 2006 and their ision is in the process of being transformed into reality by Chinese eveloper Shimao.

esign Concept

prawling landscapes and natural beauty have made it a popular tourist estination and the district has been designated as an important local nd national leisure resource. The winning concept was inspired by the tunning location and the natural environment of the rocky cliffs, waterfalls nd surrounding hills. The design of the Songjiang Quarry Hotel is meant to eflect the natural landscape of the quarry. By building the hotel where it is, esigners and developers hope to prevent further damage to the ecological nvironment around it. The Songjiang Quarry Hotel may become just about ne greenest hotel ever made.

esign Feature

he project will see the construction of a five-star hotel, built into the side f a disused, 90 meters deep, water filled quarry. Apart from the green roof, is intended that many features, ranging from geothermal energy to solar nergy utilization, are included in this project. The reasons for the green roof re two-fold: it is both for the building to fit seamlessly into the surrounding nvironment and become a "natural" part of the local topography, and also or its eco-friendly and energy saving qualities.

part from an awe-inspiring location, a cascading waterfall from the top f the quarry into the pool below it, and striking waveform architecture, he Songjiang Quarry Hotel will have plenty to offer. Also, a transparent lass "waterfall" located in the centre of the building is a major architectural eature. Combining architectural appearance with internal public space, esigners have managed to create waterfall-like lighting band which eminds of waterfall plunging down from mountain.

亚美尼亚埃里温四季帐篷大厦

All Seasons Tent Tower

设计单位：OFIS 建筑事务所

开发商：埃里温梅赛德斯奔驰有限公司

项目地址：亚美尼亚埃里温

项目面积：60 000 m²

设计团队：Rok Oman　Spela Videcnik
Robert Janez　Janez Martincic
Janja Del Linz　Katja Aljaz
Andrej Gregoric

Designed by: OFIS Arhitekti

Client: Mercedes Benz Co., City of Yerevan

Location: Yerevan, Armenia

Area: 60,000 m²

Design Team: Rok Oman, Spela Videcnik, Robert Janez, Janez Martincic, Janja Del Linz, Katja Aljaz, Andrej Gregoric

项目概况

这是 OFIS 参加在亚美尼亚埃里温举办的“梅赛德斯•奔驰”酒店设计大赛的参赛作品，设计以“四季帐篷”为主题，通过仿生的设计手法提供了一个形象生动、结构精巧、功能复合的酒店建筑方案。

设计特色

建筑构建在一个小山坡上，抬升的地形使之在场地内占据了优越的地理位置，为构建一个独特的、具有标志性意义的建筑奠定了基础。这个建筑由两个圆柱形的大楼构成，两座大楼的底部相连，这一建筑形态呼应了埃里温地区呈回旋圆周的城市肌理。建筑形态具有多个象征意义，既像游牧民族支起的帐篷，又似建筑后面绵延的山峰，也像一座绿色的攀岩墙。大楼的外层用绿色帐篷式的外壳围绕起来，它既是一种特殊的景观，也映衬了不同的季节：夏天，外壳会被绿色植物覆盖起来；冬天，外壳则会被白雪覆盖。

建筑设计

埃里温地处地震带，因此需要设计一个安全的建筑结构以防震。垂直结构用混凝土大楼核心和韧性混凝土柱来增强稳定性，可以抵抗重心荷载和地震引发的震动。较高大楼的内圈设计了两个核心来确保大楼的稳定性，而较低的大楼则设计了一个核心。建筑呈梯田式的形状，故而建筑最外层的部分都向建筑中心倾斜，而最内部的部分则是网状结构，核心墙壁的厚度和柱子直径都随着建筑的增高而减小，确保了建筑的稳定性和抗震性。

为了优化环境状况，减少大楼的能量需求，还需要做一些特殊处理，以最大限度地获取自然能源：夏天，外壳可以起到遮荫的作用，减少大厦所吸收的太阳热能；在混凝土板中安装管道系统，夏天时可以供应冷气，冬天则供应热气；大楼顶层安装的光伏板能转化太阳能，为大楼供电。

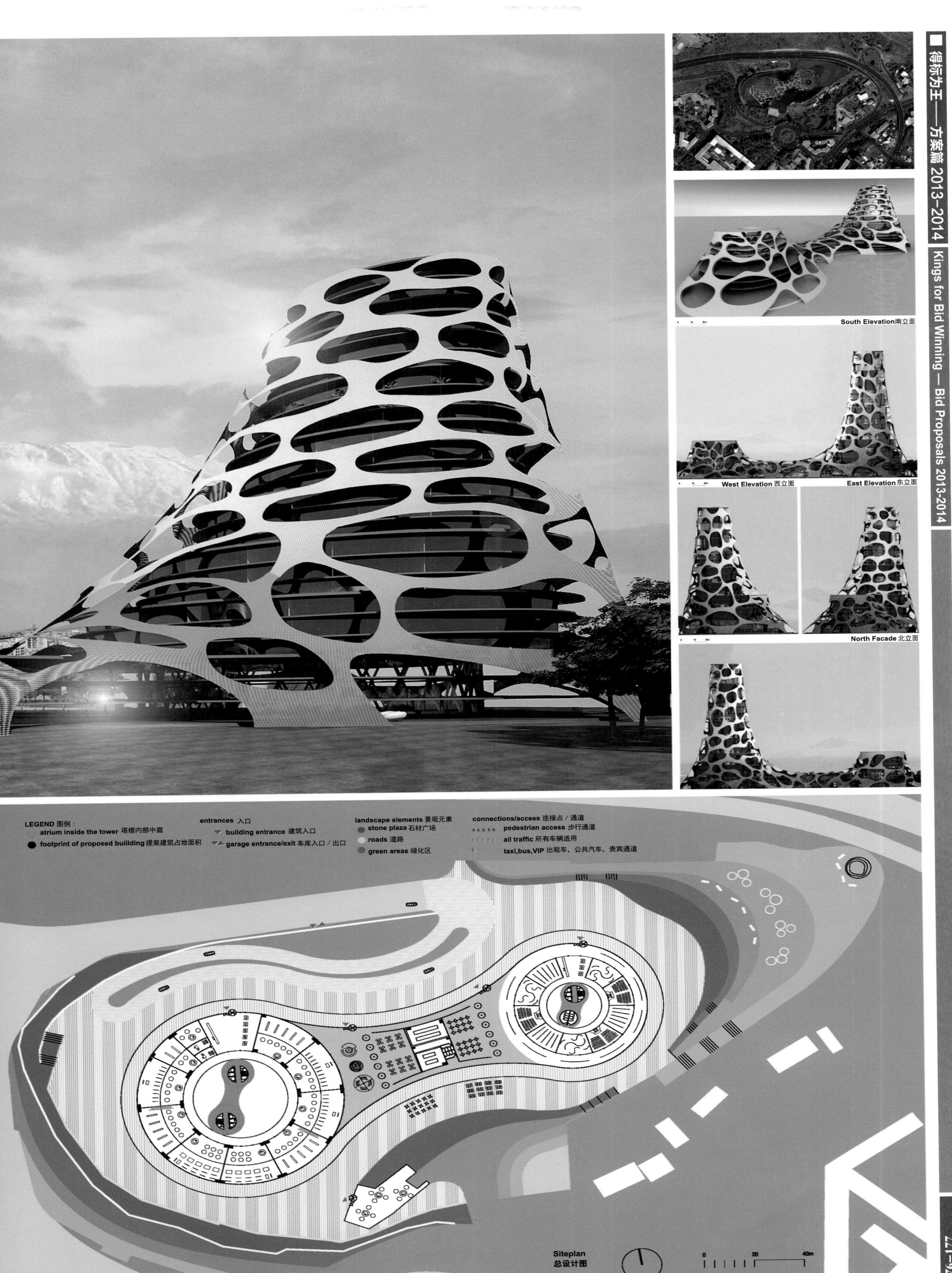
South Elevation 南立面
West Elevation 西立面
East Elevation 东立面
North Facade 北立面
LEGEND 图例：
atrium inside the tower 塔楼内部中庭
footprint of proposed building 提案建筑占地面积
entrances 入口
building entrance 建筑入口
garage entrance/exit 车库入口／出口
landscape elements 景观元素
stone plaza 石材广场
roads 道路
green areas 绿化区
connections/access 连接点／通道
pedestrian access 步行通道
all traffic 所有车辆适用
taxi,bus,VIP 出租车，公共汽车，贵宾通道
Siteplan
总设计图
0
20
40m

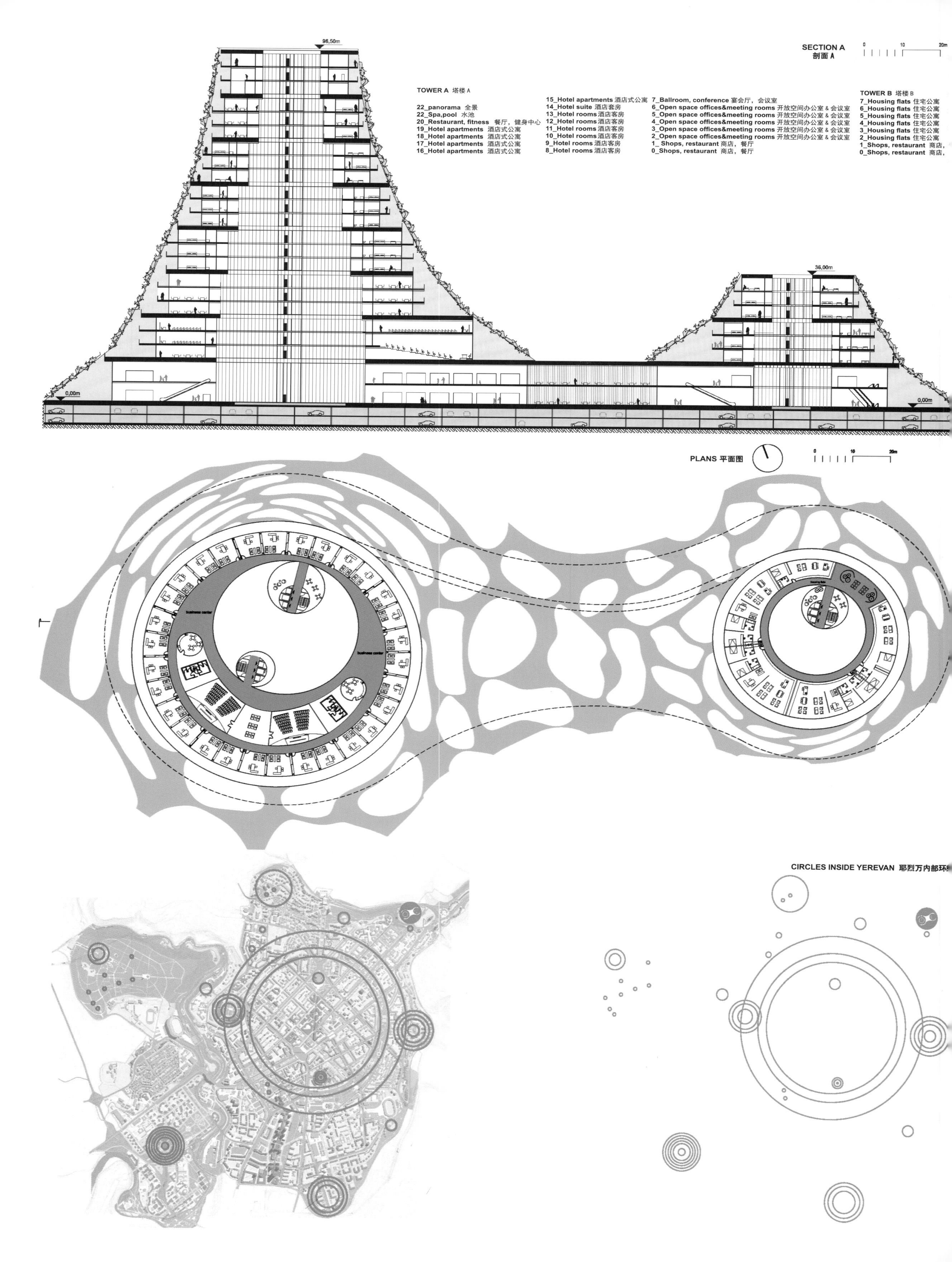

SECTION A
剖面 A
96,50m
36,00m
0,00m
TOWER A 塔楼 A
22_panorama 全景
22_Spa,pool 水池
20_Restaurant, fitness 餐厅，健身中心
19_Hotel apartments 酒店式公寓
18_Hotel apartments 酒店式公寓
17_Hotel apartments 酒店式公寓
16_Hotel apartments 酒店式公寓
15_Hotel apartments 酒店式公寓
14_Hotel suite 酒店套房
13_Hotel rooms 酒店客房
12_Hotel rooms 酒店客房
11_Hotel rooms 酒店客房
10_Hotel rooms 酒店客房
9_Hotel rooms 酒店客房
8_Hotel rooms 酒店客房
7_Ballroom, conference 宴会厅，会议室
6_Open space offices&meeting rooms 开放空间办公室 & 会议室
5_Open space offices&meeting rooms 开放空间办公室 & 会议室
4_Open space offices&meeting rooms 开放空间办公室 & 会议室
3_Open space offices&meeting rooms 开放空间办公室 & 会议室
2_Open space offices&meeting rooms 开放空间办公室 & 会议室
1_ Shops, restaurant 商店，餐厅
0_Shops, restaurant 商店，餐厅
TOWER B 塔楼 B
7_Housing flats 住宅公寓
6_Housing flats 住宅公寓
5_Housing flats 住宅公寓
4_Housing flats 住宅公寓
3_Housing flats 住宅公寓
2_Housing flats 住宅公寓
1_Shops, restaurant 商店，
0_Shops, restaurant 商店，
PLANS 平面图
CIRCLES INSIDE YEREVAN 耶烈万内部环

ofile

=IS Arhitekti has proposed "All Seasons Tent Tower" as ırt of the Mercedes Benz Hotel tower competition for e city of Yerevan, Armenia. Taking "All Seasons Tent" theme, the Company has created a vivid, exquisite, mplex functional hotel building project by biomimetic sign.

esign Feature

ıe prominent location and dominating position of the e represents a chance for a unique arrangement, with a ong identity as a symbol or landmark of contemporary chitecture in the city of Yerevan. The project consists of o cylindrical towers. It takes the concept of two terraced lindrical tents-towers connected at the ground floor. The und shape of the towers corresponds with the circular d round structures in Yerevan's urban pattern. The ape of the building represents several simbolic, formal and sustainable issues: the tent, the mountain and green climbing wall. The facade is clad in a metal mesh, which is covered in live greenery in the summer and with snow in winter.

Architectural Design

As the city of Yerevan is situated in a region prone to earthquakes, the main driver when deciding on a suitable structural system for the building was safety during earthquakes. The vertical structure, which resists gravity load and forces resulting from earthquake action, was rationalized to reinforce the tower's concrete cores and composite columns. The two cores on the inner perimeter of the taller tower are required to ensure structural stability, one core in the case of the lower tower. Due to the terraced shape of the building, the columns on the outermost perimeter are inclined towards the centre of the building. Those on the innermost perimeter have mesh forms. The core wall thickness and column dimensions are reduced with the height of the building to ensure stability and earthquake resistance.

Special care has been taken to optimize environmental conditions and minimize the energy demands of the tower. The external facades will feature a high performance skin with an adaptable external shading device to reduce solar gains in the summer. A concrete slab embedded pipe system provides cooling without draft problems and in winter comfortable heating. In the summer, the cooling of internal spaces is achieved primarily through the use of the slab system; during winter, the fresh air will be heated inside the units and distributed into the rooms using the displacement ventilation principle. Photovoltaic cells are integrated into the roof of top floor to generate power for the tower.

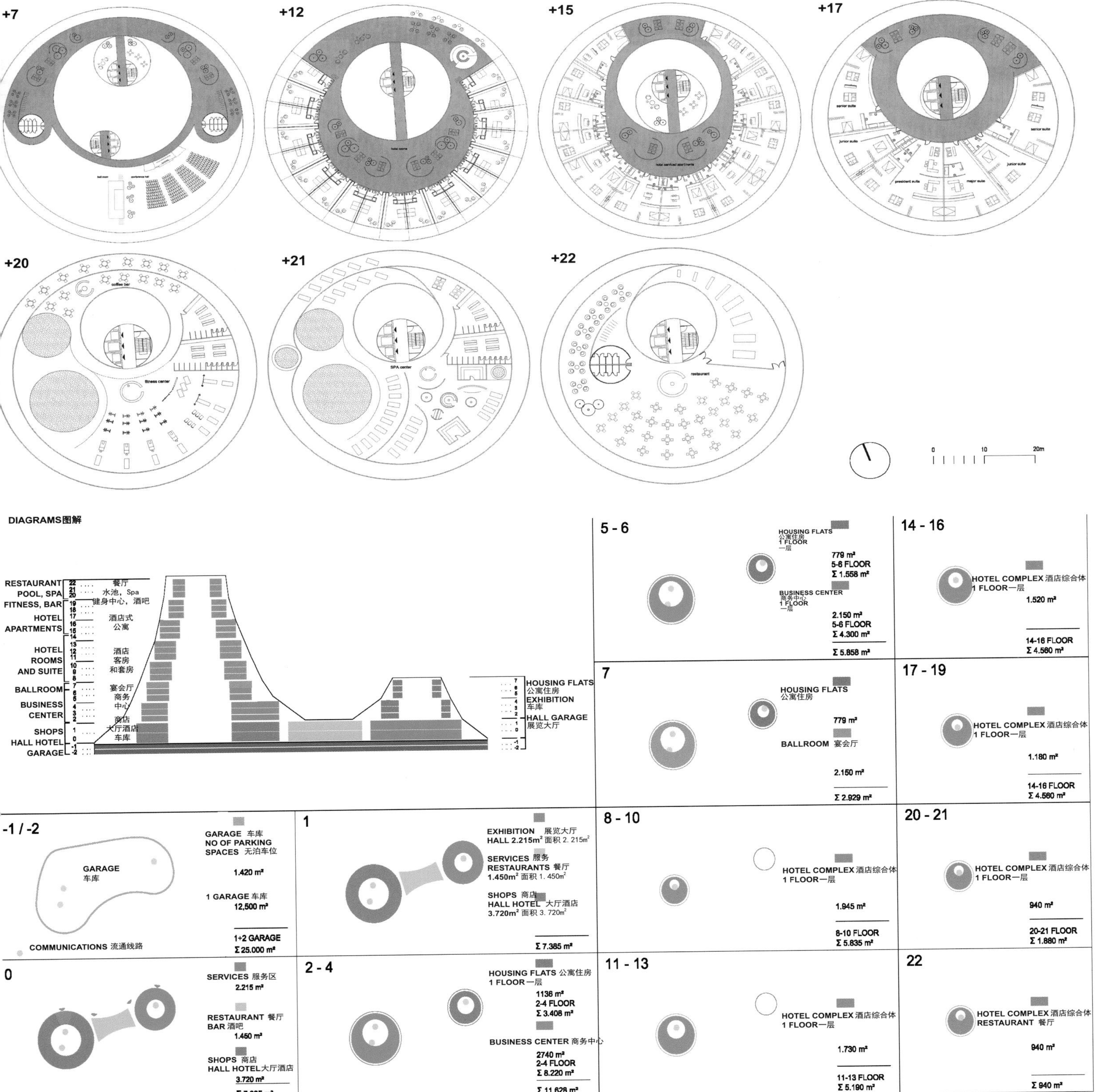

山东青岛红树林度假会展酒店综合体

Tsingtao Jiaonan Bay Mangrove Tree Resort and Conference Hotel

设计单位：ZNA 泽碧克建筑设计事务所

开发商：今典集团

项目地址：中国山东省青岛市

用地面积：296 000 m²

建筑面积：764 228 m²

Designed by: ZEYBEKOGLU NAYMAN ASSOCIATES. INC

Developer: Antaeus Group

Location: Tsingtao, Shandong, China

Site Area: 296,000 m²

Floor Area: 764,228 m²

项目概况

项目坐落于青岛胶南滨海地区，气候温和、舒适宜人。项目定位为北方最大的度假会展综合体，包含了度假酒店、演艺中心、会所、水上乐园以及德国小镇商业街等多个功能区。

设计特色

设计以海洋文化为基调，结合当地的文化特色，力图创造一处宜人、休闲的度假胜地。独具特色的建筑群落通过地下商业水街彼此串联，而广场又与海岸、下沉庭院交汇出宜人的自然景观。各具特色的空间在酒店综合体内有机共存，使项目呈现出独有的度假特性和吸引力。

分区设计

地标酒店

地标酒店是青岛红树林度假酒店建筑群落中最具特色的建筑，简洁大方的酒店外观使之成为该区域的标志性建筑。立面设计的灵感源自于贝壳及海浪的纹理，设计师将简化的贝壳弧线转化成层次交叠的轮廓，创造出度假酒店建筑群中优美的天际线；波浪起伏的竖向线条充分展现出曼妙的波动韵律，并与豪华酒店、会展酒店及家庭酒店的立面元素相互呼应，营造出浓郁的度假氛围。

家庭、会展以及豪华酒店

作为地标酒店的呼应元素，三个酒店采用了相似的建筑语汇以及色彩、材料，在体现整体性的同时，彰显了地标酒店的主导性。主体立面采用仿木褐色，而顶冠处简洁抽象的树状图案，则体现了红树林酒店的品牌内涵和识别性。

德国小镇商业街

这所小镇的设计既沿袭了青岛当地的德式建筑文化，又结合了当地滨海的地域特征，力图在体现德国异域风情建筑风格的同时，凸显本土文化特质，彰显富有活力的小镇风情。德国小镇商业街的立面设计，提取了德式传统的建筑元素、材料、色彩以及细部装饰手法，通过对其进行简化、抽象处理，应用到不同尺度的空间。

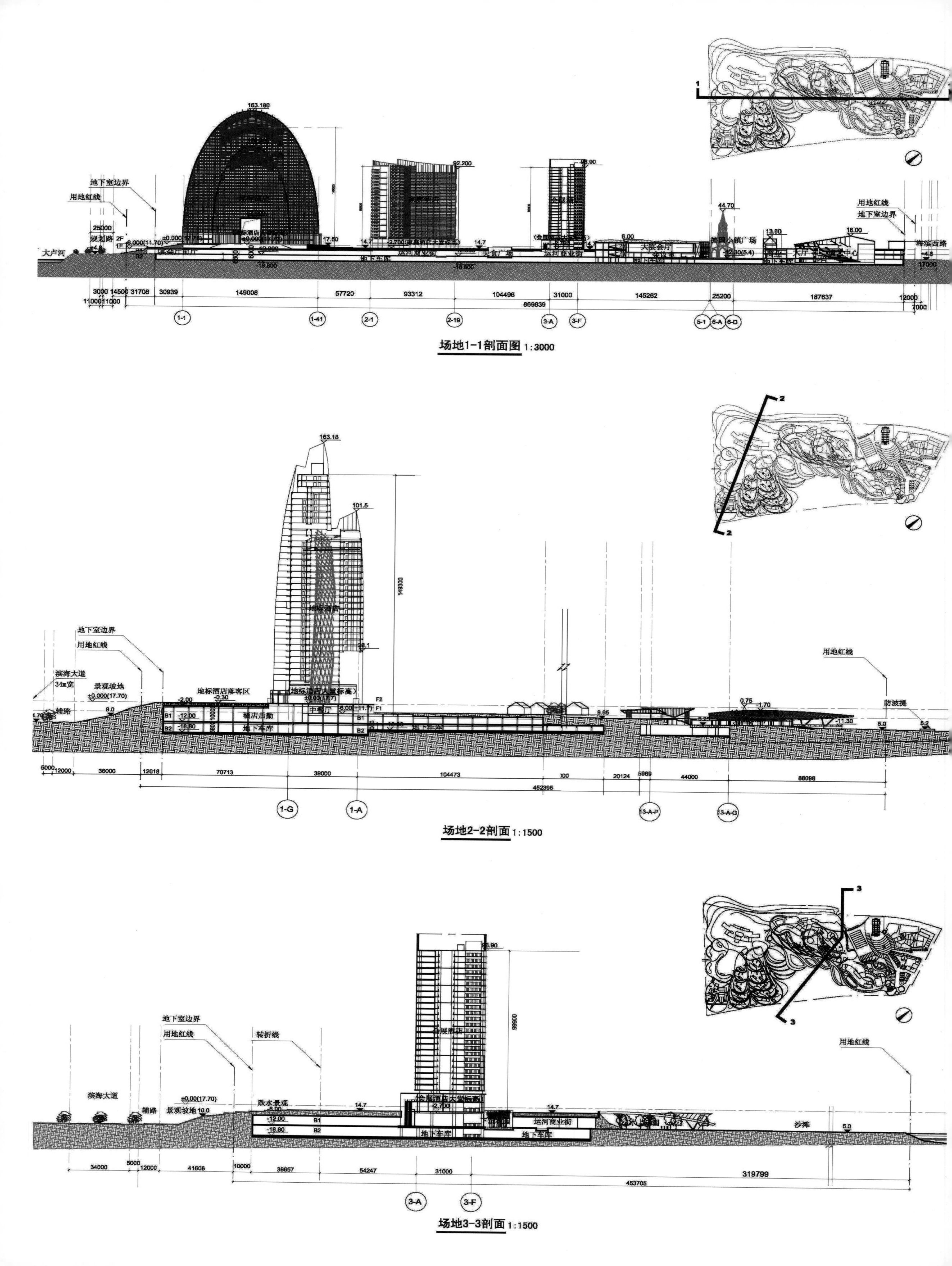

场地1-1剖面图 1:3000

场地2-2剖面 1:1500

场地3-3剖面 1:1500

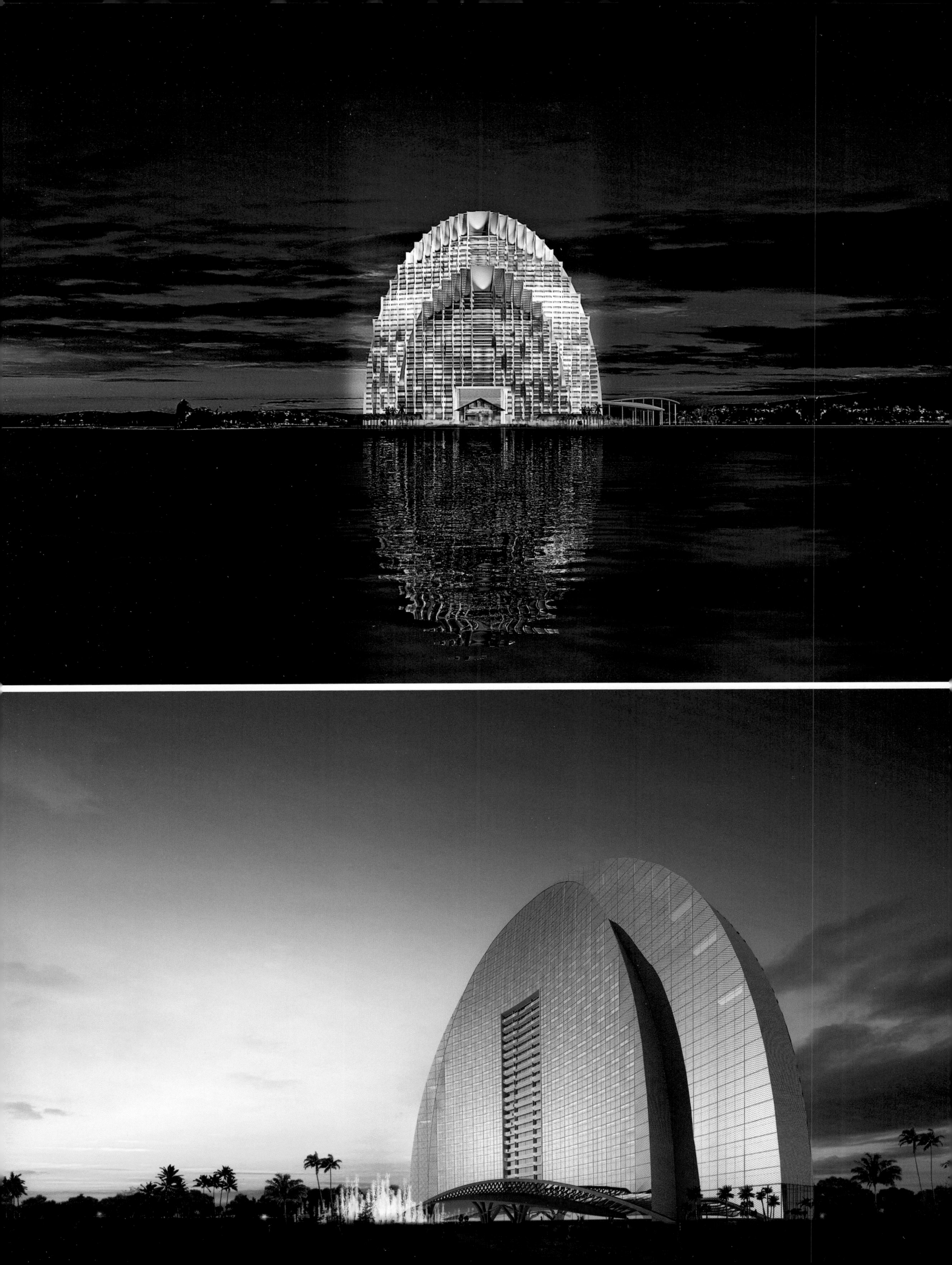

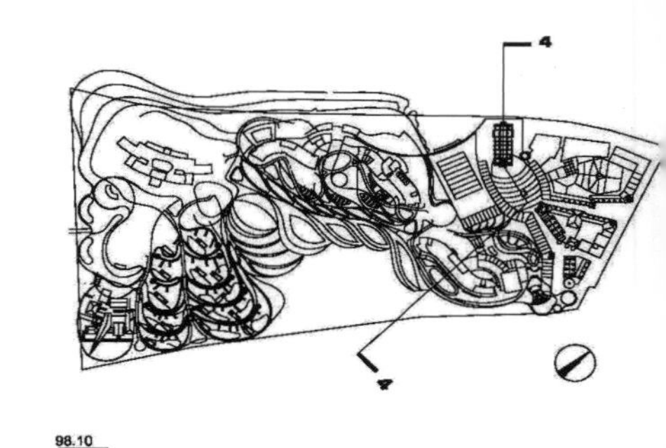

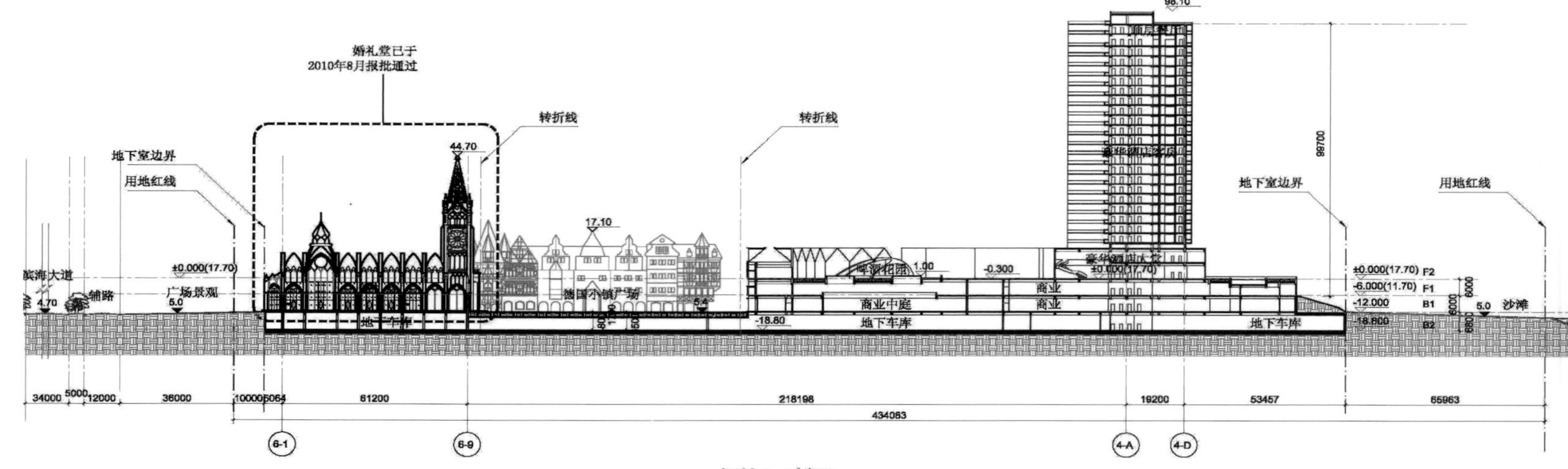

场地4-4剖面 1:1500

场地5-5剖面图 1:1000

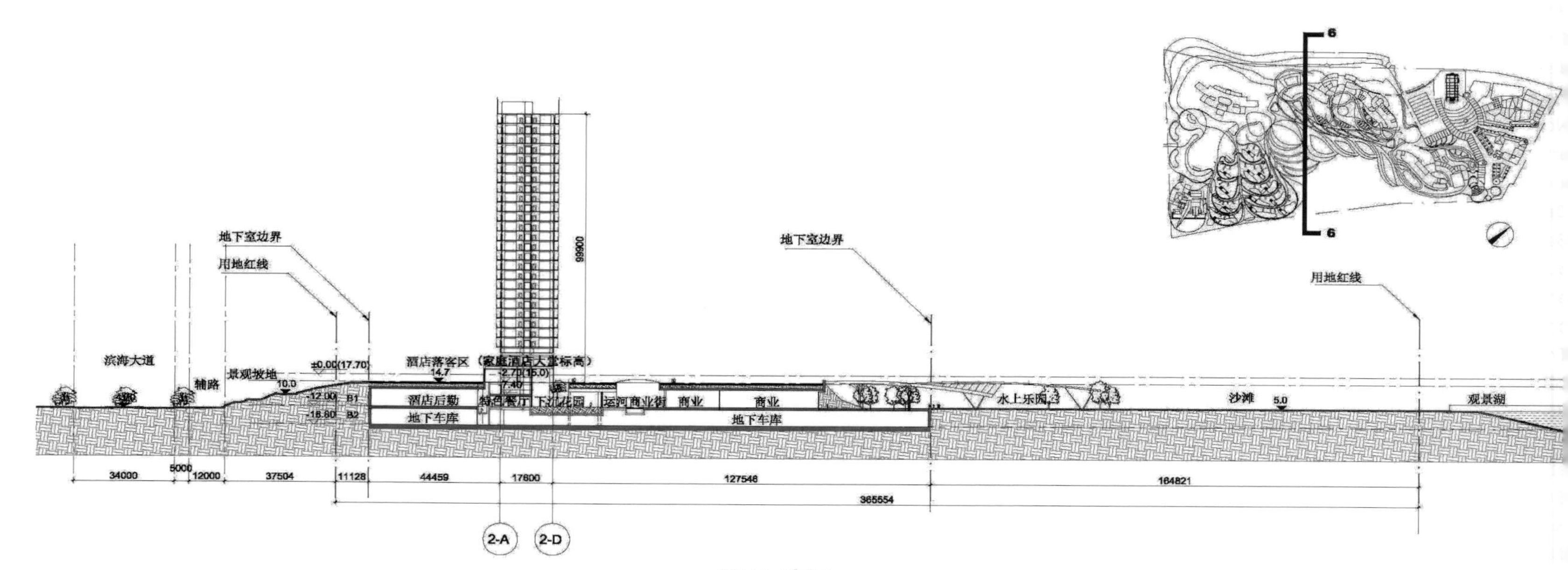

场地6-6剖面 1:1500

Profile

The project is located in Tsingtao Jiaonan waterfront district with mild pleasant climate. It's been defined as the largest recreation and convention resort in North China which includes resort hotel, performing art center, club, water park, Germantown and other functional areas.

Design Feature

Based on marine culture and combined with local cultural features, the project strives to create an enjoyable leisure resort. Unique buildings are connected by underground commercial water street. The intersection of the square, coast, and the sunken courtyard build a pleasant natural landscape. These different spaces exist organically in the synthesis of the hotel to show the characteristics and the special attraction of the resort.

Subarea Design

Iconic Hotel

The iconic hotel is the most unusual architecture among Tsingtao Jiaonan building community. The minimal elegance of the hotel exterior makes this building the landmark architecture of the region. The facade design generates from the texture of the waves and the shells. The simplified shell arc turns into an overlapping shape and creates a graceful skyline in the resort hotel community. The undulant vertical lines adequately display graceful fluctuation rhythm, which is also echoed in the façade elements of the Luxury Hotel, Convention Hotel and Family Hotel to reflect the unique atmosphere of Qingdao by the shore.

Family Hotel, Convention Hotel and Luxury Hotel

Family Hotel, Convention Hotel and Luxury Hotel emphasize wholeness as the echoes of Landmark Hotel. These three hotels use a similar architectural language, color and materials to reflect integrity and reveal the dominant Landmark Hotel. Wood-like brown color which is adopted by the main elevation, the concise and abstract tree pattern on the crown, reflect the brand recognition of the Mangrove Hotel.

Germantown Business Street

Following the influence of the German architectural culture and the coastal region characteristics of Qingdao, the project is to create a unique German inspired style and environment, and at the same time highlight native cultural characteristics, and create the lively atmosphere in the town. The facade of Germantown business street design is based on German traditional existing architectural elements, material, color and ornamental details, and creating simplified and abstract forms. Then it is used in the different scale spaces, like the street and public square space.

海南三亚海棠湾红树林度假酒店

Hainan Sanya Haitang Bay Mangrove Tree Resort Hotel

设计单位：ZNA 泽碧克建筑设计事务所

开发商：今典集团

项目地址：中国海南省三亚市

占地面积：215 547 ㎡

建筑面积：115 701 ㎡

Designed by: ZEYBEKOGLU NAYMAN ASSOCIATES. INC

Developer: Antaeus Group

Location: Sanya, Hainan, China

Site Area: 215,547 m²

Floor Area: 115,701 m²

项目概况

项目基地位于海南海棠湾中部东隅沙坝酒店地带核心位置，是一个七星级滨海度假会议酒店。建筑高度达 120 米，明显高于周边其他开发项目，在体量高度上将成为区域内独特的地标性建筑。

建筑设计

酒店主楼的入口空间为一个环形庭院，两侧的景观植栽塑造出亲切的落客空间。酒店落客雨棚以简洁优雅的弧线造型，给予游客鲜明的印象。裙房高两层，塑造了宏伟的门厅，局部设置夹层以提供更经济有效的办公空间。建筑外立面为流线型的开放观景阳台，使用了特殊的感光涂料，在不同时段的光线照射下能反射出奇幻的色彩。

中庭高近 100 米，设计师在交错分布的廊桥平台上布置植物、跌水，形成梦幻般的空中花园景观。整个中庭空间自然通风，将功能、环保节能与美观完美统一。

交通组织

设计在西侧的滨海大道上设置了三个出入口，酒店主入口居中，南侧为服务区入口，北侧则是通往总统套房、别墅会所及水疗会所的入口。水疗会所和别墅会所设有地下电瓶车车库，从这里，客人可自驾电瓶车沿基地北端的环湖路抵达水疗别墅，也可沿基地东北角的下山路进入别墅群。

水疗别墅区和海景别墅区的车行道宽度均为 4 米，可满足消防车的需要。核心景观区的景观路虽然仅宽 1.5 米，但 4 米净宽内没有种植树木或大型植被，既可满足消防车通过的需要，又可对景观区内的独立建筑施行扑救。

rofile

'he site is located in the core of hotel area of middle Haitang Bay, Hainan Province. It is a seven-star vaterfront resort & conference hotel. The building eight of 120 meters prominently higher than other evelopment projects around makes this project a nique landmark building of this area.

rchitectural Design

Entrance space of the hotel's main building forms a ircular courtyard. Landscape vegetations on both ides create intimate drop-off space. The canopy of he drop-off space offers a vivid impression to visitors vith simple elegant arc-shape. Two levels podium reates grand entrance hall. Mezzanines locally set provides more economical and efficient office space. Linear open balconies of building façade have utilized special photosensitive coating which reflects fantastic colors in the sun at different times of the day.

The atrium is about 100 meters high. Plants and waterfalls on gallery bridges create dream-like sky garden landscape. The whole atrium is naturally ventilated unifying functionality, energy saving, environment protection and aesthetics together.

Traffic Organization

There are three entrances at the west waterfront avenue, the main entrance of the hotel in the middle, the entrance of service area in the south and the entrance of presidential suite, villa club and spa club in the north. Spa club and spa club contain electromobile garage underground. From here, clients can drive their electromobiles along lakeside road of the north to spa villas. Or, they can go to villa cluster along mountain road at the northeast corner of the site.

The vehicle lanes for spa villa district and seaside villa district are four meters wide which makes fire truck access available. Even though the landscape road in central landscape area is only 1.5 meters, there are no large trees or vegetations by the roadside, fire trucks are free to pass when fire breaks out in buildings of this landscape area.

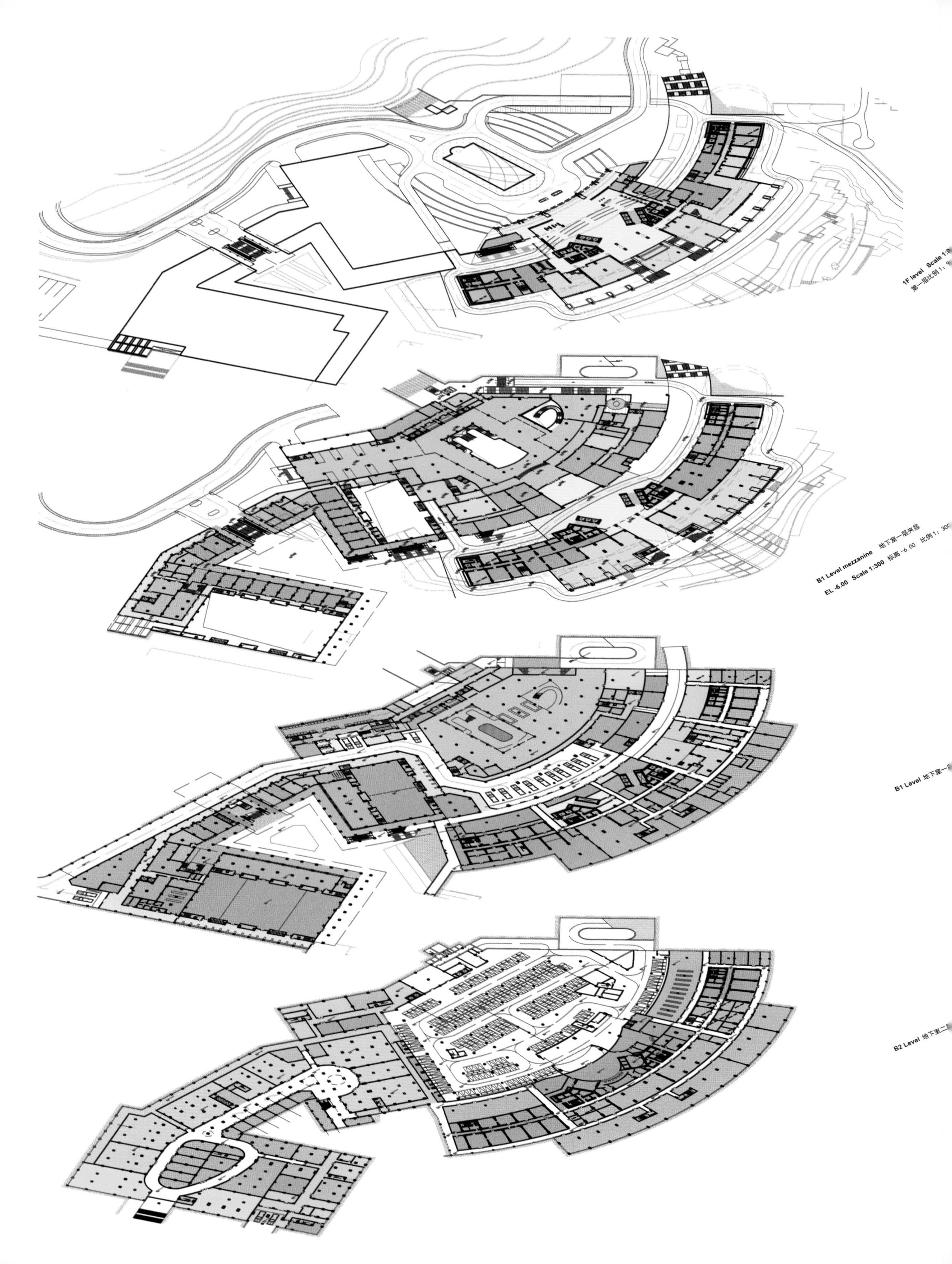
1F level Scale 1:9
第一层比例1：9
B1 Level mezzanine 地下室一层夹层
EL -6.00 Scale 1:300 标高 -6.00 比例1：30
B1 Level 地下室一
B2 Level 地下室二

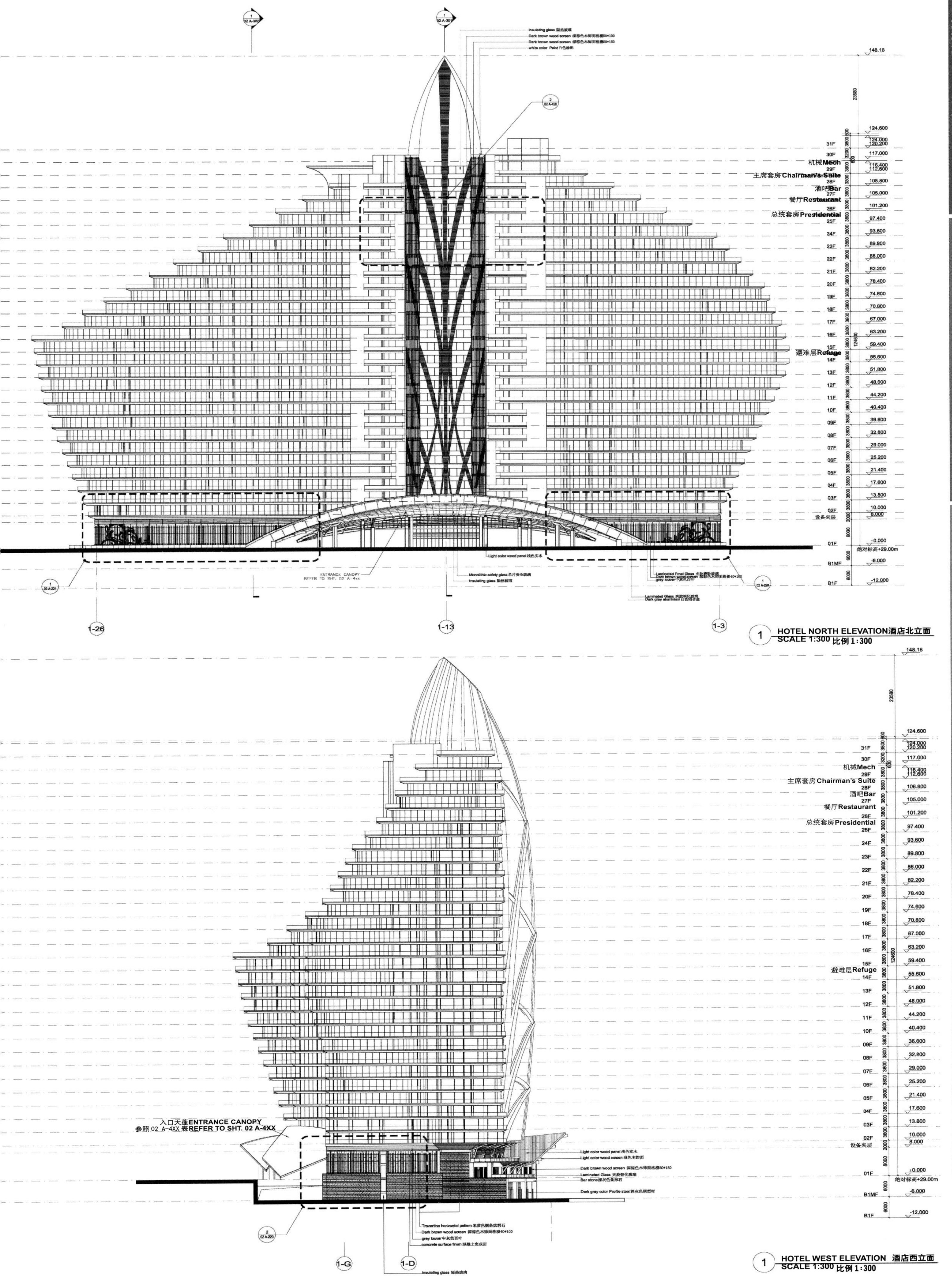

1 HOTEL NORTH ELEVATION酒店北立面
SCALE 1:300 比例 1:300

1 HOTEL WEST ELEVATION 酒店西立面
SCALE 1:300 比例 1:300

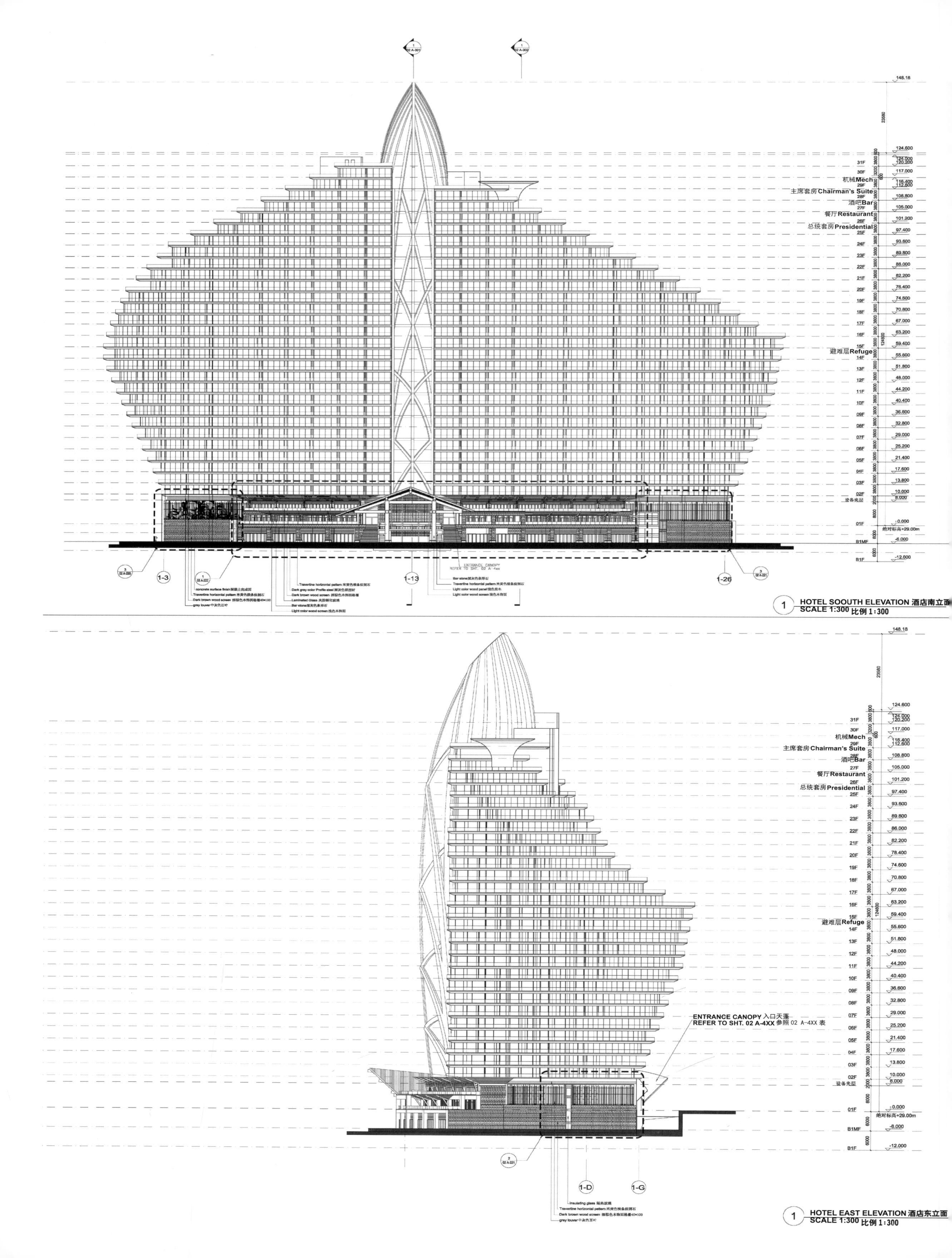
148.18
124.600
31F
124.000
120.200
30F
117.000
机械Mech
116.400
29F
112.600
主席套房Chairman's Suite
28F
108.800
酒吧Bar
27F
105.000
餐厅Restaurant
26F
101.200
总统套房Presidential
25F
97.400
24F
93.600
23F
89.800
22F
86.000
21F
82.200
20F
78.400
19F
74.600
18F
70.800
17F
67.000
16F
63.200
15F
59.400
避难层Refuge
14F
55.600
13F
51.800
12F
48.000
11F
44.200
10F
40.400
09F
36.600
08F
32.800
07F
29.000
06F
25.200
05F
21.400
04F
17.600
03F
13.800
02F
10.000
设备夹层
8.000
01F
±0.000
绝对标高+29.00m
B1MF
-6.000
B1F
-12.000
1-3
1-13
1-26
ENTRANCE CANOPY
REFER TO SHT. 02 A-4xx
concrete surface finish 混凝土完成面
Travertine horizontal pattern 米黄色横条纹洞石
Dark brown wood screen 深棕色木饰面格栅40*100
grey louver 中灰色百叶
Travertine horizontal pattern 米黄色横条纹洞石
Dark gray color Profile steel 深灰色钢型材
Dark brown wood screen 深棕色木饰面格栅
Laminated Glass 夹胶钢化玻璃
Bar stone 深灰色条形石
Light color wood screen 浅色木饰面
Bar stone 深灰色条形石
Travertine horizontal pattern 米黄色横条纹洞石
Light color wood panel 浅色实木
Light color wood screen 浅色木饰面
1 HOTEL SOOUTH ELEVATION 酒店南立面
SCALE 1:300 比例 1:300
ENTRANCE CANOPY 入口天蓬
REFER TO SHT. 02 A-4XX 参照 02 A-4XX 表
1-D
1-G
Insulating glass 隔热玻璃
Travertine horizontal pattern 米黄色横条纹洞石
Dark brown wood screen 深棕色木饰面格栅40*100
grey louver 中灰色百叶
1 HOTEL EAST ELEVATION 酒店东立面
SCALE 1:300 比例 1:300

四川遂宁圣莲岛五星级酒店

Sichuan Suining St. Lotus Island Five-Star Hotel

设计单位：OAD 欧安地建筑设计事务所
项目地址：中国四川省遂宁市
用地面积：92 493.72 m²
总建筑面积：57 254 m²
建筑密度：20%
绿地率：45%
容积率：0.5

Designed by: OAD Office for Architecture & Design
Location: Suining, Sichuan, China
Site Area: 92,493.72 m²
Floor Area: 57,254 m²
Building Density: 20%
Greening Ratio: 45%
Plot Ratio: 0.5

项目概况

该项目为总用地约为 92 000 平方米的五星级酒店，占据着圣莲岛的东南端，三面临湖，环境优越。设计旨在将其打造为西南地区首屈一指的高端旅游、度假、休闲区。

设计理念

项目坚持绿色、生态、可持续发展的基本思路，整个方案注重与周边水环境的结合，体现与自然的完美融合。同时，酒店规划体现了生态节能的理念，并将时尚现代的元素与独具遂宁特色的本土文化结合，打造一个极具地方特色的休闲度假酒店。

建筑设计

酒店分为餐饮住宿区、会议区、高端别墅区和温泉疗养区。在功能布局上，设计根据主楼的流线形走势，合理布置几大功能区，使各功能区之间建立便捷的联系。

酒店的主入口位于距地面 5 米高的绿化基座上，抬高的地势使酒店拥有了良好的观水视野的同时，也使在入口处设计出朝向水面、层层跌落的台地景观成为可能，这将丰富酒店花园的景观层次。

餐饮区位于酒店架空层和首层，模糊了内外的界线，充分展示了休闲度假的轻松自在。线性布置的各种餐厅，朝向开阔的花园和湖面，使就餐变得轻松而愉悦。

商业店铺沿用地北侧的河岸展开，一直延伸到酒店的公共空间，将精品购物和度假休闲联系起来。架空层绿化坡地的南侧设置为会议区入口，会议区配备有可以举行大型会议宴会的多功能厅和各种规模的中小型会议室。

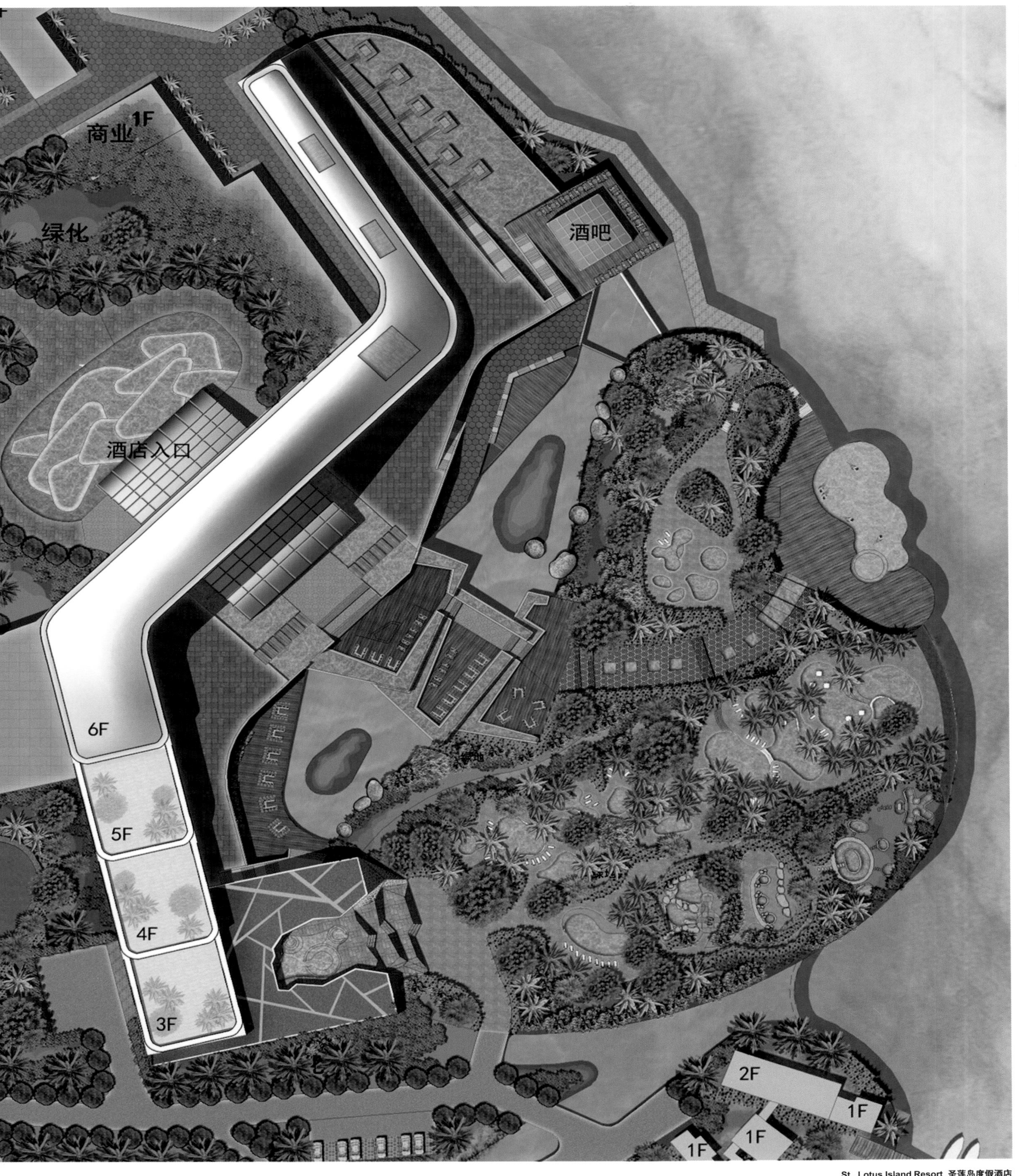

St . Lotus Island Resort 圣莲岛度假酒店

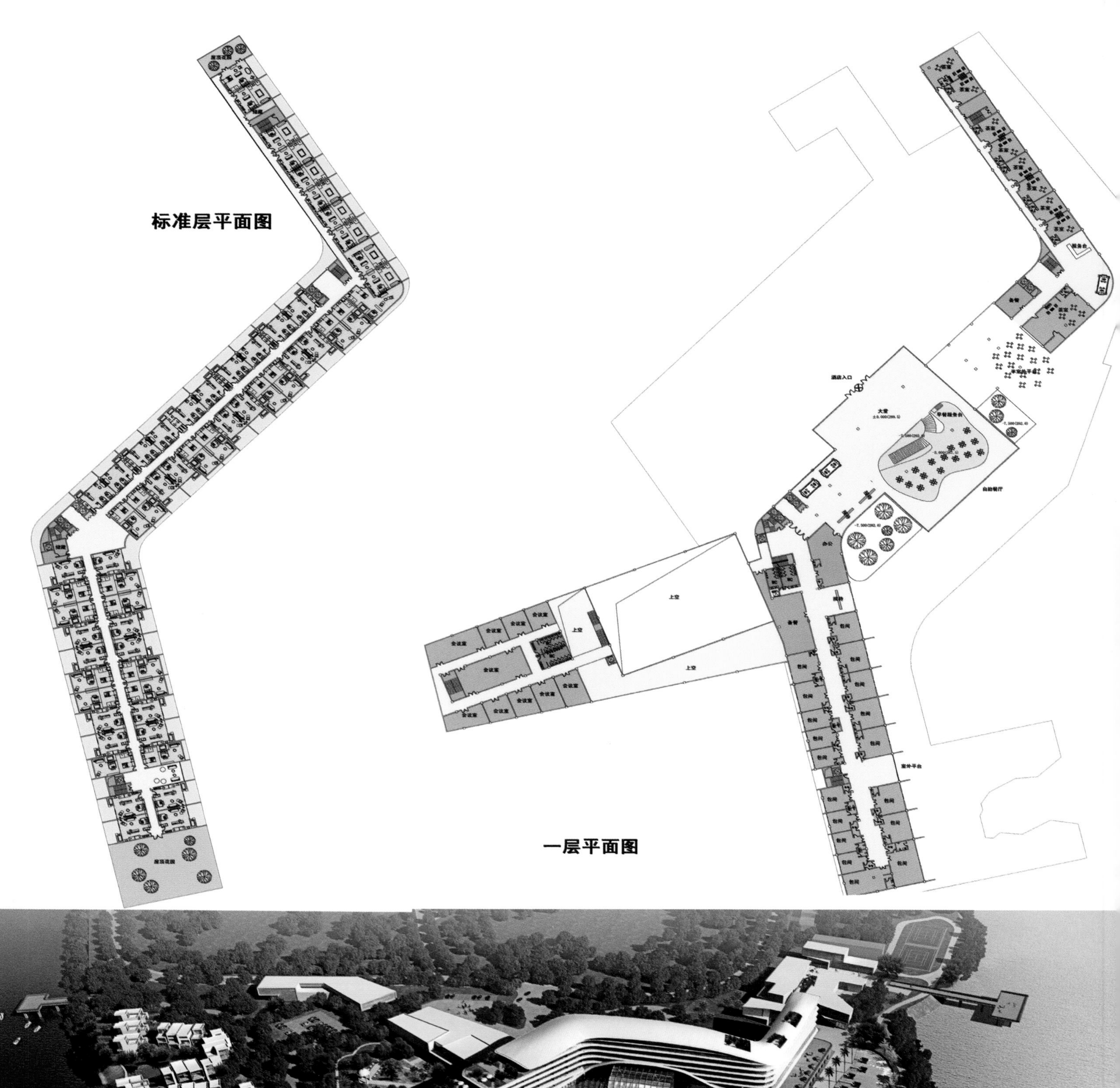
标准层平面图
一层平面图

夹层平面图

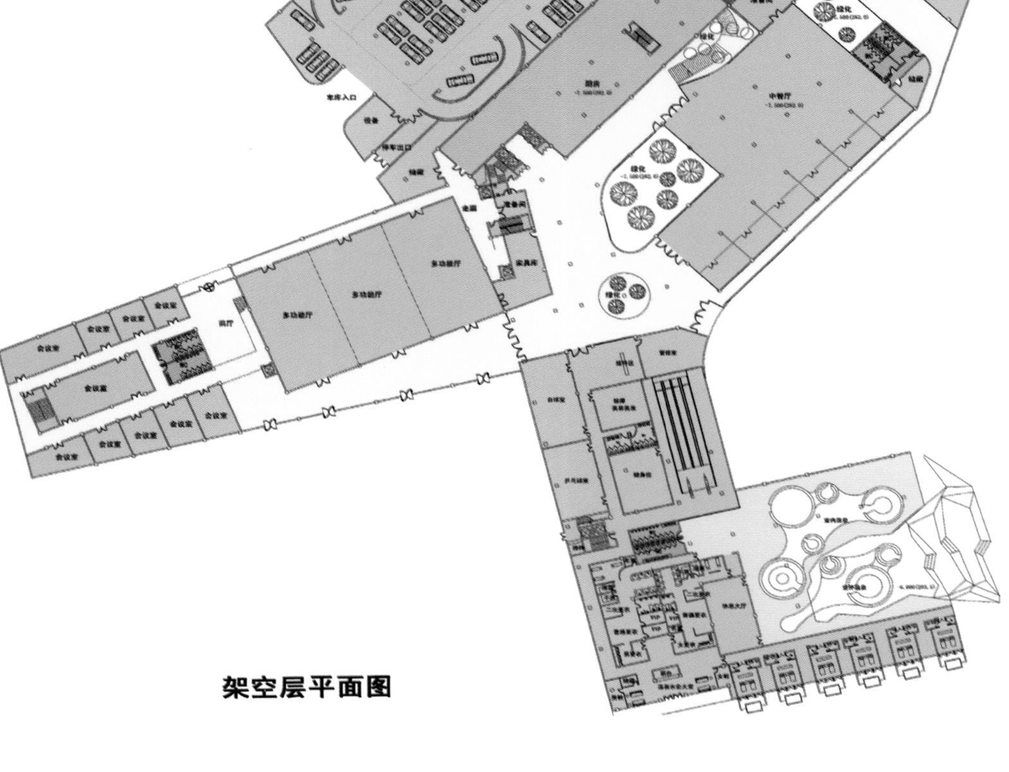

架空层平面图

总平面图

Profile

The project is a five-star hotel covering about 92,000 square meters. It is located at the southeast end of St. Lotus Island surrounded by lakes on three sides and enjoying a superior environment. The design aims to build a high-end leisure travel & holiday resort of southwest China.

Design Concept

The project adheres to a green, ecological and sustainable development. The whole scheme emphasizes the combination of buildings and surrounding water environment, which reflects a perfect integration of building and nature. In the meantime, the planning of the hotel reflects an idea of ecology and energy saving. Fashionable modern elements are combined with characteristic local culture to create this leisure resort hotel.

Architectural Design

The hotel is divided into catering & accommodation area, conference area, high-end villa area and spa area. On functions layout, the design has properly arranged several functional areas based on the main buildings' linear direction and erected convenient access between different functional areas.

The main entrance of the hotel is on a green base five meters above the ground. Raised topography offers fine waterfront views for the hotel. Meanwhile, it makes stepped terrace landscape of the entrance possible and enriches landscape layers of hotel garden.

Dining area situates in the open floor and the ground floor of the hotel. Blurring boundaries fully demonstrates the leisureliness and easiness of the hotel. Various restaurants with linear layout face to open garden and lake, which makes dining more casual and pleasant.

Stores distribute in the north of the site. The public space extending to the hotel combines shopping with holiday relaxation. The entrance of conference area is set on the south green slope of open floor. Multifunctional hall for large banquets and various dimensions of small-medium conference rooms are contained in the conference area.

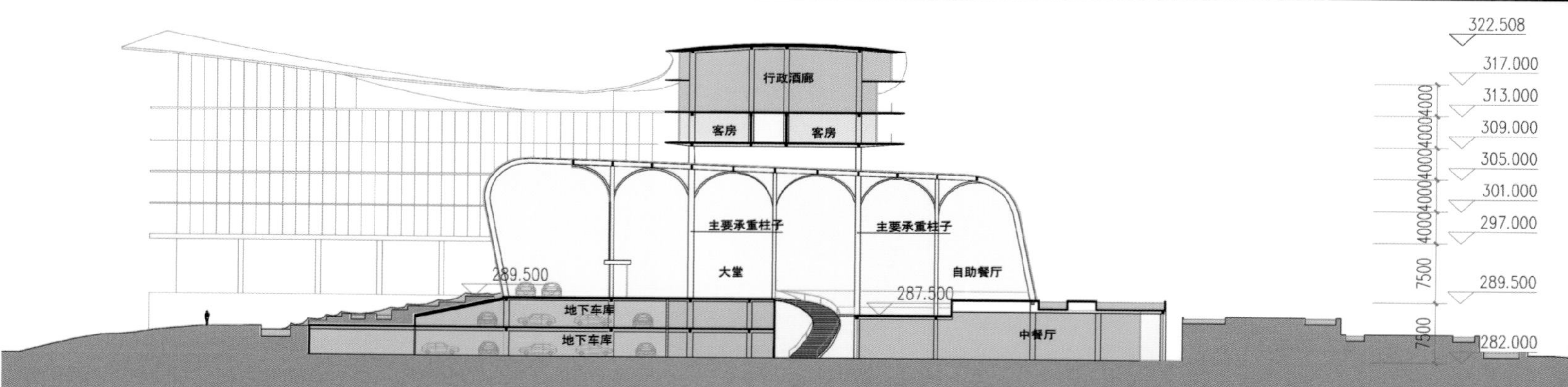

1-1 剖面图

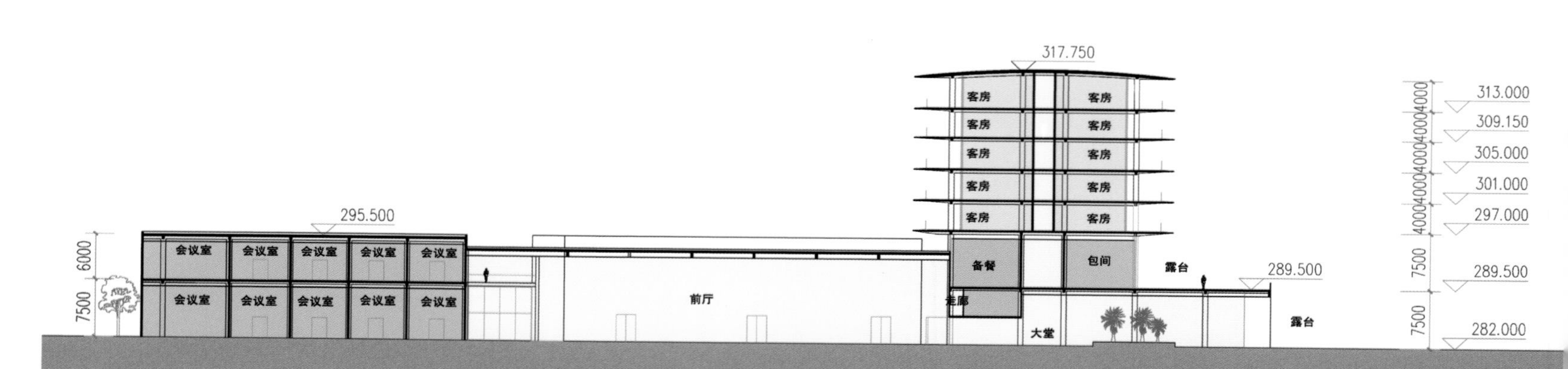

2-2 剖面图

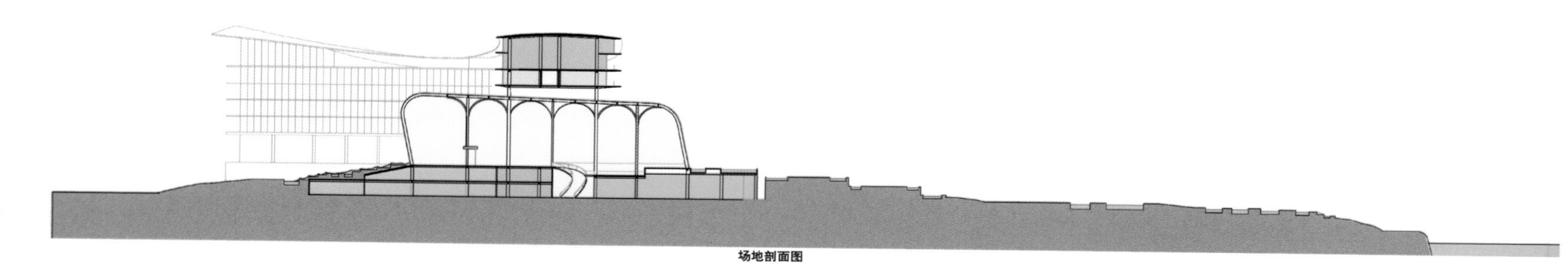

场地剖面图

①

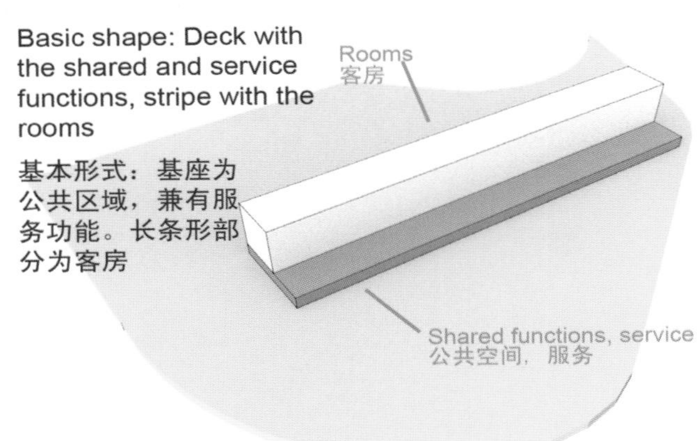

②

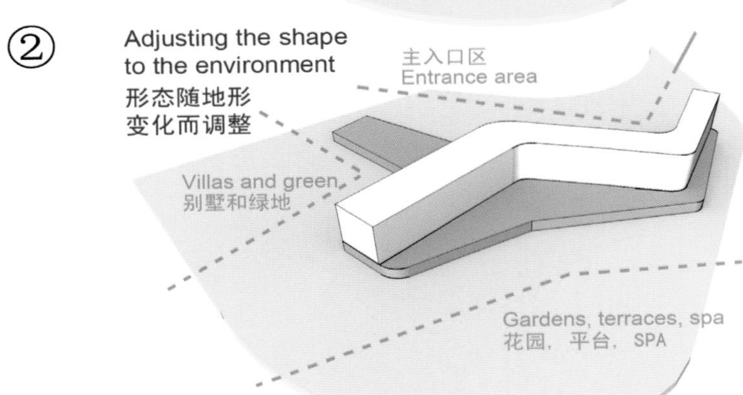

③

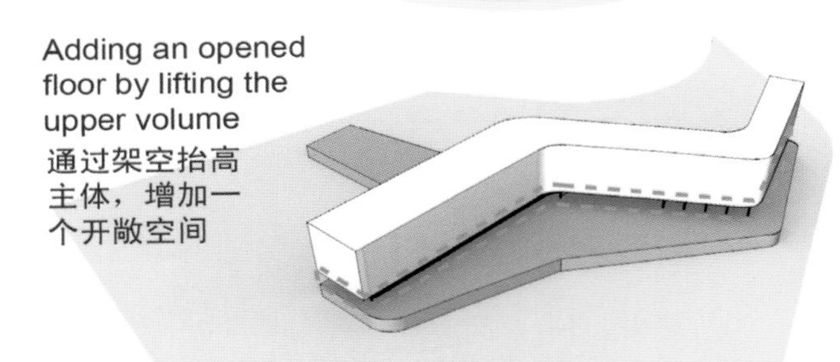

④

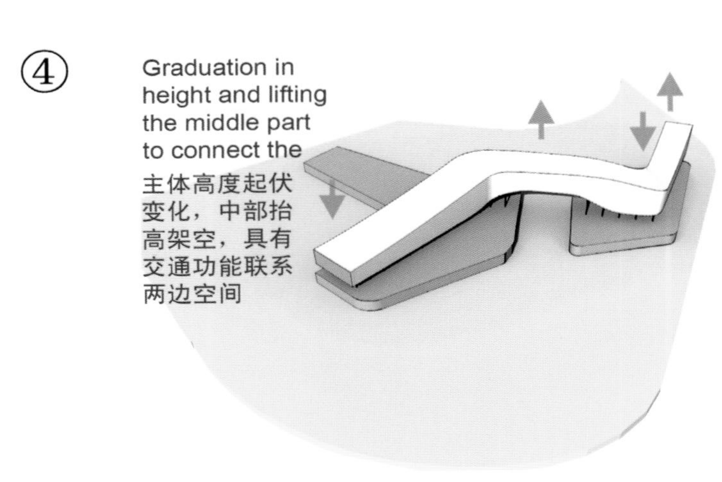

⑤

Adding volume for lobby and water architecture

中间增加水景观的大体量，作为大堂。

交通分析

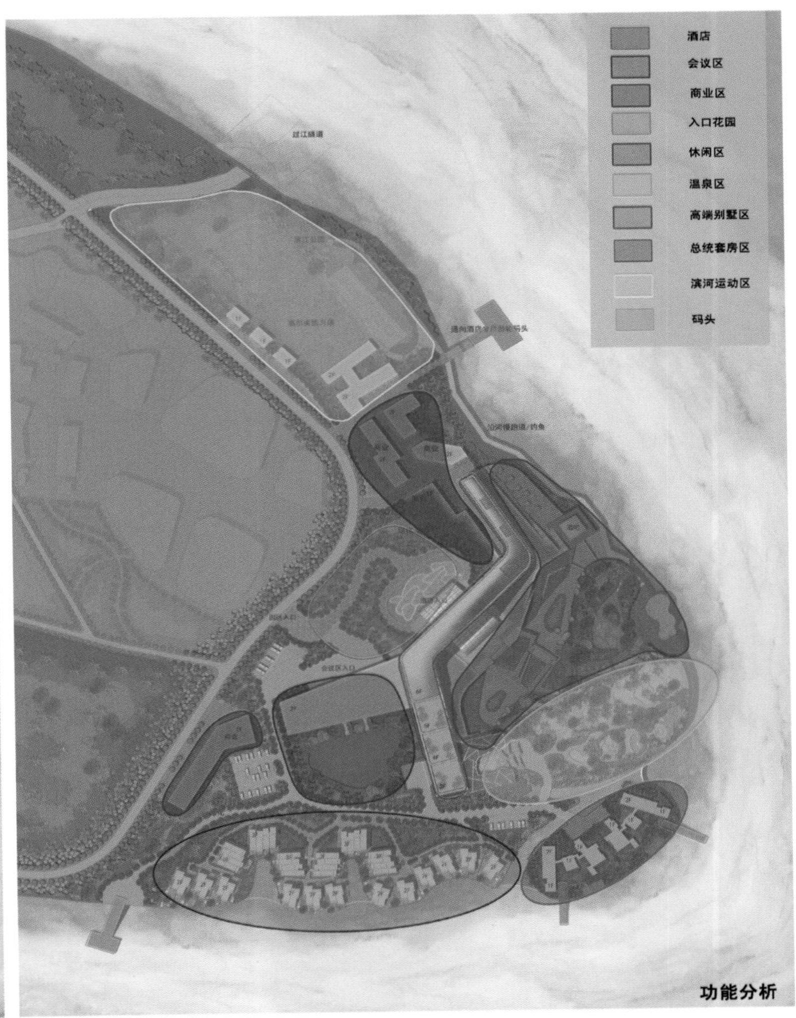

功能分析

购物中心
Shopping Center

利比亚的黎波里 Maitiga 购物中心

Maitiga Mall

设计单位：Barbosa & Guimarães
开发商：Alinmaa Construction and Real Estate
项目地址：利比亚的黎波里

Designed by: Barbosa & Guimarães
Client: Alinmaa Construction and Real Estate
Location: Tripoli, Libya

项目概况

Maitiga 购物中心是 Alinmaa 公司计划在利比亚的黎波里构建的一栋建筑。考虑到当地没有一个导向性的建筑，设计构建了一座当代的建筑和“绿洲”作为标志和指向当地景观的向导。

建筑设计

项目基地平坦，却有着不规则的形态，且高于水平面 4 米。这个购物中心的场地曾是一级方程式赛车的环道，靠近连接了的黎波里中心区和 Tajura 地区的滨海大道。

购物中心的设计概念以“汽车”为主题，设计提出的空气动力学的建筑结构也源自汽车设计概念。建筑共有三层，其地上的建筑面积达 55 000 平方米。建筑动态的外观不仅将设计理念形式化，而且确保了室内类“8”字形交通环线的高效运作，这一环线是穿过所有购物区的理想路径。

这个象征主义的商业建筑严格遵循了既定的空间格局和功能需求，在没有构建第二空间的基础上，通过连续环道上的路径，为使用者提供了一条流畅的交通流线，并将员工流线和客流路线在功能上区分开来。设计还在建筑的西侧设置了一个可容纳 750 个停车位的停车场和卡丁车赛场 。

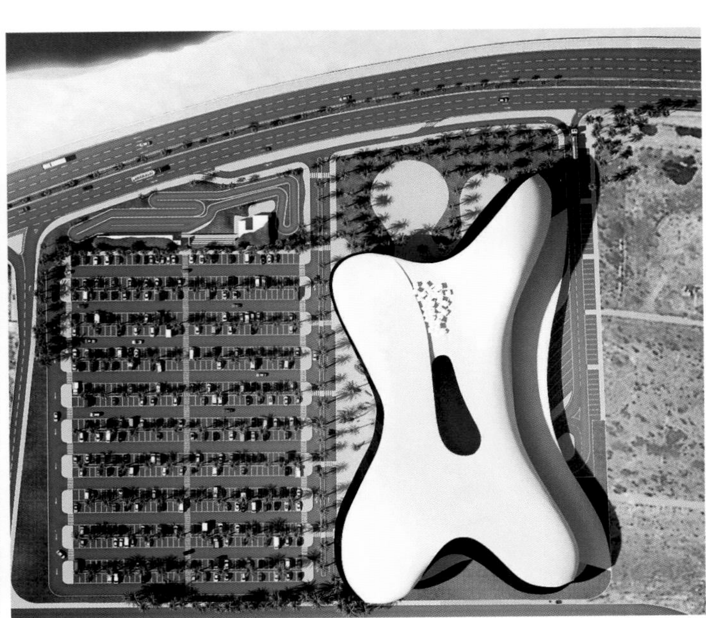

MAITIGA
MAITIGA
MAITIGA
MAITIGA
MAITIGA

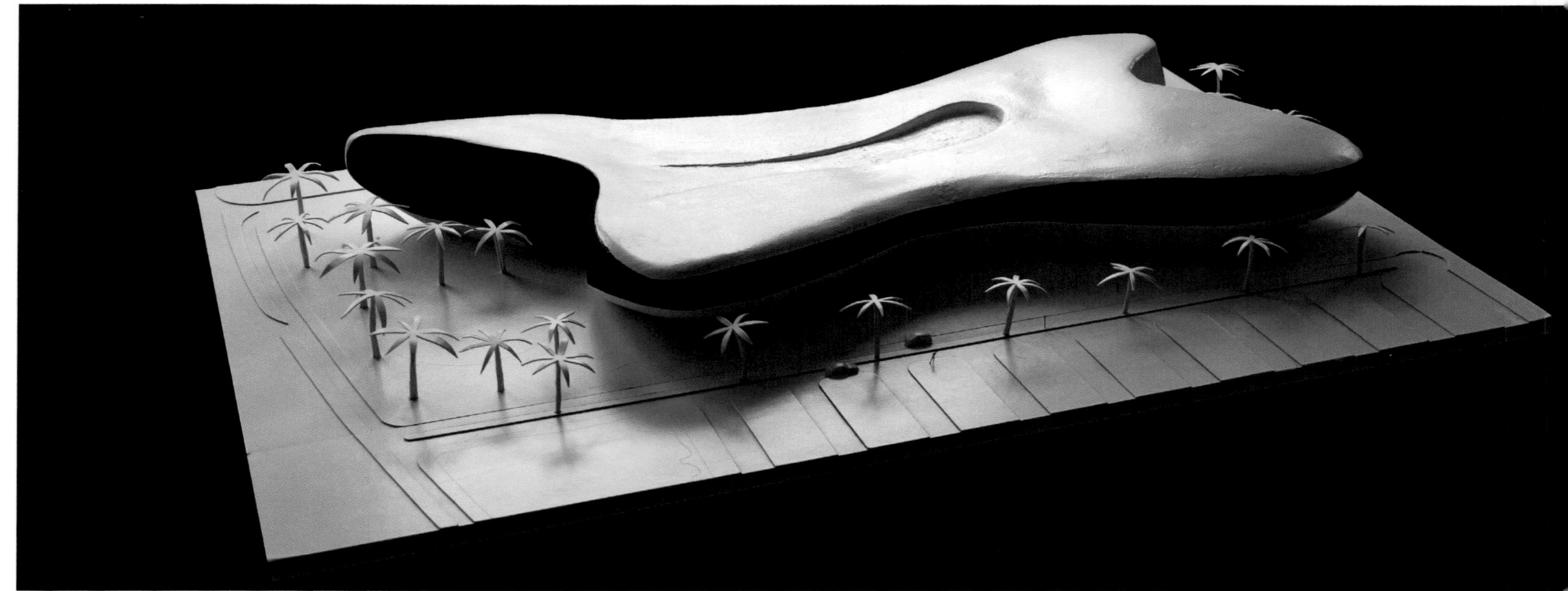

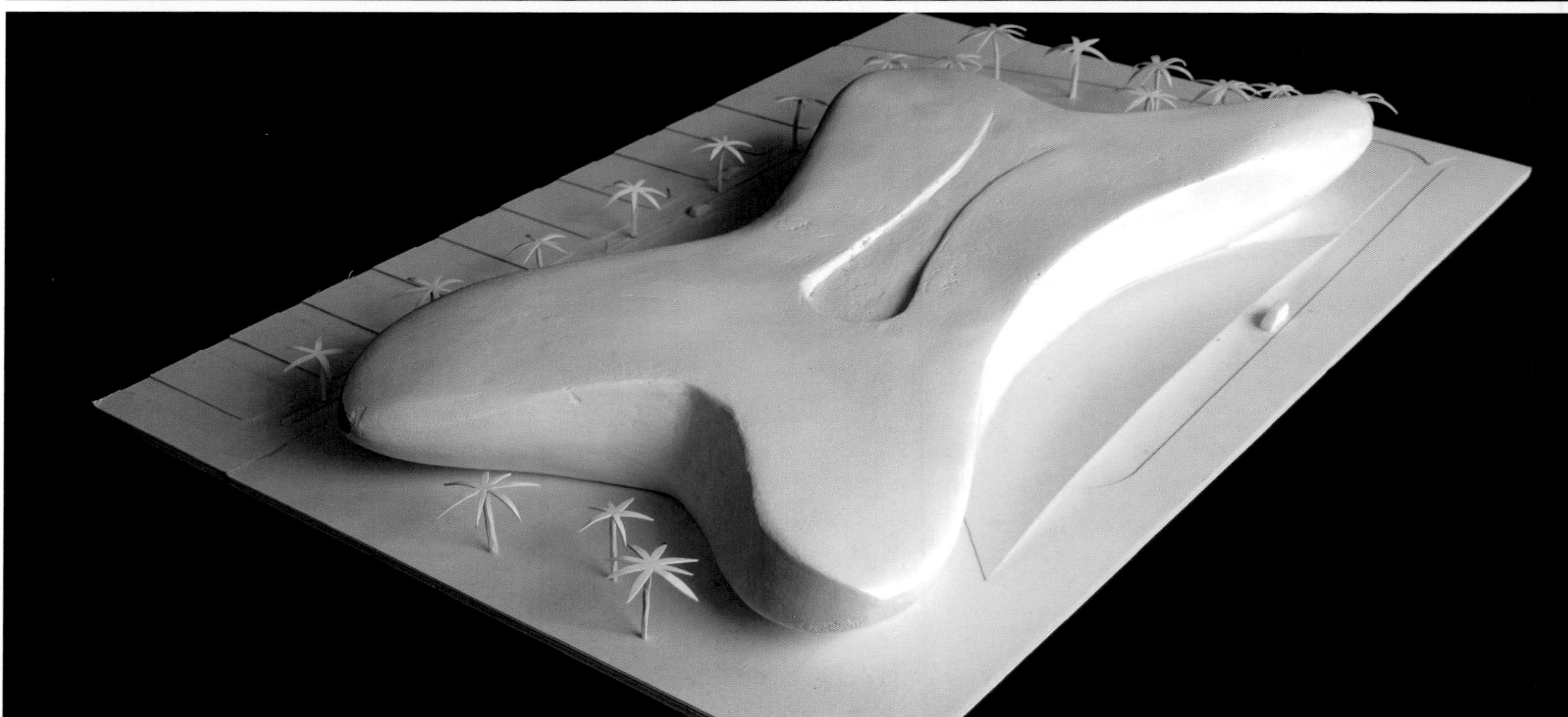

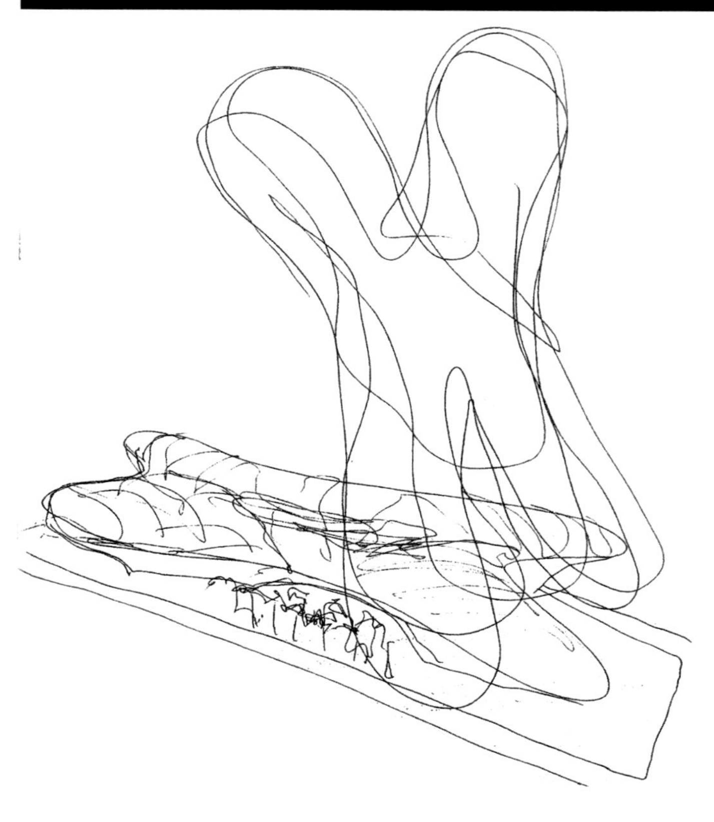

Profile

Maitiga Mall is a Shopping Center that the Alinma Company intends to build in Tripoli, Libya. Given th absence of references in the nearby surrounding designers propose to create a "green oasis" with contemporary building that works as an icon and as reference to the local landscape.

Architectural Design

The site is flat, with an irregular shape, and it is meters above the ground. The plot for the Shoppin Center, in the past, was used as a Formula On automobile circuit, next to the coastal road that links th center of Tripoli and Tajura.

The concept of the Mall is based on the motorca theme; therefore designers propose an aerodynam

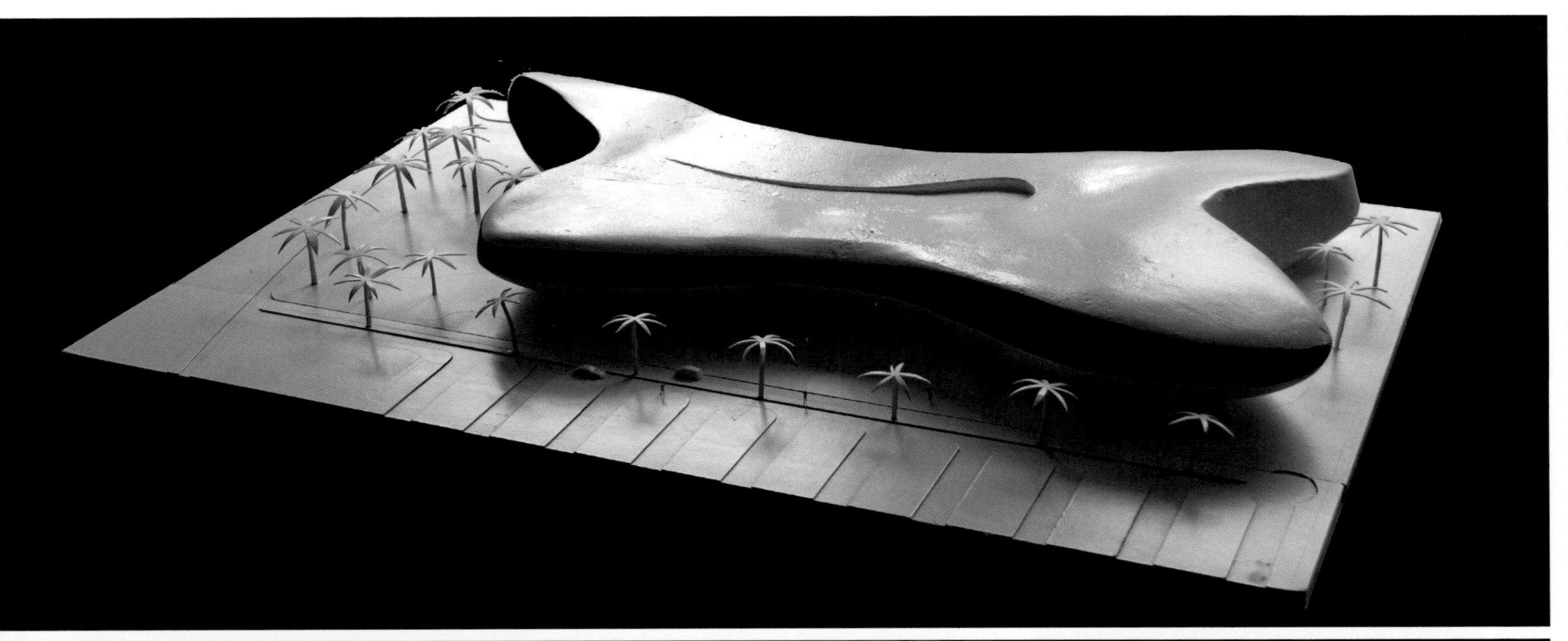

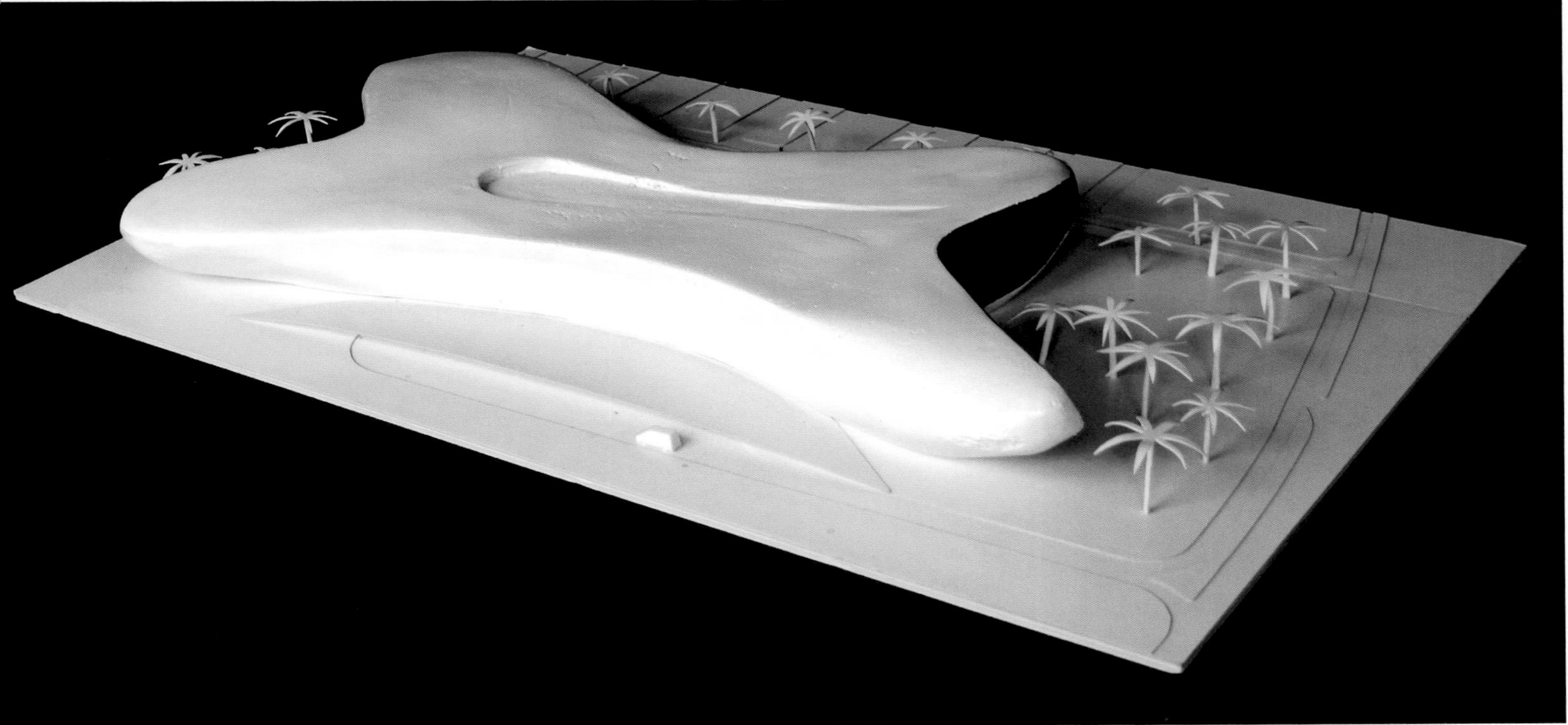

shape for the building inspired on concept cars. The three-storey building has a construction area of 55,000 square meters above the ground. The dynamic shape of the building, besides fulfill the formalization of an idea, works effectively by allowing a loop ("8") circulation in the interior, which is the ideal way to walk through shopping areas.

The mall was developed in accordance with space organization and functioning rules well defined for this typology of commercial buildings, which allow a fluid circulation of the users, without creating secondary spaces and with paths in continuous loop. Outside, on the west side of the building, there is a parking lot for 750 cars and a karting track.

瑞典马尔默 Hyllie 购物中心

Hyllie Shopping Center

设计单位：C.F.Møller Architects
开发商：Steen & Strøm
项目地址：瑞典马尔默
项目面积：120 000 ㎡

Designed by: C.F. Møller
Client: Steen & Strøm
Location: Malmo, Sweden
Area: 120,000 m^2

项目概况

Hyllie购物中心是由 C. F. Møller Architects 在马尔默新城区设计的一个极具声望的项目，包括 70 000 平方米的购物中心和 50 000 平方米的住宅和办公空间。

设计理念

设计的主要意图是构建一个有秩序的沿街序列，仿佛一条蜿蜒的河流。这一“河流”自购物中心之外的新车站延伸至西侧的主入口区，穿过购物中心后到达在其另一侧的规划湖。

设计特色

这一设想使这个购物中心形成了清晰的结构布局和流畅的空间序列，同时也赋予了建筑独特的身份象征。蛇形般蜿蜒的通道分布在各个楼层，从郁郁葱葱的绿色屋顶花园延伸至各个空间，几乎覆盖了购物中心的表面。由此形成的这一户外休闲景观可经由多个方位到达，包括整合后的多层停车场。

“河流”的物理运动通过内部通道的覆盖率来表现。如镜般的钢板保护层在抬升的街道下弯曲成变形的“镜子”，既加强了购物中心的视觉体验，同时也创造了一个独特视图的内部空间。

为了打破传统购物中心既有的结构和单调的外观，设计在与新城区主要公共空间接壤的地块设置了一系列的住宅和办公设施。设计巧妙地利用了建筑的朝向和方位，使顾客能快捷地进入中庭和屋顶花园。

Hovedindgang
Spejlbassin
Anlægt sø
Arena
Hotel
Svea Torg
Bolig vest
Bolig øst
Erhverv
Hotelgatan
Parkering 250 pladser
Parkering for Arenaen
Boliger

Hyllie
H&M

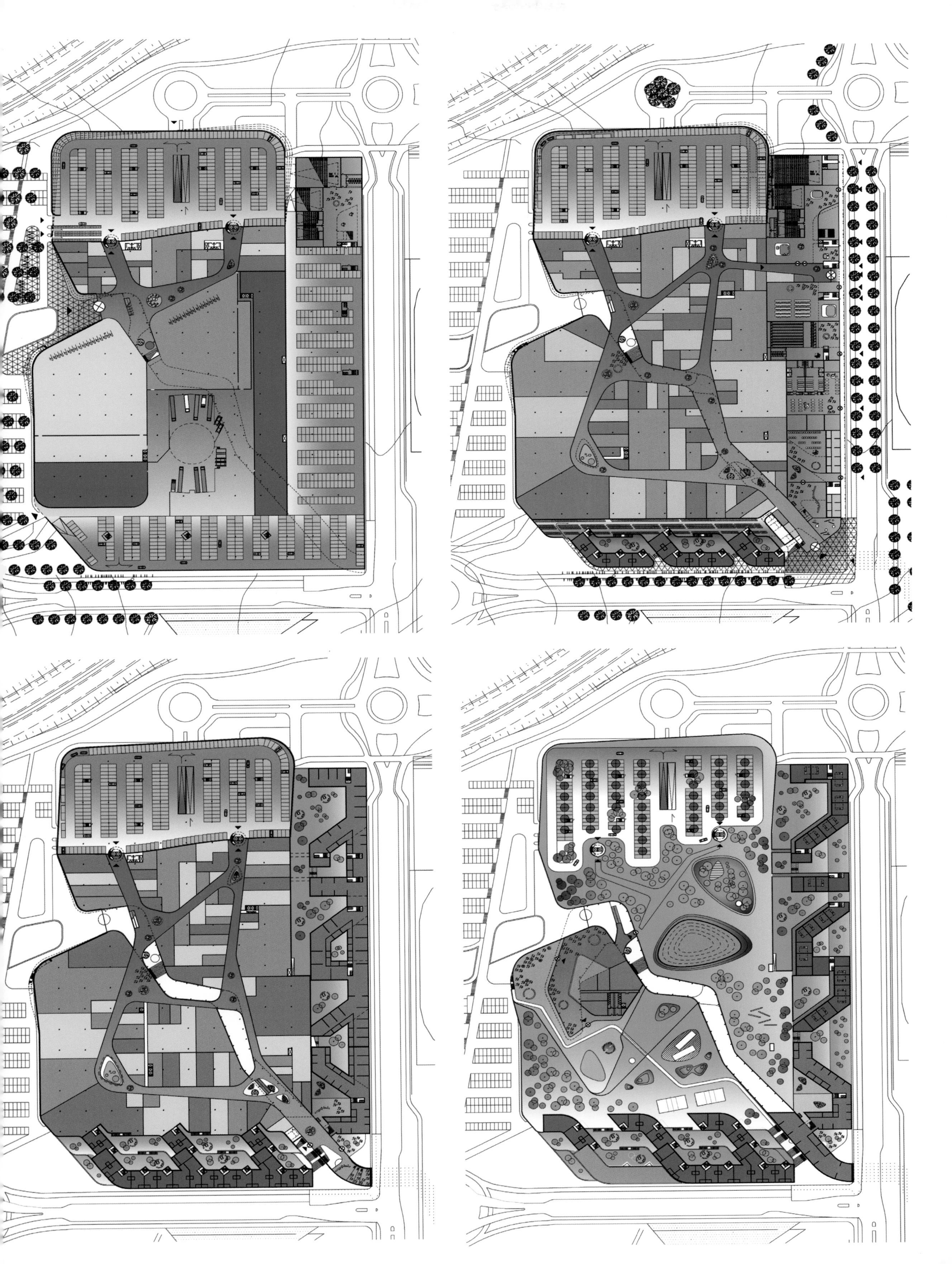

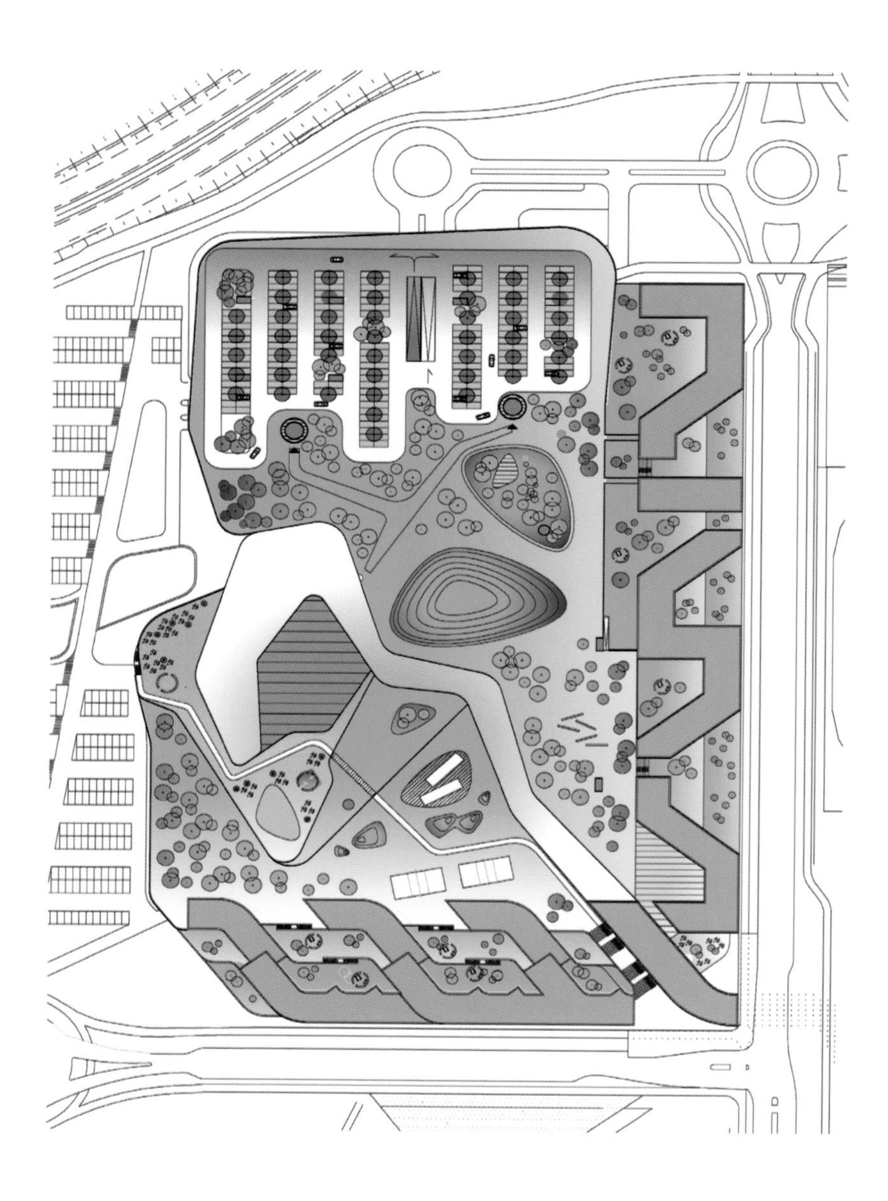

Profile

Hyllie Shopping Centre is C. F. Møller Architects' proposal in the competition for a prestige project in Hyllie, Malmo's newest urban district, a 70,000 square meter shopping centre and 50,000 square meters of housing and offices.

Design Concept

The main idea in the project proposal is a covered street sequence, which winds like a river. The source of the river is outside the centre at the new station building, from where it flows down to the centre's main entrance area to the west and thence onwards through the shopping centre to a planned lake outside the centre on the other side.

Design Feature

The idea provides the shopping centre with a clear layout, fluid space-sequences and at the same time gives identity to the centre. The snaking form of the street becomes part of the design at all levels, all through to the lush green roof-garden covering most of the centre's surface, creating an outdoor leisure-landscape accessible from numerous angles, including the integrated multi-storey car-park. The physical manifestation of the "river" is represented by the internal street's covering. The mirror-steel clad, curved underside of the raised covering becomes a distorting mirror, which manifolds the visual experience of the centre, and creates surprising perspectives of the interior.

To avoid the typically blank and introvert appearance of a traditional shopping centre, the sides bordering on the main public spaces of the new urban district are lined with a series of housing and office developments, cleverly manipulated to take advantage of the orientations and options to access courtyards and roof-gardens.

erhverv
foodcourt
foodcourt
boliger
Niv.03 - kt 36.0
Niv.02 - kt 30.0
torv
butikker
indgang
Niv.01 - kt. 24.0
SALE
butikker
lager
teknik
parkering

Alex Es
CHANCE

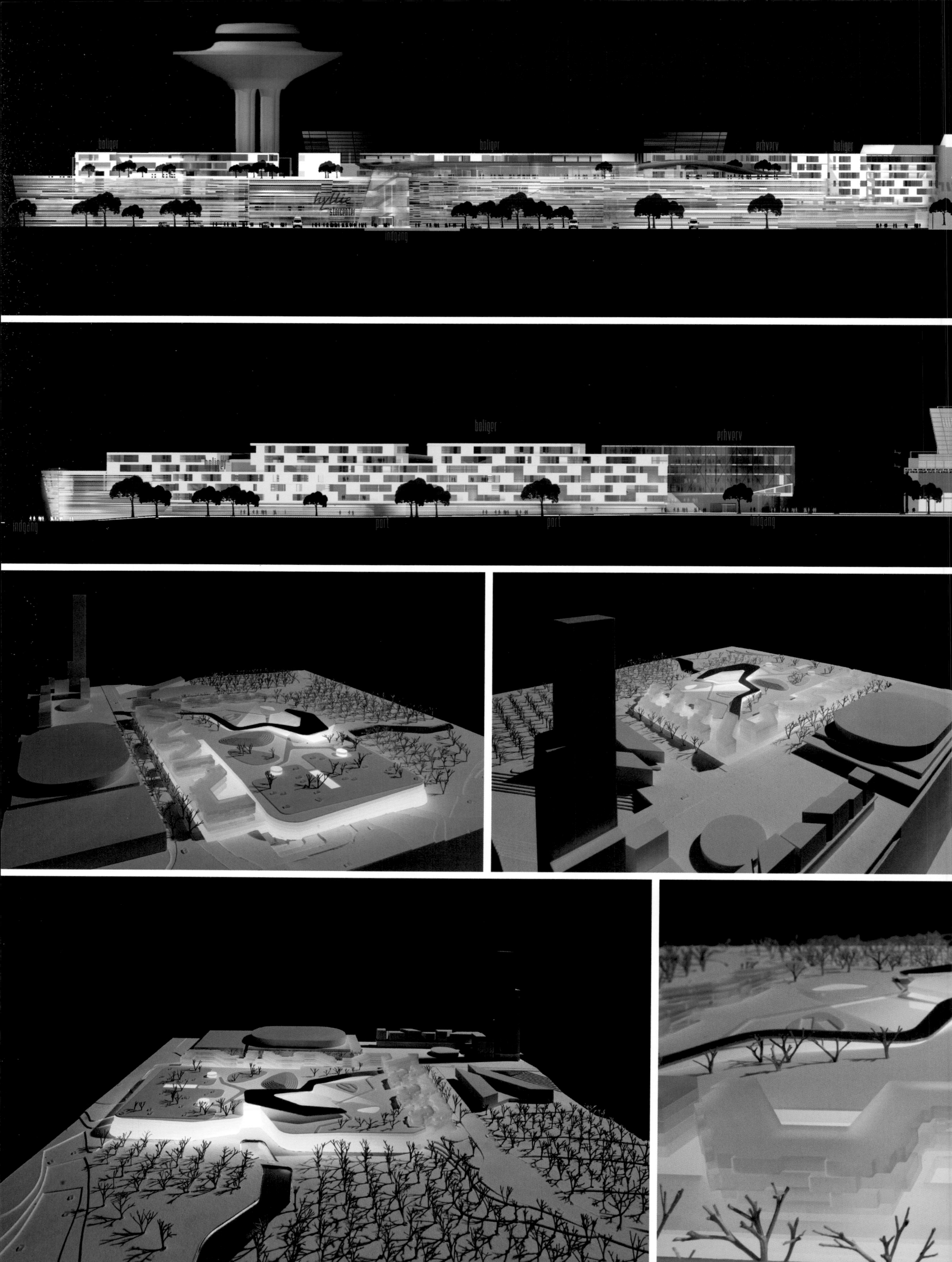
boliger
boliger
erhverv
boliger
hyllie
storcenter
indgang
boliger
erhverv
boliger
indgang
port
port
indgang

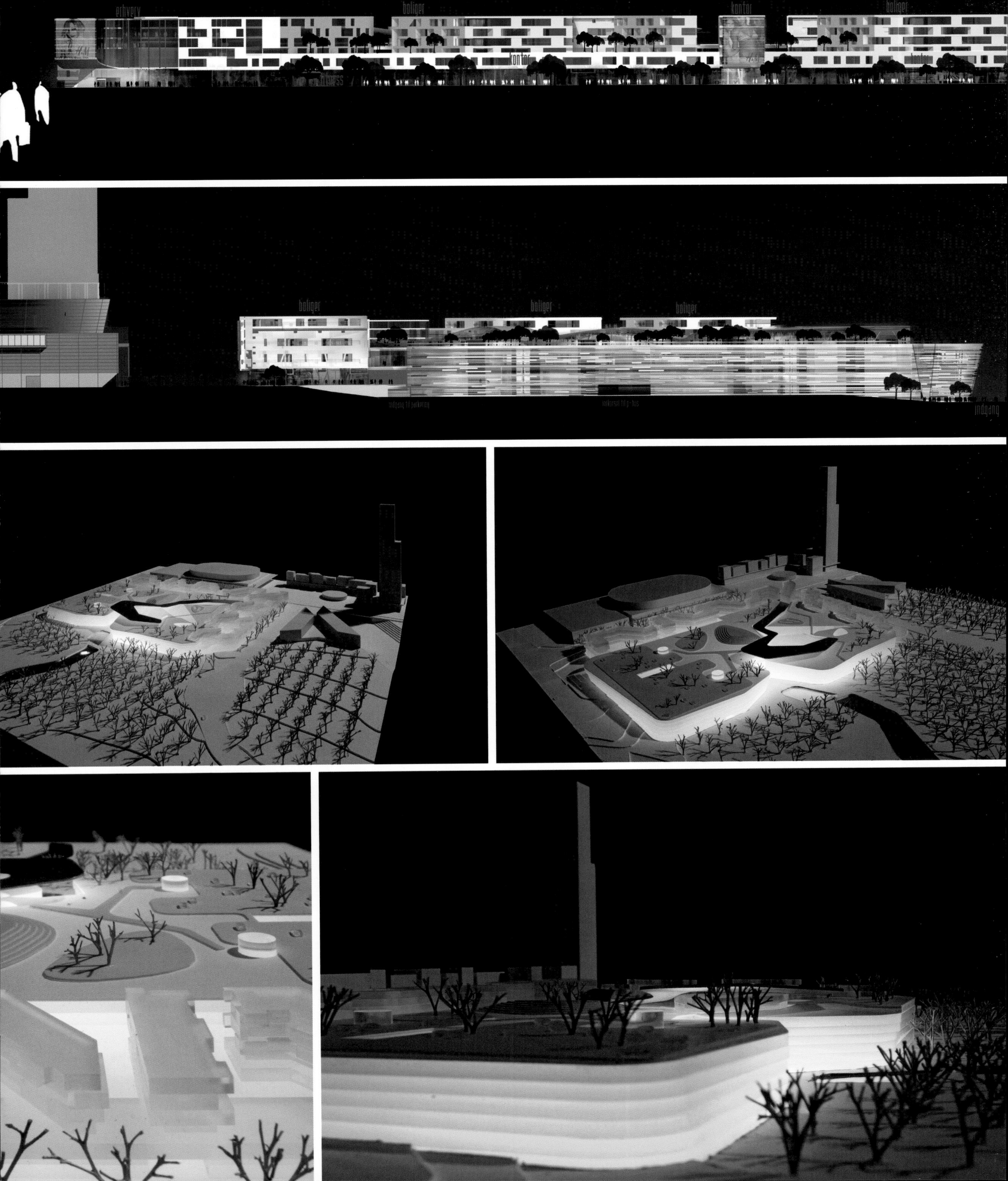
erhverv
boliger
kontor
boliger
kontor
fitness
kontor
boliger
boliger
boliger
indgang til parkering
indkørsel til p-hus
indgang

辽宁沈阳千姿汇购物中心

Liaoning Shenyang Qianzihui Shopping Center

设计单位：北京世纪安泰建筑工程设计有限公司
开发商：沈阳麟承天和置业有限公司
项目地址：中国辽宁省沈阳市
总用地面积：9 657.06 ㎡
总建筑面积：44 700 ㎡
建筑密度：62%
容积率：2.38

Designed by: Beijing SJAT Architecture & Engineering Design Company
Developer: Shenyang Lincheng Tianhe Real Estate Co., Ltd.
Location: Shenyang, Liaoning, China
Site Area: 9,657.06 m^2
Floor Area: 44,700 m^2
Building Density: 62%
Plot Ratio: 2.38

项目概况

千姿汇购物中心位于沈阳市铁西区兴工北街与虹桥路交叉口的西南角，便利的交通为这个项目创造了有利的条件。

设计理念

设计坚持了“以人为本”的设计原则，密切关注顾客的心理和行为方式，同时强调商业空间的文化性、公共性、舒适性，以创造一个富于人性化的、开放的室内外购物休闲空间。

建筑设计

千姿汇购物中心是面向所有公众开放的商业建筑，它的效益首先应建立在人流、车流、物流高效聚集的基础之上。设计经过精心组织，最大限度地减少人流与车流的交叉，将购物车流与服务车流尽量分离，使之都有各自明确通畅的流线。

作为一个商业综合体，沈阳千姿汇购物中心充分考虑了城市设计的原则。为了突出项目的地标性，设计采用了富于现代感的折面造型。主体建筑顶部标志性的 LED 旋转大屏幕和主体墙面上 800 平方米的、连续动态的 LED 显示屏，作为千姿汇购物中心的标志，无论在白天还是夜晚，都将成为建筑群中的视觉焦点。

商业设计

设计以商业内街的形式组织各功能空间，通过商业步行街、喷泉、下沉广场和人行天桥将繁杂的功能区有机地组织成一个整体，既丰富了室内外购物空间，又有效解决了交通疏散等问题。

设计充分考虑了商业建筑的文化性，设计师将 4、5 层的影院局部空间设置为徐悲鸿艺术人生画廊，使人们在购物、看电影之余，还能更全面地了解徐悲鸿大师的艺术人生。

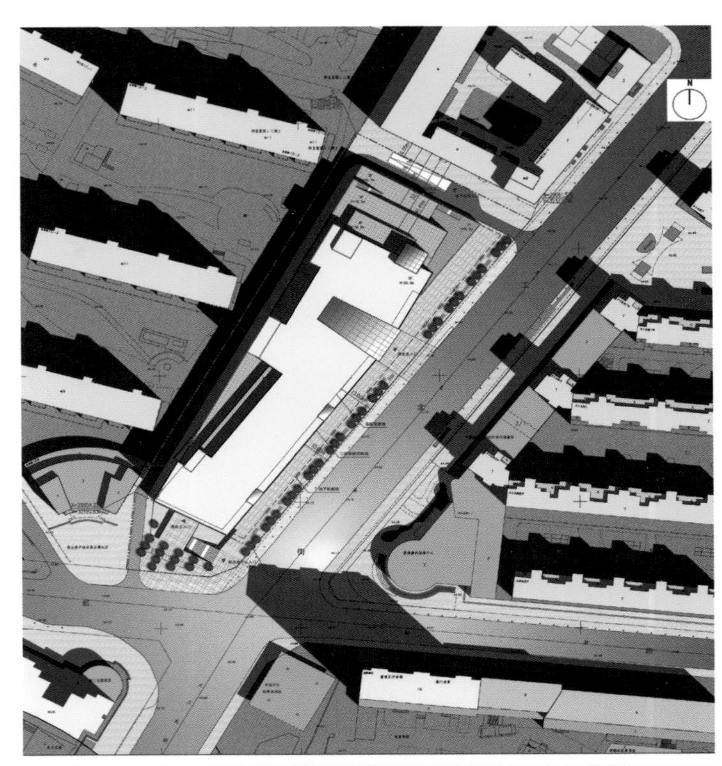

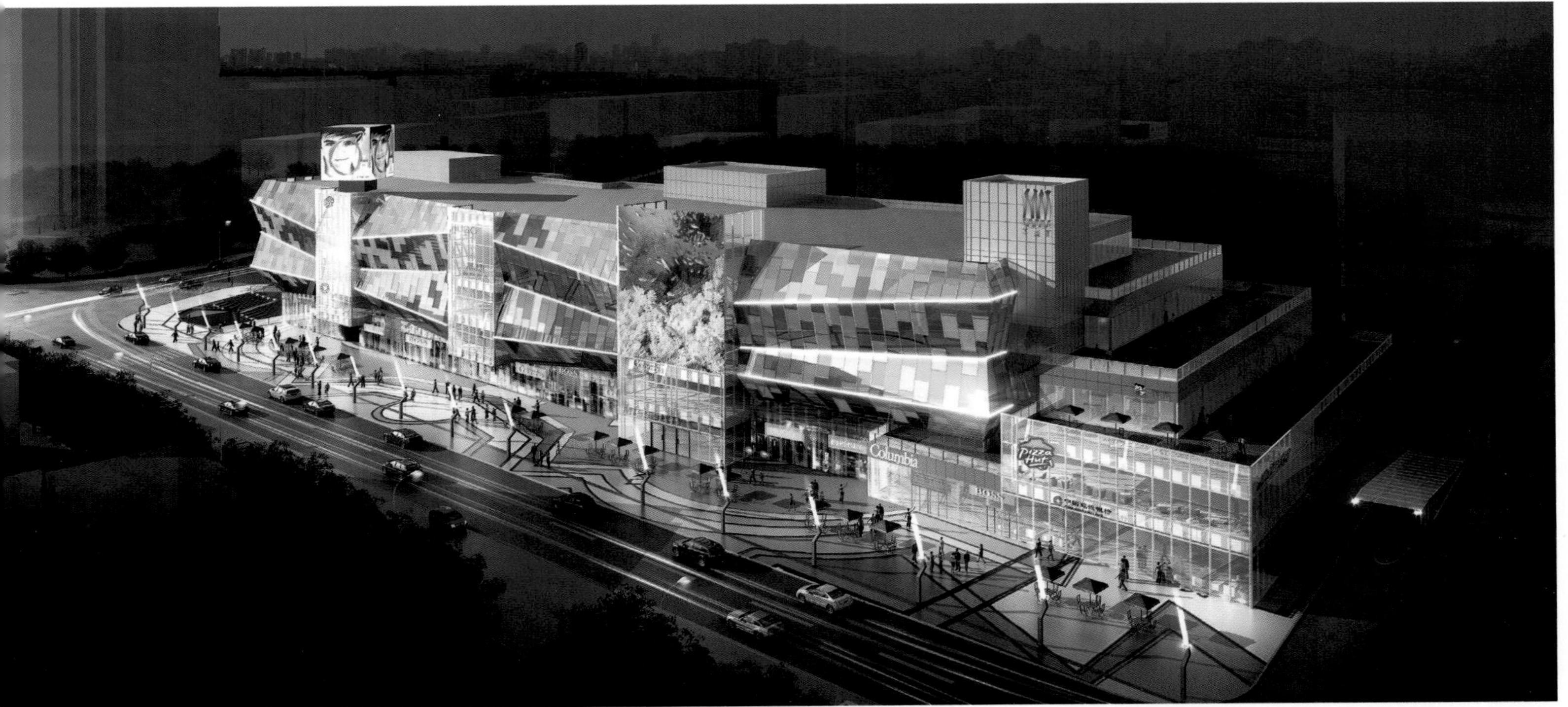

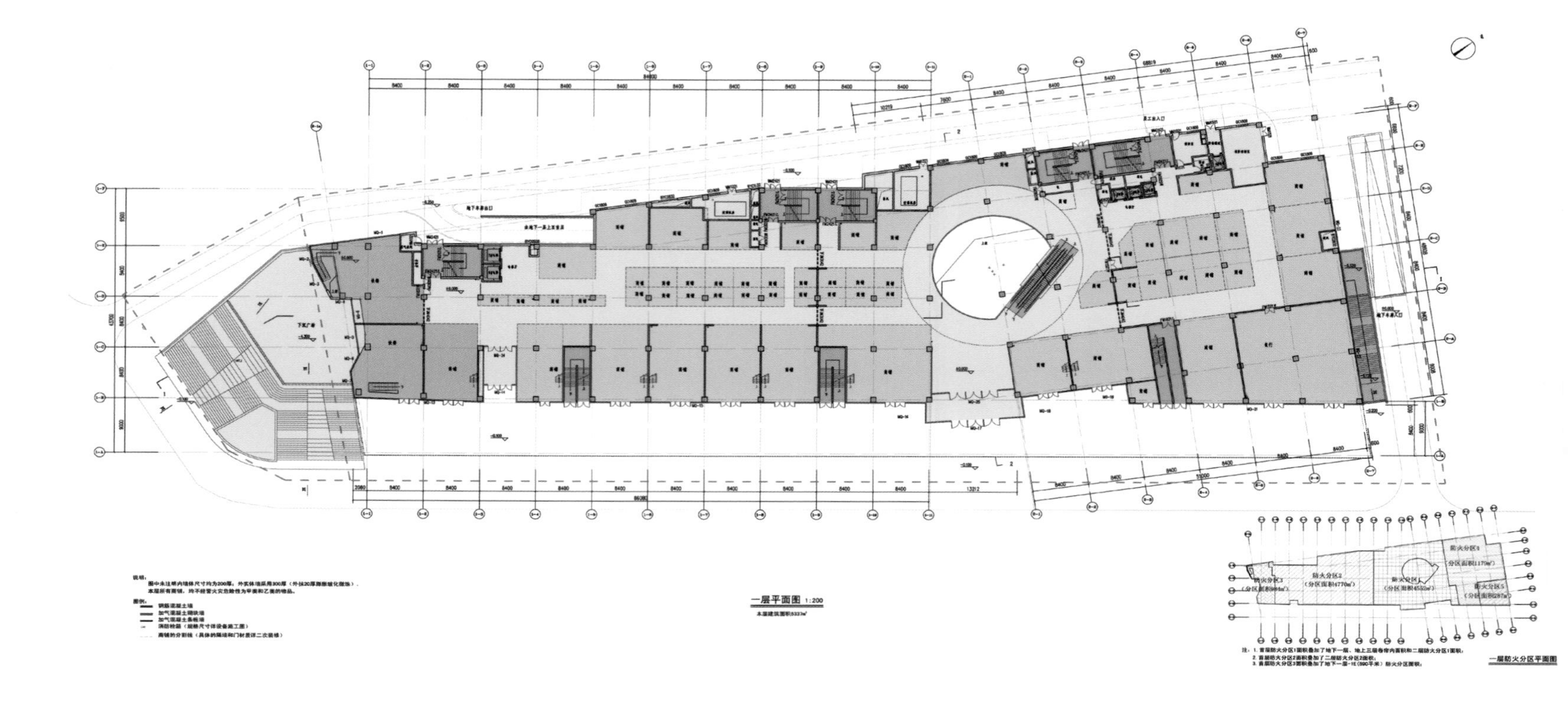
说明：
图中未注明内墙体尺寸均为200厚；外实体墙采用300厚（外抹20厚膨胀玻化微珠）。
本层所有商铺，均不经营火灾危险性为甲类和乙类的物品。
图例：
钢筋混凝土墙
加气混凝土砌块墙
加气混凝土条板墙
消防栓箱（规格尺寸详设备施工图）
商铺的分割线（具体的隔墙和门材质详二次装修）
一层平面图 1:200
本层建筑面积5337m²
防火分区3
防火分区2
（分区面积4770m²）
防火分区1
（分区面积4552m²）
防火分区4
（分区面积1179m²）
防火分区5
（分区面积287m²）
注：1. 首层防火分区1面积叠加了地下一层、地上三层卷帘内面积和二层防火分区1面积；
2. 首层防火分区2面积叠加了二层防火分区2面积；
3. 首层防火分区3面积叠加了地下一层-1E（590平米）防火分区面积；
一层防火分区平面图

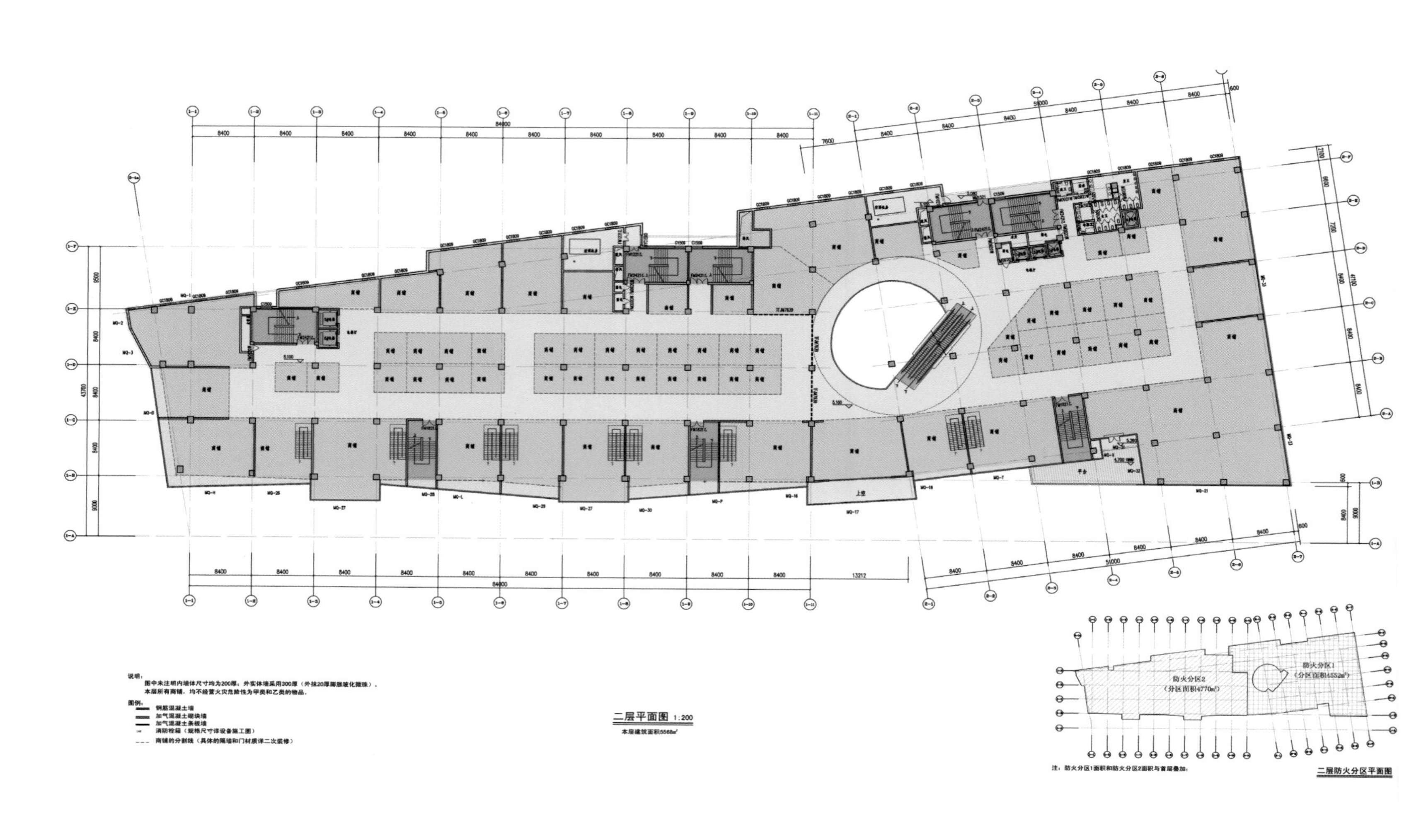
说明：
图中未注明内墙体尺寸均为200厚；外实体墙采用300厚（外抹20厚膨胀玻化微珠）。
本层所有商铺，均不经营火灾危险性为甲类和乙类的物品。
图例：
钢筋混凝土墙
加气混凝土砌块墙
加气混凝土条板墙
消防栓箱（规格尺寸详设备施工图）
商铺的分割线（具体的隔墙和门材质详二次装修）
二层平面图 1:200
本层建筑面积5568m²
防火分区2
（分区面积4770m²）
防火分区1
（分区面积4552m²）
注：防火分区1面积和防火分区2面积与首层叠加；
二层防火分区平面图

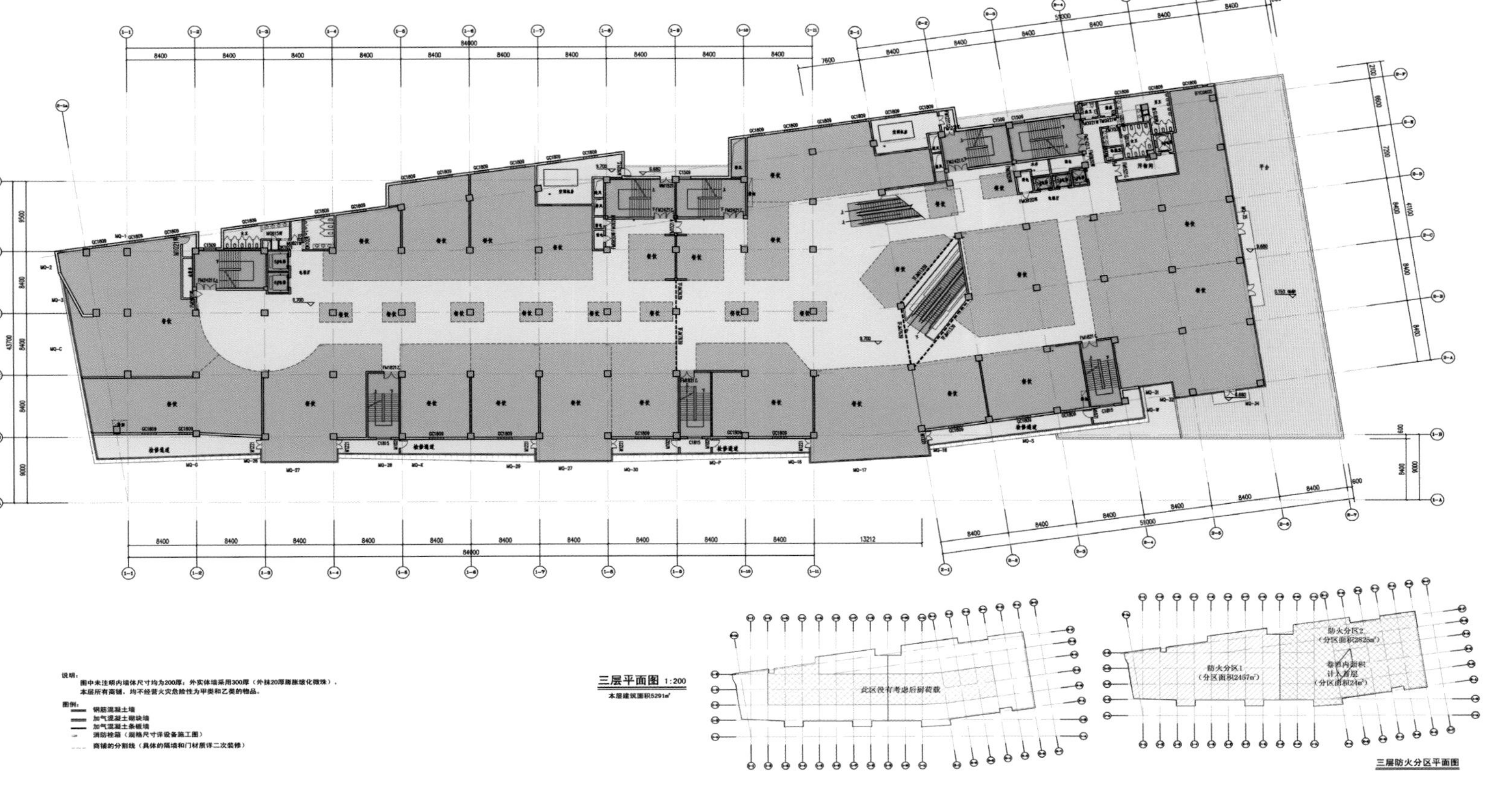

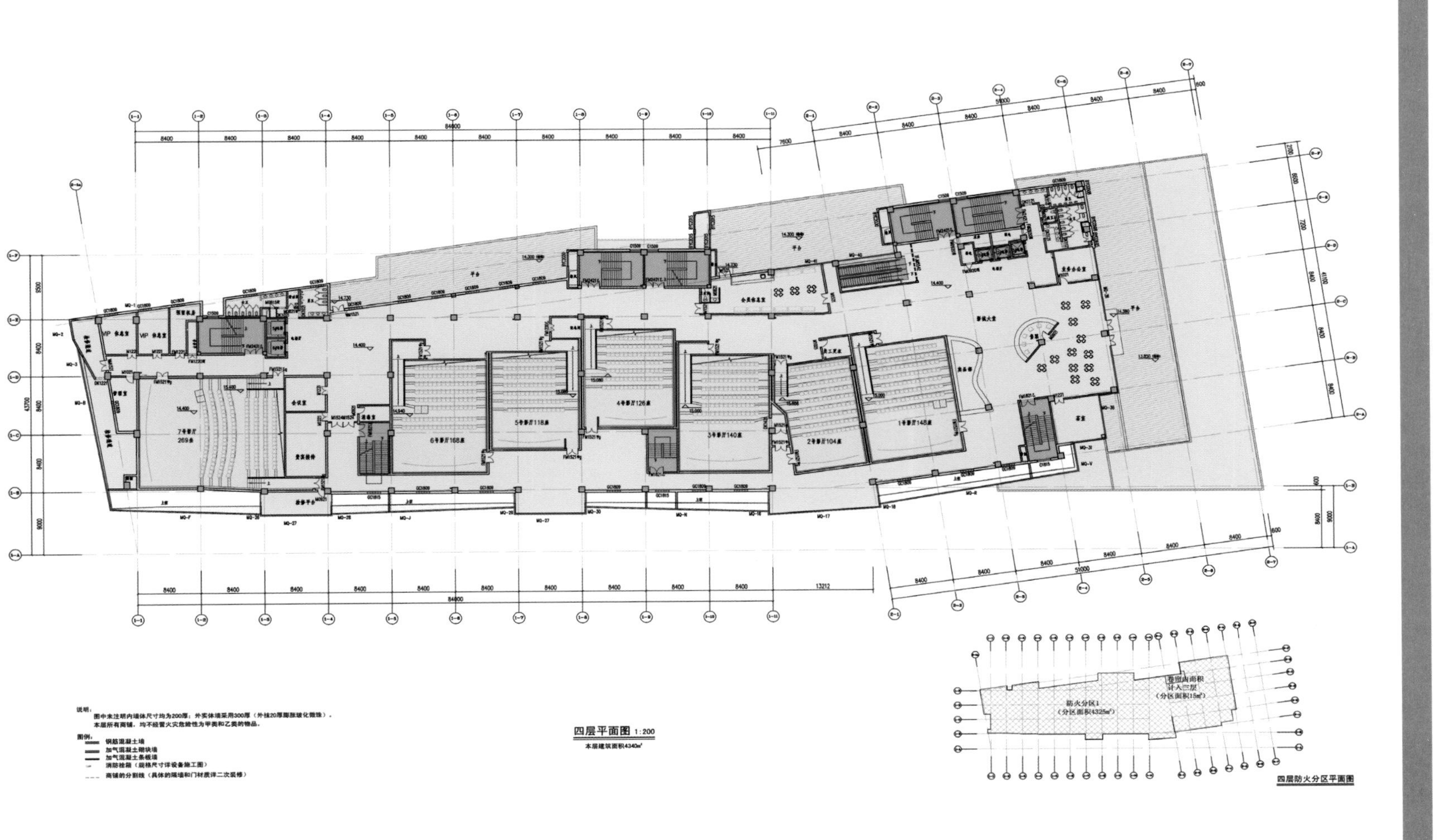

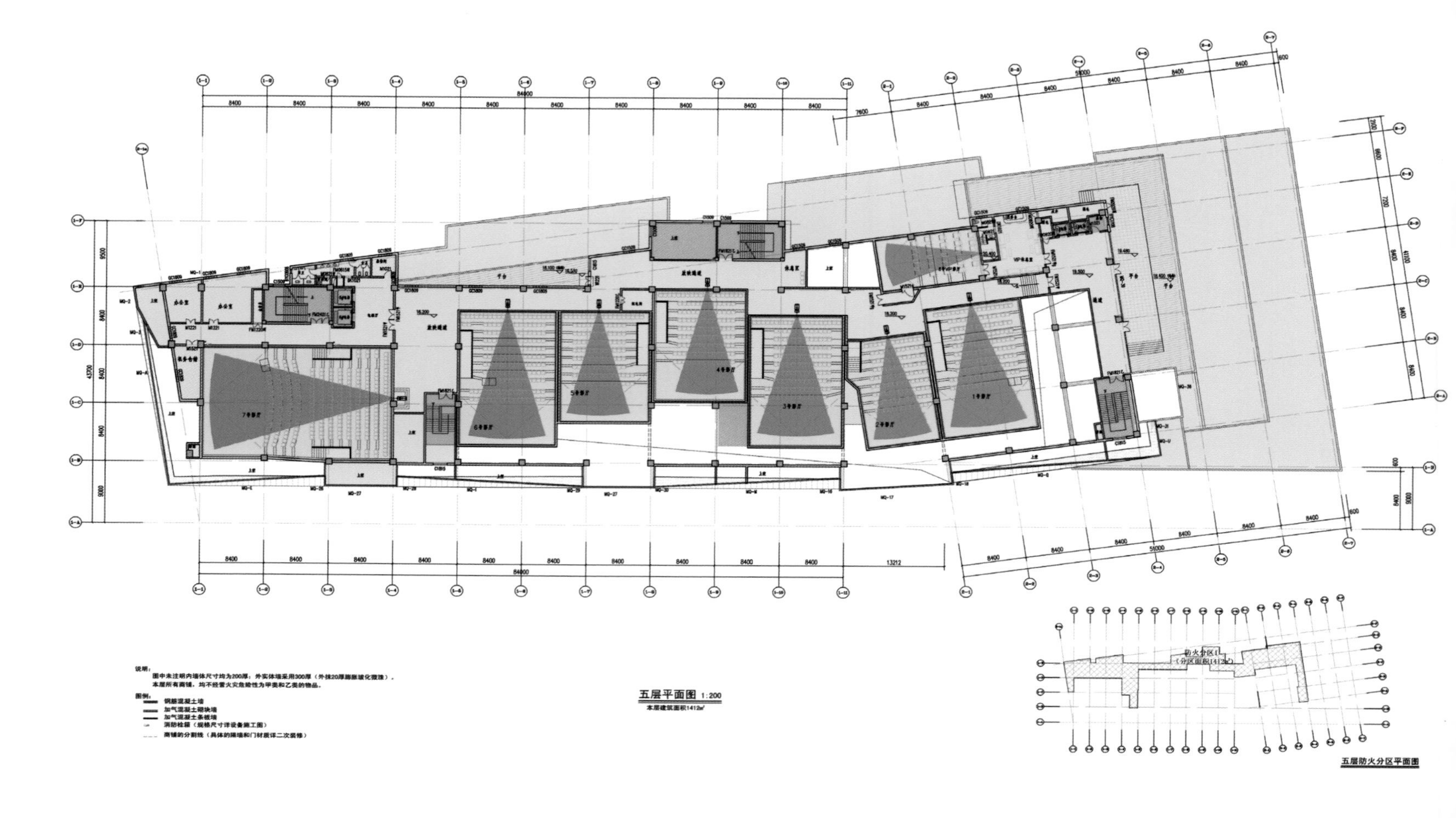
说明：
图中未注明内墙体尺寸均为200厚；外实体墙采用300厚（外抹20厚膨胀玻化微珠）。
本层所有商铺，均不经营火灾危险性为甲类和乙类的物品。
图例：
钢筋混凝土墙
加气混凝土砌块墙
加气混凝土条板墙
消防栓箱（规格尺寸详设备施工图）
商铺的分割线（具体的隔墙和门材质详二次装修）
五层平面图 1:200
本层建筑面积1412m²
五层防火分区平面图

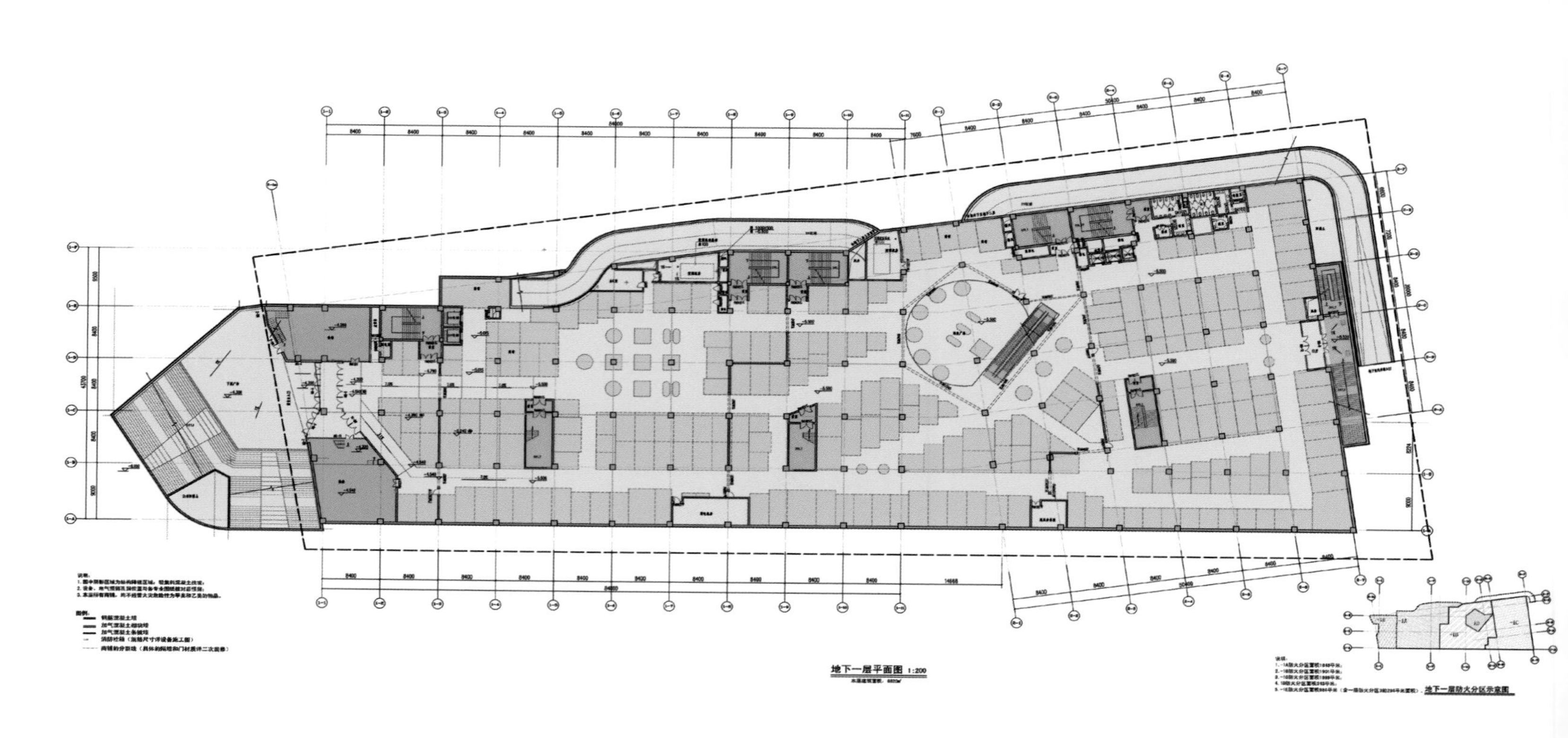
地下一层平面图 1:200
地下一层防火分区示意图

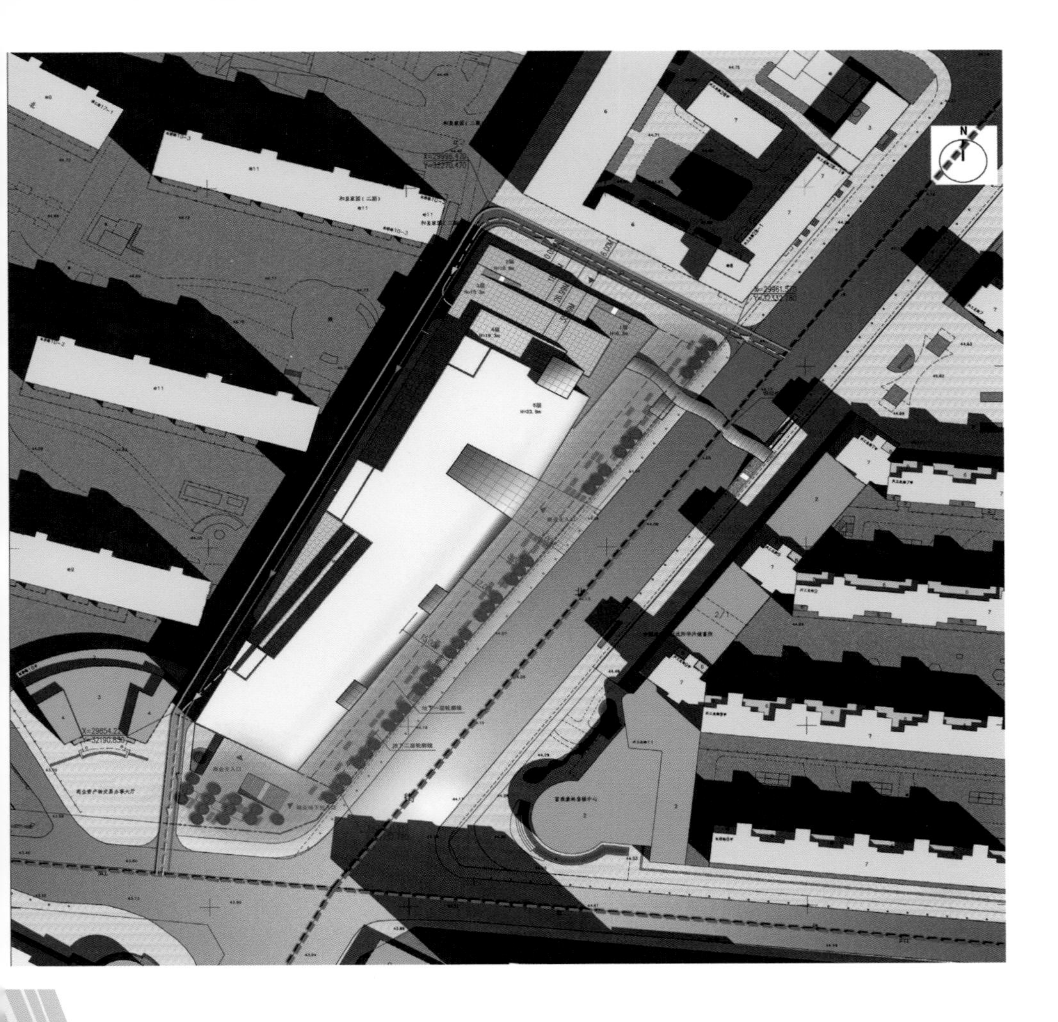

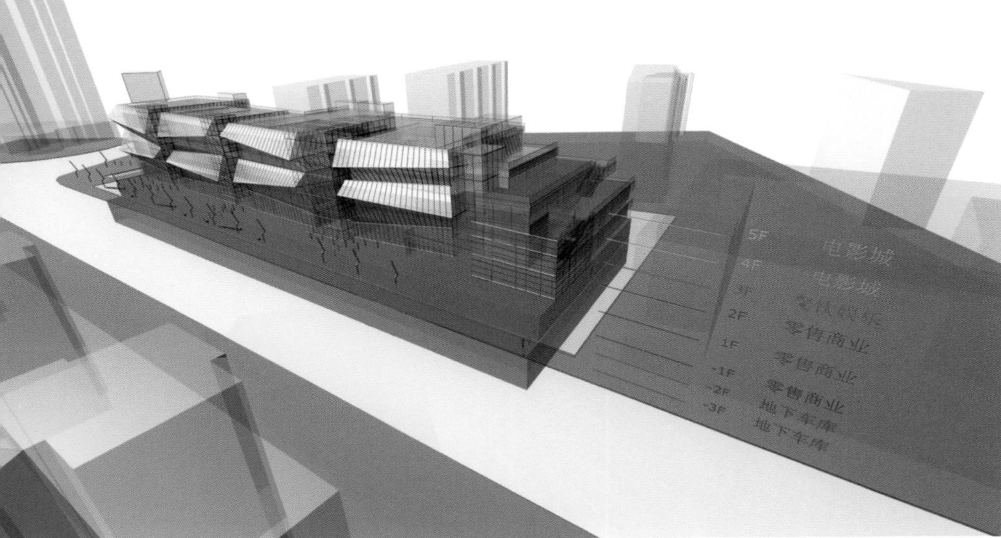

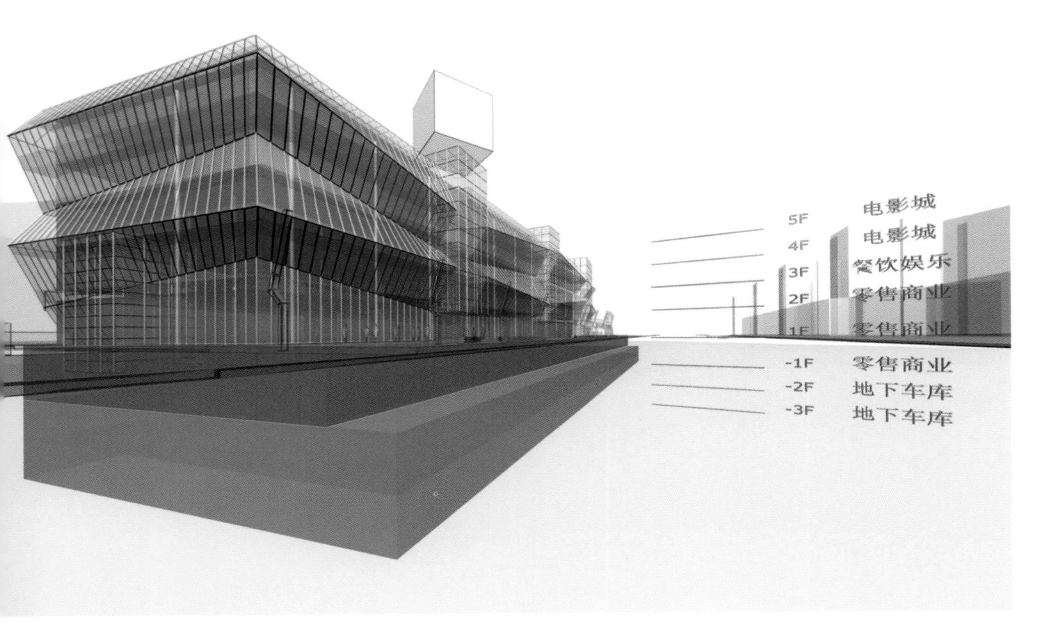

Profile

Qianzihui Shopping Center is located southwest of the intersection of Xinggong North Road and Hongqiao Road of Tiexi Area in Shenyang City. Convenient transportation creates favorable conditions for this project.

Design Concept

The design adheres to a design concept of "People Oriented". It pays close attention to customers' psychology and behavior. Culture, publicity and comfort of this commercial space are taken into account to create a humanized open outdoor leisure shopping space.

Architectural Design

Qianzihui Shopping Center is a commercial building open to all the public. Its benefits should be based on effective gathering of pedestrians, vehicles and logistics. Exquisite arrangement of the design has maximally reduced the overlapping of pedestrians and vehicles. Shopping traffic and service traffic are separated as much as possible, which ensures clear smooth traffic flow.

As a commercial complex, Shenyang Qianzihui has fully respected the principles of urban design. To highlight the project as a landmark, the design has created a modern folding façade. Iconic LED rotating screen at the top of the main building and an 800-meter continuous dynamic LED screen on the wall become symbol of Qianzihui Shopping Center. It is a visual focus in midst of the building complex whether by day or at night.

Commercial Design

Commercial inner street is applied to organize the whole functional space. An integration of commercial pedestrian street, fountain, sunken plaza and footbridge organically combines complicated functional districts, which has enriched interior and outdoor shopping spaces as well as effectively evacuated traffic.

Taking the "Cultural Characteristics" of the commercial building into account, designers have locally set Xu Beihong Artistic Life Gallery on the fourth and fifth floors' theater space. In this way, visitors shall get to know about Xu Beihong's artistic life besides shopping and watching movies.

天津天佑城

Tianjin Lucky City

设计单位：广州市纬纶建筑设计公司
开发商：佛奥集团
项目地址：中国天津市
总用地面积：17 515 m²
总建筑面积：173 676 m²
设计团队：许佐锐 朱溢钊 杜淑莹
郑燕畅 江启冬 张健玮

Designed by: Win-land Architecture Design Co., Ltd
Developer: Foao Group
Location: Tianjin, China
Land Area: 17,515 m²
Floor Area: 173,676 m²
Design Team: Xu Zuorui, Zhu Yizhao, Du Shuying,
Zheng Yanchang, Jiang Qidong, Zhang Jianwei

项目概况

项目位于天津南开区西马路与南马路的交汇处，是一个集购物、游览、美食、娱乐、休闲、商务、广告、信息、展览、康体等多功能于一体的大型商业地产项目。这个规模宏大、功能齐全的现代型综合购物中心，在当地开创了一种全新的消费概念，成为市民和旅游宾客购物、观光、休闲的首选场所。

建筑设计

在建筑原有的基础上，设计师在 1–5 楼增设了两个中庭。设计将 3 个中庭重新布局，将原来方正的建筑线条改为富有动感的弧线，使商业动线连续流畅且富有生气。中庭使顾客的视线上下贯通，也增加了商场的通透感。

设计调整了原有电动扶梯的位置，并新增了电动扶梯和跨层扶梯，引导人流到达各层商场。设计师特别设计了一条从 1 楼到 3 楼的直达手扶天梯，使顾客能够更加方便快捷地到达商场的中层。

原本的商场外立面过于素净，也没有预留广告位，不利于吸引顾客，因此，设计师重新设计了外立面，加强了广告、橱窗展示空间的设计，为商场提供了一个展示其现代、高档新形象的平台。

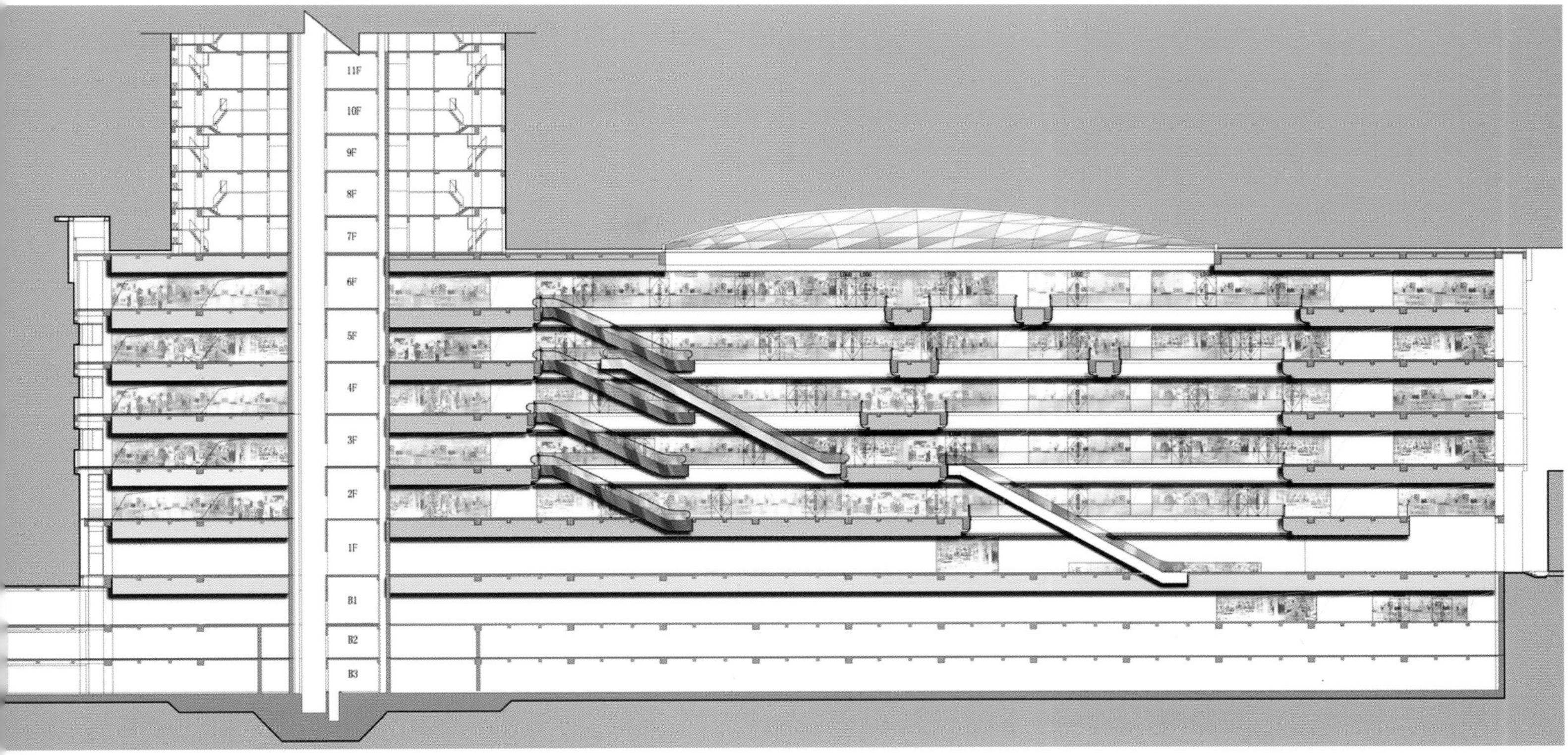
11F
10F
9F
8F
7F
6F
5F
4F
3F
2F
1F
B1
B2
B3

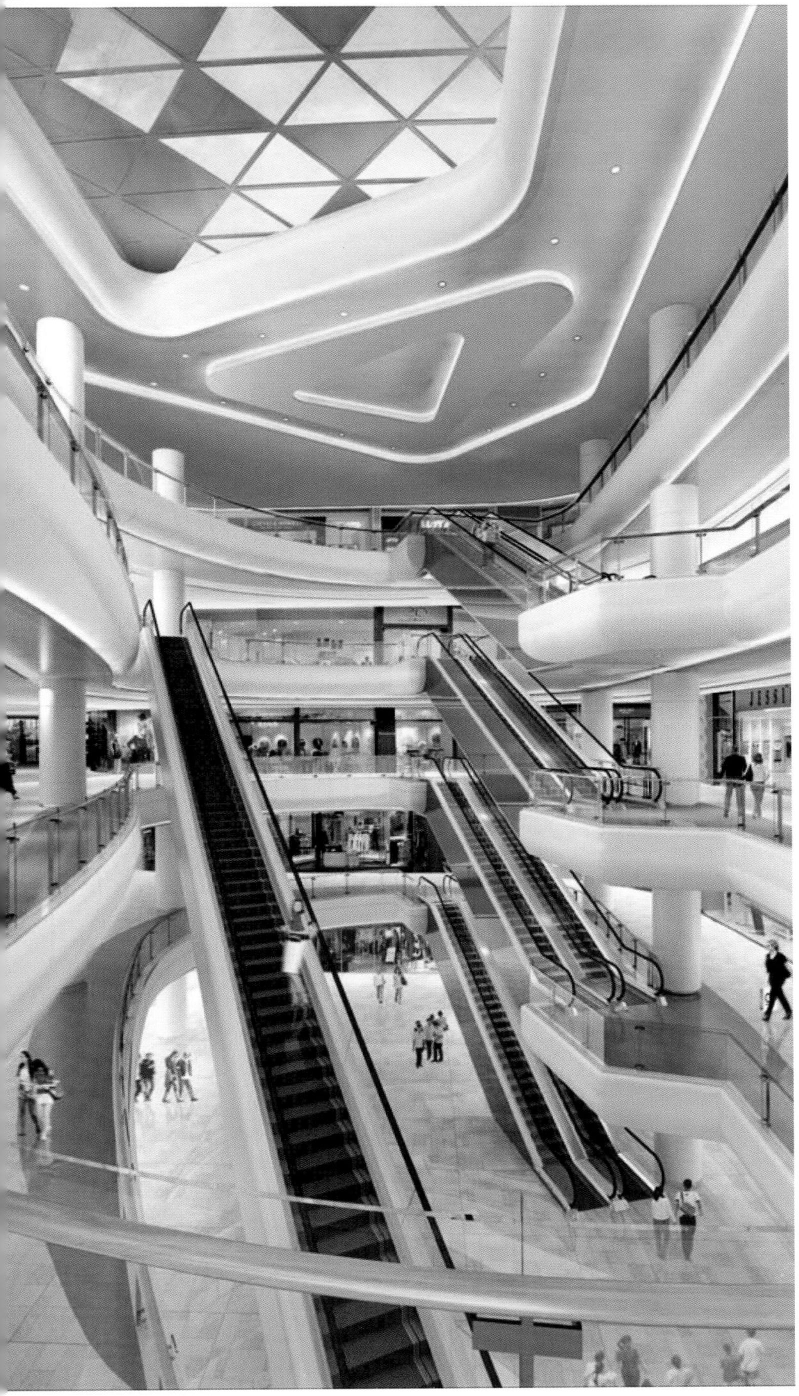

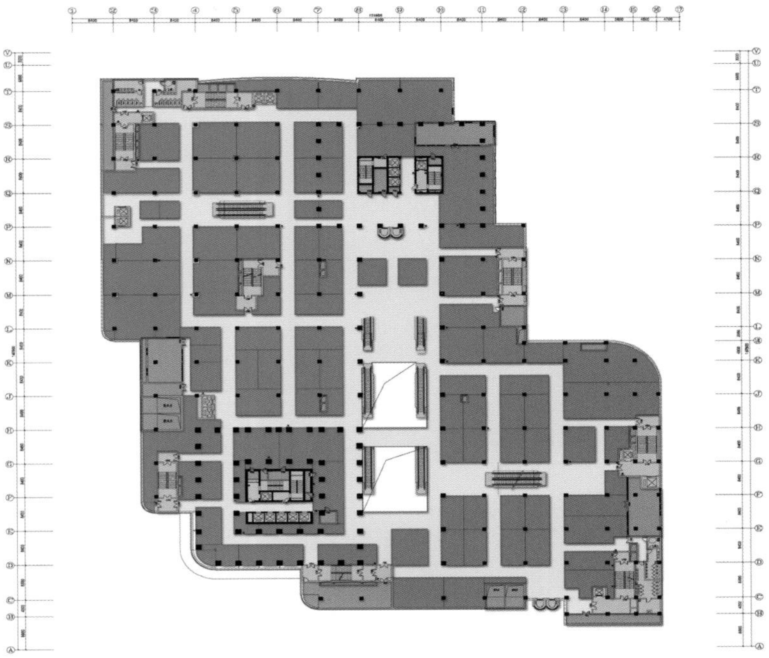

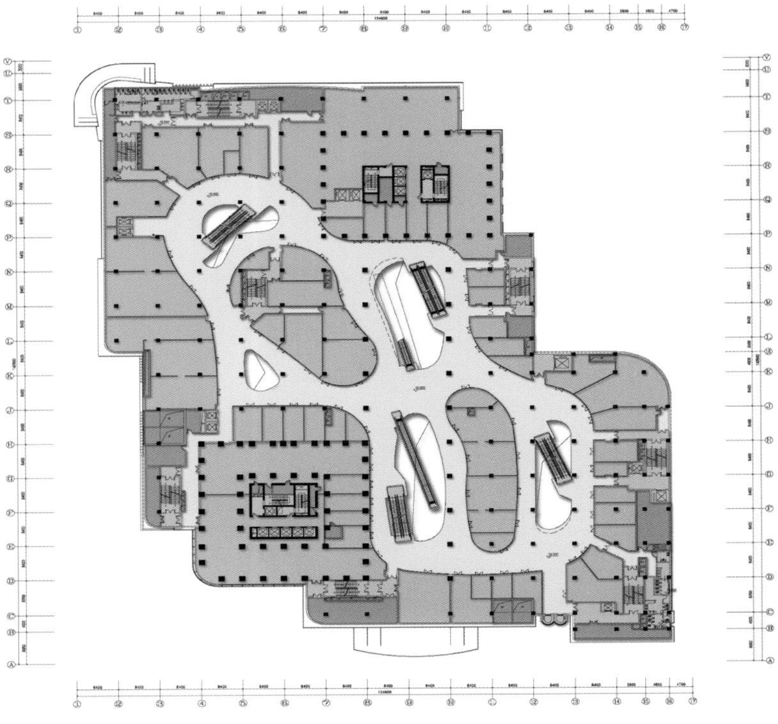

Profile

The project located at the intersection of Xima Road and Nanma Road in Nankai District, Tianjin is a large commercial real estate project integrating shopping, tourism, restaurant, entertainment, recreation, business, advertising, information, exhibition and fitness together. It has initiated a brand new consumption concept locally and becomes a preferred place for shopping and sightseeing.

Architectural Design

Based on existing buildings, designers have added two atriums for 1-5 floors. The three atriums are re-arranged, turning rigid architectural lines into dynamic arcs and making commercial lines continuous, smooth and lively. Atriums open up views top down and increase sense of penetrating of the Mall as well.

The design has reset the position of the existing escalators and added new escalators as well as cross-floor escalators, leading visitors to the Mall on each floor. The non-stop escalator from the First Floor to the Third Floor is especially designed to bring customers to the middle floor.

The original facade of the Mall is unsatisfying as it is too plain and has no room for advertising. Therefore designers have re-designed the facades to enhance advertising and window display space and to provide a platform for displaying modern high-end presence of the Mall.

LIZZY

星光天地

星光天地

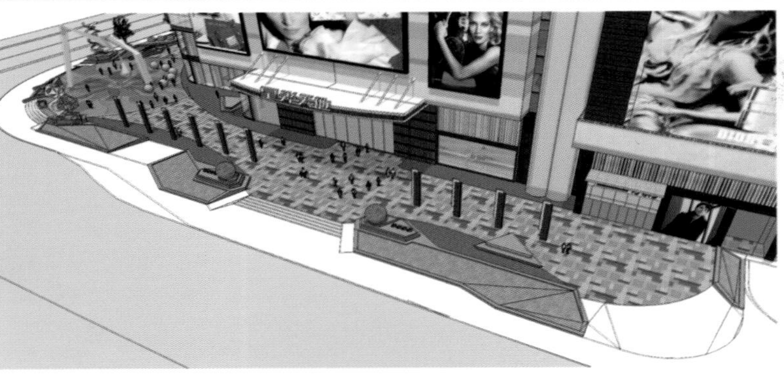

北京赛特购物中心

Beijing Scitech Redevelopment

设计单位：UNStudio
开发商：春天百货集团有限公司
项目地址：中国北京市
场地面积：26 300 ㎡
总建筑面积：250 000 ㎡
设计团队：Ben van Berkel Caroline Bos
Astrid Piber Hannes Pfau
Jürgen Heinzel Jörg Lonkwitz
Shuyan Chan Shuojiong Zhang
Qiyuan Ding Craig Yan
David Chen Stella Shen

Designed by: UNStudio
Client: PCD Stores (Group) Limited
Location: Beijing, China
Site Area: 26,300 m²
Gross Floor Area: 250,000 m²
Design Team: Ben van Berkel, Caroline Bos
Astrid Piber, Hannes Pfau, Jürgen Heinzel
Jörg Lonkwitz, Shuyan Chan, Shuojiong Zhang
Qiyuan Ding, Craig Yan, David Chen, Stella Shen

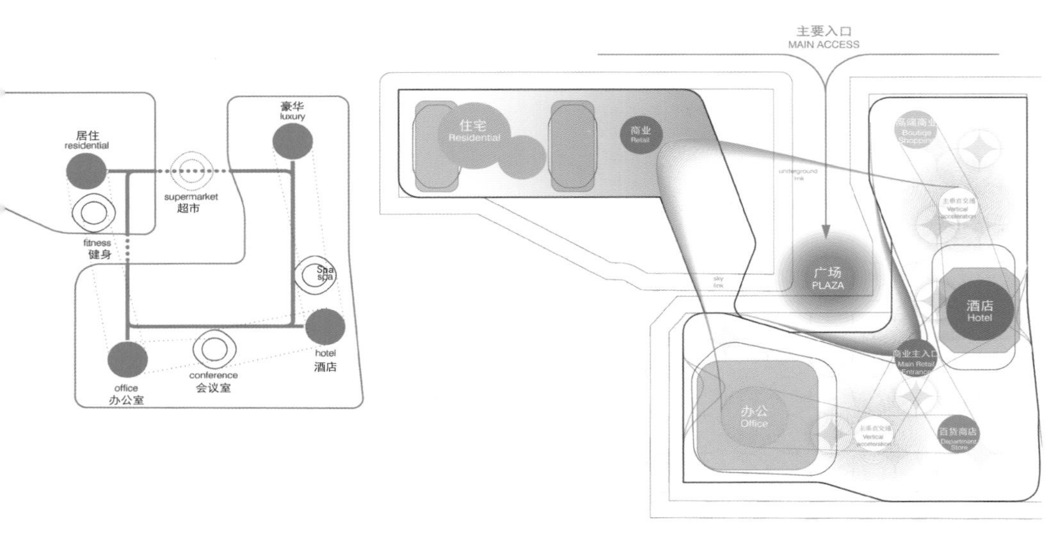

项目概况

这是由 UNStudio 设计的赛特购物中心大型综合体项目，项目位于北京传统和现代发展的交界处，毗邻通往天安门广场的城市东西中轴线。

建筑设计

设计的重点在于将多功能的重建项目与城市结合起来，并将密集的低层建筑和高层建筑整合起来。随着项目的进行，周边的零售商场也将被纳入设计范围。

在新建建筑的低层部分，一系列相互连接的庭院使不同的功能区间整合成一条“城市地毯”，建筑内部和外部的连接则为各功能区的整合提供了室内和室外空间。5 至 7 层高的墩座墙被设计成起伏的形态，既使周边的车流井然有序，同时也为行人提供了安全的绿色通道。

高层塔楼从低层塔楼中生出，为酒店和办公室提供了场地。两者与墩座墙的交汇处，是低层区和高层区的过渡空间，融合了两个区域的功能。

裙房主要为零售商业而设计，地下 5 层，地上 8 层，也包括上层水疗中心、会议设施以及屋顶景观。地下 3 层用于停车，其他的两层设置为超市、食品店和为了保持地面高度而设计的下沉区域。

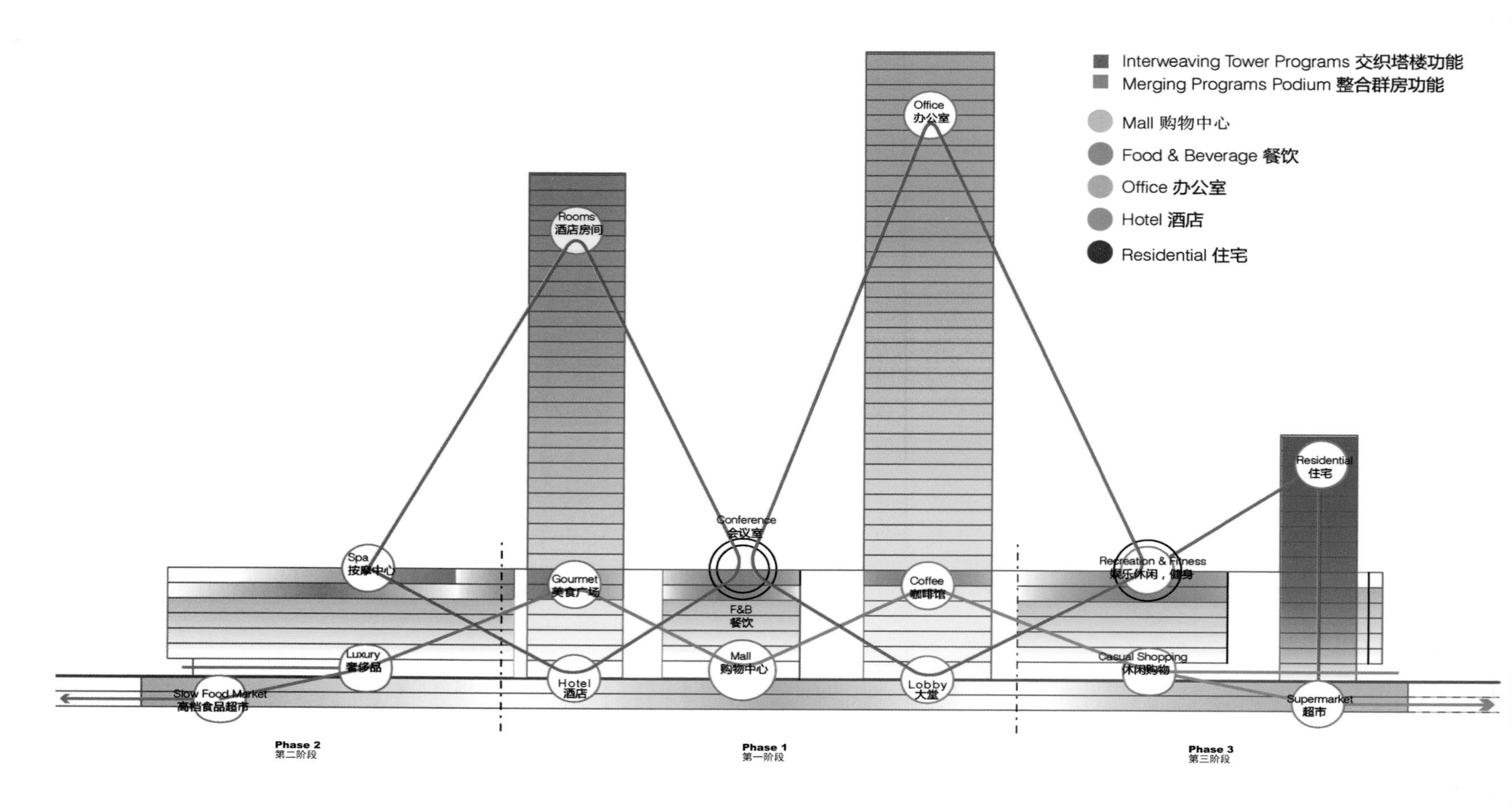
Interweaving Tower Programs 交织塔楼功能
Merging Programs Podium 整合群房功能
Mall 购物中心
Food & Beverage 餐饮
Office 办公室
Hotel 酒店
Residential 住宅
Office
办公室
Rooms
酒店房间
Residential
住宅
Conference
会议室
Spa
按摩中心
Gourmet
美食广场
F&B
餐饮
Coffee
咖啡馆
Recreation & Fitness
娱乐休闲，健身
Luxury
奢侈品
Mall
购物中心
Casual Shopping
休闲购物
Slow Food Market
高档食品超市
Hotel
酒店
Lobby
大堂
Supermarket
超市
Phase 2
第二阶段
Phase 1
第一阶段
Phase 3
第三阶段

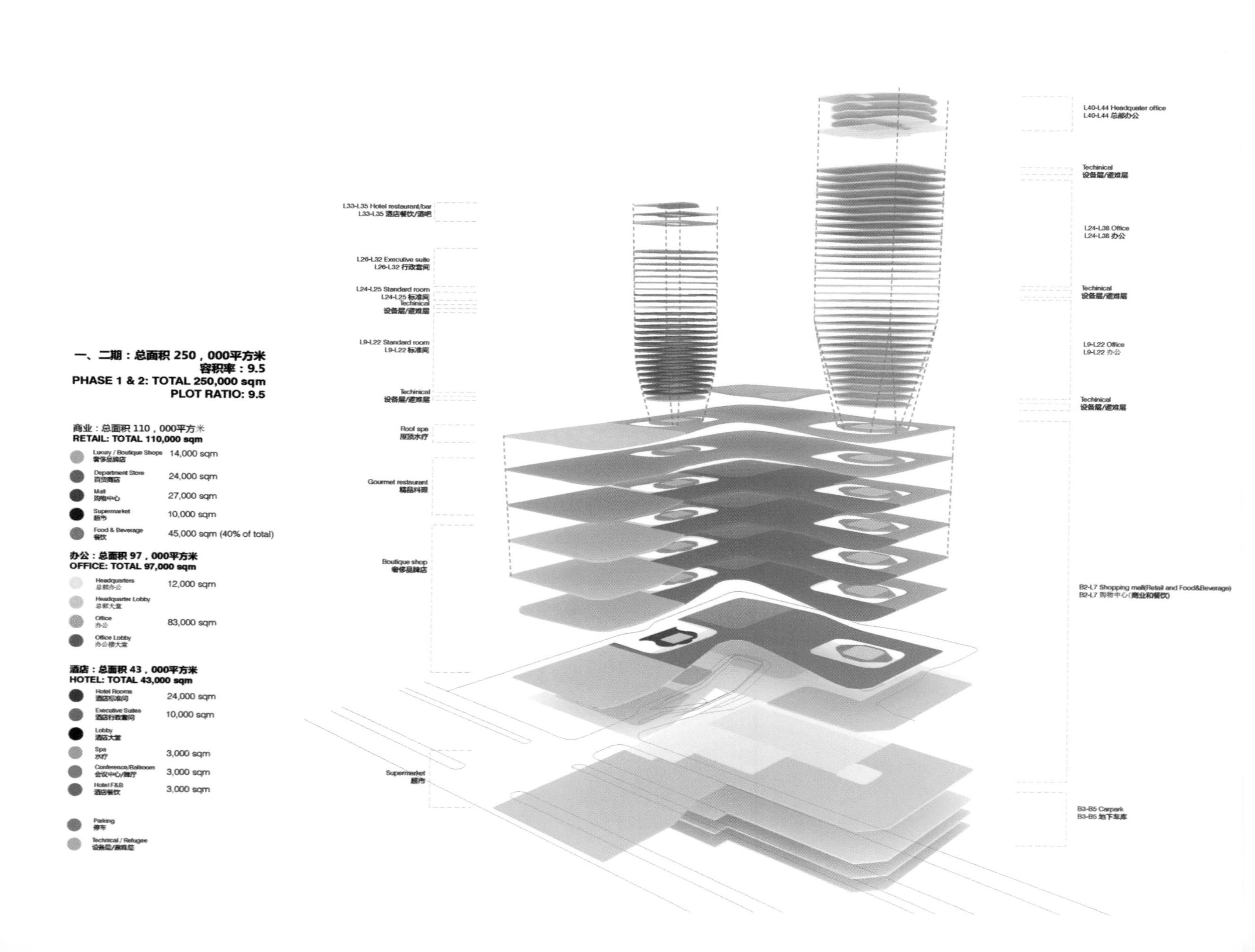
L33-L35 Hotel restaurant/bar
L33-L35 酒店餐饮/酒吧
L26-L32 Executive suite
L26-L32 行政套间
L24-L25 Standard room
L24-L25 标准间
Techinical
设备层/避难层
L9-L22 Standard room
L9-L22 标准间
Techinical
设备层/避难层
Roof spa
屋顶水疗
Gourmet restaurant
精品料理
Boutique shop
奢侈品牌店
Supermarket
超市
L40-L44 Headquater office
L40-L44 总部办公
Techinical
设备层/避难层
L24-L38 Office
L24-L38 办公
Techinical
设备层/避难层
L9-L22 Office
L9-L22 办公
Techinical
设备层/避难层
B2-L7 Shopping mall(Retail and Food&Beverage)
B2-L7 购物中心(商业和餐饮)
B3-B5 Carpark
B3-B5 地下车库
一、二期：总面积 250，000平方米
容积率：9.5
PHASE 1 & 2: TOTAL 250,000 sqm
PLOT RATIO: 9.5
商业：总面积 110，000平方米
RETAIL: TOTAL 110,000 sqm
Luxury / Boutique Shops 奢侈品牌店 14,000 sqm
Department Store 百货商店 24,000 sqm
Mall 购物中心 27,000 sqm
Supermarket 超市 10,000 sqm
Food & Beverage 餐饮 45,000 sqm (40% of total)
办公：总面积 97，000平方米
OFFICE: TOTAL 97,000 sqm
Headquarters 总部办公 12,000 sqm
Headquarter Lobby 总部大堂
Office 办公 83,000 sqm
Office Lobby 办公楼大堂
酒店：总面积 43，000平方米
HOTEL: TOTAL 43,000 sqm
Hotel Rooms 酒店标准间 24,000 sqm
Executive Suites 酒店行政套间 10,000 sqm
Lobby 酒店大堂
Spa 水疗 3,000 sqm
Conference/Ballroom 会议中心/舞厅 3,000 sqm
Hotel F&B 酒店餐饮 3,000 sqm
Parking 停车
Technical / Refugee 设备层/避难层

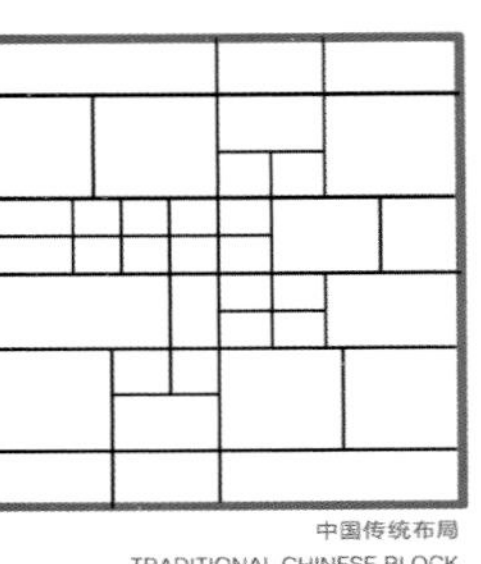

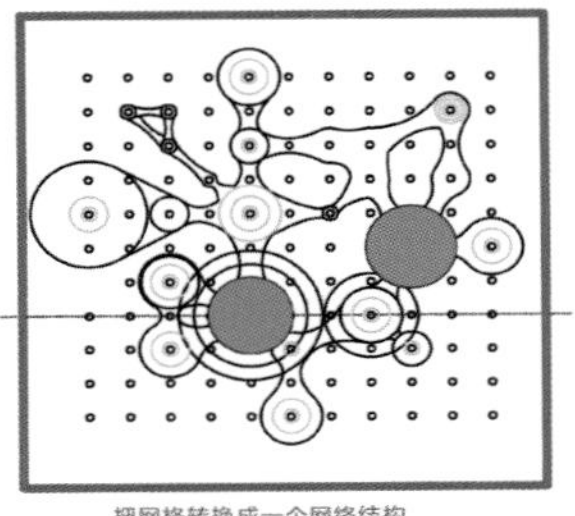

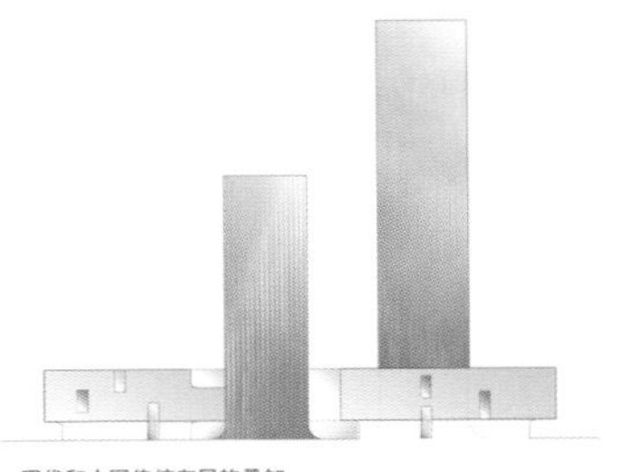

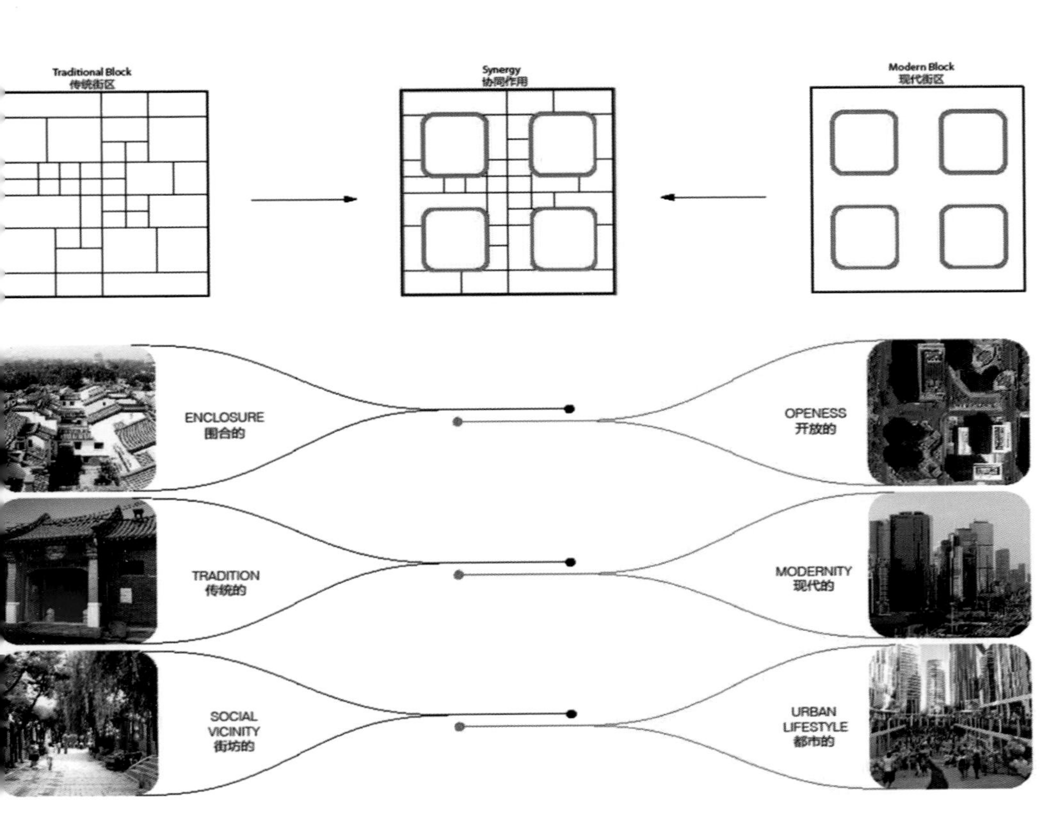

Profile

The Scitech mixed-use redevelopment designed by UNStudio is a large comprehensive project. It is located on a crossing point of traditional and modern developments in Beijing. The site is adjacent to the city's east-west central axis leading to the Tiananmen Square.

Architectural Design

The massing strategy for the plot focuses on creating optimal links between the mixed-use programming of the redevelopment, whilst interweaving a dense low rise development with a high rise component. Simultaneously, through phasing the redevelopment, the existing retail mall is integrated into the design from the outset.

In the low rise portion of the new development, a series of connected courtyards organize the different programmes around the user flows to form an urban carpet. This provides outdoor and indoor spaces which combine programmes through internal and external links. This five to seven storey high podium is designed as an undulating landscape that organizes the traffic flows on its perimeters, whilst providing interior green and sheltered spaces for pedestrian access.

The high rise towers emerge from the low rise development and provide singular usage for hotel and office premises. In the intersection with the low rise podium, functions are allocated accordingly to create maximum synergy between activities in the low rise zone and the towers.

The podium is primarily dedicated to retail function with 5 levels underground and 8 levels above ground, including spa and conference facilities located within the upper floor and roofscape. Three underground levels provide parking facilities, whilst the remaining two house a supermarket, food court and programmatically arranged drop offs to sustain the ground level for pedestrians.

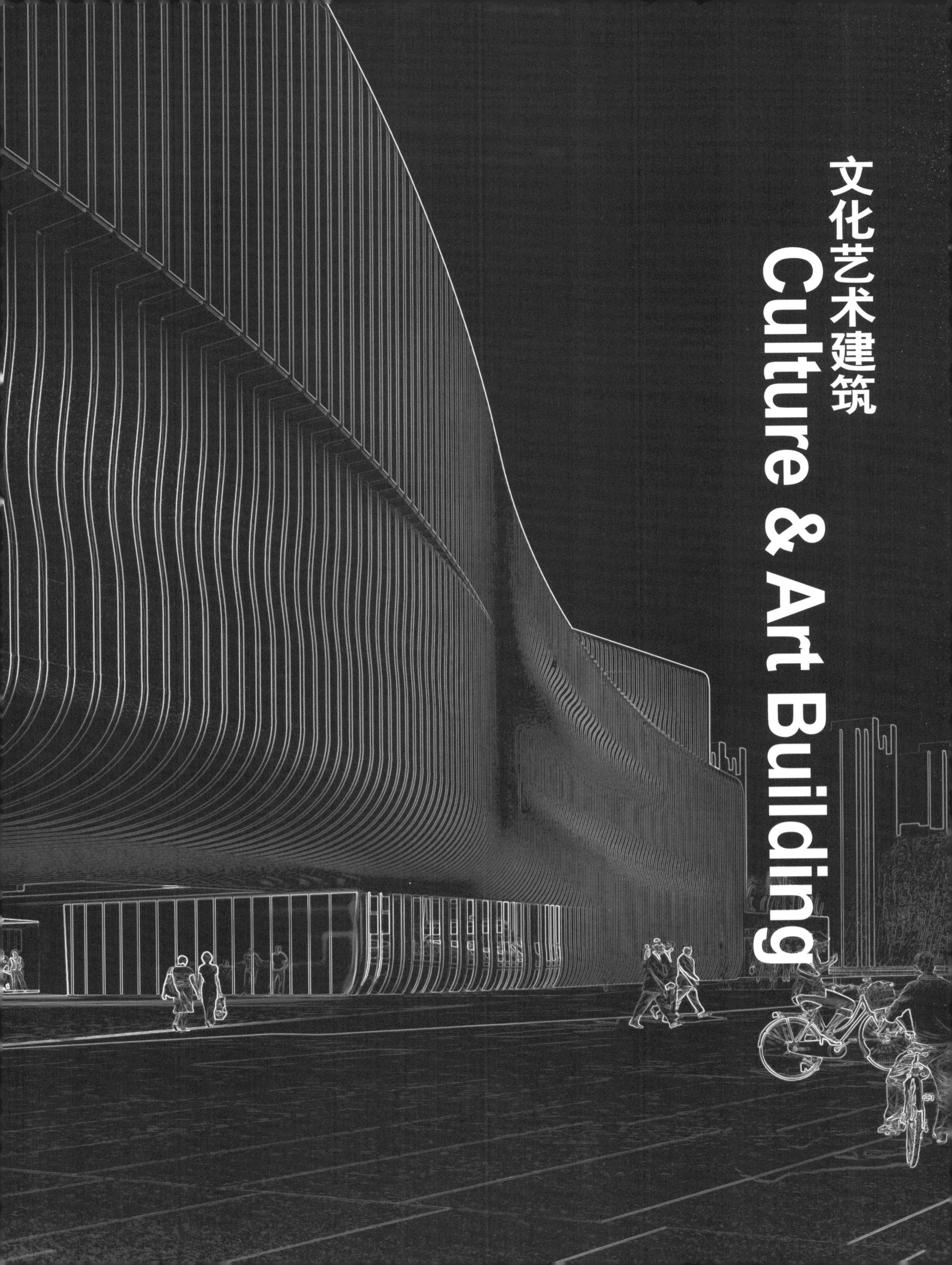
文化艺术建筑
Culture & Art Building

克罗地亚萨格勒布 Kajzerica 城市体育馆

Kajzerica City Stadium

设计单位：3LHD
开发单位：Grad Zagreb
项目地址：克罗地亚萨格勒布
设计团队：Sasa Begovic　Marko Dabrovic
Silvije Novak　Tatjana Grozdanic Begovic
Tin Kavuric　Irena Mazer
Krunoslav Szorsen　Dragana Simic
Matija Crnogorac　Ljiljana Dordevic
Vibor Granic　Jelena Mance
Ana Safar
摄影：Studio HRG d.o.o.　Sandro Lendler

Designed by: 3LHD
Client: Grad Zagreb
Location: Zagreb, Croatia
Project Team: Sasa Begovic, Marko Dabrovic, Silvije Novak, Tatjana Grozdanic Begovic, Tin Kavuric, Irena Mazer, Krunoslav Szorsen, Dragana Simic, Matija Crnogorac, Ljiljana Dordevic, Vibor Granic, Jelena Mance, Ana Safar
Photography: Studio HRG d.o.o., Sandro Lendler

目概况

这个能容纳 50 000 位观众的体育馆位于萨格勒布地区一块被抬高的高地上，靠近附近赛马场的运动和休息区。

计理念

这个圆形的竞技场的设计灵感源自于“马格努斯效应”和“Kajzerica”这一名字。“马格努斯效应”是足球比赛中最难也是最引人注目的动作之一，这种效应在球改变其路径的时候发生，设计师依据这一运动轨迹，构思了这一圆形的建筑形态。

“Kajzerica”是临近区域的地名，也指代一种特定种类的面包，设计师从一句有关面包和马戏的古语中获取灵感，将面包的形态融入体育馆的建筑形态中。

区设计

项目分为三个部分，北区即一期建筑区，它与周边的居民区和萨格勒布博览会独立开来，避免了对体育馆的干扰。中区即项目二期，是整个体育馆开发的关键，涵盖了服务区、公共车库、公共汽车停车场等公共空间。南区涵盖了商业楼、住宅大楼以及一个新商业中心，这些元素给这一地区的发展注入了活力。

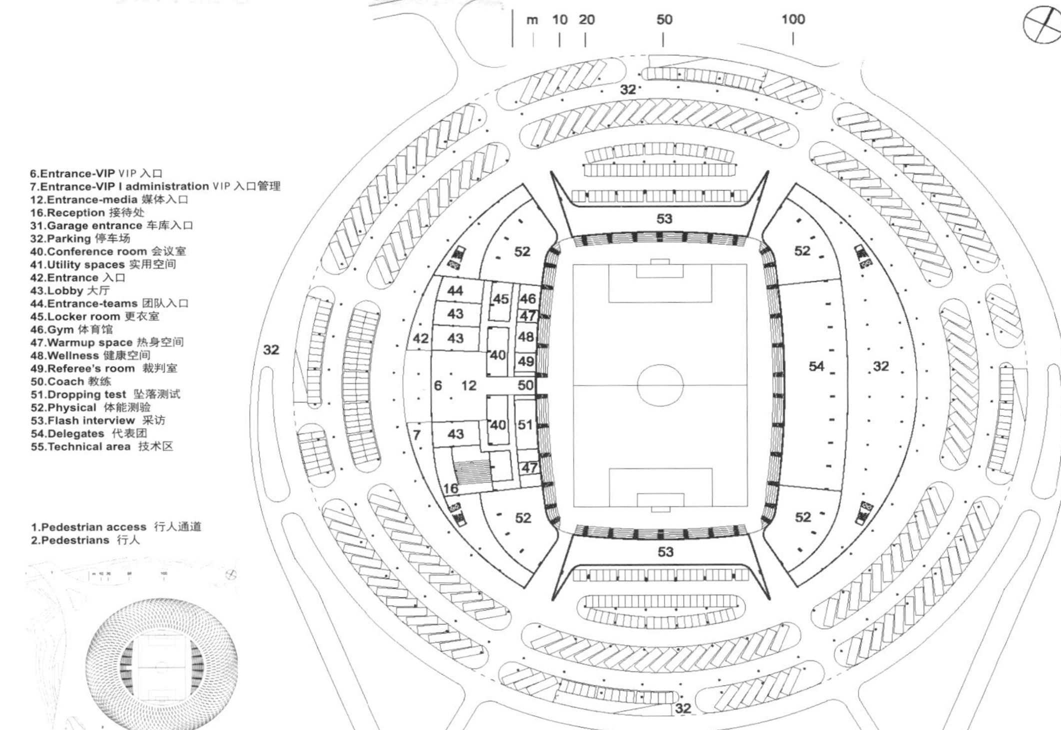

18.Conference room 会议室
20.WC 卫生间
31.Garage entrance 车库入口
32.Parking 停车场
33.Lobby 大堂
34.Press-welcome zone 媒体接待区
35.VIP-welcome zone VIP 接待区
36.Media rooms 媒体室
37.Web editing 网站编辑
38.Photografs 摄影
39.Press 新闻
40.Conference room 会议室
41.Utility room 杂物间

1.Pedestrian access 行人通道
2.Pedestrians 行人
3.Entrance upper stand 入口上方看台
4.Entrance-VVIP VVIP 入口
5.Entrance-staff 员工入口
6.Entrance-VIP VIP 入口
7.Entrance administration 入口管理
8.Media 媒体
9.Entrance 入口
10.Lower stand entrance 较低看台入口
11.Utility room 杂物间
12.Entrance-media 媒体入口
13.Caffe 咖啡区
14.Space for disabled visitors 残疾访客空间

Profile

The stadium, which seats more than 50,000 spectators, is located on an elevated plateau next to the sports and recreational zone of the racetracks.

Design Concept

The inspiration for the circular arena came from several suitable forms: the so-called Magnus' Effect, one of the most difficult and attractive moves in soccer which occurs when a ball changes its course; and the name of the neighborhood "Kajzerica", which is a name for a certain kind of a bun whose shape is reflected in the stadium. In the context of a contemporary sports arena, this design evokes the saying from ancient times about "Bread and Circuses".

Subarea Design

The coverage area is divided into three zones. The northern zone, the first building zone, enables the undisturbed building of the stadium, independent of the surrounding settlements, parcels and Zagrebački Velesajam. The middle zone, the second building phase is crucial for the development of the stadium because this is where different services will be situated, as well as the infrastructural elements used by all three zones including public garages, bus parking and similar. The southern zone is for the construction of new businesses, commercial and residential buildings as well as a new commercial center. These elements add vitality to the development of this region.

6.Entrance-VIP VIP 入口
7.Entrance-administration 管理区入口
13.Caffe 咖啡区
15.Lounge 休息室
16.Reception 接待处
17.VIP wardrobe VIP 衣橱
18.Conference room 会议室
19.Business lounge 商务休息室
20.WC 卫生间
21.Staff wardrobe 员工衣橱
22.Office 办公室
23.Restaurant 餐厅

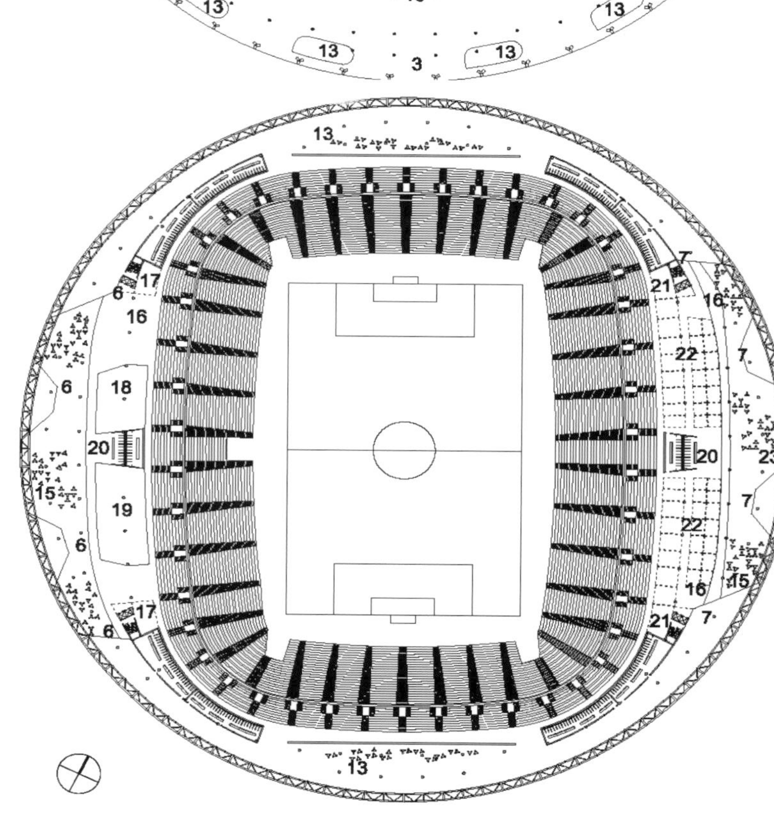

m 10 20 50 100

6.Entrance-VIP VIP 入口
14.Space for disabled visitors 残疾访客空间
24.VVIP bar VVIP 酒吧
25.VIP restaurant VIP 餐厅
26.VIP VIP
27.Concert mixing stage 音乐会混合型舞台
28.Sky studio 空中工作室

13.Caffe 咖啡区
20.WC 卫生间

29.Commentators 解说员
30.Police 警察

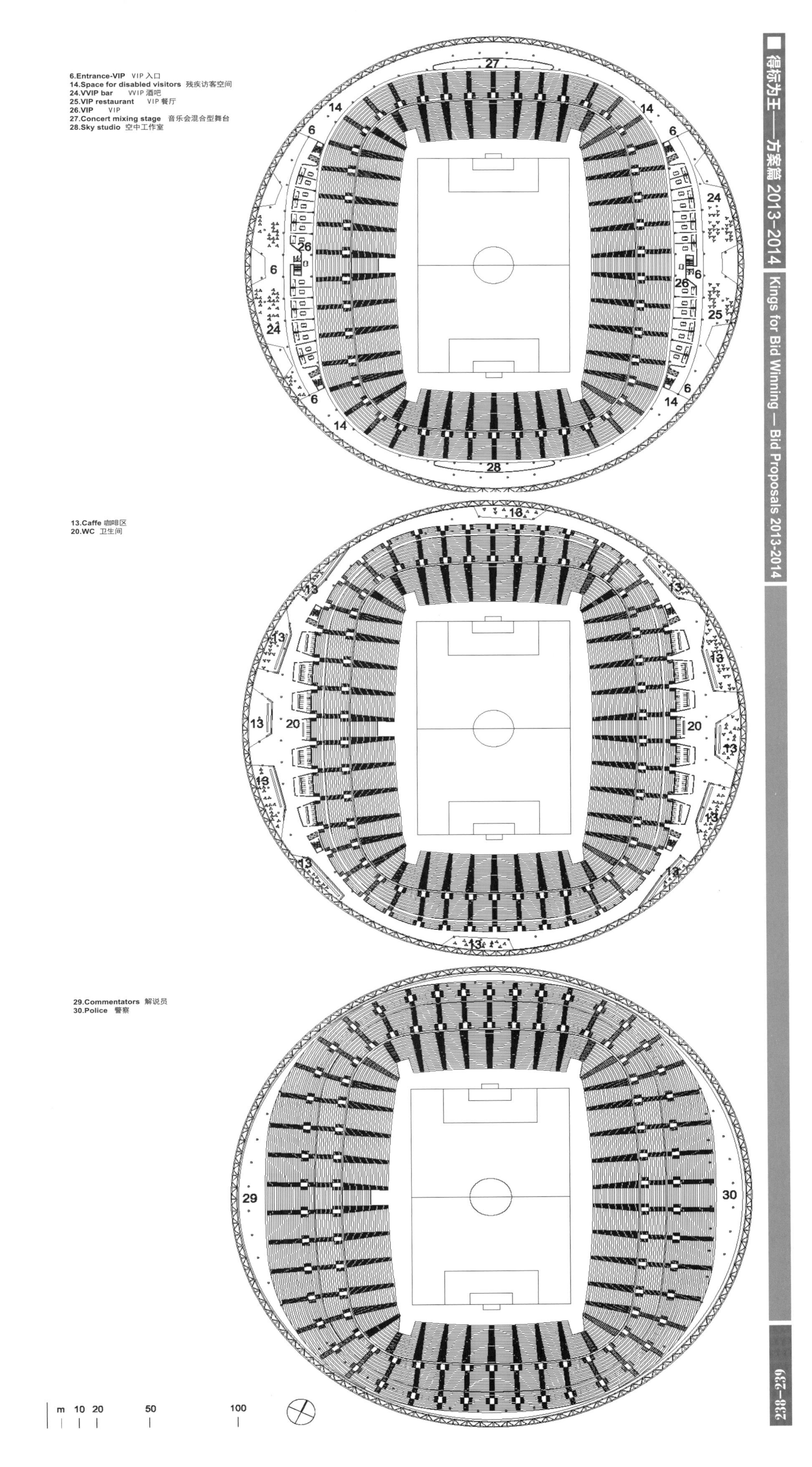

m
10
20
50
100
m
10
20
50
100

50
100
m
10
20
50
100
50
100
m
10
20
50
100

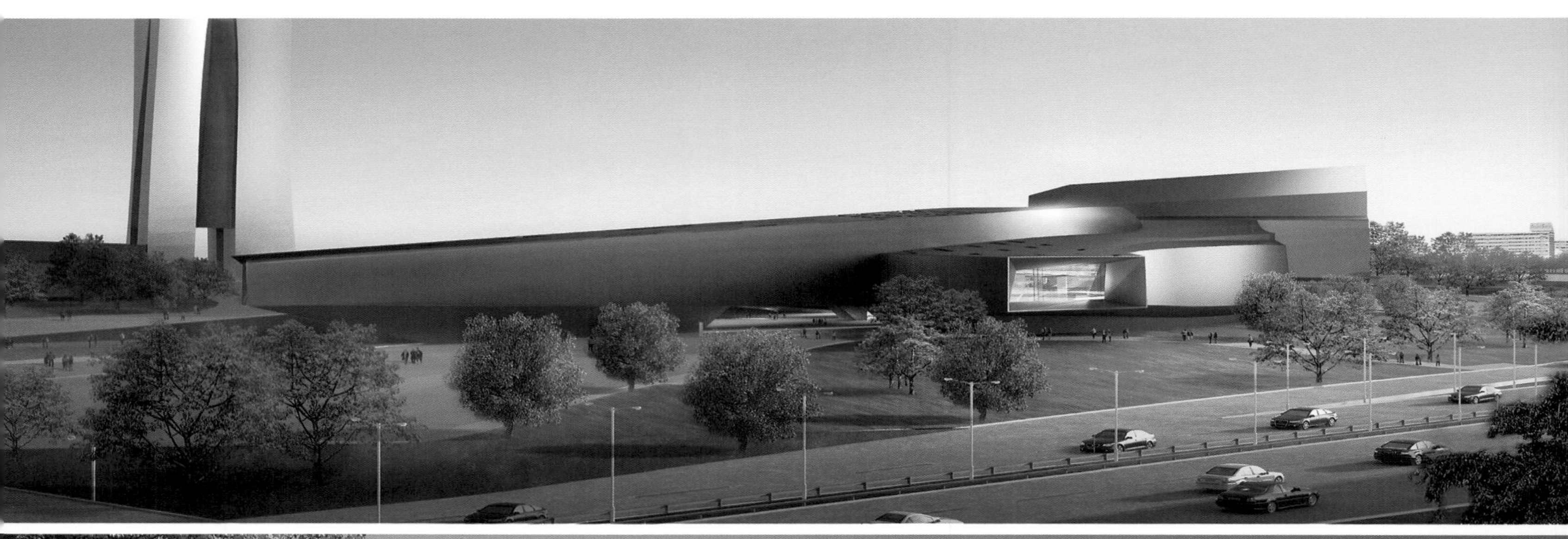

塞尔维亚贝尔格莱德科学宣传中心 39 号楼

Block 39 Center for Promotion of Science

设计单位：RTA-Office 建筑事务所
Santiago Parramón
开发单位：科学技术发展部
项目地址：塞尔维亚贝尔格莱
基地面积：116 600 ㎡
总建筑面积：13 998 ㎡
摄影：RTA-Office 建筑事务所

Designed by: RTA-Office; Santiago Parramón
Client: Ministry of Science and Technological Development
Location: Belgrade, Republic of Serbia
Site Area: 116,600 m^2
Gross Building Area: 13,998 m^2
Photography: RTA-Office

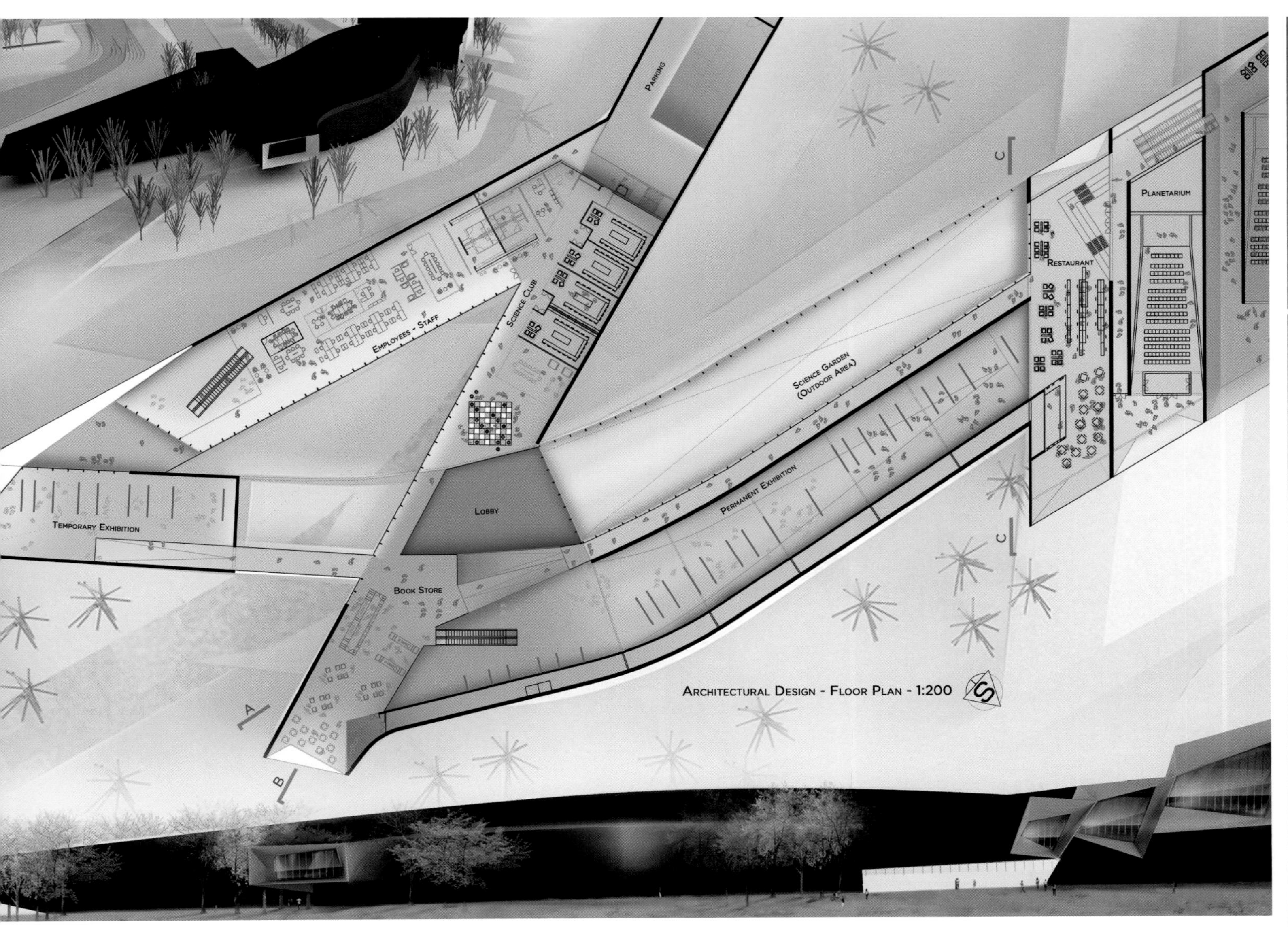

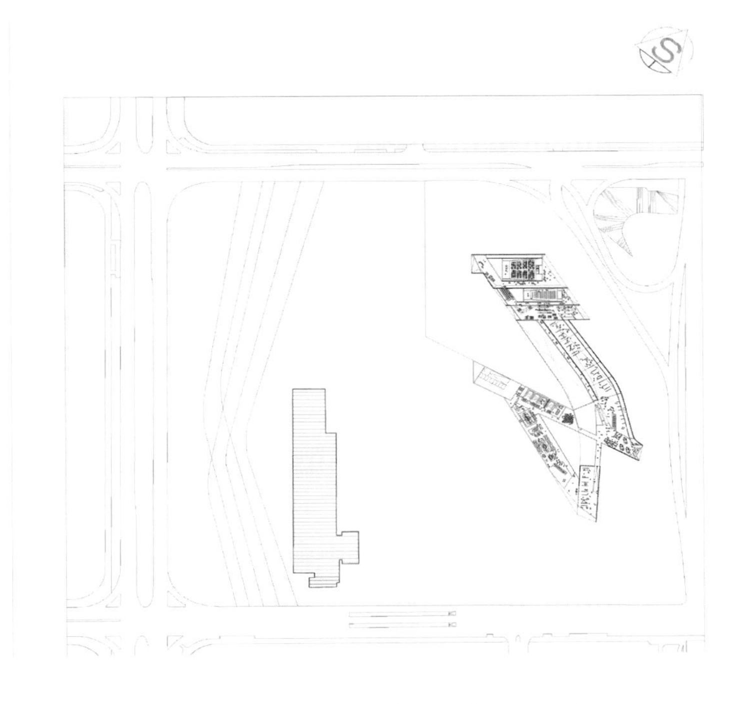

项目概况

项目位于塞尔维亚贝尔格莱德，方案旨在创造一个国际化的科学宣传中心，提供一个聚集和信息交流的大型空间，并通过这个创新、独特的建筑，为这个区域带来生机与活力。

设计理念

设计在顺应场地条件的基础上，对场地进行改造，构成新的流通系统，打破僵硬的既定城市结构，还城市以生机和活力。设计创造了新的公共空间、文化空间和大学空间，呈现了具有国际表现力的城市元素。

空间设计

设计师将“公共空间”与“私密空间”进行转化，并相互穿插。“私密空间”随着城市环境打开，将建筑融入城市环境中。设计增加了一种流畅、开放、融合的空间，这是一种充满感官体验的空间，动感而富于变化。

建筑设计

建筑矗立在场地上，好似一个由高科技打造的精致雕塑，融入城市肌理中。建筑以纵向排布的方式横铺在场地上，并向远方延展，享有两座大桥之间的视野。这个大型的流线型建筑，不仅将改善周围的环境，也将成为一个独特的城市空间。

设计统一采用了一种可回收利用的黑色钢铁，既赋予了建筑优雅生动的质感，强化了建筑的视觉效果，同时也是对可持续设计理念的回应。

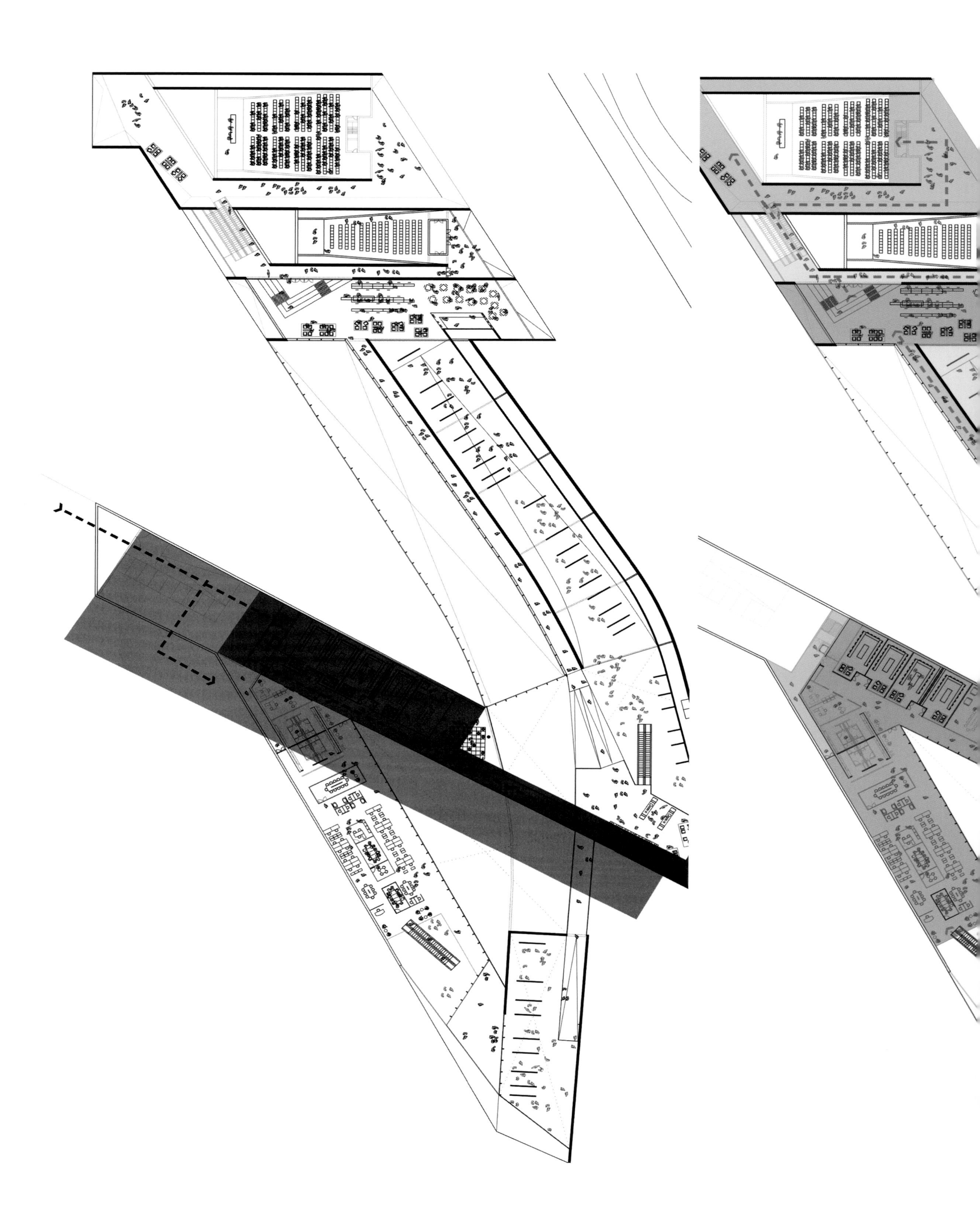

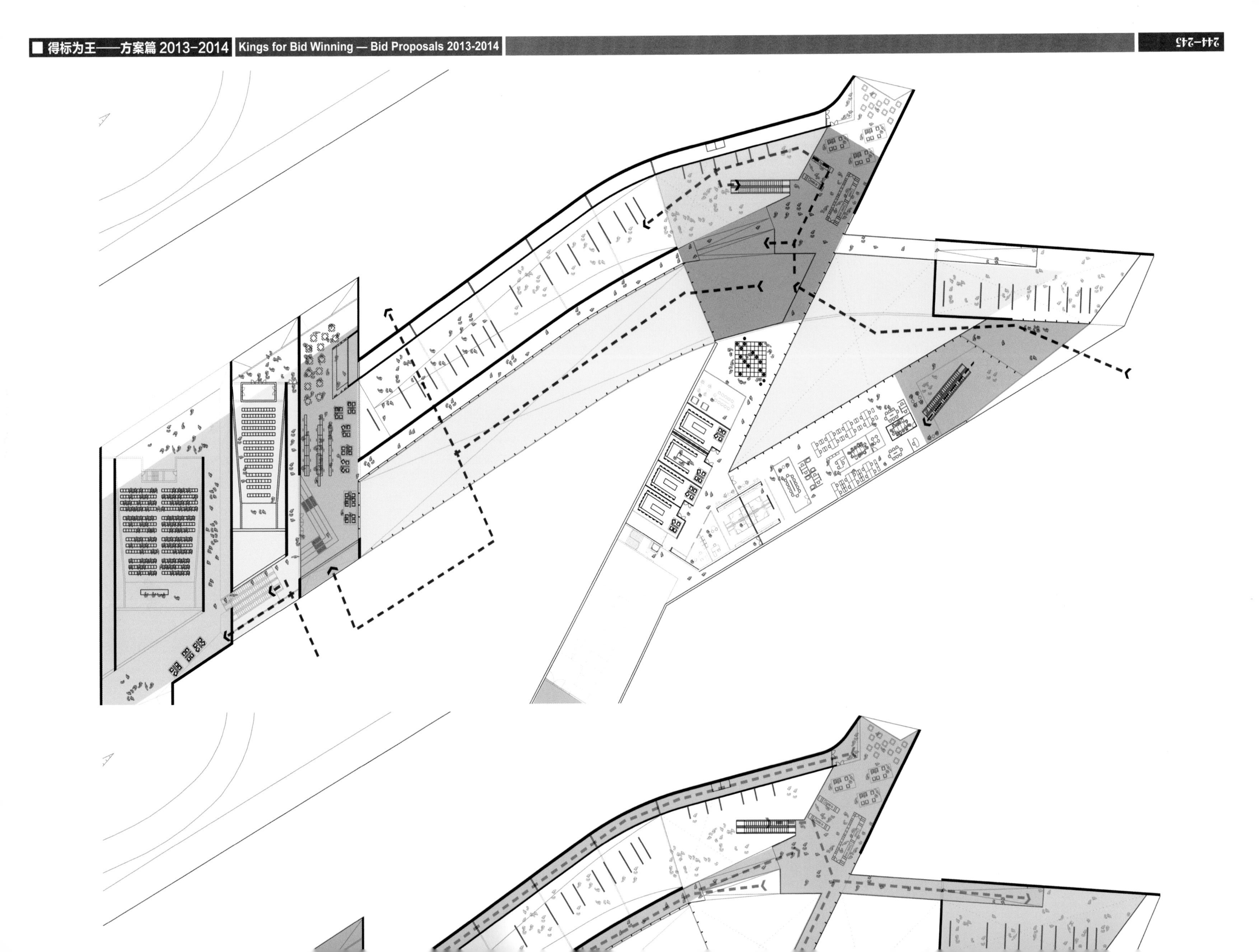

Views from Campus

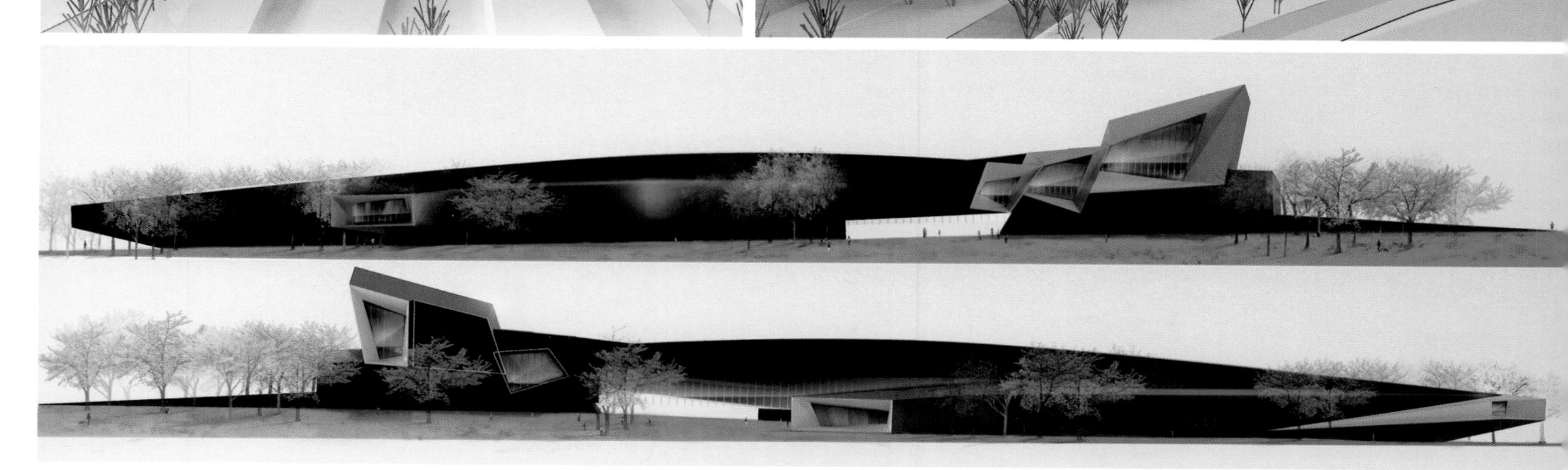

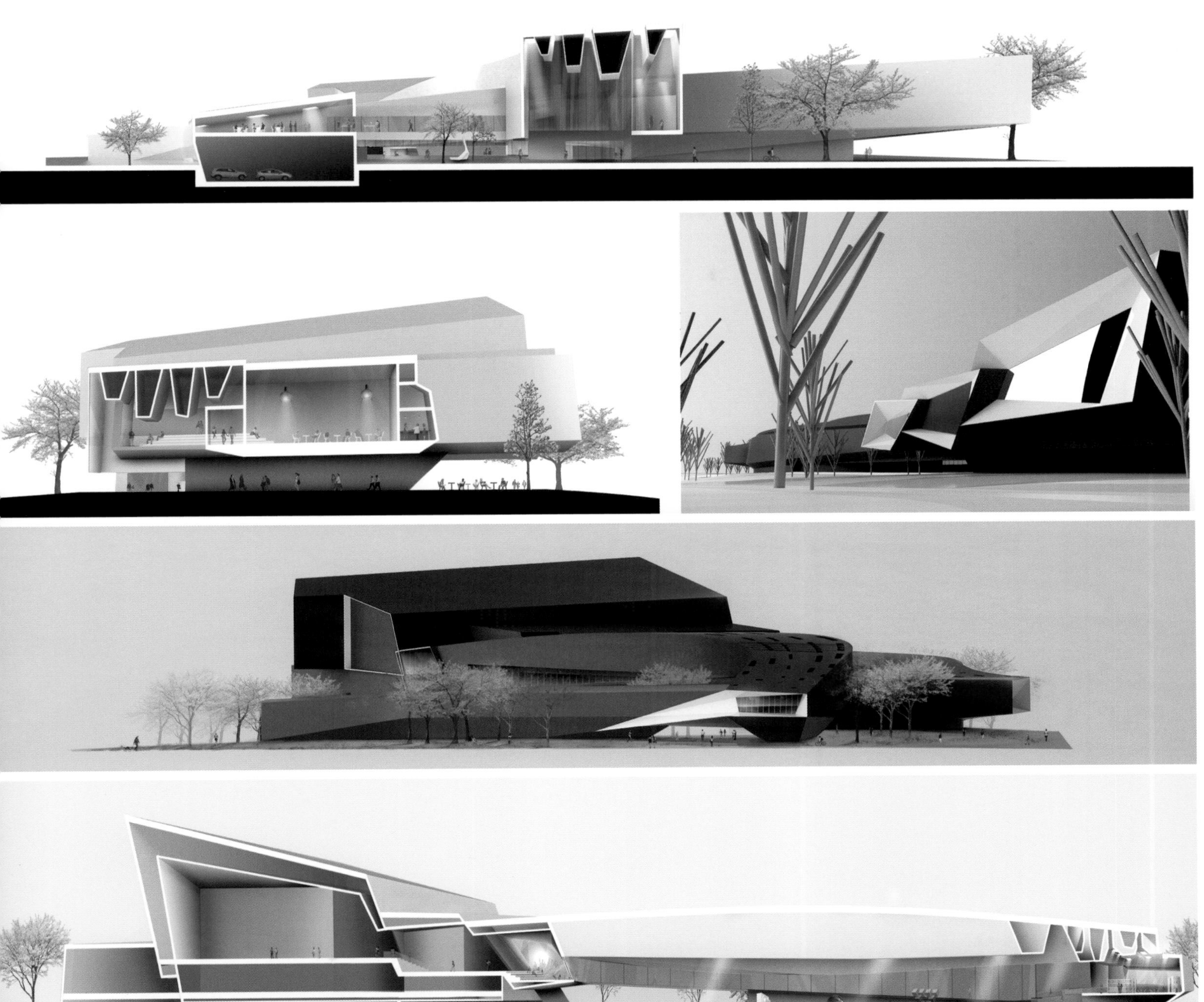

Profile

The project is located in Belgrade, Republic of Serbia. It aims to become a global calling for the promotion of science, to become a place for meeting and exchange of information in a larger scale and to bring vitality and life to this area through this creative and innovative building.

Design Concept

Based on existing conditions of the site, the design has carried out transformation to the site composing new circulation system, releasing the rigid structures of the actual urban organization and bringing the city vitality and vigor. The design has created new public space, cultural space and college space, presenting international expressive urban elements.

Space Design

Designers transform "Public Space" into "Private Space" and vise versa. "Private Space" opens to urban environment and brings buildings into the city. The design adds a kind of smooth, open and integrated space which filled with sensory experience and rich in dynamics and changes.

Architectural Design

The new building stands in the site as a delicate piece shaped by a technological sculptor. The building is a piece that is displayed longitudinally. It extends afar enjoying views between the two bridges. The large linear building has not only improved surrounding environment but also become a unique urban space. The use of a single material, recycled black steel endows the building with graceful vivid texture, enhanced visual effect and also responds to its sustainable design concept.

安徽省艺术博物馆

Anhui Provincial Art Museum

设计单位：RTA-Office 建筑事务所
Santiago Parramón
项目地址：中国安徽省合肥市
总用地面积：1 214 056.93 ㎡
总建筑面积：36 000 ㎡
设计团队：Margherita Filpi　Isabel Granell
Mariana Rapela　Kelly Sadikin
Arsenii Shiianov　Lorena Trinidad
Miguel Vilacha　Fei Fei Zhan

Designed by: RTA-Office; Santiago Parramón
Location: Hefei, Anhui, China
Total Land Area: 1,214,056.93 m^2
Total Building Area: 36,000 m^2
Design Team: Margherita Filpi, Isabel Granell, Mariana Rapela,Kelly Sadikin, Arsenii Shiianov, Lorena Trinidad, Miguel Vilacha, Fei Fei Zhan

项目概况

这个由 RTA-Office 建筑事务所负责设计的安徽省艺术博物馆位于一个大型的绿化区，设计方案特别注重空间布局及是否有足够的空间举办国际艺术活动等问题。

建筑设计

场地位于该市新的政治文化区，处于公园的西南角。设计旨在实现建筑与周围环境的完美融合，通过现代的建筑理念，构建一个简约、高雅、多功能的艺术空间，展示安徽省及其他省市的文化艺术作品和文化艺术成就。

设计灵感来源于安徽崎岖的山路、自然元素以及丝绸绘画和木雕等艺术品。设计规划了两个独立的结构，利用统一的外立面将两个单独的建筑联系起来。

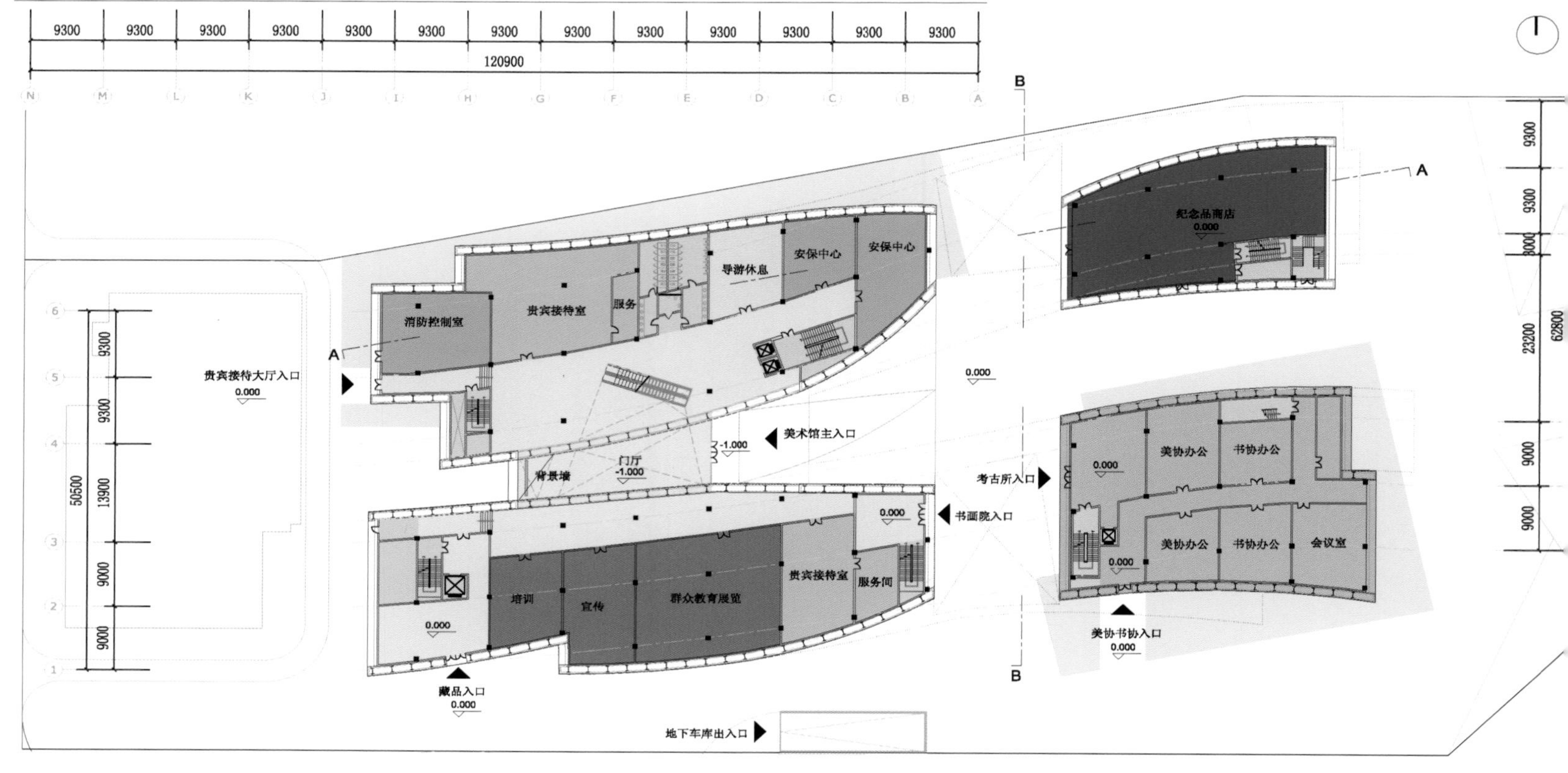
9300
120900
消防控制室
贵宾接待室
服务
导游休息
安保中心
安保中心
贵宾接待大厅入口
美术馆主入口
门厅
背景墙
纪念品商店
美协办公
书协办公
会议室
考古所入口
书画院入口
培训
宣传
群众教育展览
贵宾接待室
服务间
藏品入口
美协书协入口
地下车库出入口
50500
62800

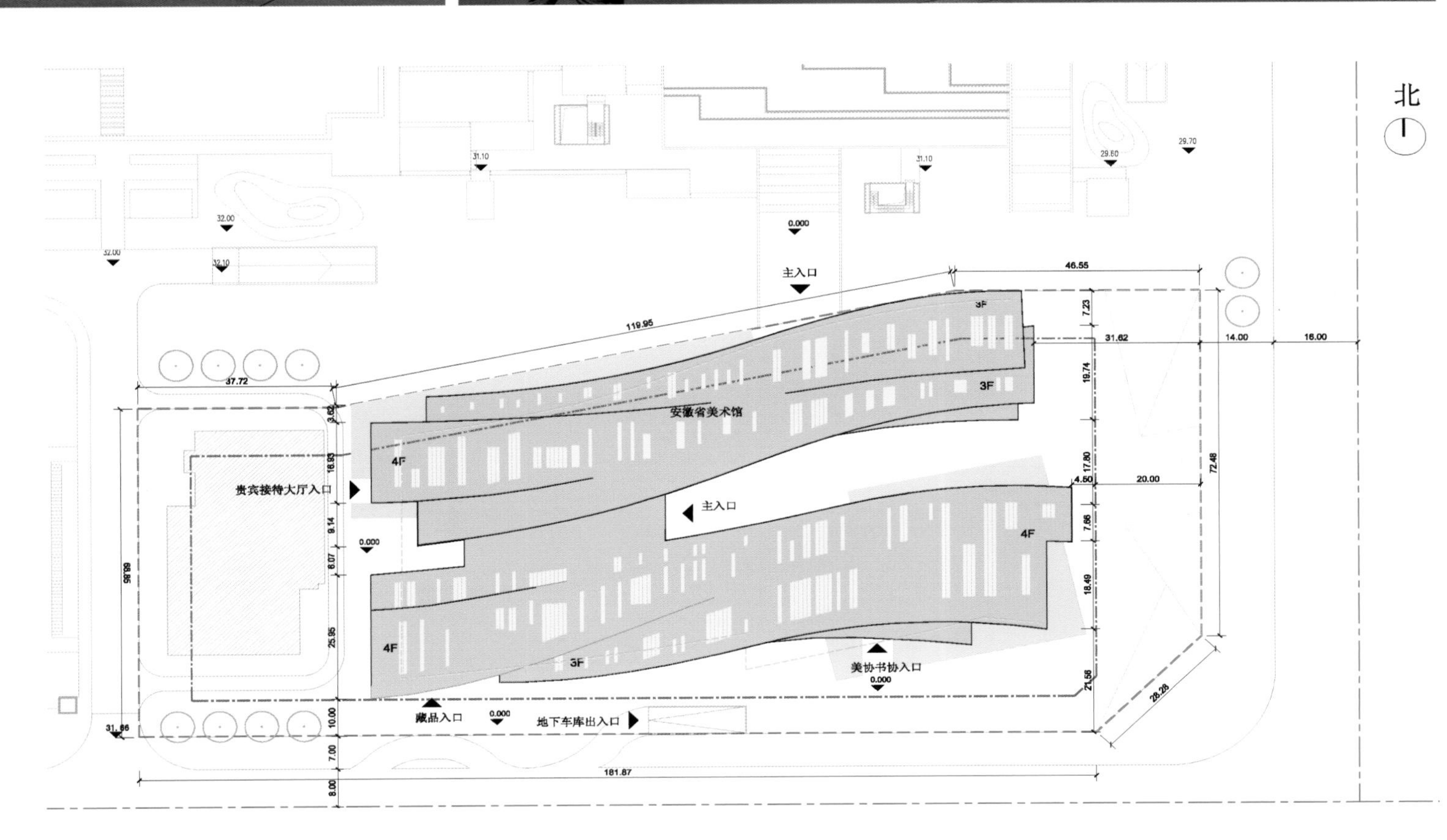
北
主入口
安徽省美术馆
贵宾接待大厅入口
主入口
美协书协入口
藏品入口
地下车库出入口
4F
3F
181.87
119.95
46.55

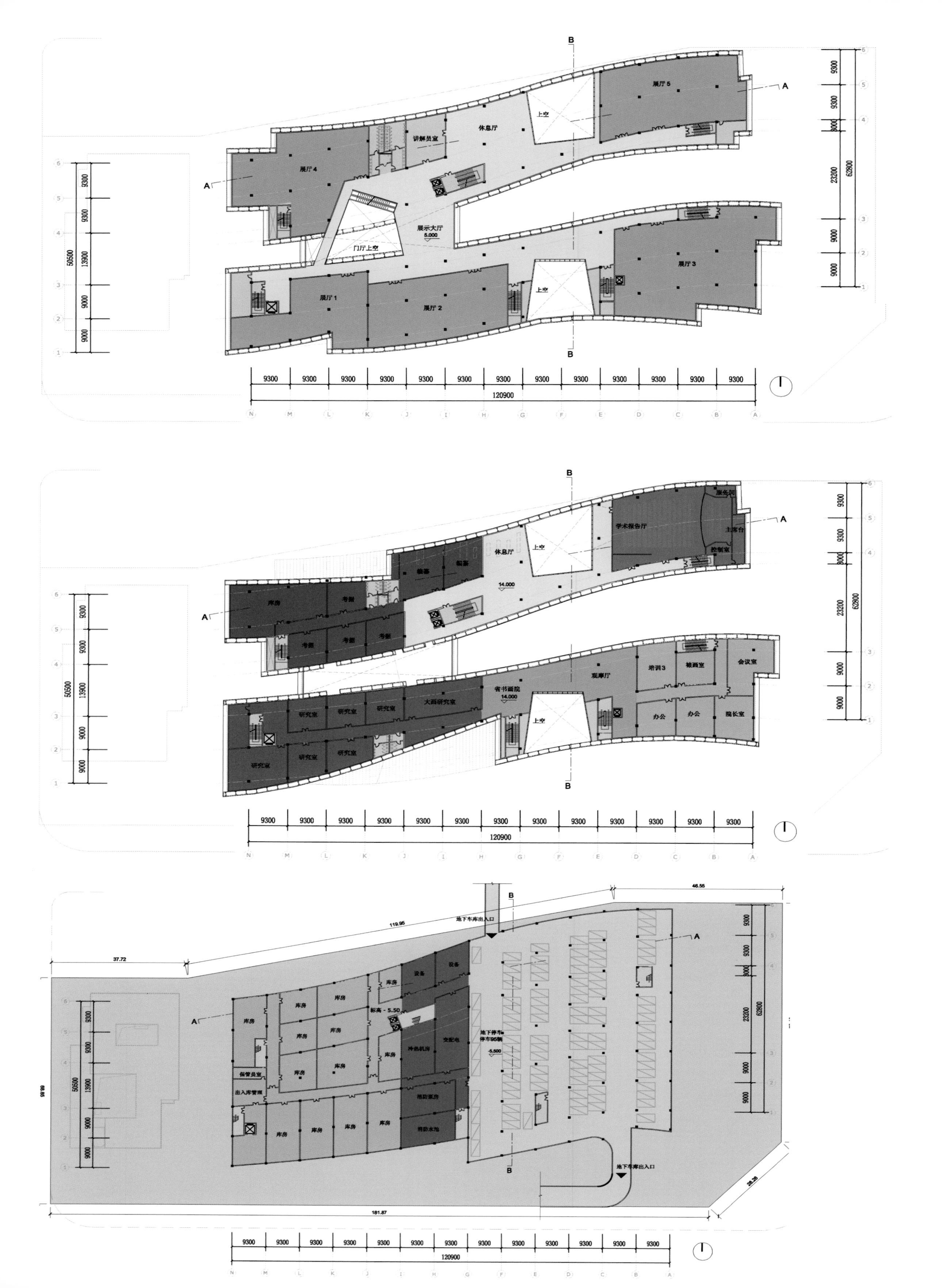

展厅 5
上空
休息厅
讲解员室
展厅 4
展示大厅
5.000
门厅上空
展厅 3
展厅 1
展厅 2
120900
学术报告厅
服务间
主席台
控制室
库房
考据
研究室
大画研究室
省书画院
14.000
观摩厅
培训3
接待室
会议室
办公
院长室
地下车库出入口
标高 - 5.50
设备
库房
冷热机房
变配电
保管员室
出入库管理
消防泵房
消防水池
地下停车
停车95辆
-5.500

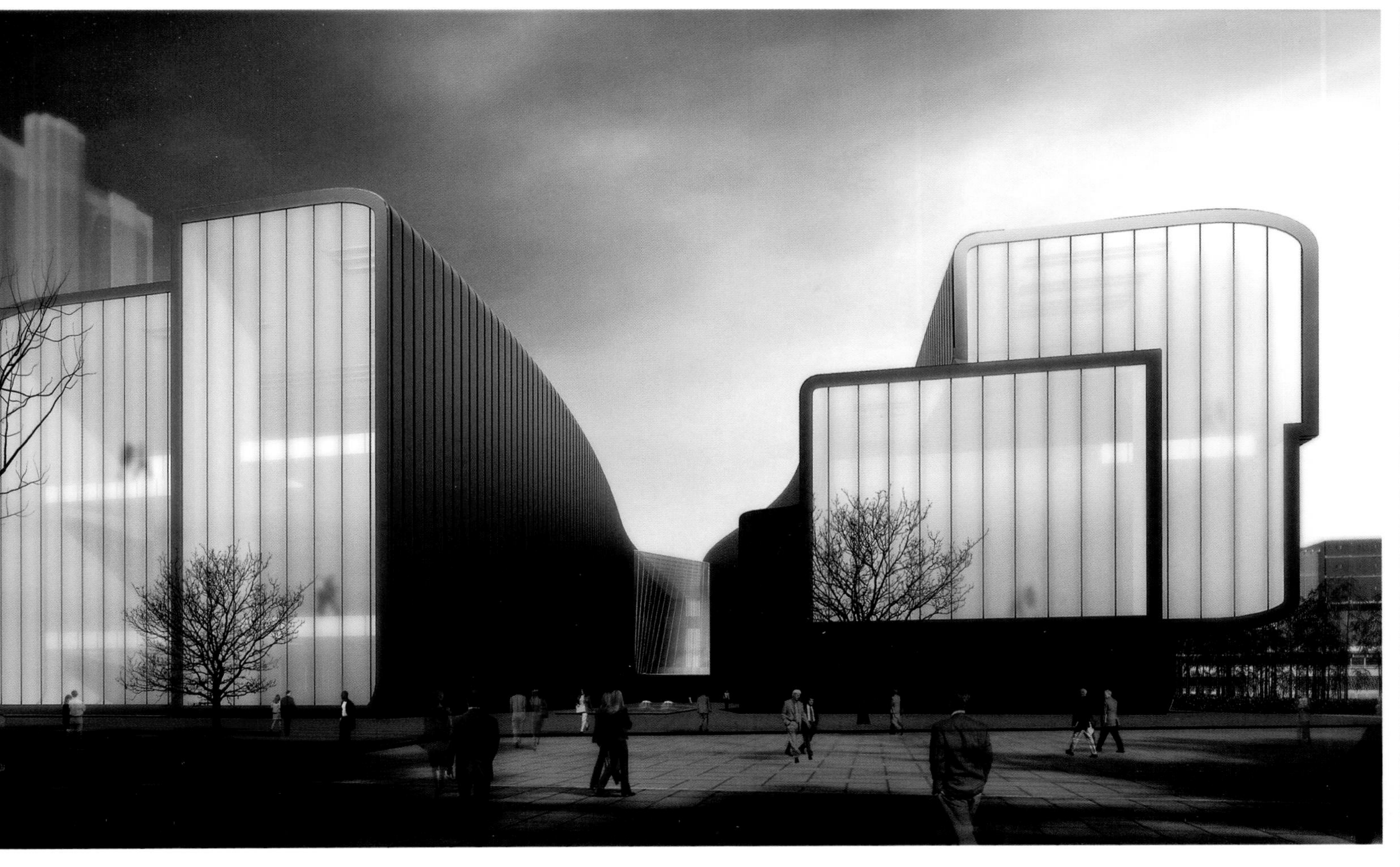

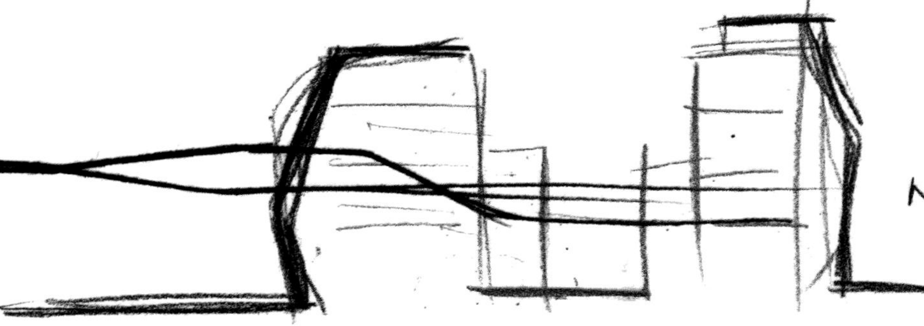

Profile

Anhui Provincial Art Museum developed by RTA-Office is located in a big green area. The proposal pays special attention to spaces' distribution and is designed with the aim of holding international art activities.

Architectural Design

The plot is located in a new political and cultural district. The new building will occupy the southwest corner of the park. The client asked for a design in accordance with the surrounding environment, for a modern architectural concept, simple and elegant, at the same time for a design with complete functions. Exhibition reflects the history of art development in Anhui Province and outside through artists' works of art, literature, etc.

The project finds inspiration from the rugged mountain landscapes, the natural elements, and the silk painting and wood carving — unique of Anhui Province. The proposed design organizes the program into two separate structures which are connected by unified architectural skin.

韩国釜山歌剧院

Busan Opera House

设计单位：Henning Larsen 建筑事务所
Tomoon 建筑事务所
合作单位：奥雅纳工程顾问公司
开发单位：Busan Metropolitan City
项目地址：韩国釜山
建筑面积：47 100 ㎡
摄影：Henning Larsen 建筑事务所

Designed by: Henning Larsen Architects; Tomoon Architects
Collaboration: ARUP
Client: Busan Metropolitan City
Location: Busan, Korea
Gross Floor Area: 47,100 m^2
Photography: Henning Larsen Architects

项目概况

釜山这个美丽的城市位于高山与峡湾之间，起伏的丘陵为城市提供了优美的天际线。釜山歌剧院是釜山展示的舞台，让所有人都能以独一无二的姿态参与其中。

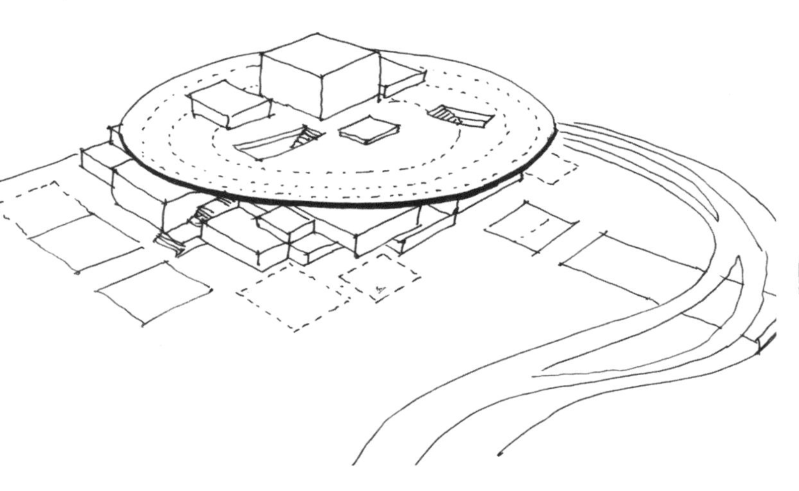
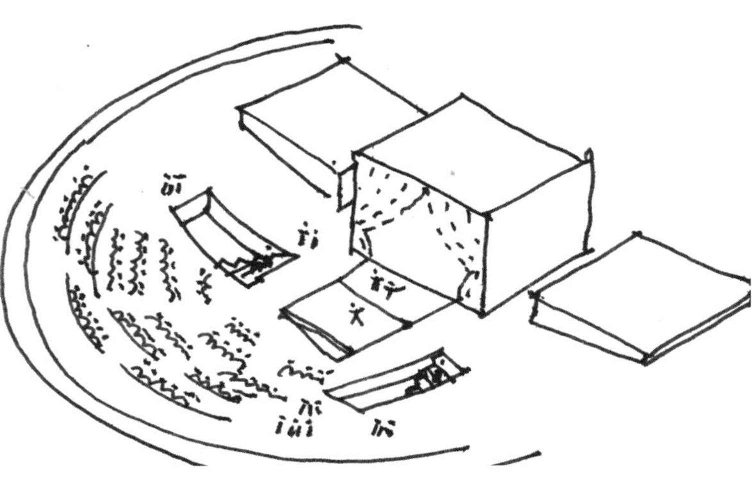
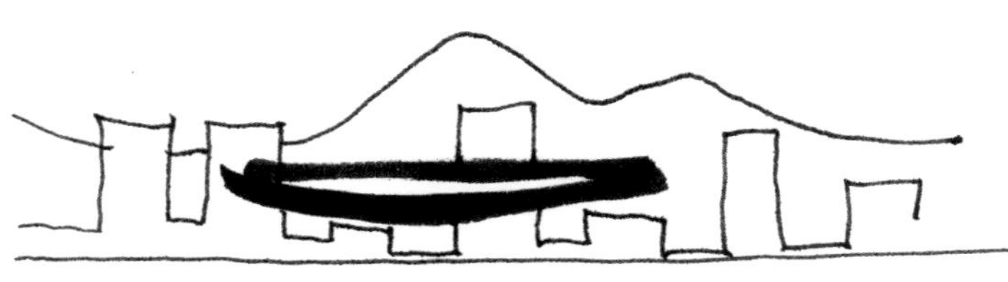

350 VEHICLE PARKING LOT
350 车位的停车场
Staff Core
员工核心
Parking for Performers/Staff 20 Lot
表演嘉宾 / 员工停车场，共 20 个车位
Exhibition Audience Hall
展览会大厅
STAGE SUPPORT AREA
舞台配套区
Background Warehouse
后台仓库
HALL 大厅
Chiller Plant 制冷设备
Sprinkler Tank
喷水灭火系统水箱
EXIT 出口
PARKING 停车场
Electrical LV Room
低电压房
Generator Room
发电机房
MAIN STAGE PIT
主舞台
Sprinkler Pumps
喷淋泵
Transformer Room
变电室
Domestic Hot Water Generation Plant
生活热水生产间
Electrical MV Room
多电压房
Cold Water Tank
冷水箱
PRACTICE ROOM
Conference Audience Hall
会议室大厅
OPERA SUB ENT
歌剧院次入口
Chair wagon storage
座椅储藏室
REHEARSAL ROOM FOR ORCHESTRA
管弦乐队排练室
Media room
媒体室
BUILDING MAINTENANCE 建筑维修
General Stage Warehouse
普通舞台仓库
Multi-purpose Repair Room
多用途维修间
Box office
售票处
AHU Room
空气处理机室
Warehouse
Dressing Room for Orchestra
Lounge for Orchestra
Instrument Warehouse
Piano Warehouse

DAILY KISS AND RIDE 日常临停接送区

PUBLIC FRONT ARRIVAL PLAZA 公共前抵达广场

PUBLIC UNDERGROUND PARKING 公共地下停车场

通往屋顶的入口和公共通道
ENTRANCE AND PUBLIC ACCESS TO ROOF

BASIN 水池

BASIN 水池

BUS PARKING 巴士停车场

SUPERVISED BACKSTAGE AREA 后台区

CAFE RESTAURANT 餐厅

咖啡厅 CAFE

厨房

舞台配套区 STAGE SUPPORT AREA

LOADING 装载

EXHIBITION SUPPORT SPACE

后台走廊 BOH CORRIDOR

舞台配套区 STAGE SUPPORT AREA

BACK STAGE 后台

歌剧排练室 REHEARSAL ROOM FOR OPERA

PERFORMER & STAFF ENT.

BACK STAGE PARKING 后台停车场

CAFE PAVILION

出口停车场 EXIT PARKING

CONFERENCE HALL 会议厅

侧舞台左侧 SIDE STAGE LEFT

OPERA HOUSE 歌剧院 MAIN STAGE

侧舞台右侧 SIDE STAGE RIGHT

BASIN 水池

BASIN 水池

EXIT PARKING

CONFERENCE HALL ENT.

VIP ENT.

特殊活动临停接送区 SPECIAL EVENT KISSAND RIDE

POSSIBLE BOAT DROP OFF

ENTRANCE 入口

整个公共大堂 ENTIRE PUBLIC LOBBY

KISS & RIDE VIP ENT. 临停接送区 VIP入口

KISS AND RIDE VIP ENTRANCE

BASIN 水池

OPERA MAIN FOYER 歌剧大堂

SUB ENT. 次入口

ENTRANCE 入口

CAFE PAVILION 咖啡馆

CAFE RESTAURANT 咖啡厅餐厅

MOVABLE WALLS FOR WAVE PROTECTION 可移动墙体

BASIN 水池

CAFE PAVILION

POSSIBLE VIP BOAT DROP OFF

设计构思

釜山歌剧院位于城市与水滨之间，其设计灵感来源于釜山特殊的地理位置。建筑的总体形态源于韩国的传统哲学——天地交泰、地水相连。设计巧妙地运用了弧形结构，微妙的弧线形既在观念上加强了这种传统哲学意识，又将各种本不相关联的因素在特定的条件下整合在一起，成为包容个体动态活动的强烈象征。

设计特色

设计旨在构建一个艺术与歌剧之乡，碎木箱似的外壳将建筑包裹住，为建筑提供了荫蔽。内凹的建筑形态为圆形剧场提供了向心感，也形成了一个既可遮风避雨，又可隔离噪音的空间。

圆形的屋顶以圆润的曲线结构拥抱天空，在棱角分明的城市建筑与周围的山峰之间凸显出来，同时又与海洋和天际线形成呼应。大屋顶既具备剧场的功能，同时也是一个宏伟的观景平台，于此可将周围的高山、峡湾等自然风光与城市景观纳入眼中。

礼堂是建筑的中心，就像一颗镶嵌在木箱之间巨大的珍珠。其内部沿用了经典的马蹄形状，设计了一个适于人们聚会庆典的公共场所。沿着墙壁表面的木板可以根据不同演出类型的需要转换角度以保证良好的视听效果。

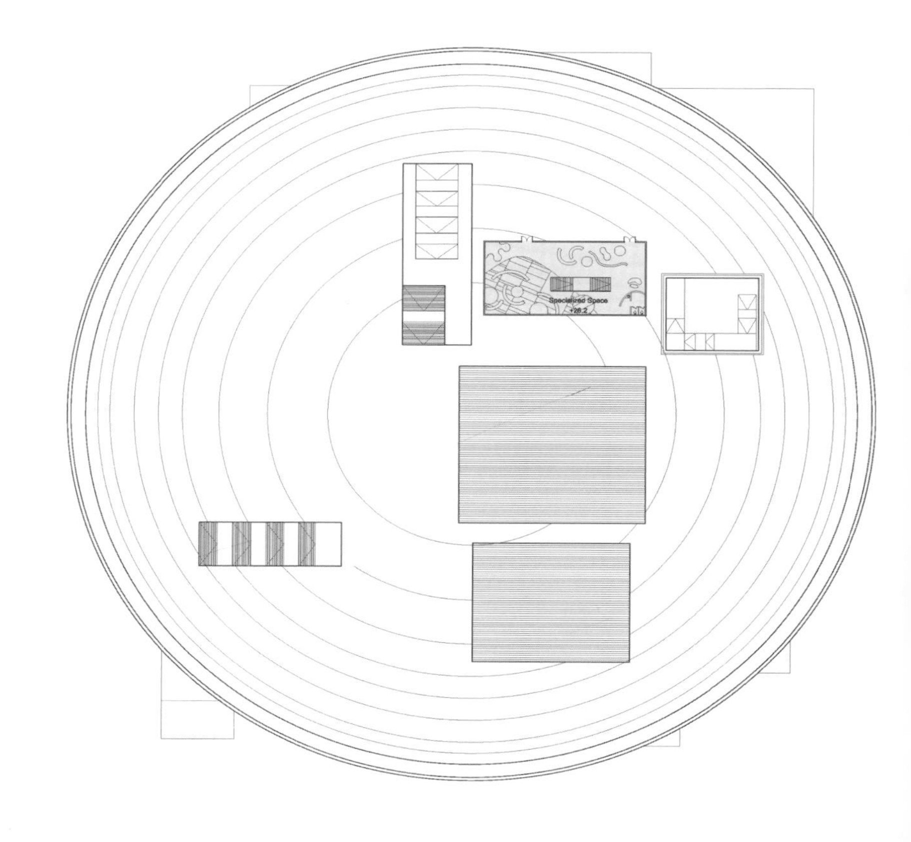

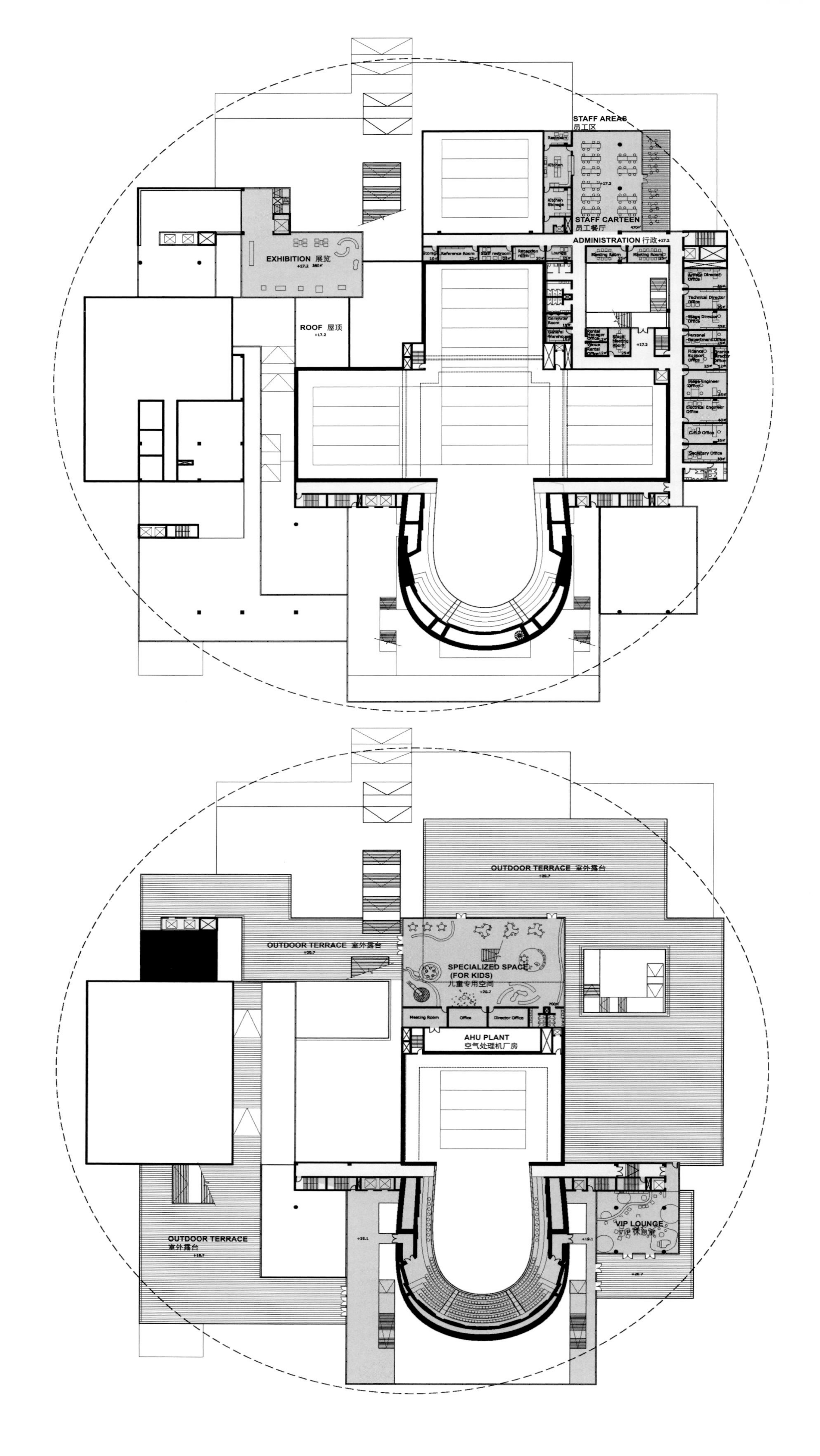
STAFF AREAS
员工区
STAFF CARTEEN
员工餐厅
ADMINISTRATION 行政
EXHIBITION 展览
ROOF 屋顶
OUTDOOR TERRACE 室外露台
OUTDOOR TERRACE 室外露台
SPECIALIZED SPACE
(FOR KIDS)
儿童专用空间
Meeting Room
Office
Director Office
AHU PLANT
空气处理机厂房
OUTDOOR TERRACE
室外露台
VIP LOUNGE
VIP 休息室

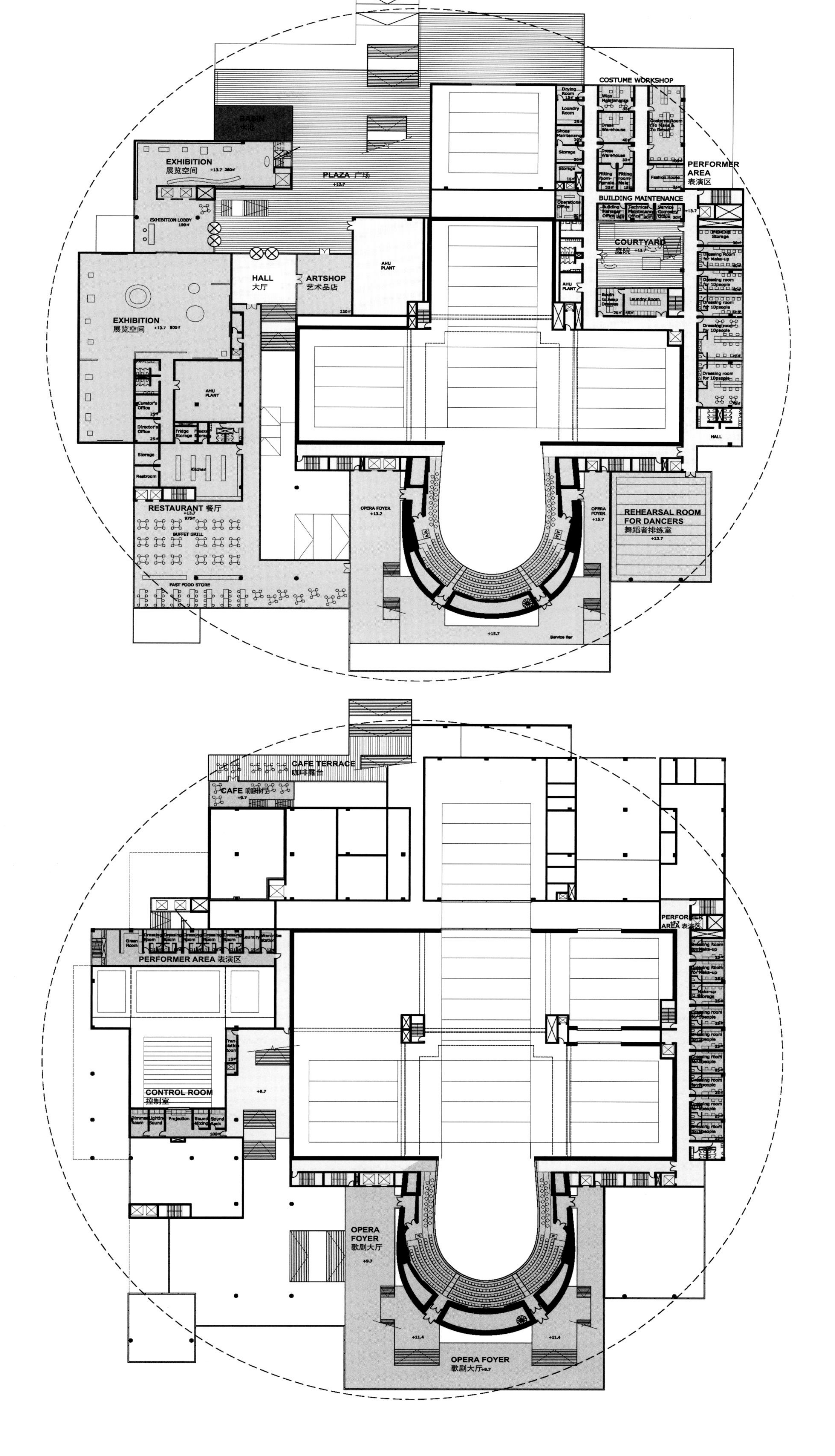
COSTUME WORKSHOP
PERFORMER AREA
表演区
EXHIBITION
展览空间
PLAZA 广场
BUILDING MAINTENANCE
COURTYARD
庭院
HALL
大厅
ARTSHOP
艺术品店
EXHIBITION
展览空间
AHU PLANT
RESTAURANT 餐厅
BUFFET GRILL
FAST FOOD STORE
OPERA FOYER
REHEARSAL ROOM
FOR DANCERS
舞蹈者排练室
CAFE TERRACE
咖啡露台
CAFE 咖啡厅
PERFORMER AREA 表演区
CONTROL ROOM
控制室
OPERA
FOYER
歌剧大厅
OPERA FOYER
歌剧大厅

Profile

Busan has a beautiful location between high mountains and long fjords. The hilly terrain provides gorgeous skyline for this city. The new opera is the stage of Busan, letting everyone become part of and follow the unique life at this special place.

Design Concept

The proposal for the Busan Opera House connects the citizens of Busan to the waterfront. The architecture takes it's inspiration from the particular location in Busan. The building's overall shape is originated from Korea's traditional philosophy: celestial and terrestrial forces in harmony, the earth connecting with water. Arc-shaped structure is intelligently utilized in this design. Subtle arch shape heightens this traditional philosophy while at the same time integrates irrelevant factors to a whole which becomes an impressive symbol for individual activities.

Design Feature

The architecture envisions an "Art or Opera Village" of fragile wooden chests that are sheltered from rain and sun. Internally protruded building offers center-oriented feeling as well as forms a space for sheltering from wind & rain and isolating from noise. The clear shape of the grand roof creates a magnificent icon in the vertical context of high-rises and mountains. At the same time it creates a subtle unity with the ocean and the horizon. The roof functions as an amphitheatre and constitutes a grand new place capable of setting in scene the unique location with the sea, mountains and city of Busan.

The auditorium is the symbolic heart of the house conceived as a giant pearl set between the wooden boxes. The inside of the auditorium follows the classic horseshoe form creating a festive social space. Panels along wall surface can convert angles according to requirements of different performances so as to enhance acoustic effect and ensure favorable visual & audio effect.

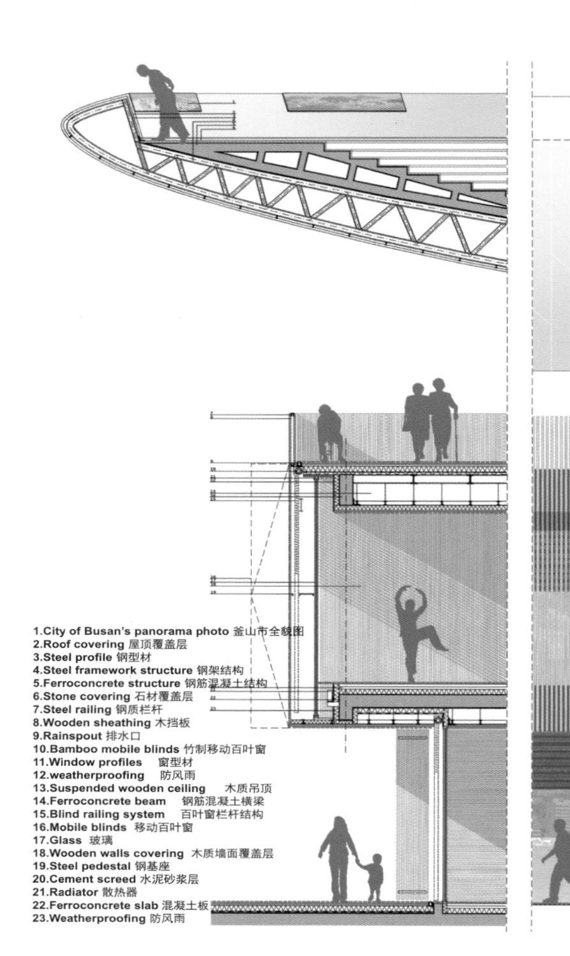

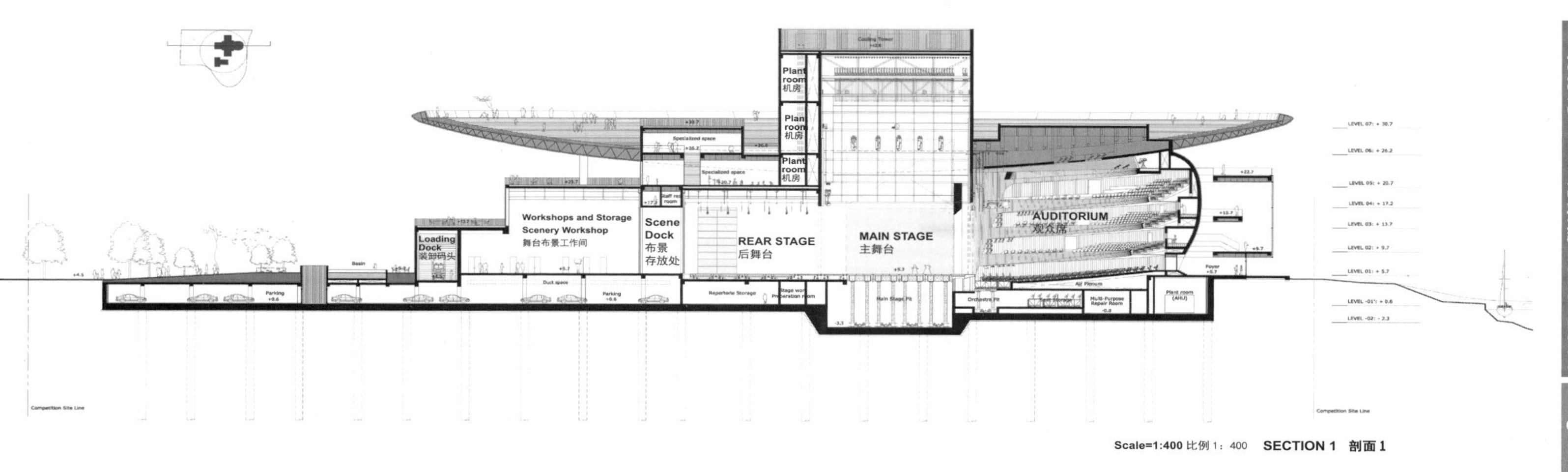

Scale=1:400 比例 1：400 **SECTION 1 剖面1**

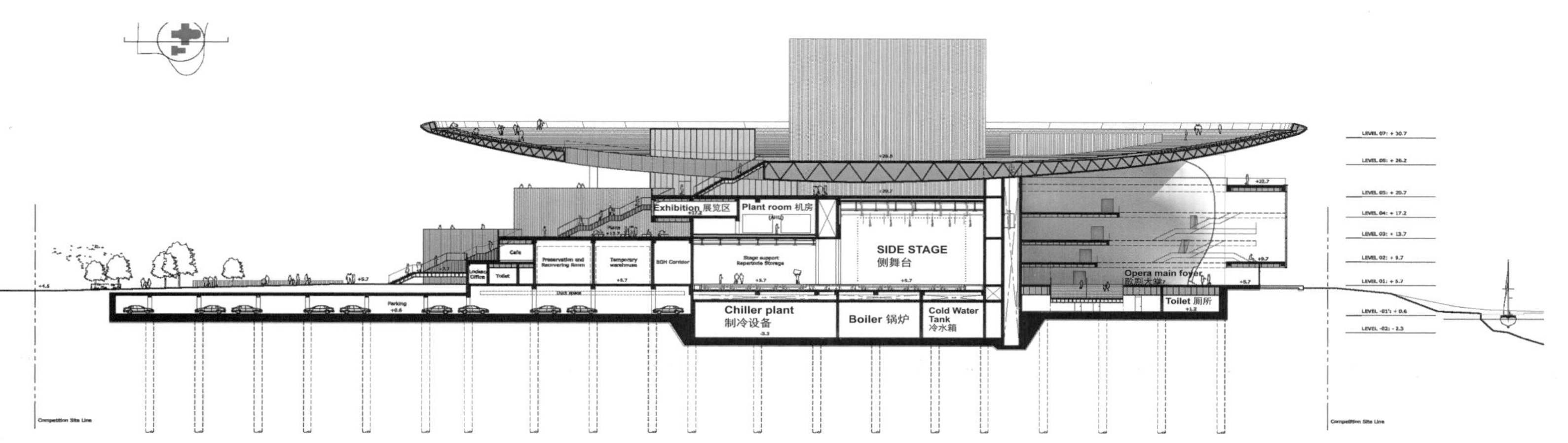

Scale=1:400 比例 1：400 **SECTION 2 剖面2**

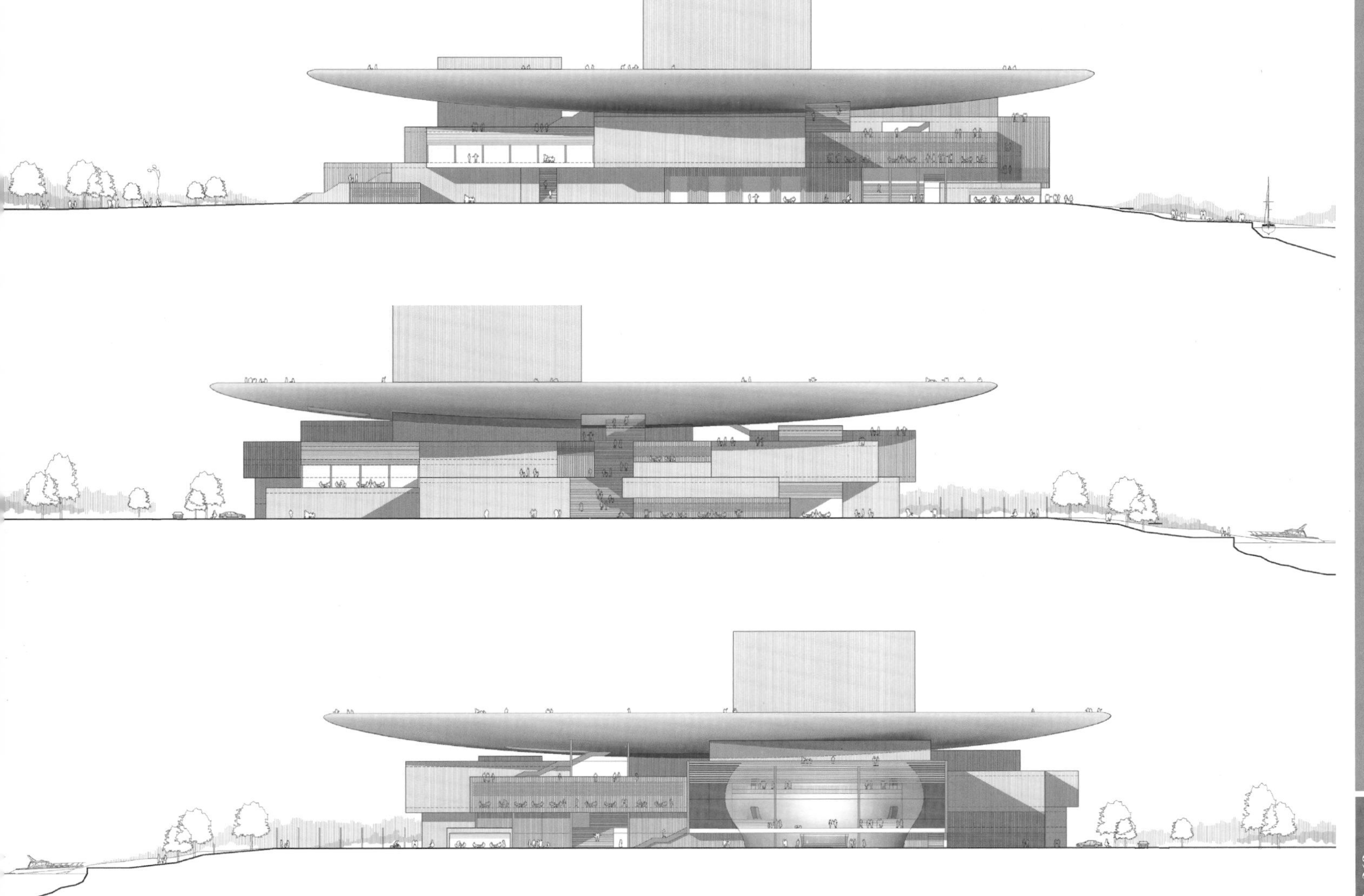

波兰克拉科夫欧洲－远东画廊

Europe – Far East Gallery, Krakow, Poland

设计单位： Ingarden & Ewy 建筑师事务所
开发单位： 克拉科夫日本艺术科技中心 Manggha 博物馆
项目地址： 波兰克拉科夫
项目面积： 2 000 m²
设计团队： Jacek Dubiel　Piotr Hojda
Tomasz Żeludziewicz　Bartosz Haduch
Piotr Urbanowicz　Sebastian Machaj
Jakub Wagner　Bogdan Blady
Enio Ferreira

Designed by: Ingarden & Ewy
Client: Manggha Museum of Japanese Art and Technology, Krakow
Location: Krakow, Poland
Area: 2,000 m²
Design Team: Jacek Dubiel, Piotr Hojda, Tomasz Żeludziewicz, Bartosz Haduch, Piotr Urbanowicz, Sebastian Machaj, Jakub Wagner, Bogdan Blady, Enio Ferreira

项目概况

欧洲－远东画廊是日本 Manggha 艺术中心的一个附属楼，旨在展现远至印度半岛的东南亚文化。这一展览空间将组织古典与现代的东方艺术展览，收藏的远东艺术品也将保存在特定的收藏空间。

建筑设计

画廊有两个展览厅，分别位于建筑的第一层和第二层，可以灵活地展现不同风格的传统和现代艺术品。建筑还包含了储存空间、办公空间、会议厅以及一些必备的备用设施，其中具有行政和办公职能的秘书处、会议室以及配有储物间的办公室都设置在第二层。

无论是在形态还是功能上，建筑都延续了日本 Manggha 艺术中心的风格和主要特征，其规模和建筑布局从属于 Manggha 艺术中心，好像 Manggha 的一道背景。

建筑的位置偏离了 Manggha 中心的开放空间，两者围合形成的内部公共空间得到扩展，丰富了建筑的自我表现形式。

CENTRUM SZTUKI I TECHNIKI JAPOŃSKIEJ "MANGGHA"

PAWILON HERBACIANY

SZKOŁA JAPOŃSKA

GALERIA EUROPA-DALEKI WSCHÓD

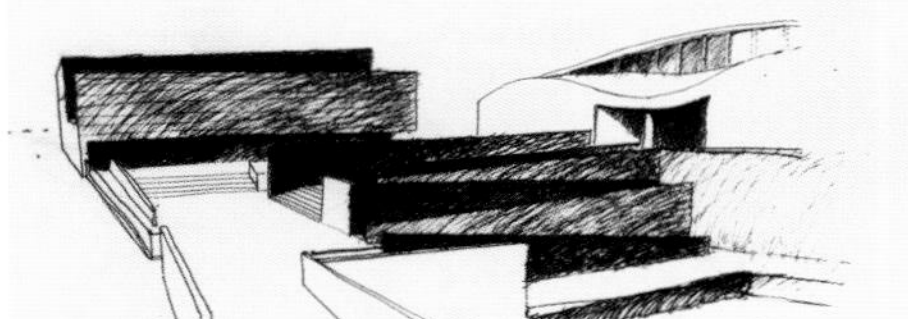

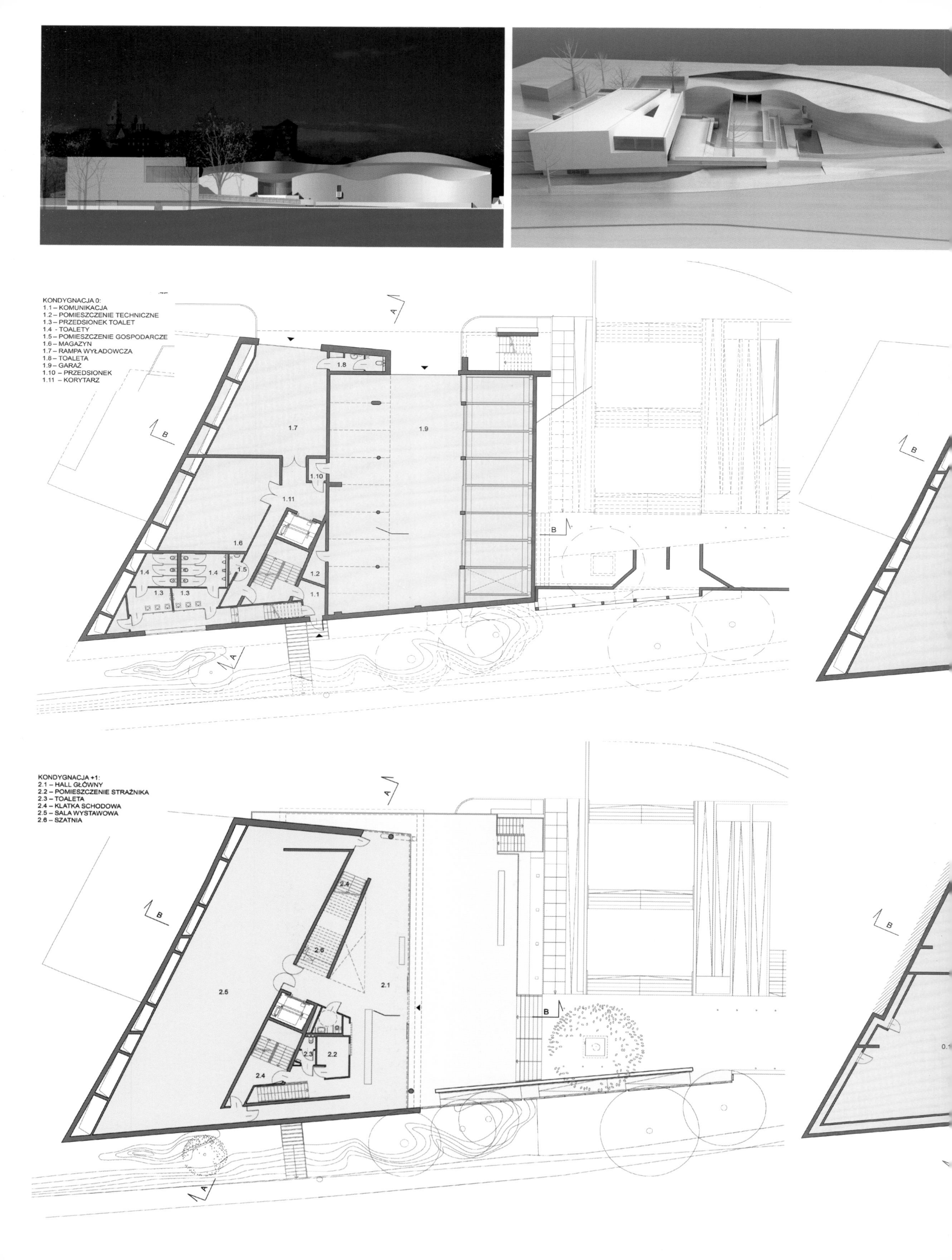
KONDYGNACJA 0:
1.1 – KOMUNIKACJA
1.2 – POMIESZCZENIE TECHNICZNE
1.3 – PRZEDSIONEK TOALET
1.4 - TOALETY
1.5 – POMIESZCZENIE GOSPODARCZE
1.6 – MAGAZYN
1.7 – RAMPA WYŁADOWCZA
1.8 – TOALETA
1.9 – GARAŻ
1.10 – PRZEDSIONEK
1.11 – KORYTARZ
KONDYGNACJA +1:
2.1 – HALL GŁÓWNY
2.2 – POMIESZCZENIE STRAŻNIKA
2.3 – TOALETA
2.4 – KLATKA SCHODOWA
2.5 – SALA WYSTAWOWA
2.6 – SZATNIA

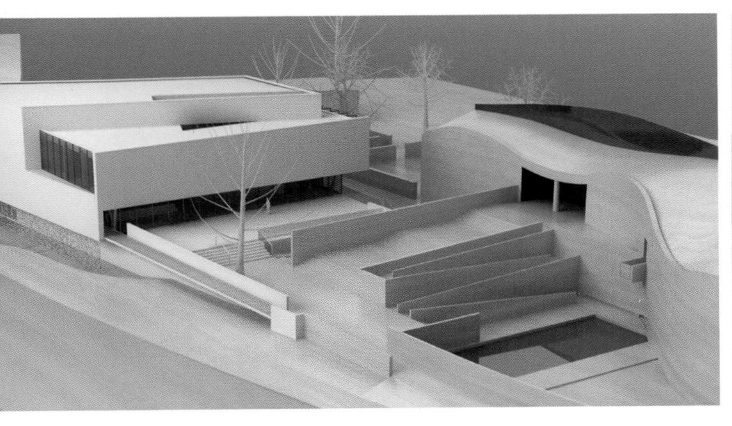

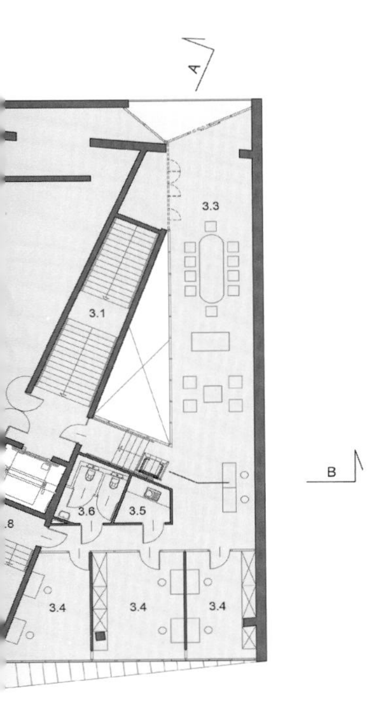

KONDYGNACJA +2:
3.1 – KLATKA SCHODOWA
3.2 – SALA WYSTAWOWA
3.3 – SALA KONFERENCYJA + BIURO
3.4 – BIURA
3.5 – POMIESZCZENIE SANITARNE
3.6 – TOALETY
3.7 – ZAPLECZE
3.8 – KLATKA SCHODOWA

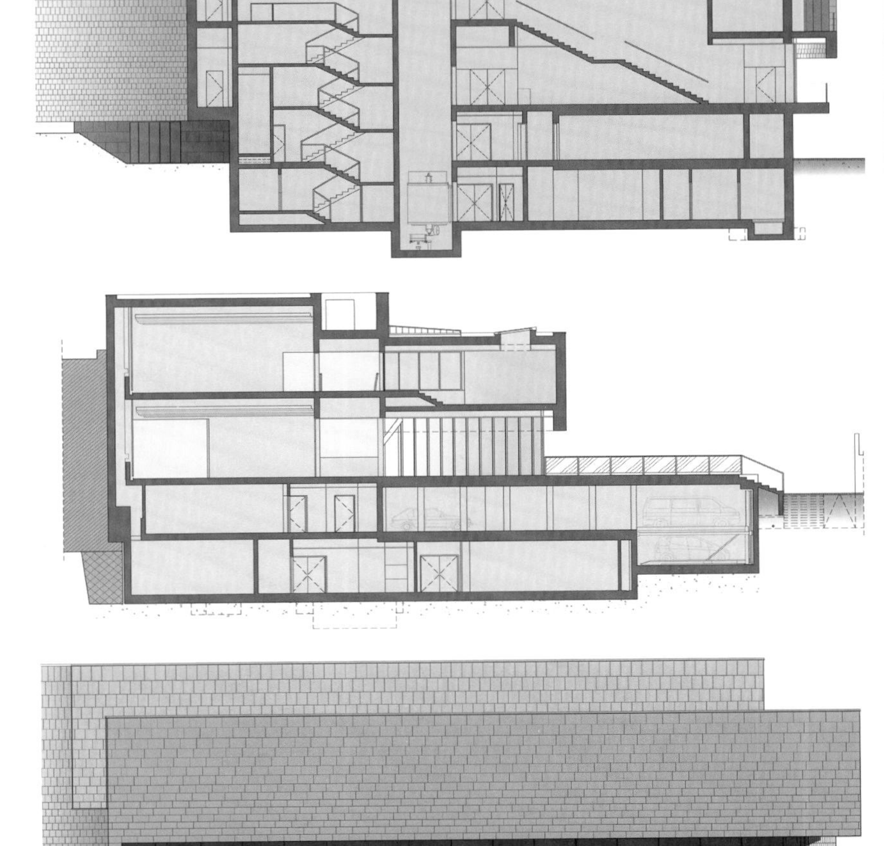

KONDYGNACJA -1:
0.1 – MAGAZYN
0.2 – WENTYLATORNIA
0.3 – PRZYŁĄCZ C.O.
0.4 – ROZDZIELNIA
0.5 – POMIESZCZENIE HYDROFORNI
0.6 – MAGAZYN
0.7 – MAGAZYN
0.8 – POMIESZCZENIE POMOCNICZE
0.9 – POMIESZCZENIE POMOCNICZE
0.10 – SZYB WINDOWY
0.11 – OBNIŻONA CZĘŚĆ GARAŻU

Profile

The "Europe – Far East" Gallery has been designed as an annex to the Center for Japanese Art and Technology Manggha. It is meant to present the culture of the countries of South-Eastern Asia, situated as far as the Indian Peninsula. The exhibition program is to include the organization of diverse exhibitions of old and contemporary oriental art and the artifacts from the Far East acquired by the Center shall be kept in the storage space of the designed facility.

Architectural Design

The gallery has two exhibition halls located on the first and the second floor, which enables flexible presentation of various forms of art, both traditional and modern. Apart from that the building shall have the necessary backup facilities, storage space, office space and a conference hall. The administrative and office functions, i.e. the secretariat, the conference hall and the offices with the utility rooms have been located on the second floor level.

The predominant feature, both from the point of view of the form and of the function, shall still be vested in the neighboring building, namely the Center for Japanese Art and Technology Manggha to which the scale of the new building and its urban and architectural composition have been subordinated.

The gallery has been removed from the open area near the Manggha Center, which has expanded internal public space delineated by the elevations of both buildings and enriched self-expression of the buildings.

波兰克拉科夫会议中心

Congress Center in Krakow, Poland

设计单位：Ingarden & Ewy 建筑师事务所
概念设计：Piotr Urbanowicz Sebastian Machaj
Jakub Wagner Piotr Hojda
Bartosz Haduch Tomasz Koral
J. Kossowicz T.Babicz
P.Redmierski Tomasz Trzebunia-Niebies
开发单位：克拉科夫市议会
项目地址：波兰克拉科夫
总建筑面积：37 200 ㎡
奖项：国际建筑设计竞赛一等奖

Designed by: Ingarden & Ewy
Concept Design Team: Piotr Urbanowicz, Sebastian Machaj,
Jakub Wagner, Piotr Hojda,
Bartosz Haduch, Tomasz Koral,
J. Kossowicz, T.Babicz,
P.Redmierski, Tomasz Trzebunia-Niebies
Client: Krakow City Council
Location: Krakow, Poland
Total Floor Area: 37,200 m^2
Award: 1st Prize in International Architectural Competition

项目概况

项目位于波兰克拉科夫 Grunwaldzki Roundabout 地块，处于汽车、公共汽车和电车道路交汇处，地理位置优越。波兰克拉科夫会议中心的建成将成为维斯瓦河滨海区域建筑设计的参照，也将是克拉科夫片区的一个地标性建筑。

设计特色——材料的混搭构建多彩立面

立面设计参照瓦维尔宫，陶瓷面板、花岗岩、石灰岩、砂岩等经典建筑材料通过个性化的手法填充进玻璃钛钢结构，不同类型的材料和玻璃互相映衬，形成多彩镶嵌的立面，丰富了建筑的表现力，强化了建筑的张力。

建筑设计

项目基地毗邻古城区卡奇米日和普格兹广场区，设

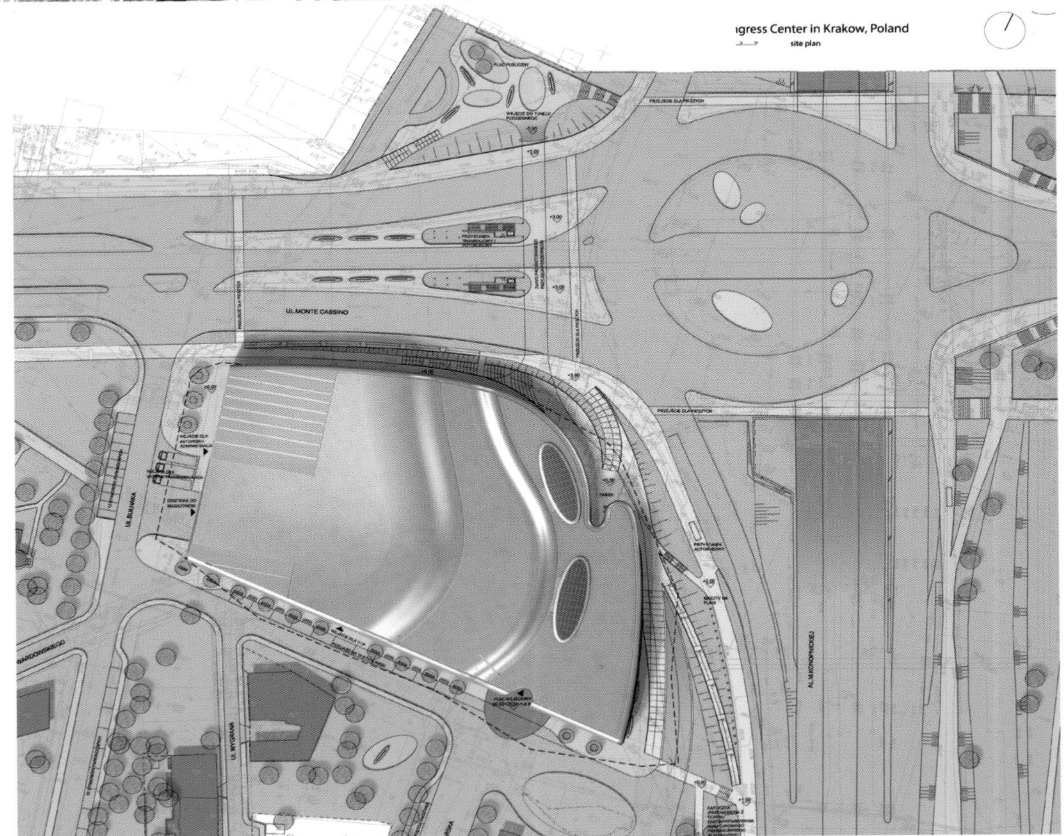

计根据项目所处位置和环境状况以及建筑布局和功能需求，将大面积公共空间面向城市主要遗迹，将建筑融入周边环境中。由于周边绿色林荫大道对这一建筑飞线的影响，使这一立方体的规格沿着维斯瓦河的方向逐渐减弱。

建筑外观在各类材料的有机组合下形成的不规则形状，与覆盖有玻璃的、壮观的克拉科夫门厅形成对比。3 层高的门厅覆盖着玻璃，形成一种“看与被看”的视觉感受，当人们处于瓦维尔宫的观景平台，便可看到这一建筑的全貌。

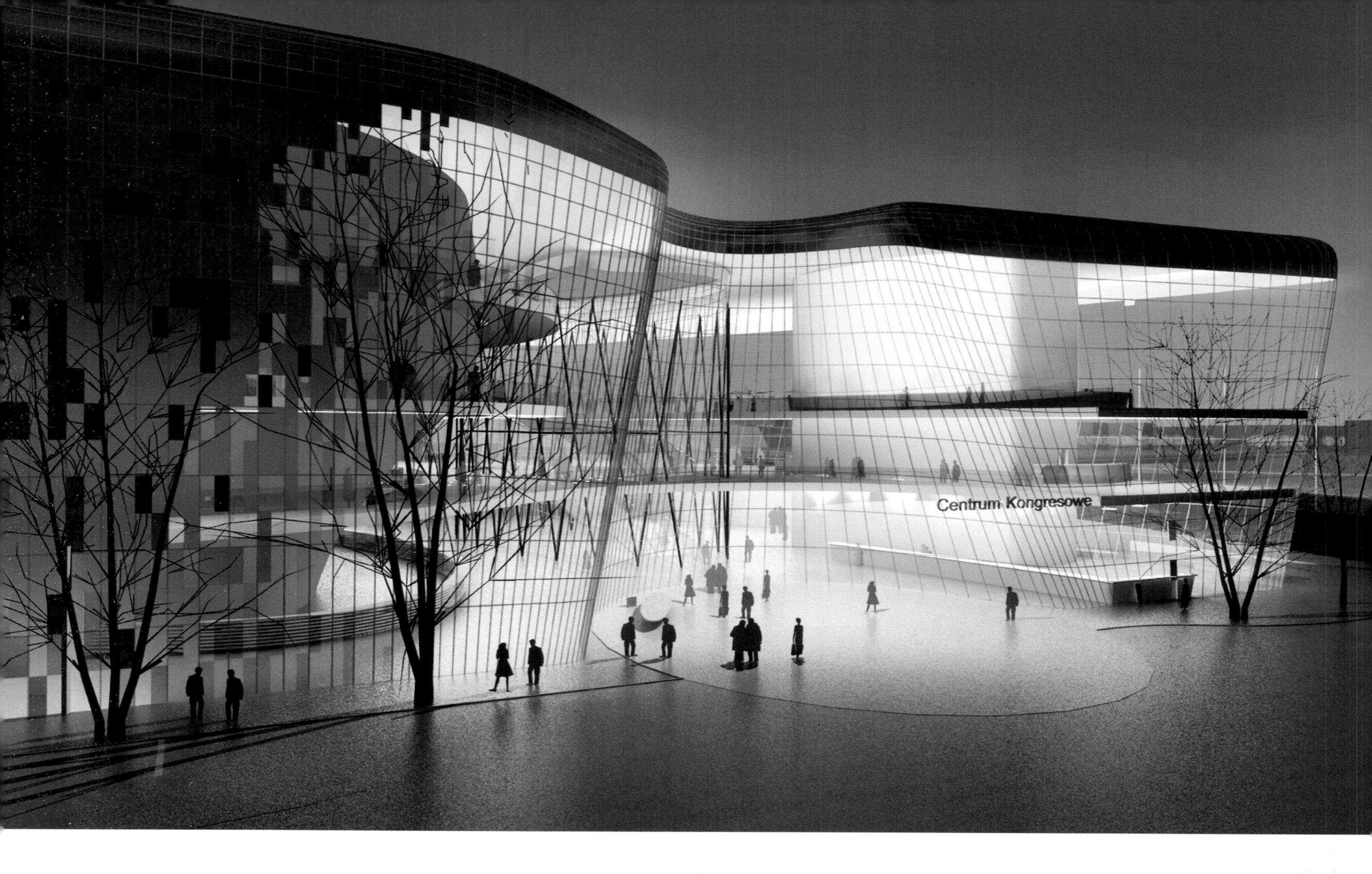

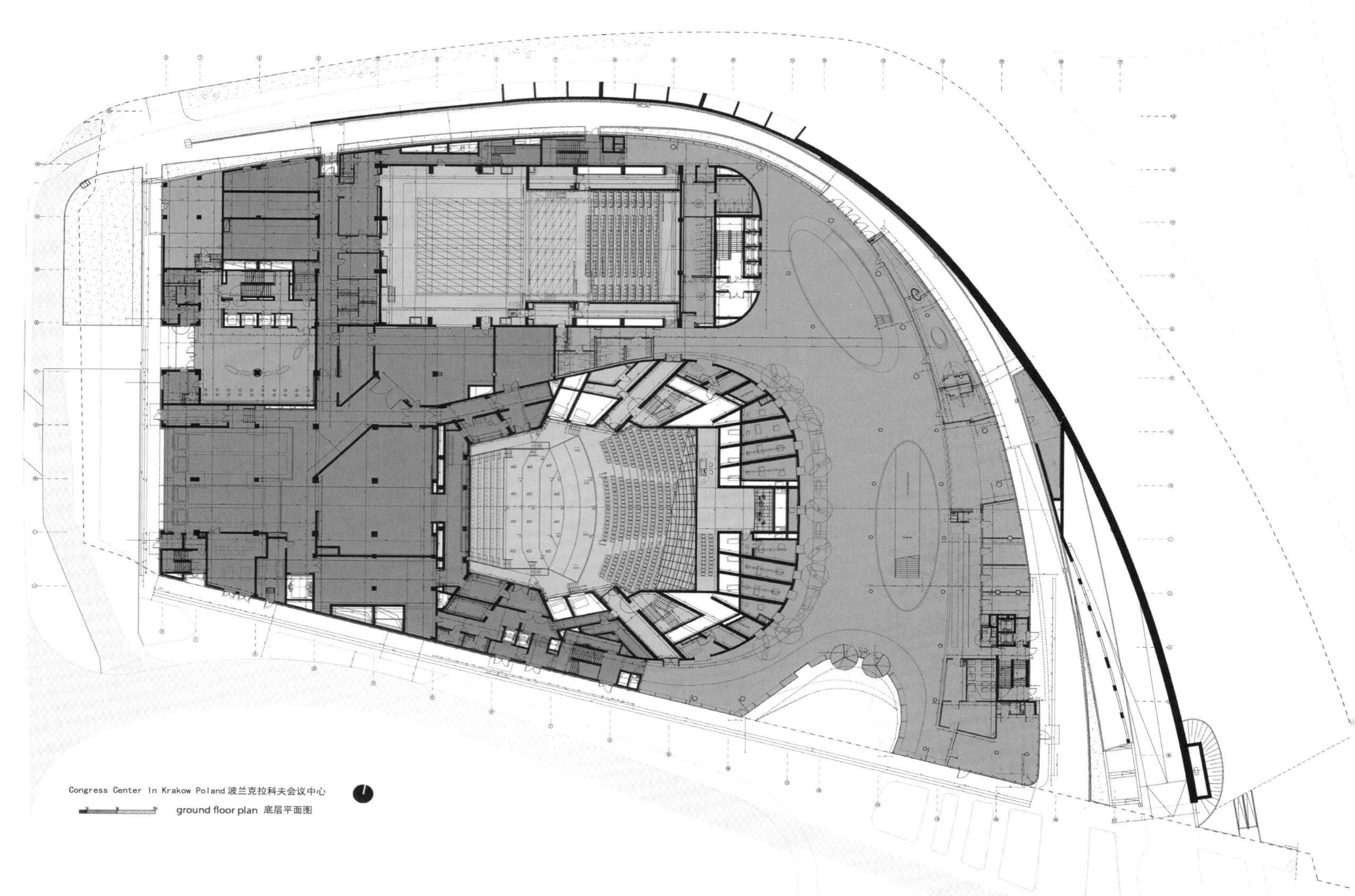

Congress Center In Krakow Poland 波兰克拉科夫会议中心

ground floor plan 底层平面图

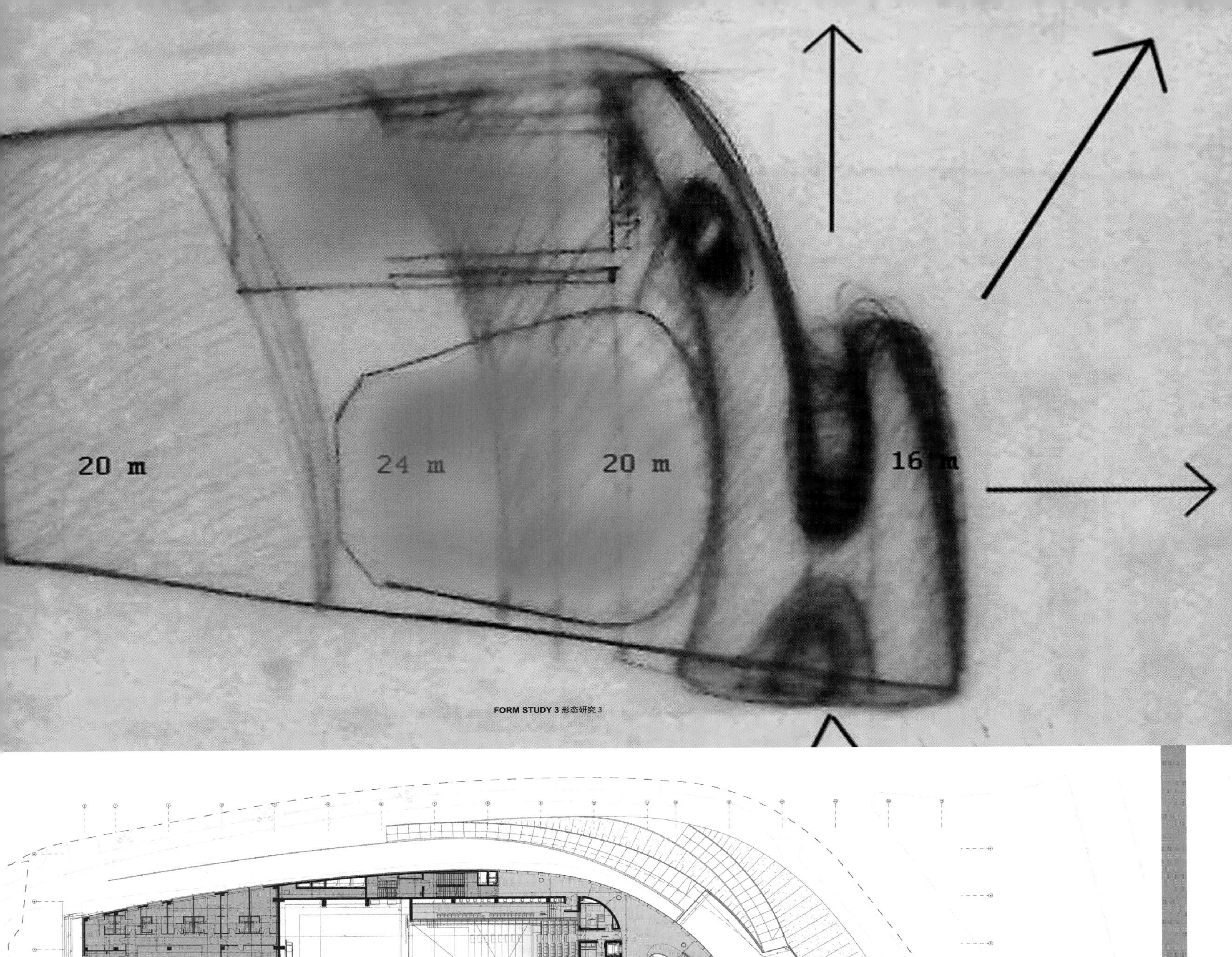

FORM STUDY 3 形态研究 3

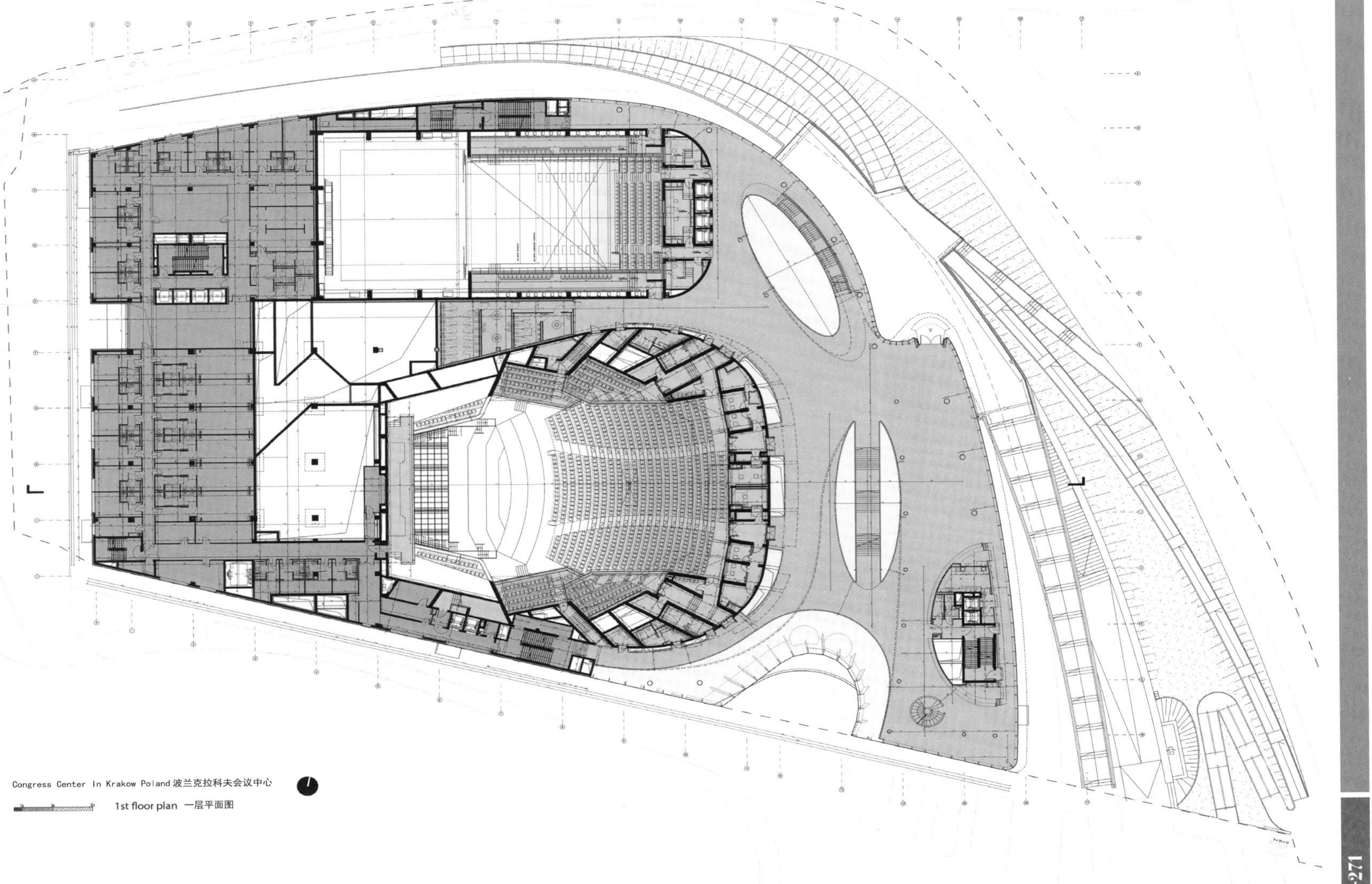

Congress Center In Krakow Poland 波兰克拉科夫会议中心

1st floor plan 一层平面图

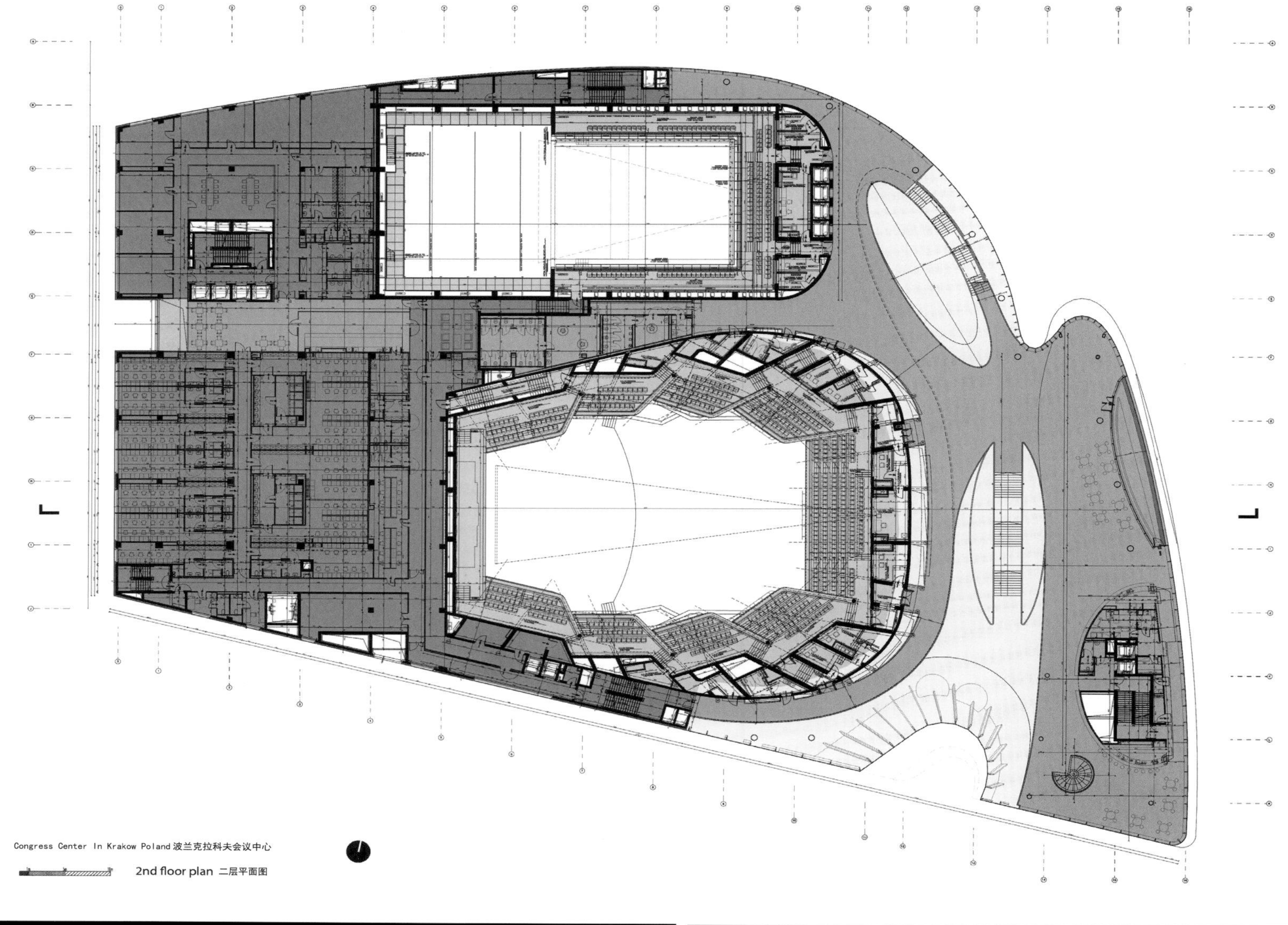

Congress Center In Krakow Poland 波兰克拉科夫会议中心

2nd floor plan 二层平面图

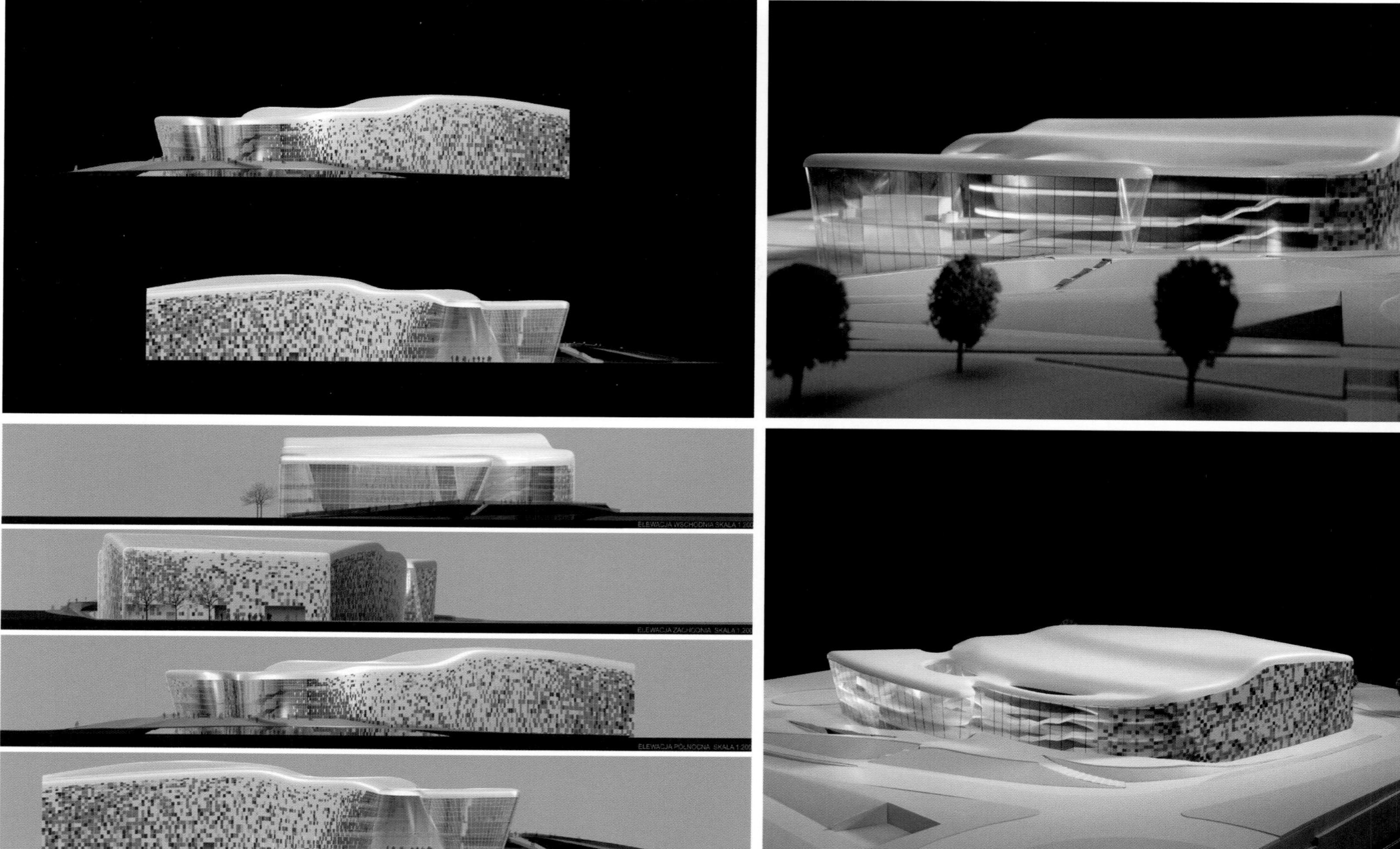

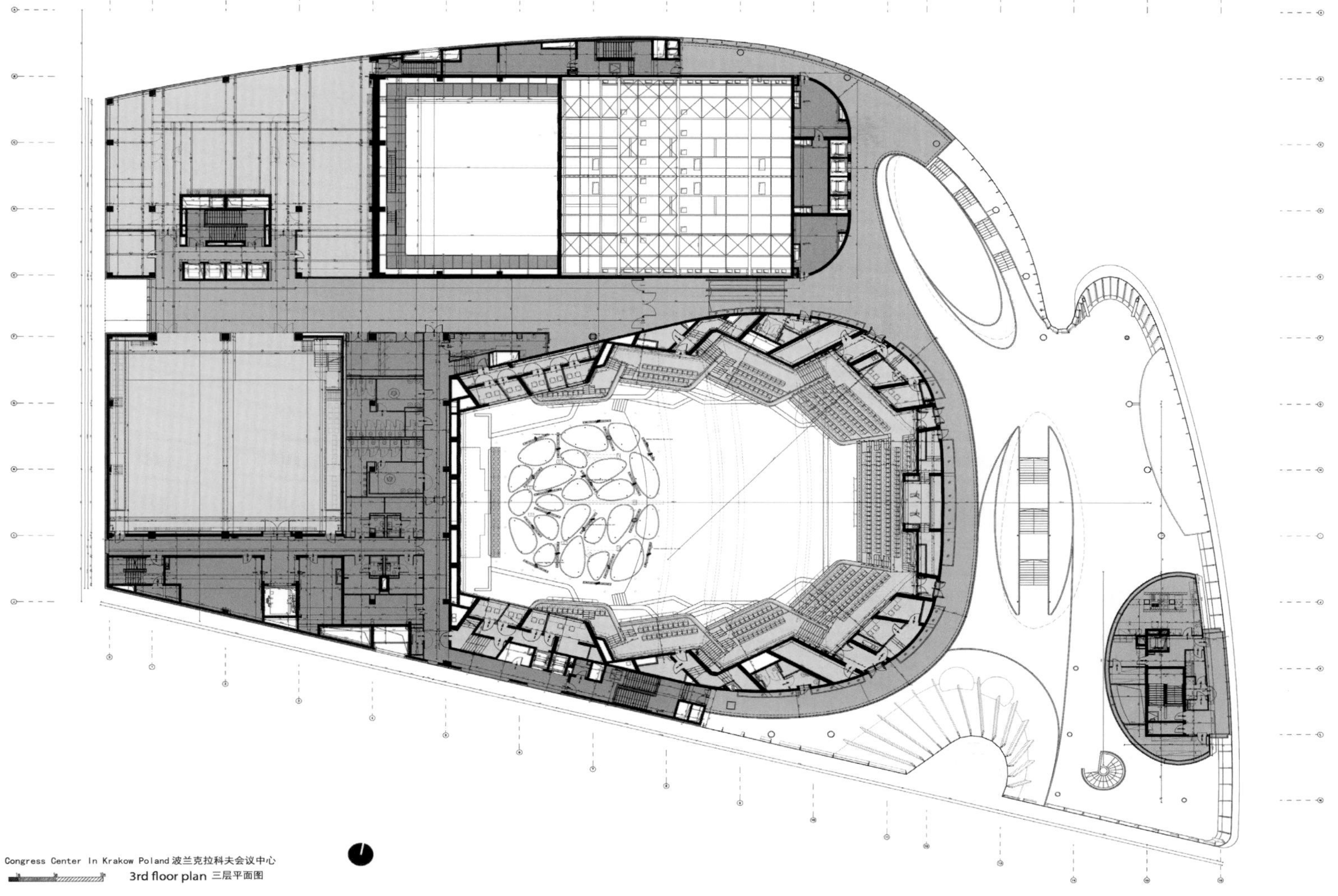

Congress Center In Krakow Poland 波兰克拉科夫会议中心

3rd floor plan 三层平面图

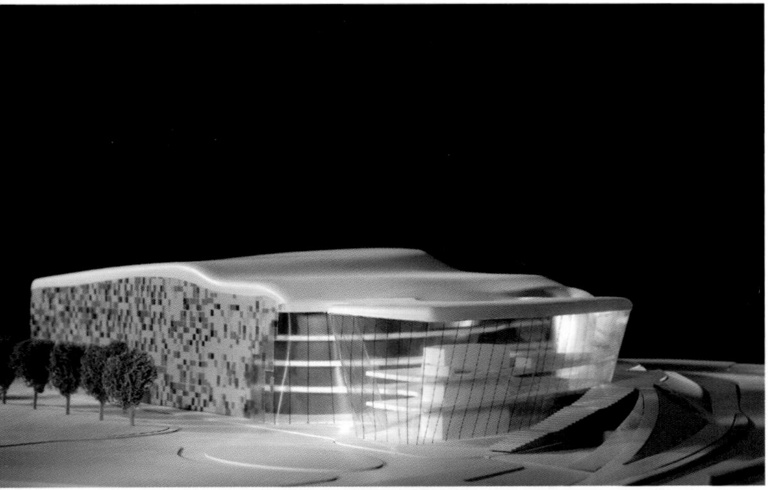

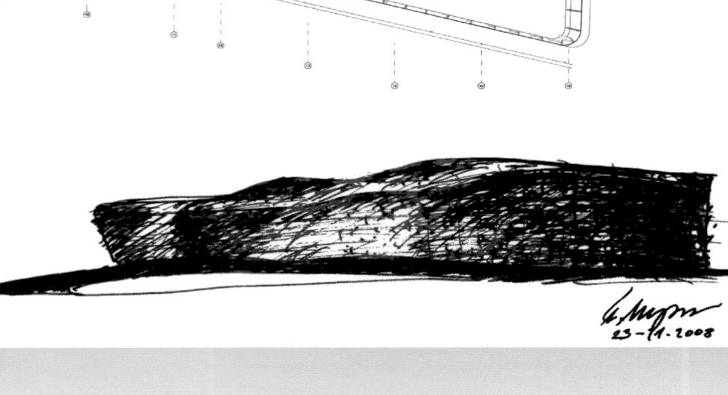

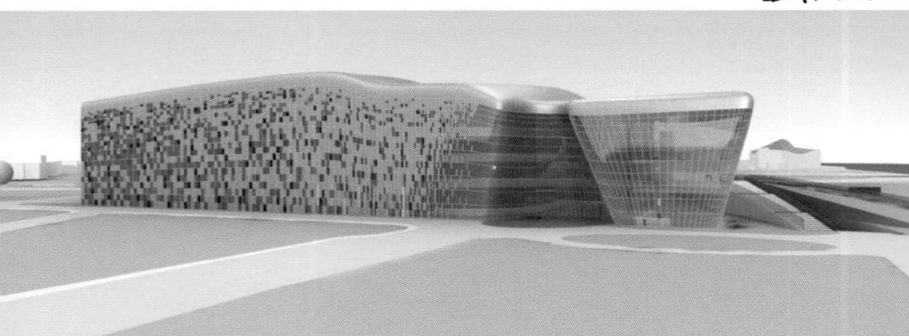

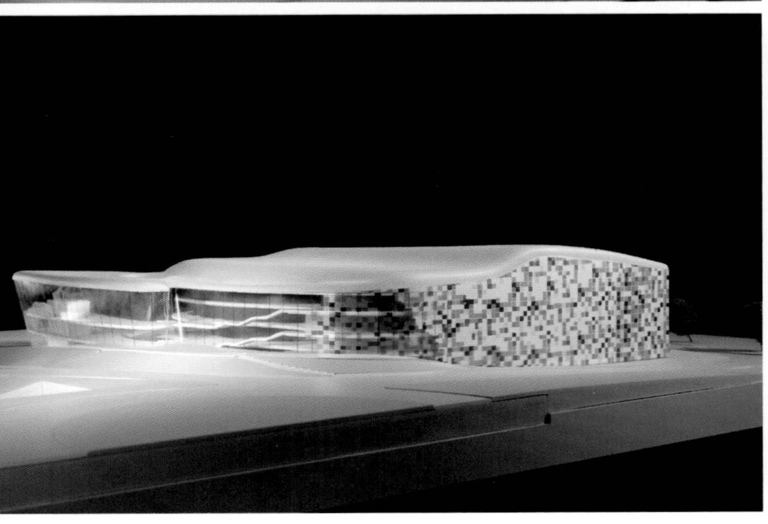

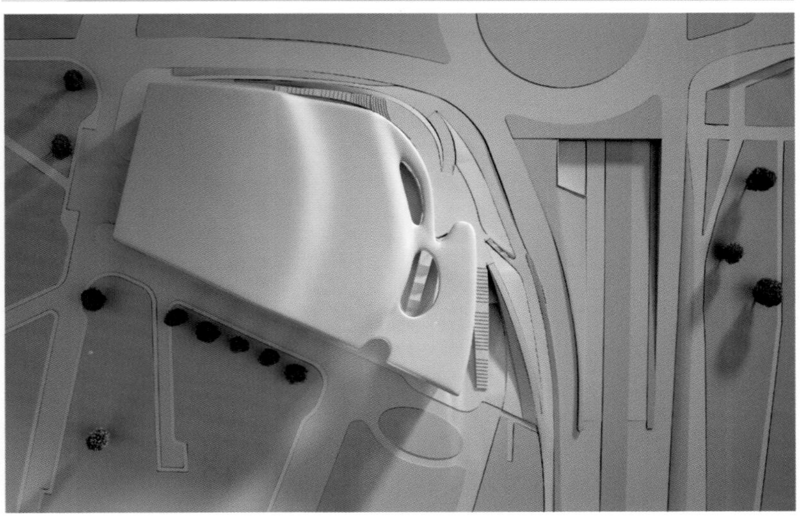

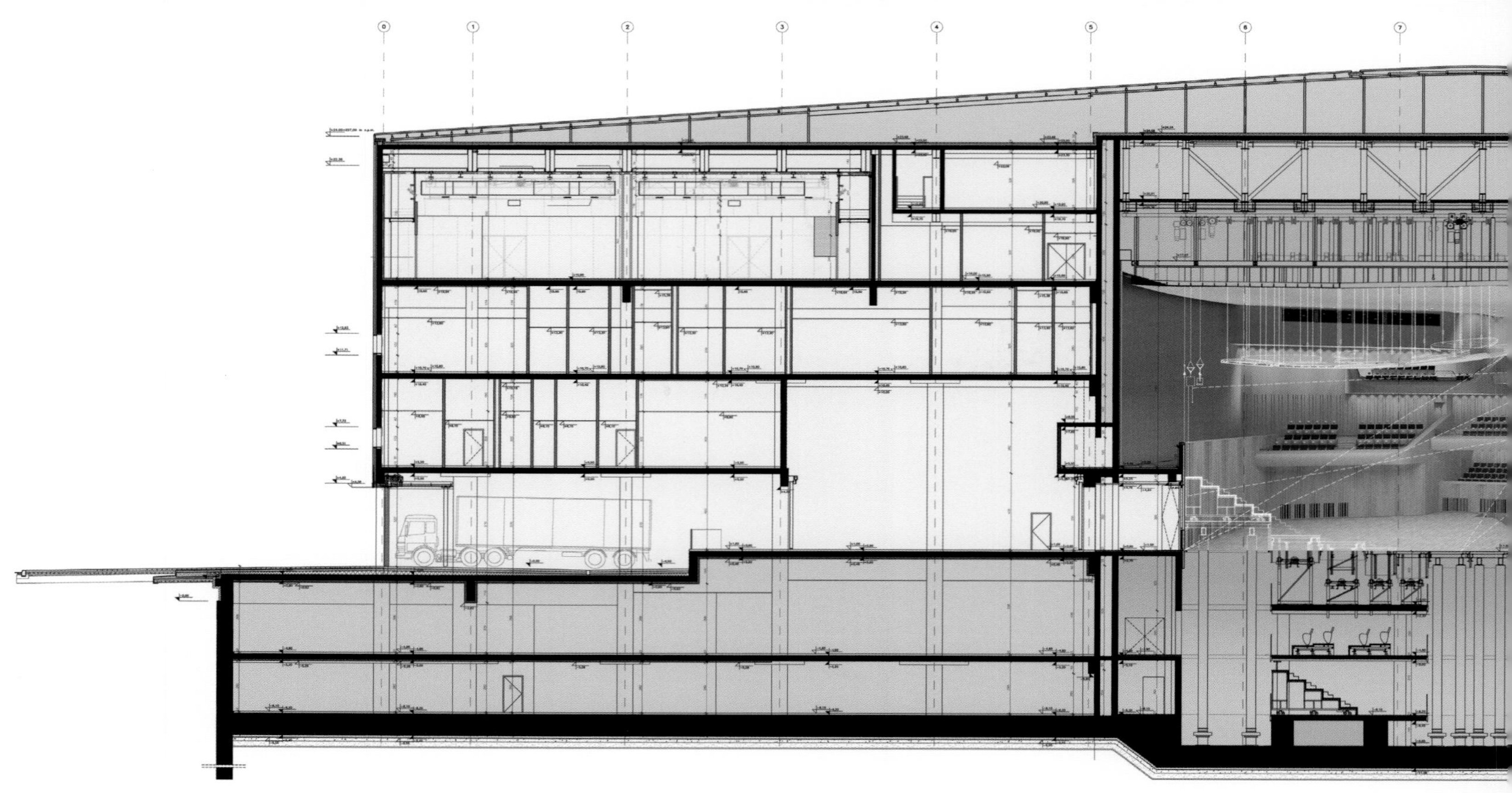

Congress Center In Krakow Poland 波兰克拉科夫会议中心

Section 剖面

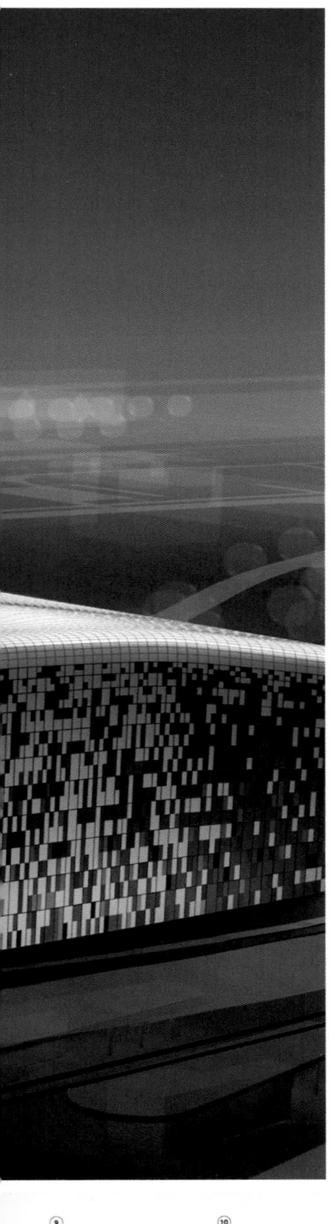

Profile

New Congress Center occupies a corner of the plot nearby Grunwaldzki Roundabout an important car, bus and tram road junction. Congress Center will create a new architectural reference to a fragment of the Krakow City nearby Vistula River waterfront. It will also become a landmark building of Krakow.

Design Feature — Collocation of Materials Compose Colorful Facades

Elevation design of the project refers to Wawel Castle. Ceramic panels, granite, limestone, sandstone and other typical historical materials are filled into glass and titanium steel structure. Composition of differentiated materials juxtaposed with glass creates a colorful mosaic on elevations which has greatly enriched architectural expression of the building and enhanced the tension of the building.

Architectural Design

The site is adjacent to the prestigious historical and contemporary context of the Old City, Kazimierz and Podgorze districts. Based on project's location, environment conditions and buildings' layout and function demands, extensive public spaces are opened to the city's main historical landmarks while integrates buildings with surrounding environment. The solid dimension is diminished in direction of Vistula River due to optical reduction of a large building scale in context of green boulevards nearby. Contemporary countenance of the building is formed by organic irregular shapes and differentiation of materials contrasted with spectacular foyer covered with glazing opened up to a great view of Krakow panorama. Glazed, three story high foyer allows visitors "to see and to be seen". Building solid will be visible from Wawel Castle view terrace.

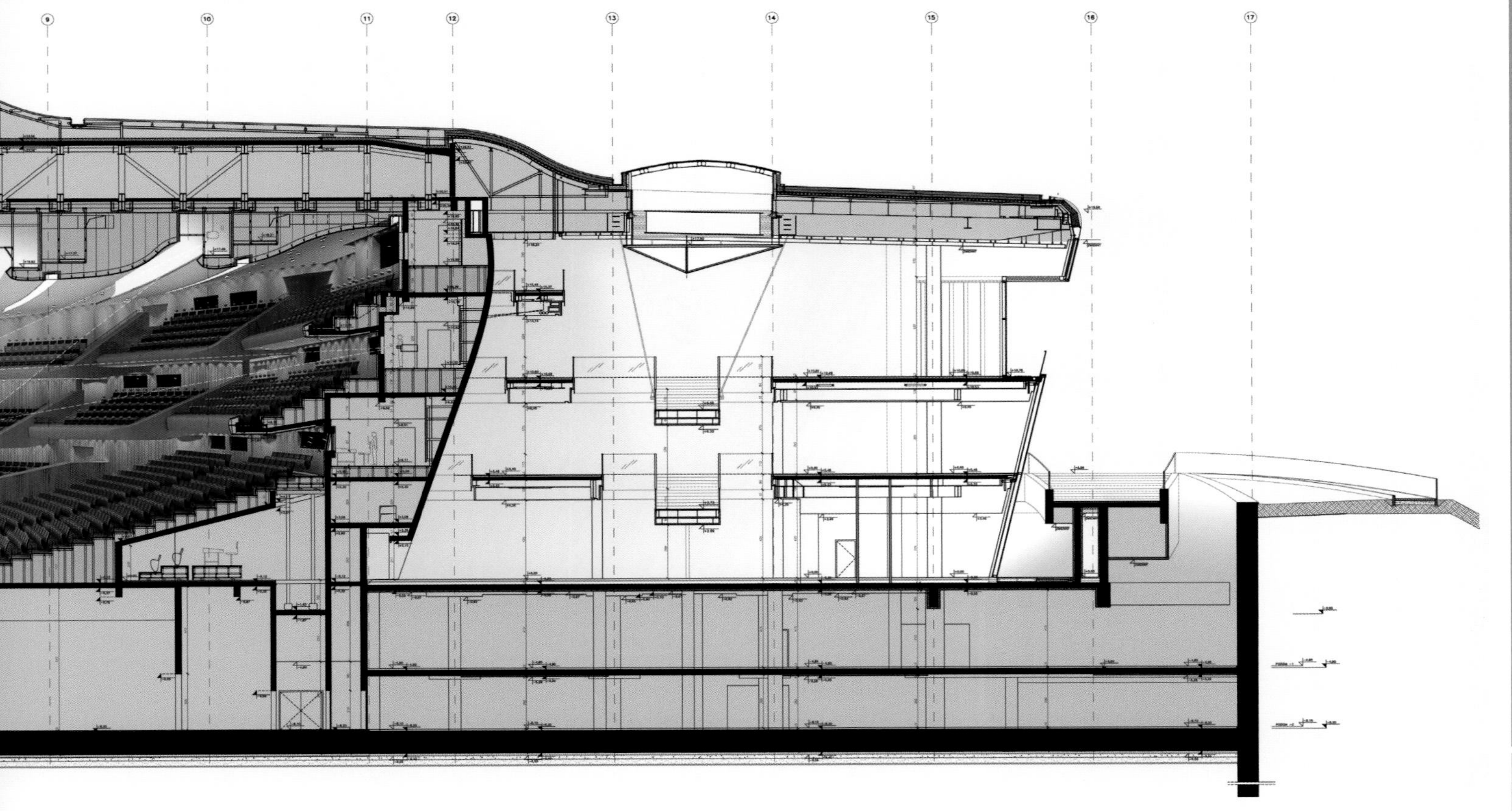

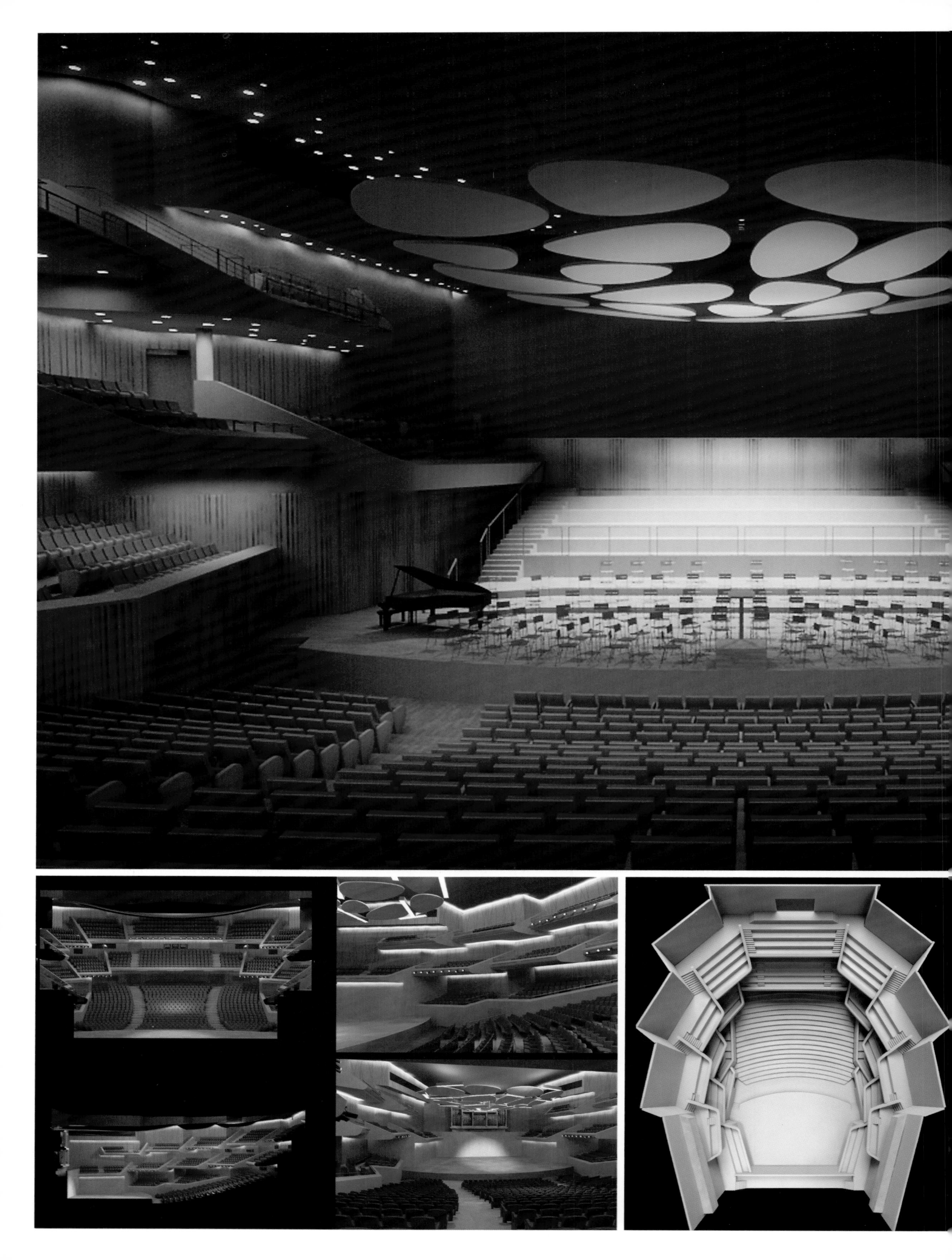

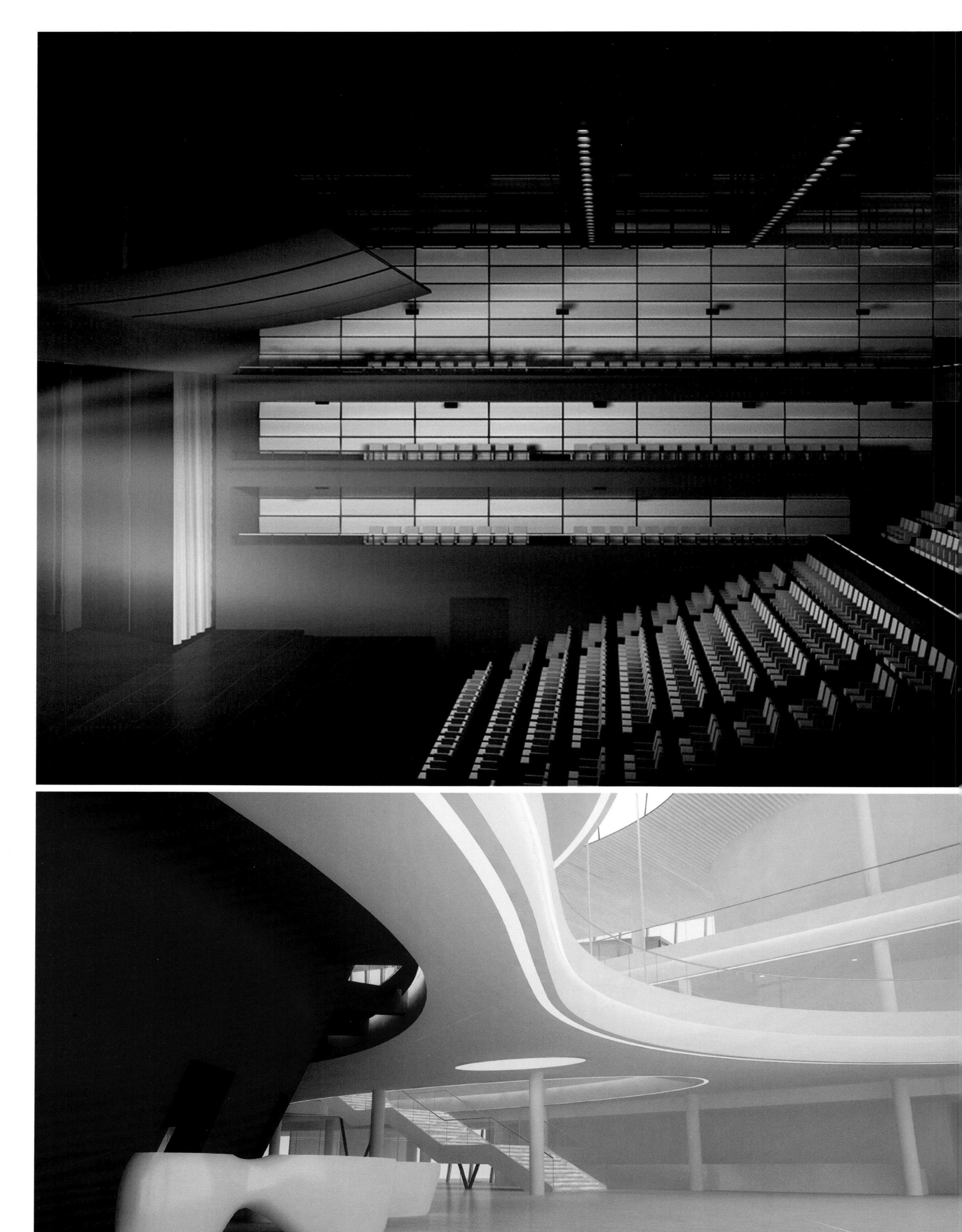

美国纽约坎贝尔体育中心

Campbell Sports Center

设计单位：斯蒂文·霍尔建筑师事务所
开发单位：哥伦比亚大学
项目地址：美国纽约
建筑面积：4 459 m²
设计团队：Steven Holl　Chris McVoy
Olaf Schmidt　Marcus Carter
Christiane Deptolla　Peter Englaender
Runar Halldorsson　Jackie Luk
Filipe Taboada　Dimitra Tsachrelia
Ebbie Wisecarver
摄影：Andy Ryan

Designed by: Steven Holl Architects
Client: Columbia University
Location: New York, NY, United States
Building Area: 4,459 m²
Design Team: Steven Holl, Chris McVoy, Olaf Schmidt, Marcus Carter, Christiane Deptolla, Peter Englaender, Runar Halldorsson, Jackie Luk, Filipe Taboada, Dimitra Tsachrelia, Ebbie Wisecarver
Photography: Andy Ryan

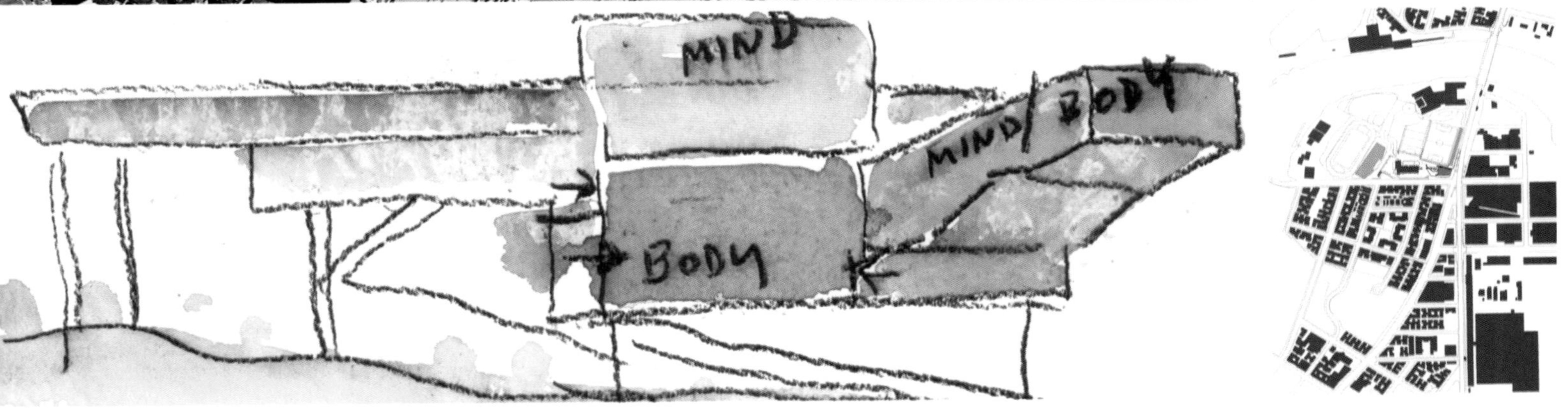

项目概况

坎贝尔体育中心坐落在纽约第 218 街和百老汇街的拐角处，这一建筑将成为贝克田径综合大楼具有吸引力的新“大门”，同时也将成为一栋具有可持续发展和生态创新功能的建筑。

设计目标

坎贝尔体育中心的主要设计目标是：提高哥伦比亚田径场的标识性；使之成为贝克田径综合大楼一个极具吸引力的“大门”；通过内部流通系统营造艺术运动空间；利用阶梯将场地扩展到建筑之中；构建可持续和生态创新的建筑。

建筑设计

方案以“地上的点，空中的线”为理念展开设计，建筑从位于些微倾斜的场地上的基点上延展，形成物理意义上的“推”和“拉”。作为“空中的线”，外部楼梯和露台延展了运动场地，并将之纳入建筑中来。

CAMPBELL SPORTS CENTER

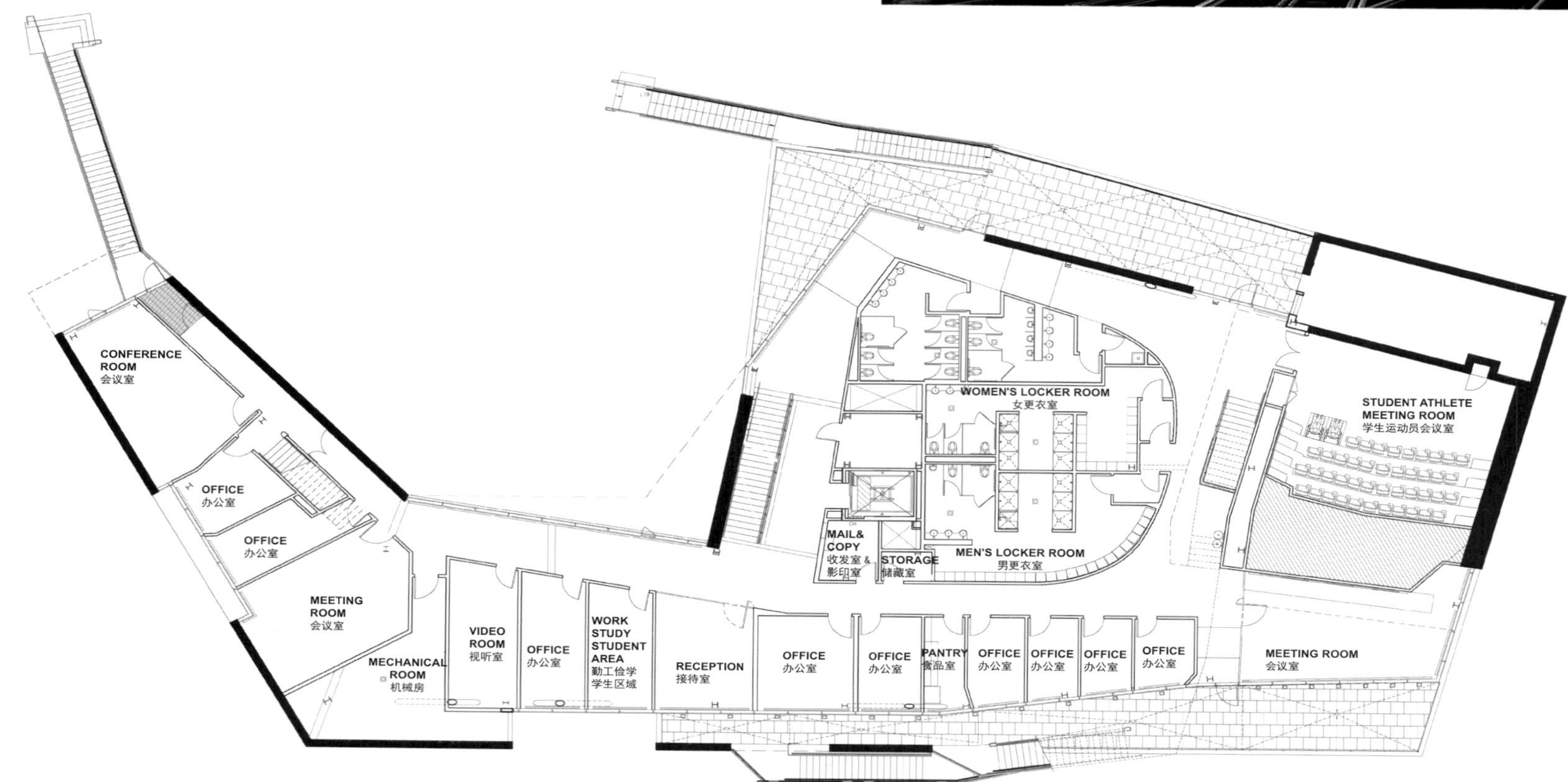
CONFERENCE ROOM
会议室
OFFICE
办公室
OFFICE
办公室
MEETING ROOM
会议室
MECHANICAL ROOM
机械房
VIDEO ROOM
视听室
OFFICE
办公室
WORK STUDY STUDENT AREA
勤工俭学学生区域
RECEPTION
接待室
OFFICE
办公室
OFFICE
办公室
PANTRY
食品室
OFFICE
办公室
OFFICE
办公室
OFFICE
办公室
OFFICE
办公室
MEETING ROOM
会议室
MAIL& COPY
收发室&影印室
STORAGE
储藏室
WOMEN'S LOCKER ROOM
女更衣室
MEN'S LOCKER ROOM
男更衣室
STUDENT ATHLETE MEETING ROOM
学生运动员会议室

FIRE PUMP
消防泵

CUSTODIAL STORAGE
保管仓库

SHAFT OPENING ABOVE
垂直通道向上打开

TANK (SPRINKLER)
水箱（喷水器）

ELECTRICAL ROOM
电气室

VEHICLE STORAGE HEATED
车辆仓库加热

VEHICLE STORAGE UNHEATED/UNCLOSED AREA
车辆仓库未加热／未闭合区域

VERIZON
威瑞森无线通讯

ELEV.ROOM
电气房

AIR HANDLER UNITS PUMPS, BOILER ROOM
空气处理器单元泵，锅炉房

SHOP
商铺

FIRE SERVICE
消防设施

DOMESTIC WATER
生活用水

GAS
煤气

OFFICE
办公室

PANTRY
食品室

LACTATION
哺育室

STORAGE
储藏室

CONFERENCE ROOM
会议室

RECEPTION VARSITY SUITE
接待大学生代表队的套间

MAIL© ROOM
收发室 & 影印室

STUDENT-ATHLETE LOUNGE
学生运动员休息室

HOSPITALITY SUITE
接待室

STUDENT-ATHLETE STUDY CENTER
学生运动员学习中心

STORAGE
储藏室

PANTRY
食品室

STUDENT ATHLETE MEETING ROOM
学生运动员会议室

MECHANICAL ROOM
机械房

Profile

The new Campbell Sports Center at the corner of 218th Street and Broadway will form an inviting new gate for Baker Athletics Complex and become a sustainable ecological and innovative building.

Design Goal

The main design goal of Campbell Sports Center is to create new visibility for Columbia Athletics, to create an inviting new gate for Baker Athletics Complex, to shape state of the art athletic spaces with interconnecting flow, to extend field onto and into the building with stepped ramps, and to build sustainable, ecological and innovative building.

Architectural Design

The design concept "Points on the Ground, Lines in Space" develops from point foundations on the sloping site. The building's elevations push and pull in space. External stairs, "Lines in Space," and terraces extend the field up and into the building.

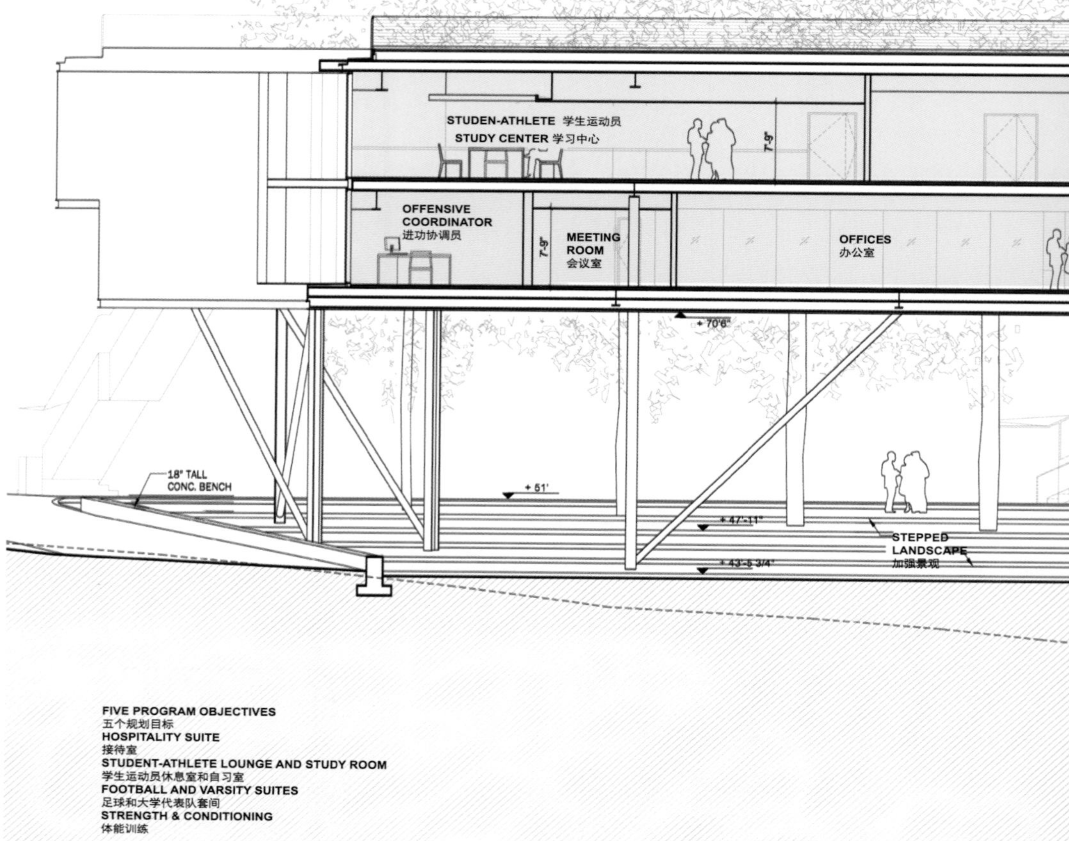

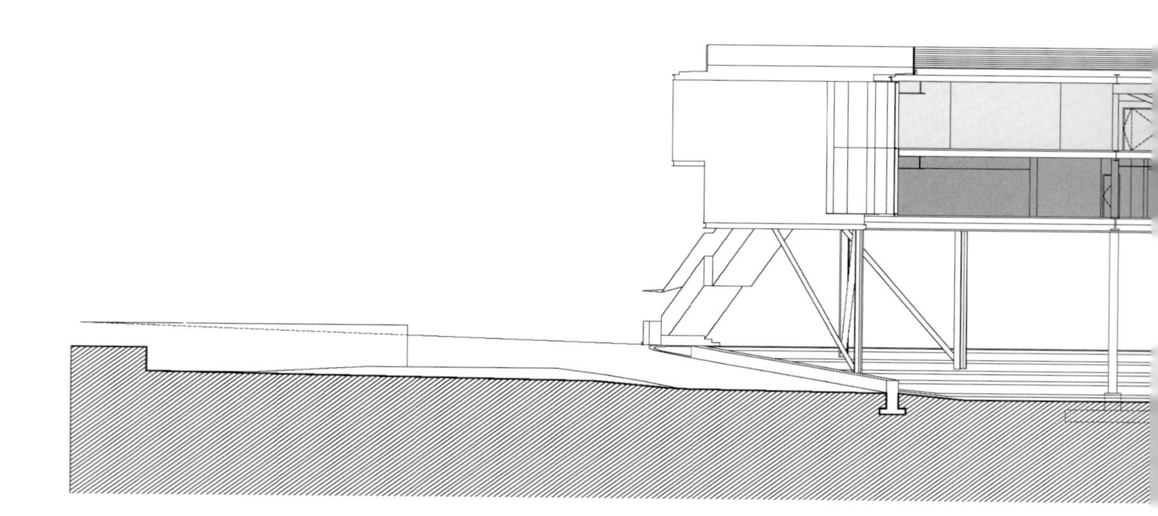

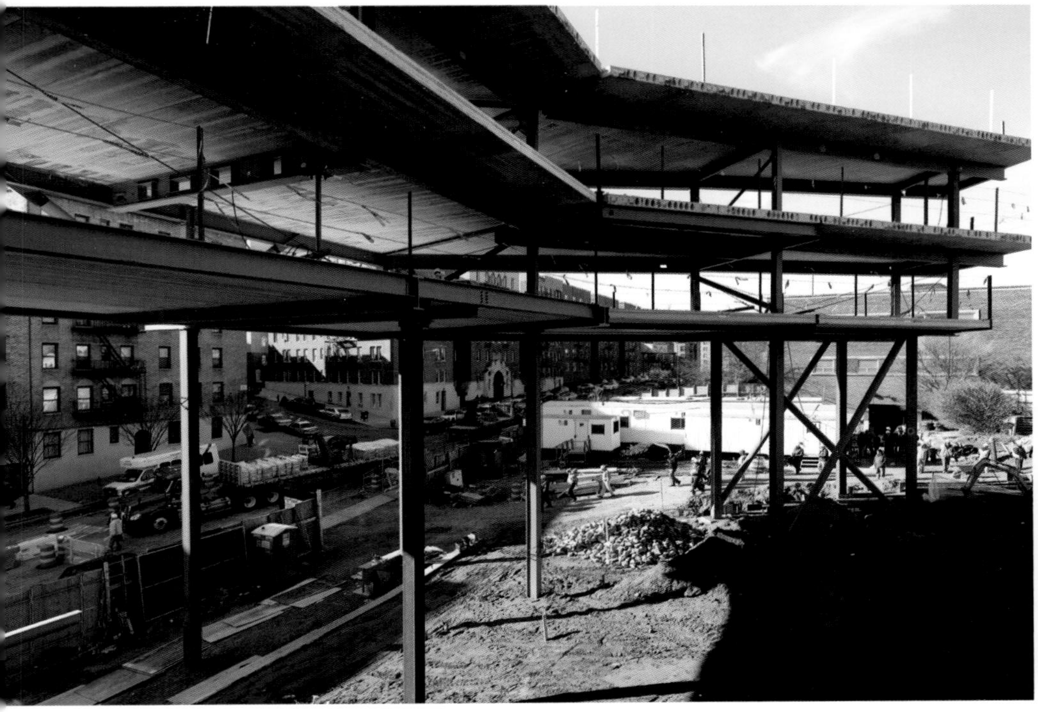

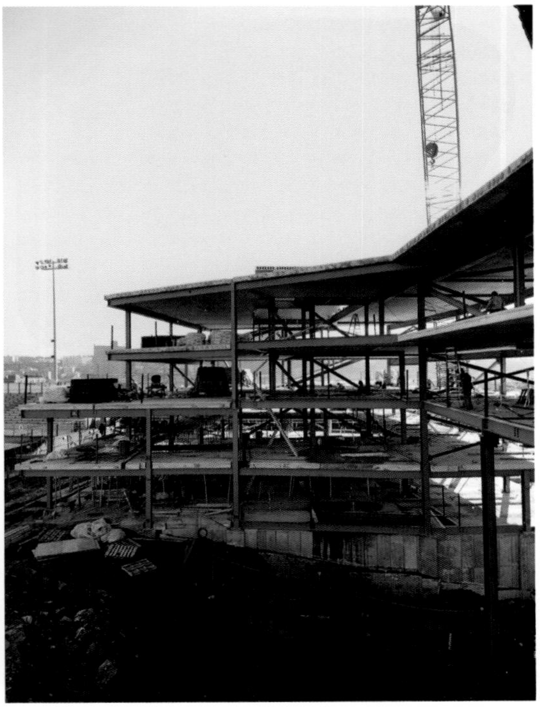

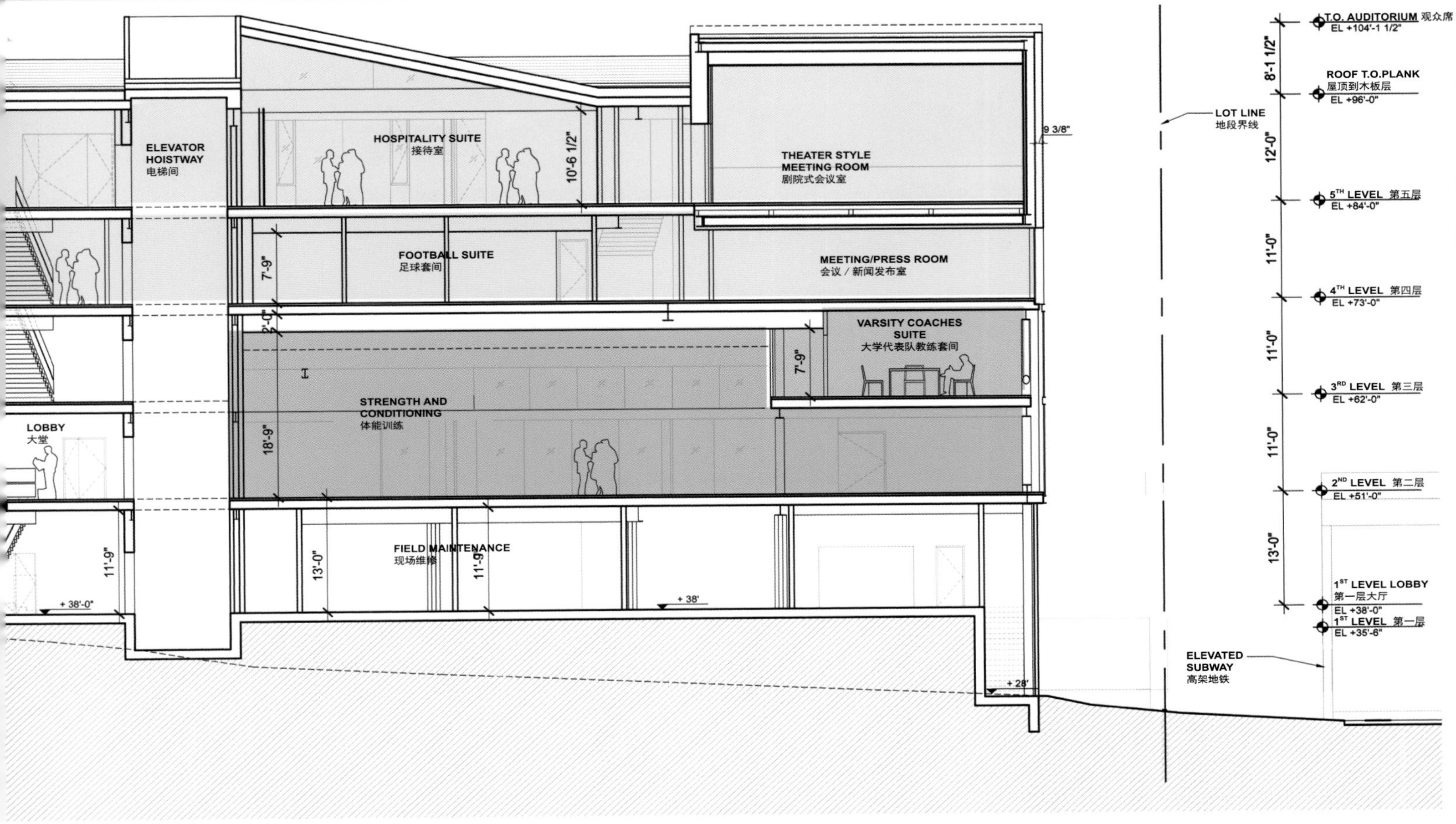

T.O. AUDITORIUM 观众席
EL +104'-1 1/2"
ROOF T.O.PLANK
屋顶到木板层
EL +96'-0"
LOT LINE
地段界线
ELEVATOR HOISTWAY
电梯间
HOSPITALITY SUITE
接待室
THEATER STYLE MEETING ROOM
剧院式会议室
5TH LEVEL 第五层
EL +84'-0"
FOOTBALL SUITE
足球套间
MEETING/PRESS ROOM
会议／新闻发布室
4TH LEVEL 第四层
EL +73'-0"
VARSITY COACHES SUITE
大学代表队教练套间
STRENGTH AND CONDITIONING
体能训练
3RD LEVEL 第三层
EL +62'-0"
LOBBY
大堂
2ND LEVEL 第二层
EL +51'-0"
FIELD MAINTENANCE
现场维修
1ST LEVEL LOBBY
第一层大厅
EL +38'-0"
1ST LEVEL 第一层
EL +35'-6"
ELEVATED SUBWAY
高架地铁

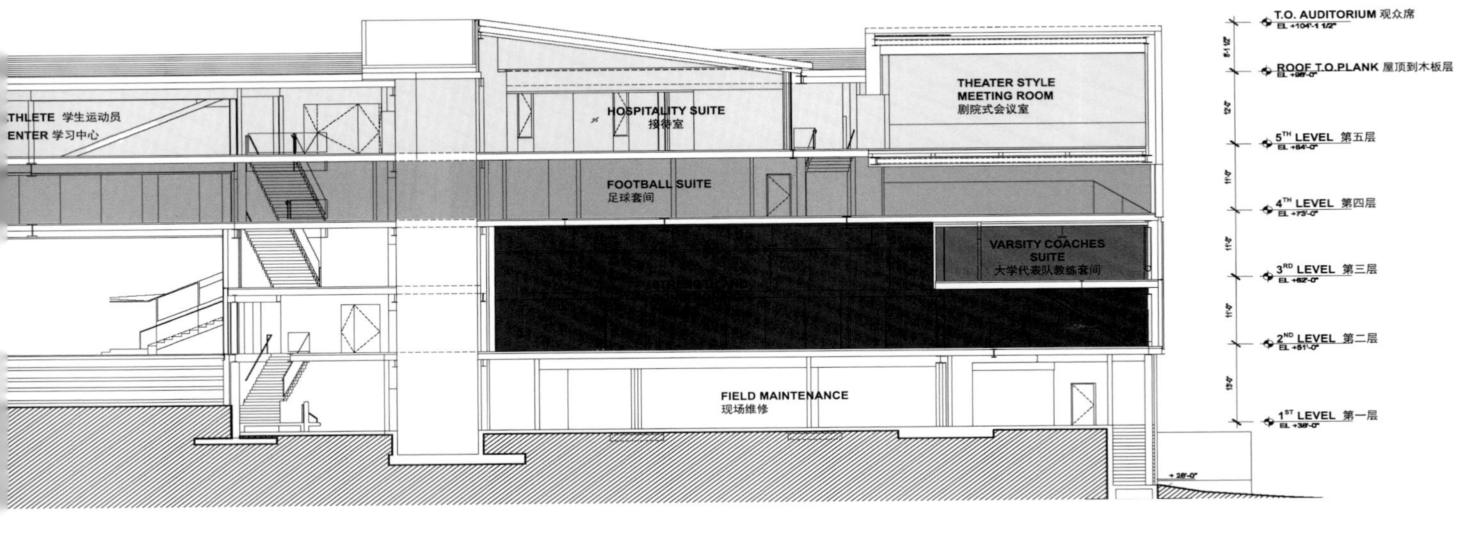

T.O. AUDITORIUM 观众席
ROOF T.O.PLANK 屋顶到木板层
ATHLETE 学生运动员
ENTER 学习中心
HOSPITALITY SUITE
接待室
THEATER STYLE MEETING ROOM
剧院式会议室
5TH LEVEL 第五层
FOOTBALL SUITE
足球套间
4TH LEVEL 第四层
VARSITY COACHES SUITE
大学代表队教练套间
3RD LEVEL 第三层
2ND LEVEL 第二层
FIELD MAINTENANCE
现场维修
1ST LEVEL 第一层

克罗地亚希贝尼克 Draga 城市中心

Draga City Center

设计单位： 3LHD
项目地址： 克罗地亚希贝尼克
基地面积： 46 229 m²
设计团队： Sasa Begovic　Marko Dabrovic
Silvije Novak　Tatjana Grozdanic Begovic
Zorislav Petric　Vibor Granic
Sanja Jasika　Josko Kotula
Ines Vlahovic

Designed by: 3LHD
Location: Šibenik, Croatia
Site Area: 46,229 m²
Project Team: Sasa Begovic, Marko Dabrovic, Silvije Novak, Tatjana Grozdanic Begovic, Zorislav Petric, Vibor Granic, Sanja Jasika, Josko Kotula, Ines Vlahovic

项目概况

Draga 位于希贝尼克的城市中心区，区域内涵盖了众多行政、文化、商业等公共设施，具备了开发一个新的行政和文化中心的潜力。同时，高密度的开发模式，将这个沿海城市向海洋敞开，为构建新的滨海区创造了先决条件。

设计理念

沿交通干线开发、遵循场地的地形结构和城市线型特征布局是 Draga 城市中心项目的主题和设计灵感，这为城市在更宽广的区域实现持续发展创造了条件。

结构布局

依据沿交通干线开发的主题定位，设计师加强了对密集城市结构与新开发区域结构及两者关系的处理。设计通过衔接和延续东西方向已有街道的主要线型结构，构建了一个新的网格结构。Draga 城市中心的设计遵循自然地貌结构，下降至海岸线，融入已有城市结构中，并与步行横轴、小巷和广场相连，形成新的公共空间。

建筑设计

五栋大楼延续了希贝尼克的建筑传统，建筑造型和立面设计受街道和旧城区结构的启发，形成瀑布般流畅的韵律感。线型体量的建筑有两个临街立面，自然地从较高平台过渡到较低平台。

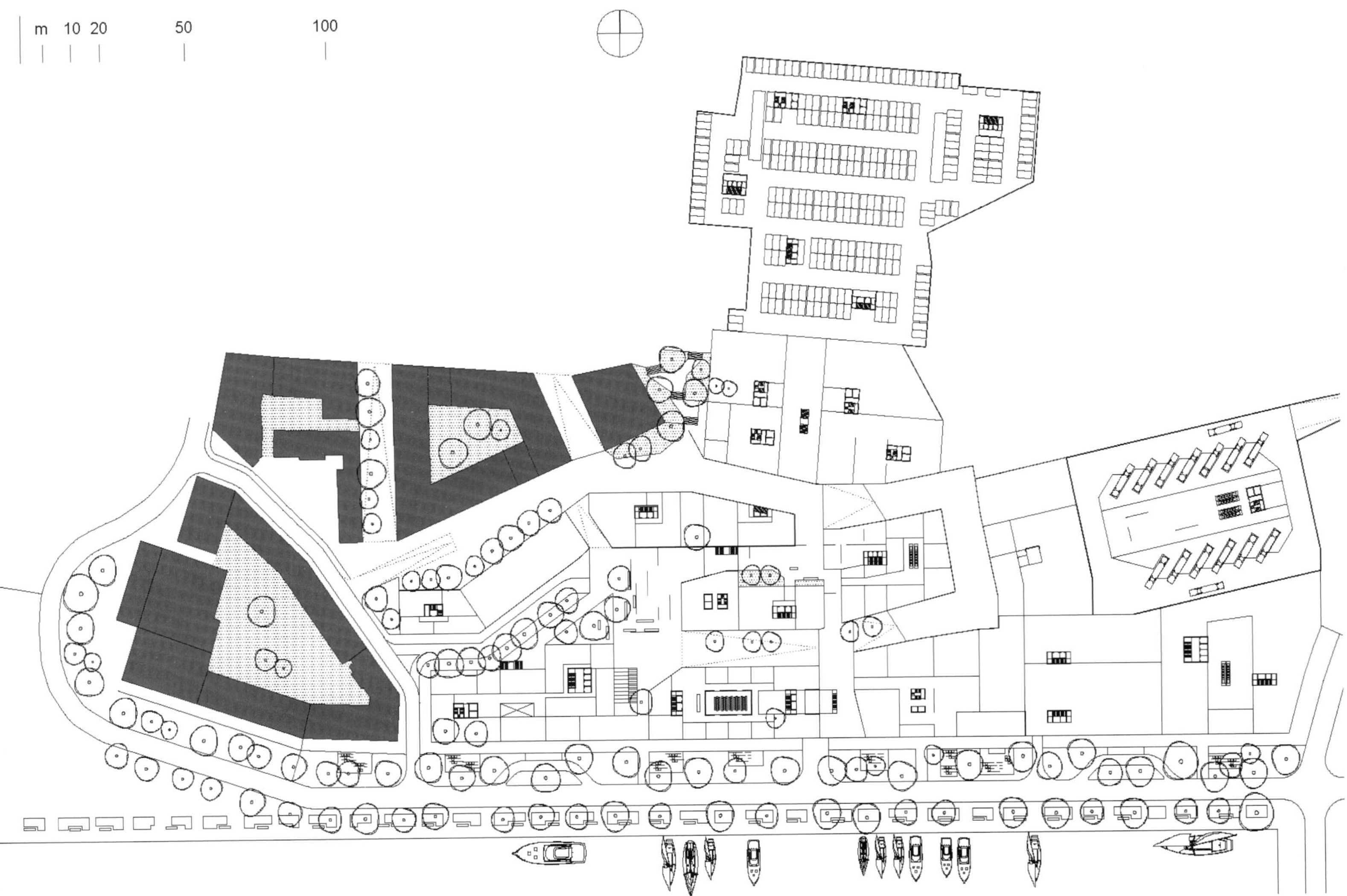
m
10
20
50
100

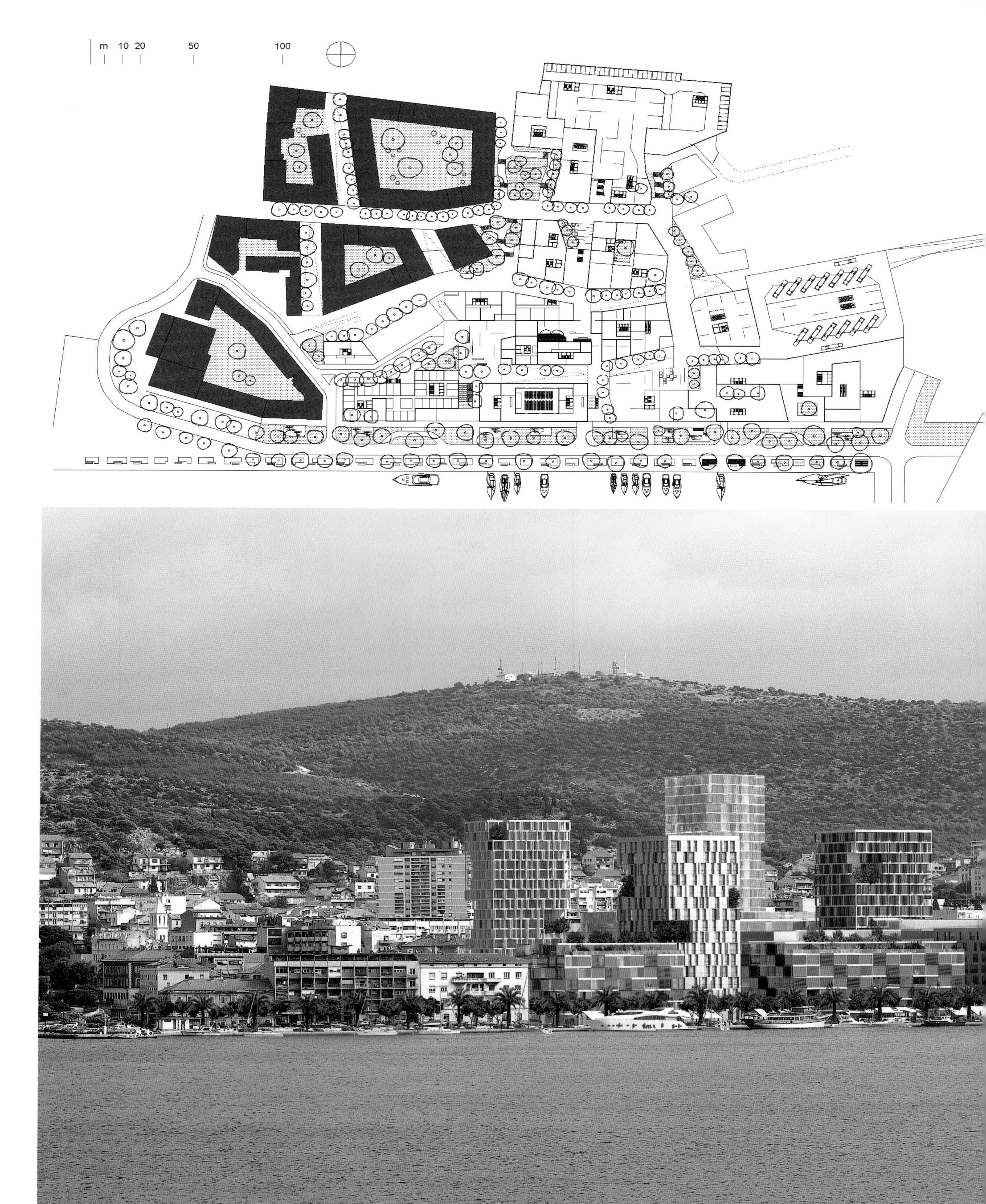
m 10 20 50 100

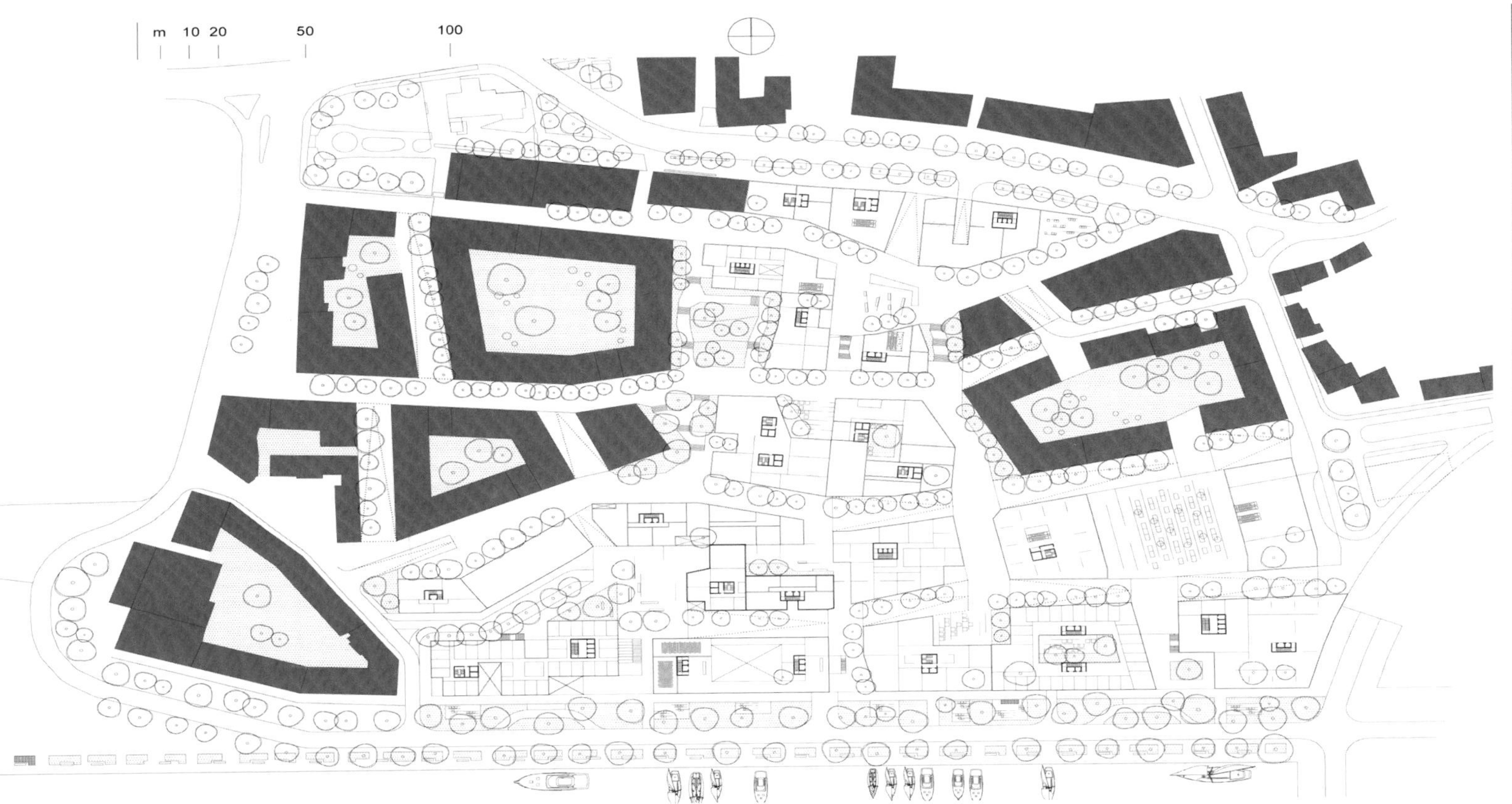

Profile

Draga is located in the central area of Šibenik accommodating a number of administrative, cultural and commercial facilities and as such represents a potential for developing a new administrative and cultural city center. This high density is also a prerequisite for the development of the current coast which will open the city to the sea and become a new urban waterfront.

Design Concept

Identification of elements and recognition of the city's extremely linear character, which has developed alongside of dominant traffic parts by following the terrain's configuration, are motifs and inspirations used for the new project of Draga in Šibenik.

Structural Layout

Since the recent construction follows the main traffic arteries, the treatment of dense urban structure, the new eastern section structure and their relationship is intentionally strengthened. By connecting and naturally continuing the main linear directions of existing streets from west and east, designers create a new grid. Urban terraces, used for developing new blocks, are created by following the natural terrain configuration and descent to the sea line. They are woven in to the city's existing structure, connected to pedestrian transverses, small alleys and squares that make up the new public space.

Architectural Design

Five verticals continue Šibenik's tradition of verticals. The diversity in façade treatment and volume shaping stimulated and inspired by the pace of the streets and Šibenik's old town core, contributes to a harmonious volume, street fronts, passages and cascades rhythm. The proposed linear elongated blocks of buildings are located on the borders of the transition from higher to lower plateau / terrace, primarily with two street facades.

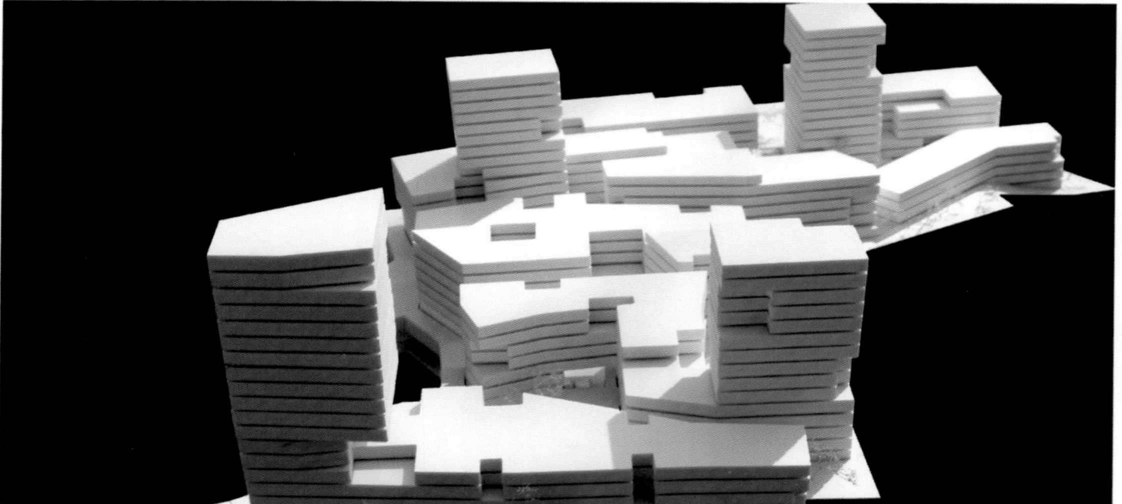

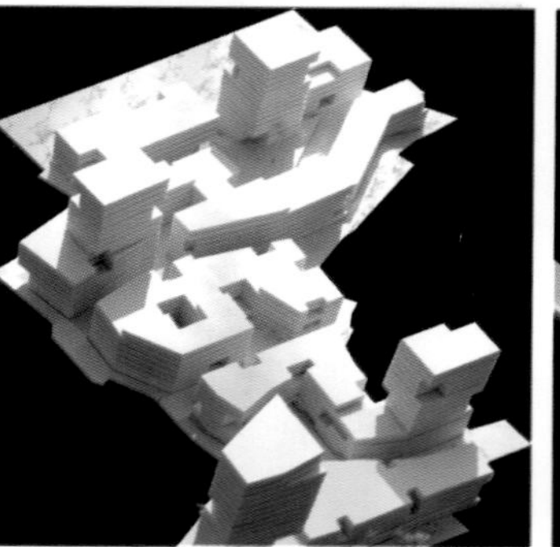

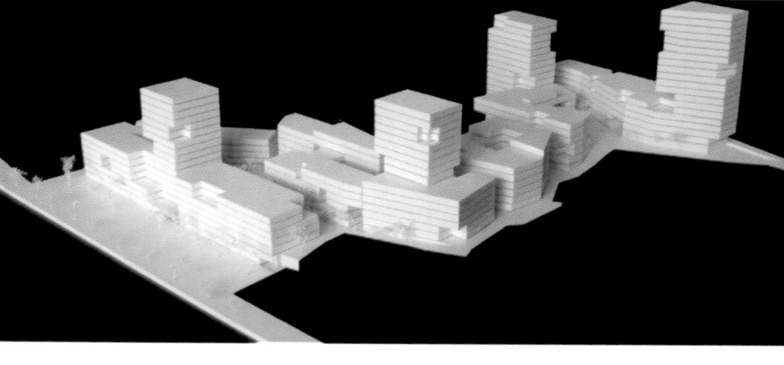

m 10 20 50

m 10 20 50 100

m 10 20 50 100

m 10 20 50

m 10 20 50 100

美国弗吉尼亚联邦大学现代艺术研究院

Institute for Contemporary Art, Virginia Commonwealth University

设计单位：斯蒂文·霍尔建筑师事务所
合作单位：BCWH Architects
项目地址：美国弗吉尼亚州里士满
建筑面积：2 973 m²
设计团队：Steven Holl　Chris McVoy
Dimitra Tsachrelia　Garrick Ambrose
Rychiee Espinosa　Scott Fredricks
Gary He　Christina Yessios

Designed by: Steven Holl Architects
Collaboration: BCWH Architects
Location: Richmond, VA, United States
Building Area: 2,973 m²
Design Team: Steven Holl, Chris McVoy, Dimitra Tsachrelia, Garrick Ambrose, Rychiee Espinosa, Scott Fredricks, Gary He, Christina Yessios

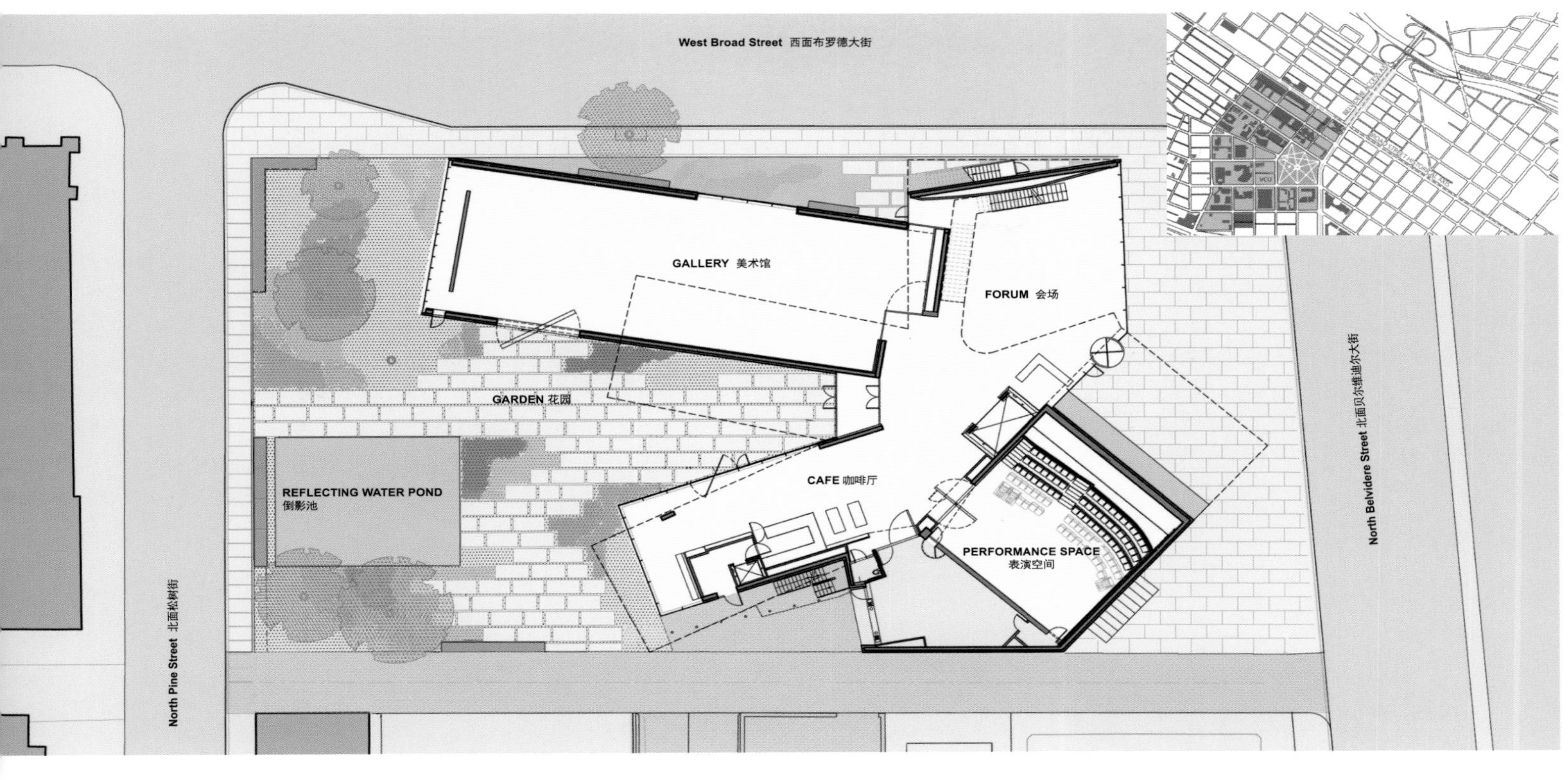

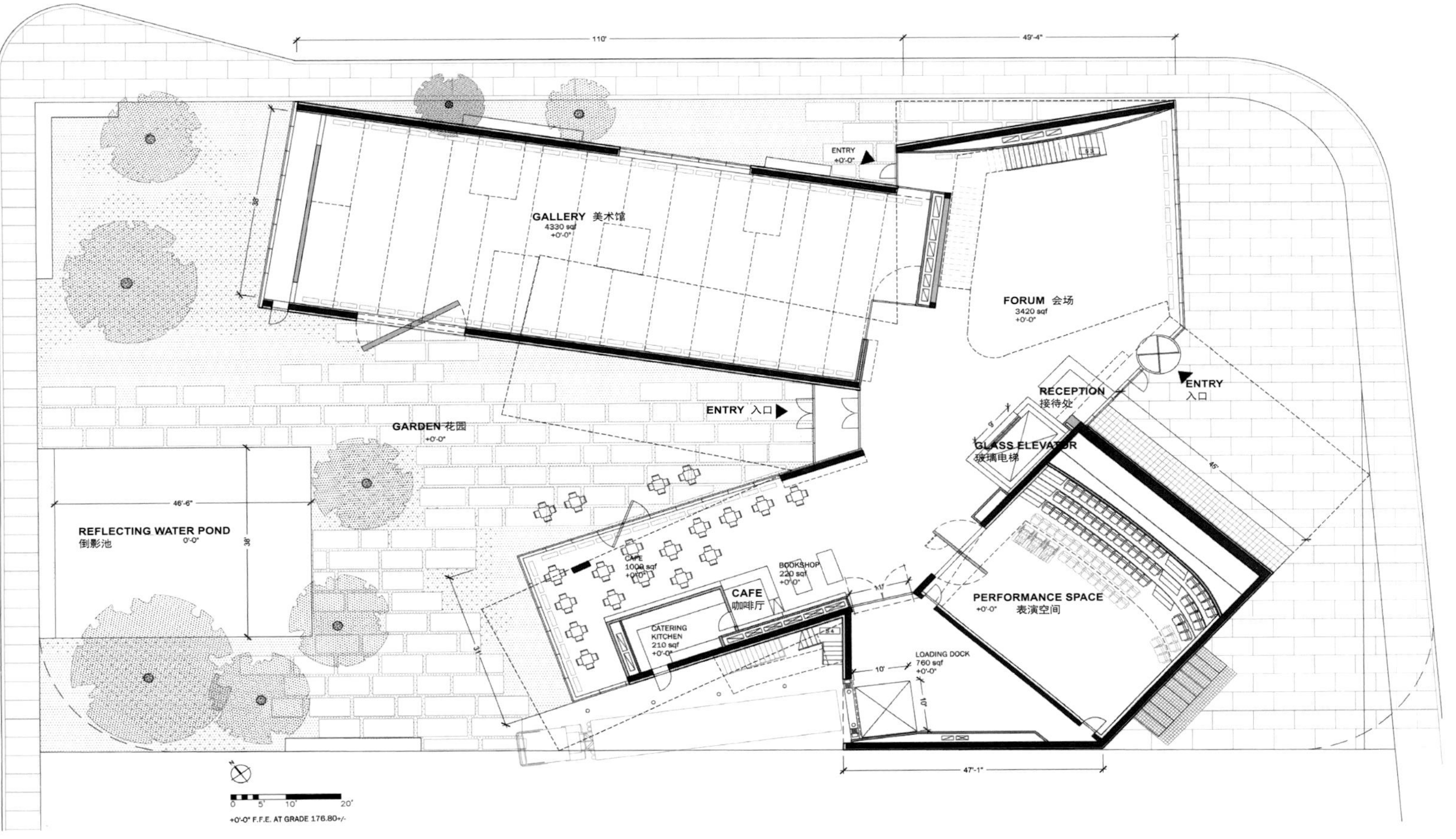

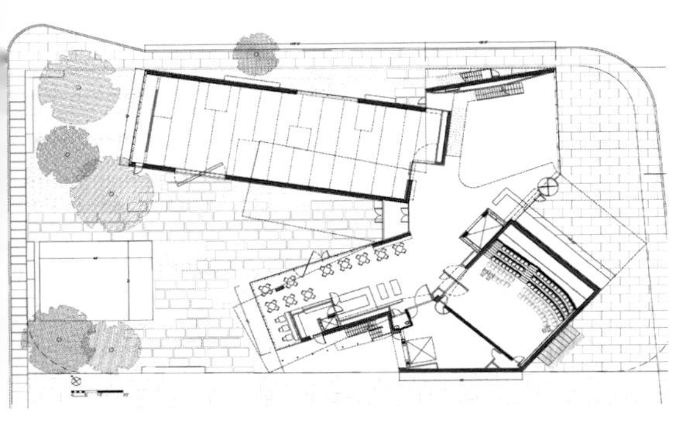

项目概况

项目是由斯蒂芬·霍尔建筑师事务所设计的弗吉尼亚联邦大学现代艺术研究院，涵盖了视觉艺术中心、画廊、影院、音乐厅、舞蹈室以及演艺教学公演空间。

建筑设计

项目位于两条道路的交汇处，形成连接大学校园与里士满市的一个“大门”，吸引了人们的视线。表演空间与广场交汇，形成了建筑的主入口，从而在两者交汇处增添了一个垂直的“临时界面”。“临时界面”连接了美术馆、表演空间、雕塑花园和广场。沿着这一建筑长廊，人们可以在变化的视野中体验多种元素融合之美。

新的艺术研究院分为4个美术馆，每个美术馆都表现出不同的特色和特征。场馆设计十分灵活，这使4个场馆能够进行各自的展览活动，也可以将几个场馆组合起来举办一个连续的展览。画廊空间布局灵活，展品展览方式也变得多样，参展的艺术品可以以悬挂的方式展现，也可以固定在楼板上。

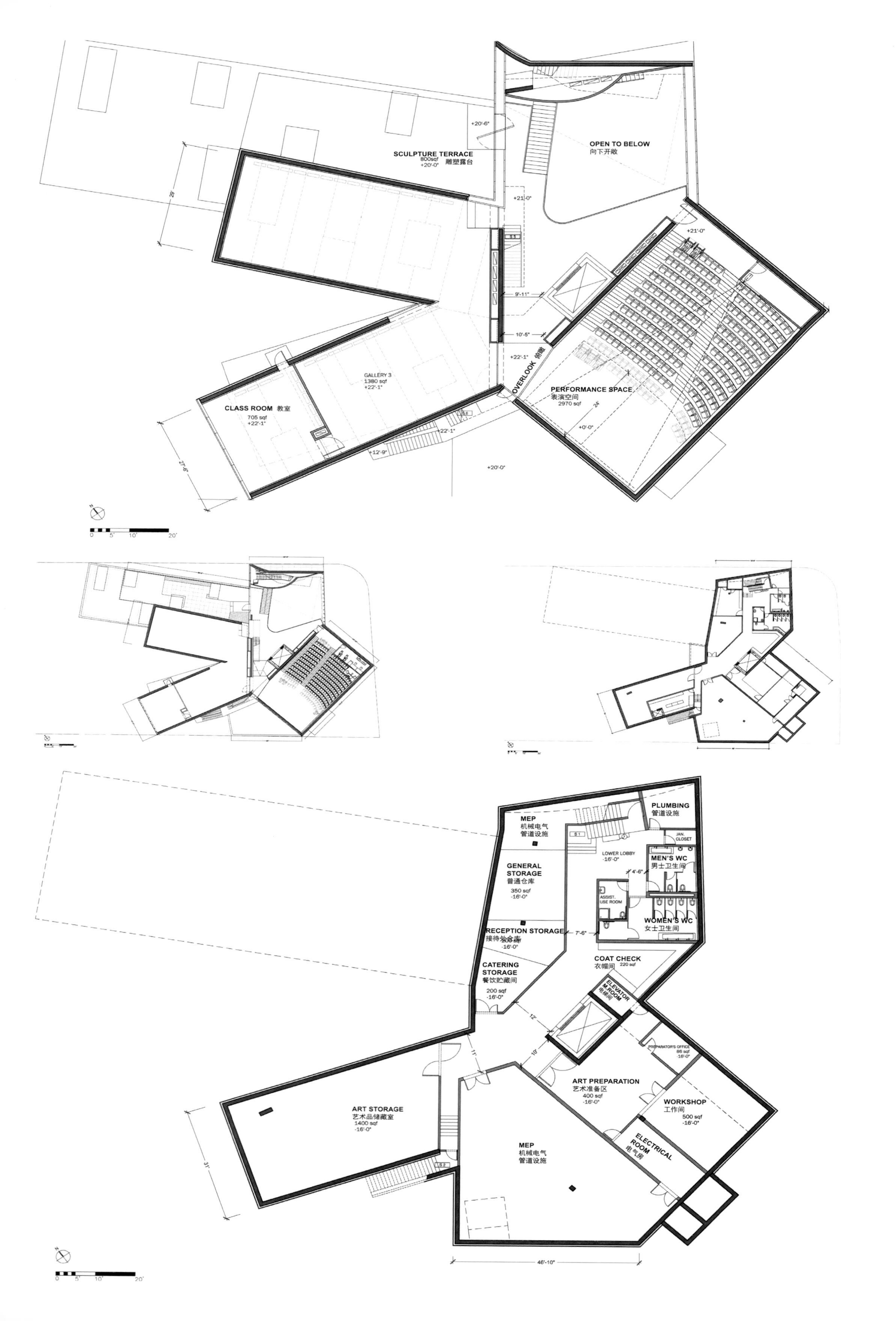

SCULPTURE TERRACE
800sqf
雕塑露台
+20'-0"
OPEN TO BELOW
向下开敞
+21'-0"
9'-11"
10'-5"
+22'-1"
OVERLOOK 俯瞰
GALLERY 3
1380 sqf
+22'-1"
CLASS ROOM 教室
705 sqf
+22'-1"
PERFORMANCE SPACE
表演空间
2970 sqf
+0'-0"
+12'-9"
+20'-0"
MEP
机械电气
管道设施
PLUMBING
管道设施
JAN. CLOSET
LOWER LOBBY
-16'-0"
MEN'S WC
男士卫生间
GENERAL
STORAGE
普通仓库
350 sqf
-16'-0"
ASSIST. USE ROOM
RECEPTION STORAGE
-16'-0"
WOMEN'S WC
女士卫生间
CATERING
STORAGE
餐饮贮藏间
200 sqf
-16'-0"
COAT CHECK
衣帽间
220 sqf
ELEVATOR M.ROOM
PREPARATOR'S OFFICE
-16'-0"
ART PREPARATION
艺术准备区
400 sqf
-16'-0"
WORKSHOP
工作间
500 sqf
-16'-0"
ART STORAGE
艺术品储藏室
1400 sqf
-16'-0"
MEP
机械电气
管道设施
ELECTRICAL
ROOM
电气房
46'-10"

GALLERY 4 美术馆 4
1900 sqf
+32'-6"
OPEN TO BELOW
向下开敞
KITCHENETTE
小厨房
90 sqf
MEETING ROOM
480 sqf
+32'-6"
会议室
MEETING ROOM
200 sqf
WC 洗手间
OPEN TO
BELOW
向下开敞
ADMINISTRATION 行政室
+32'-6"
OPEN TO BELOW
向下开敞
0 5' 10' 20'

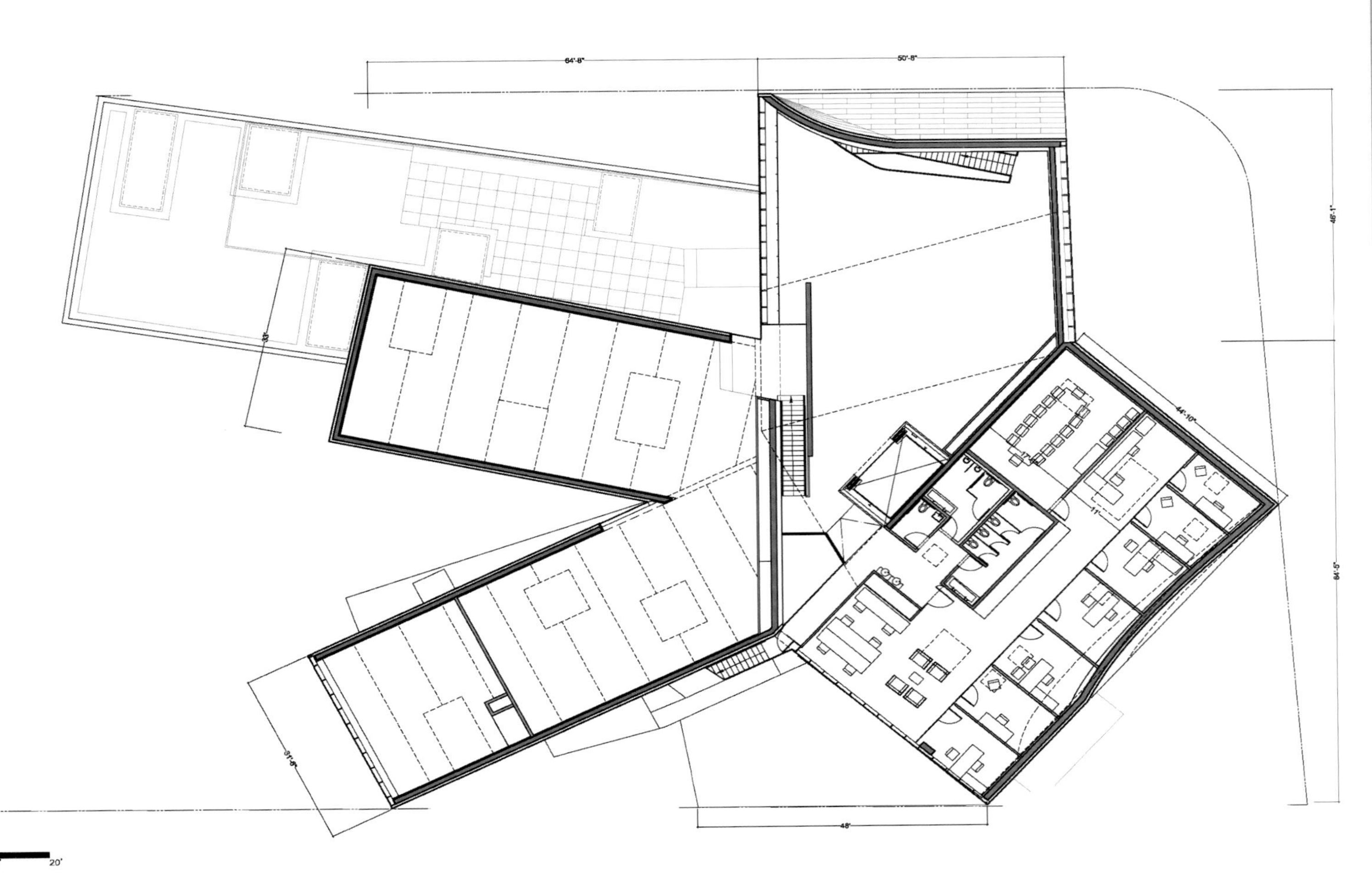
64'-8"
50'-8"
48'
0 5' 10' 20'

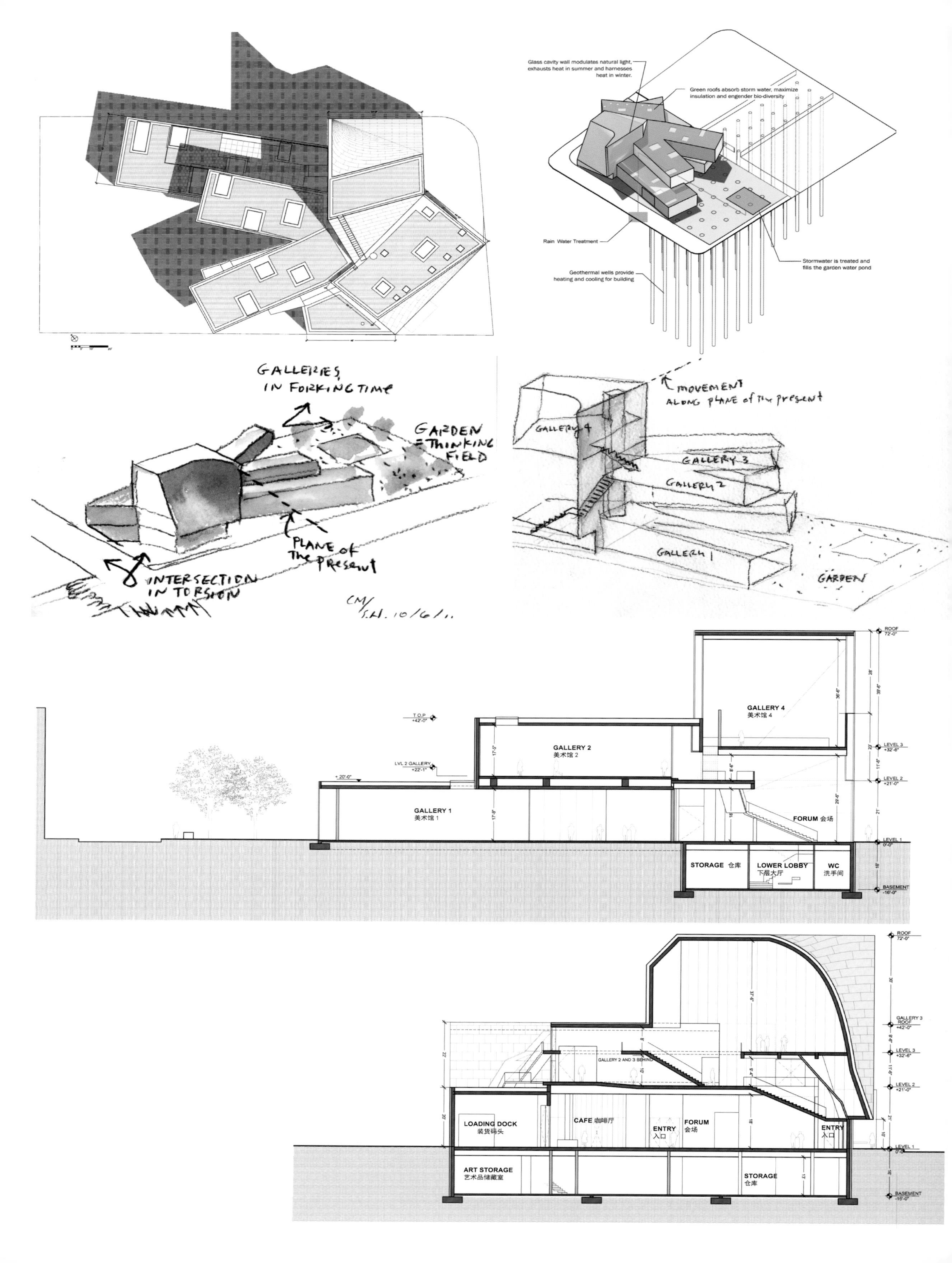
Glass cavity wall modulates natural light, exhausts heat in summer and harnesses heat in winter.
Green roofs absorb storm water, maximize insulation and engender bio-diversity
Rain Water Treatment
Stormwater is treated and fills the garden water pond
Geothermal wells provide heating and cooling for building
GALLERIES IN FORKING TIME
GARDEN = THINKING FIELD
PLANE of The Present
INTERSECTION IN TORSION
MOVEMENT ALONG PLANE of the present
GALLERY 4
GALLERY 3
GALLERY 2
GALLERY 1
GARDEN
ROOF 72'-0"
T.O.P. +42'-0"
GALLERY 4 美术馆 4
GALLERY 2 美术馆 2
LVL 2 GALLERY +22'-1"
+20'-0"
GALLERY 1 美术馆 1
LEVEL 3 +32'-6"
LEVEL 2 +21'-0"
FORUM 会场
LEVEL 1 0'-0"
STORAGE 仓库
LOWER LOBBY 下层大厅
WC 洗手间
BASEMENT -16'-0"
ROOF 72'-0"
GALLERY 3 ROOF +42'-0"
LEVEL 3 +32'-6"
GALLERY 2 AND 3 BEHIND
LEVEL 2 +21'-0"
LOADING DOCK 装货码头
CAFE 咖啡厅
ENTRY 入口
FORUM 会场
ENTRY 入口
LEVEL 1 0'-0"
ART STORAGE 艺术品储藏室
STORAGE 仓库
BASEMENT -16'-0"

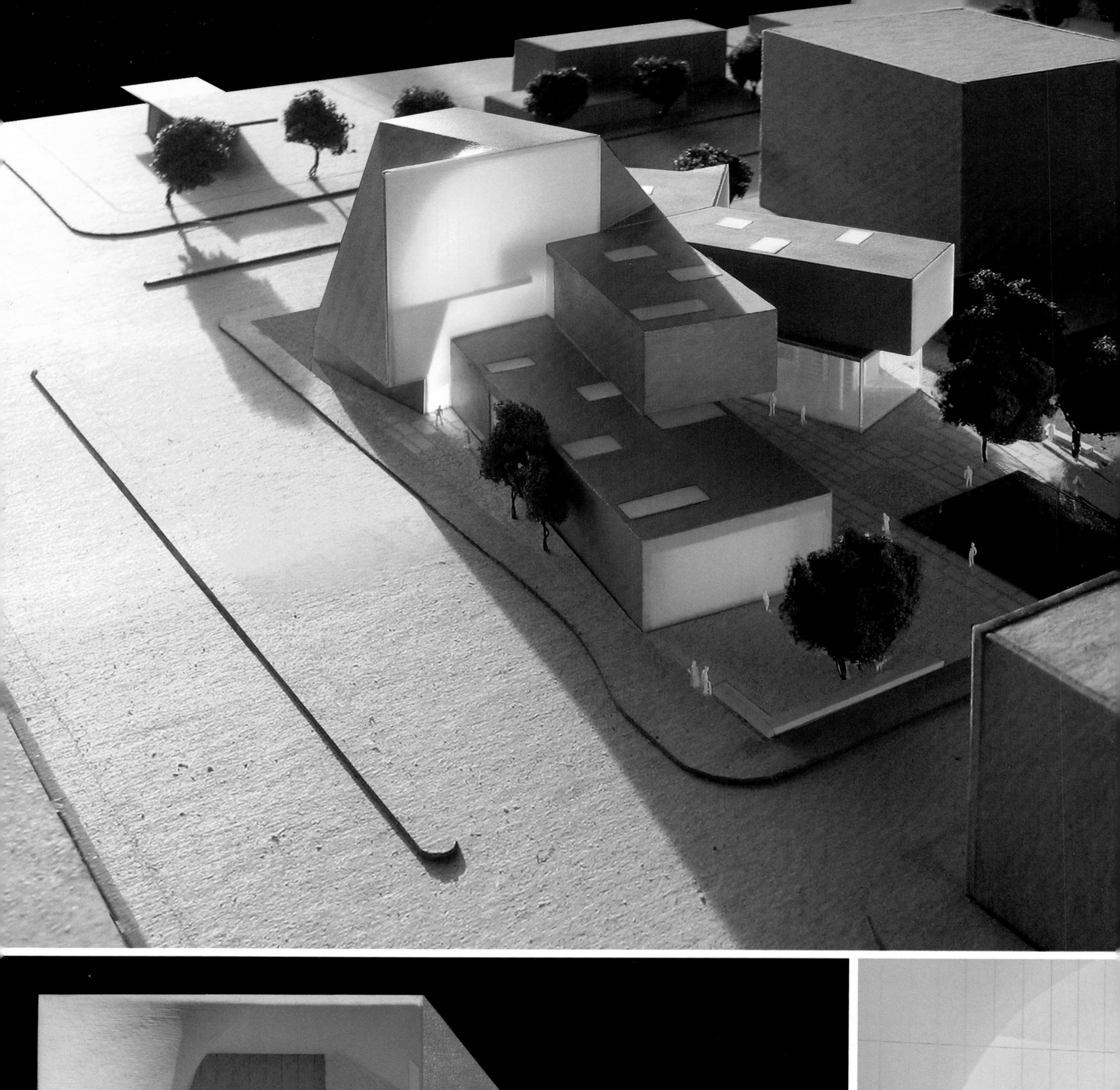

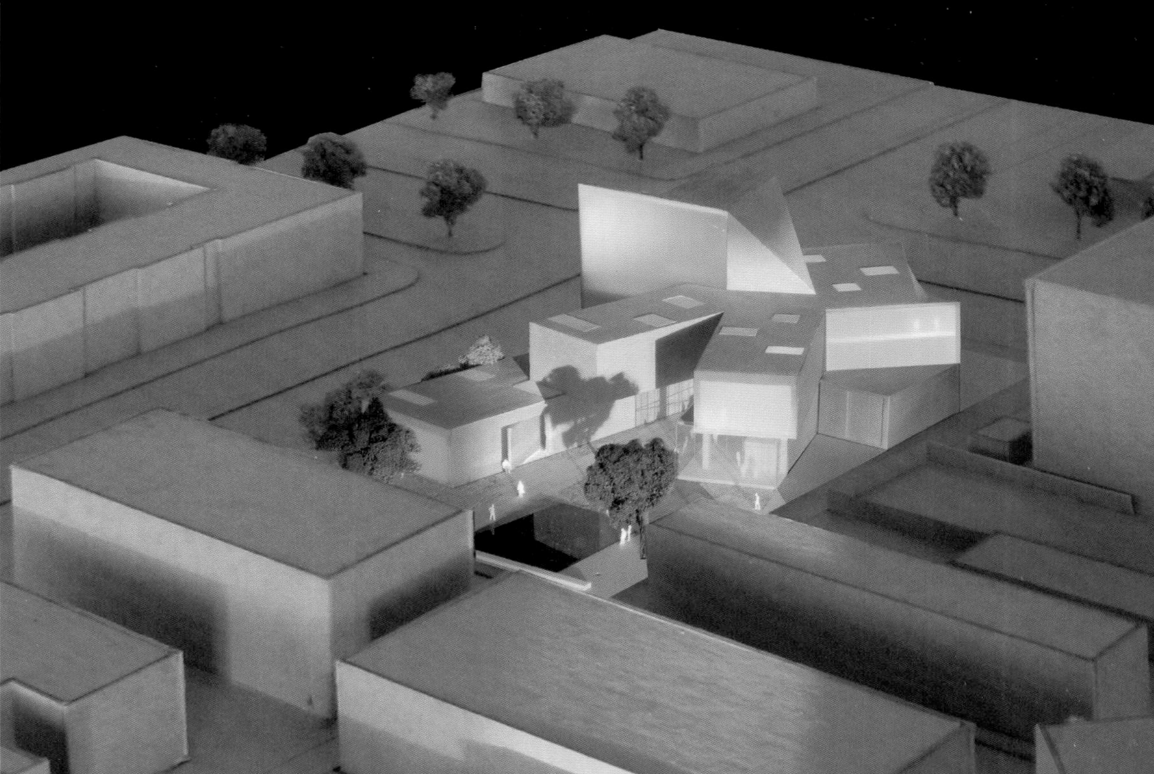

Profile

The project, designed by Steven Holl Architects, is an Institute for Contemporary Art of Virginia Commonwealth University. It covers visual arts center, gallery, cinema, music hall, dancing room and performance space.

Architectural Design

On the busiest intersection of Richmond at Broad and Belvidere Streets, the building will form a gateway to the University with an inviting sense of openness. The performance space and forum form the main entrance at their intersection which adds a vertical "Plane of the Present" there. Vertical movement along the "Plane of the Present" links the galleries, the performance space, sculpture garden, and forum. Along this architectural promenade, the integration of all the building elements can be experienced in changing views.

The new Institute for Contemporary Art is organized in four galleries, each with a different character. Flexibility allows for four separate exhibitions, one continuous exhibition, or combinations. As flexible spaces, the galleries can accept suspended art or projects anchored to the floor slab.

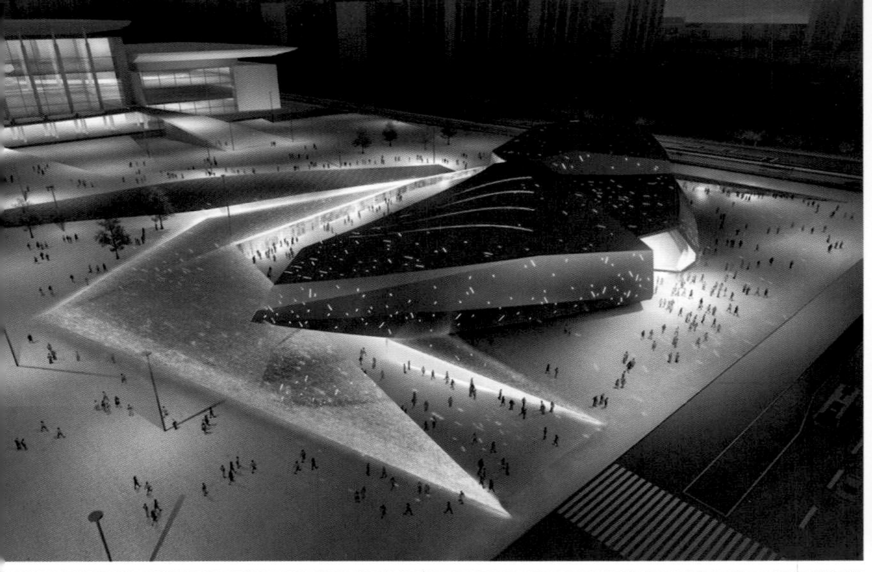

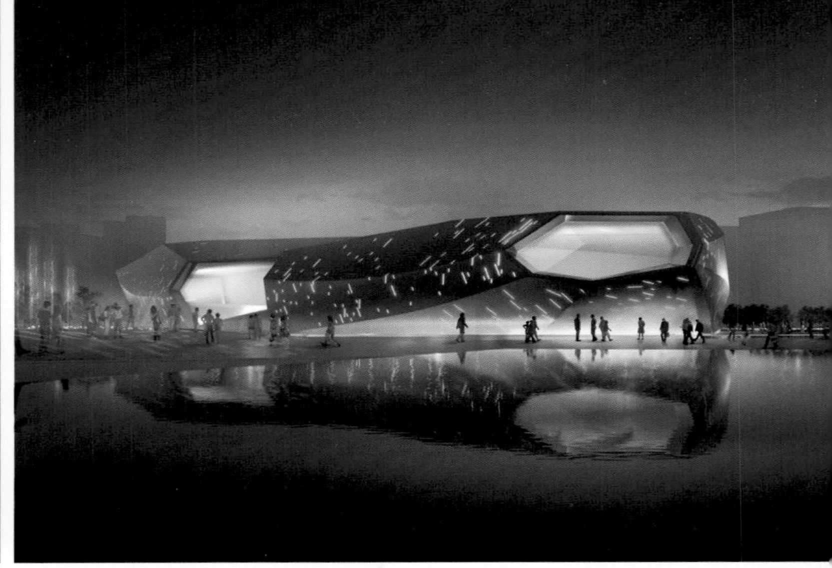

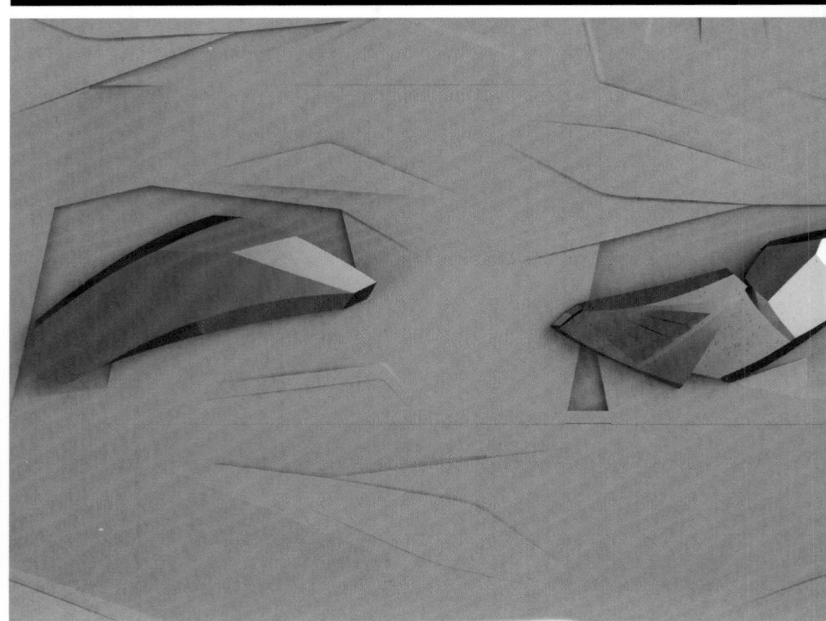

山东济南高新区科技文化中心

Jinan High-tech Science and Technology Cultural Center

单位设计：RTA-Office 建筑事务所
CCTN 中联建筑
项目地址：中国山东省济南市
总用地面积：17 000 m²
总建筑面积：21 655 m²
摄影：RTA-Office 建筑事务所

Designed by: RTA-Office; CCTN. ARCH China United Zhujing Architecture Design Co., Ltd.
Location: Jinan, Shandong, China
Total Land Area: 17,000 m²
Total Building Area: 21,655 m²
Photography: RTA-Office

项目概况

科技文化中心位于济南高新区，设计采用最先进的技术创造了一个 3D 模型，通过现代的表达方式展现这一区域的文化特质和传统韵味。

设计构思

济南是齐鲁文化的中心，也是中华文明最重要的发祥地之一，拥有着深厚的文化底蕴和浓郁的艺术氛围，为项目提供了潜在的优势和发展机遇。建筑师认为，这些建筑物需要反映济南的文化韵味，因此，建筑师提出了一个独特的设计方案，即通过软件语言铸就这个现代建筑的灵魂，展示其所有的现代特征。

设计特色

整个项目由两个建筑物和一个方形的广场构成。两个建筑物既是独立的，也是相互连接的，它们就像是两块在舞台上（广场中）舞动的“石头”。设计采用了轴线对称的布局方式，“舞蹈石”位于轴线的两端，通过广场连接在一起。这一布局方式源自济南的城市布局特征，广场象征着济南的城市中心，对称的舞蹈石呼应了济南的东城与西城。

为了突出广场作为城市舞台的地位和作用，设计师对广场进行了复兴重建。除了北面中心的楼梯被保留作为舞台包厢外，其余的楼梯都被拆除了，并以斜坡的形态展现出来，便于人们自由移动。

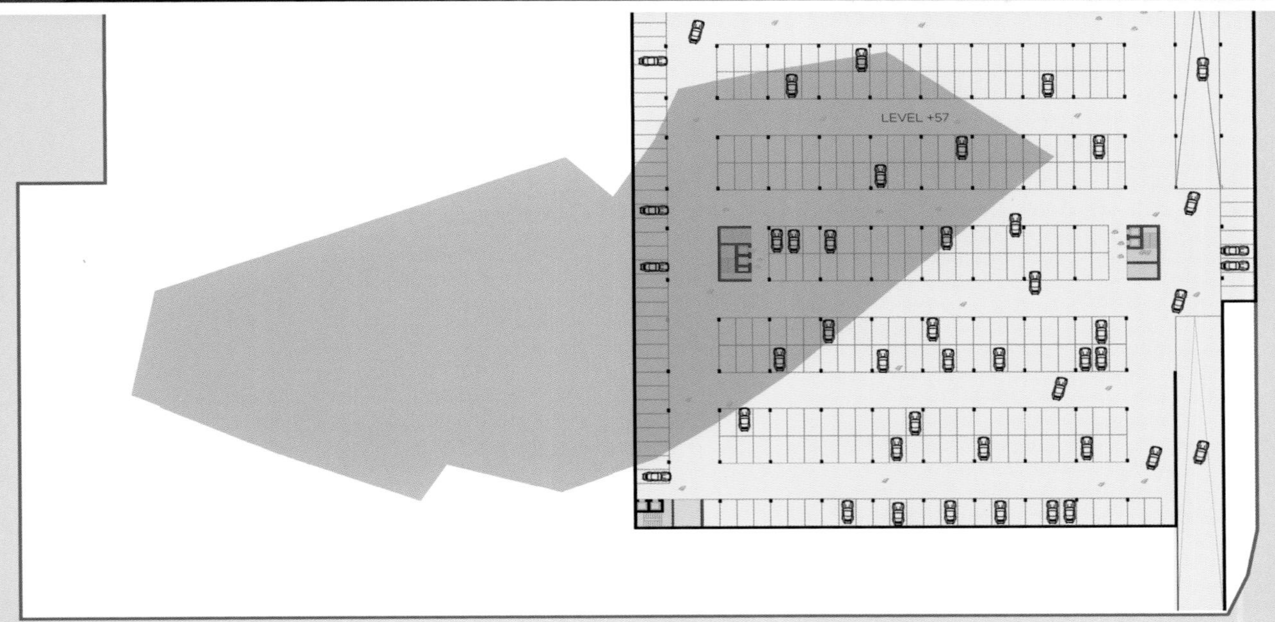

地下二层停车场 UNDERGROUND -2 PARKING LEVEL +57
图例 LEGEND

停车场 Parking	6.829 m2
洗手间 Toilet	19 m2
坡道 Ramps	527 m2
分布空间 Distribution Space	127 m2
地下二层总面积 TOTAL UNDERGROUND -2	7.502 m2

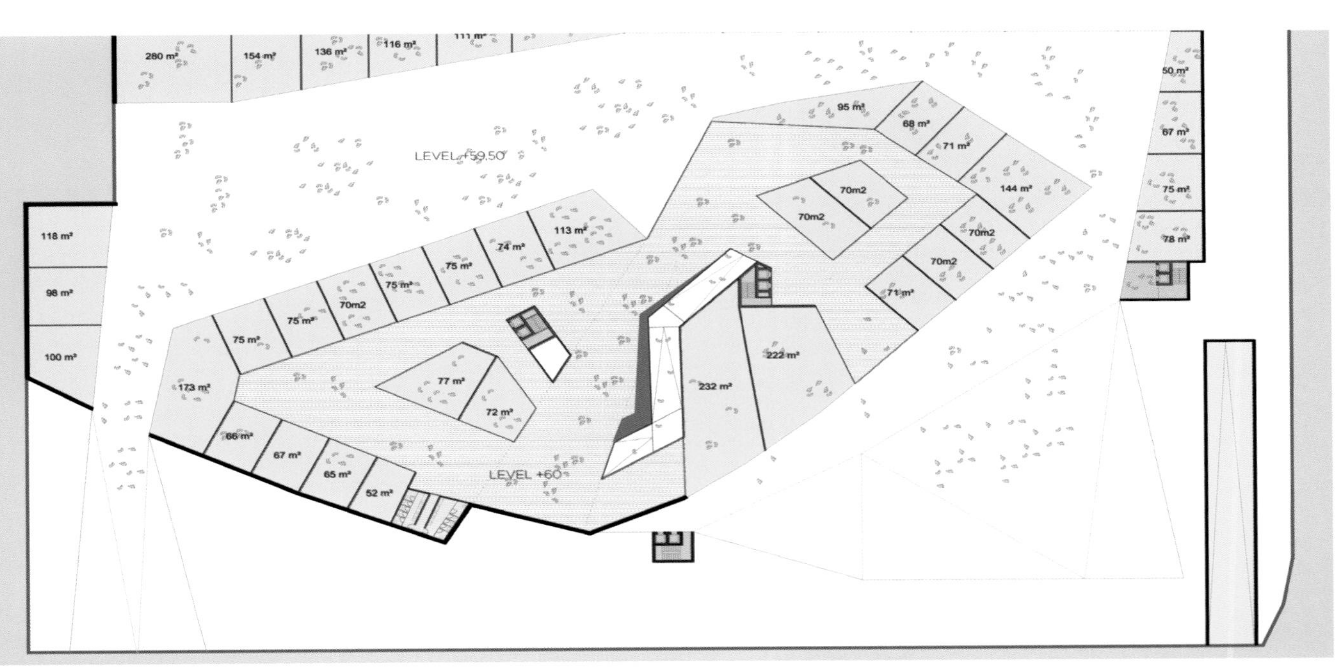

地下一层 UNDERGROUND -1 LEVEL +60
图例 LEGEND

商业 Commercial	4.256 m2
卫生间 Toilets	77 m2
室内分配空间 Interior Distribution Space	2.200 m2
坡道 Ramp	170 m2
二层总面积	

Exterior Part

外面空间 Exterior Space	4.178 m2
停车场入口 Parking Access	

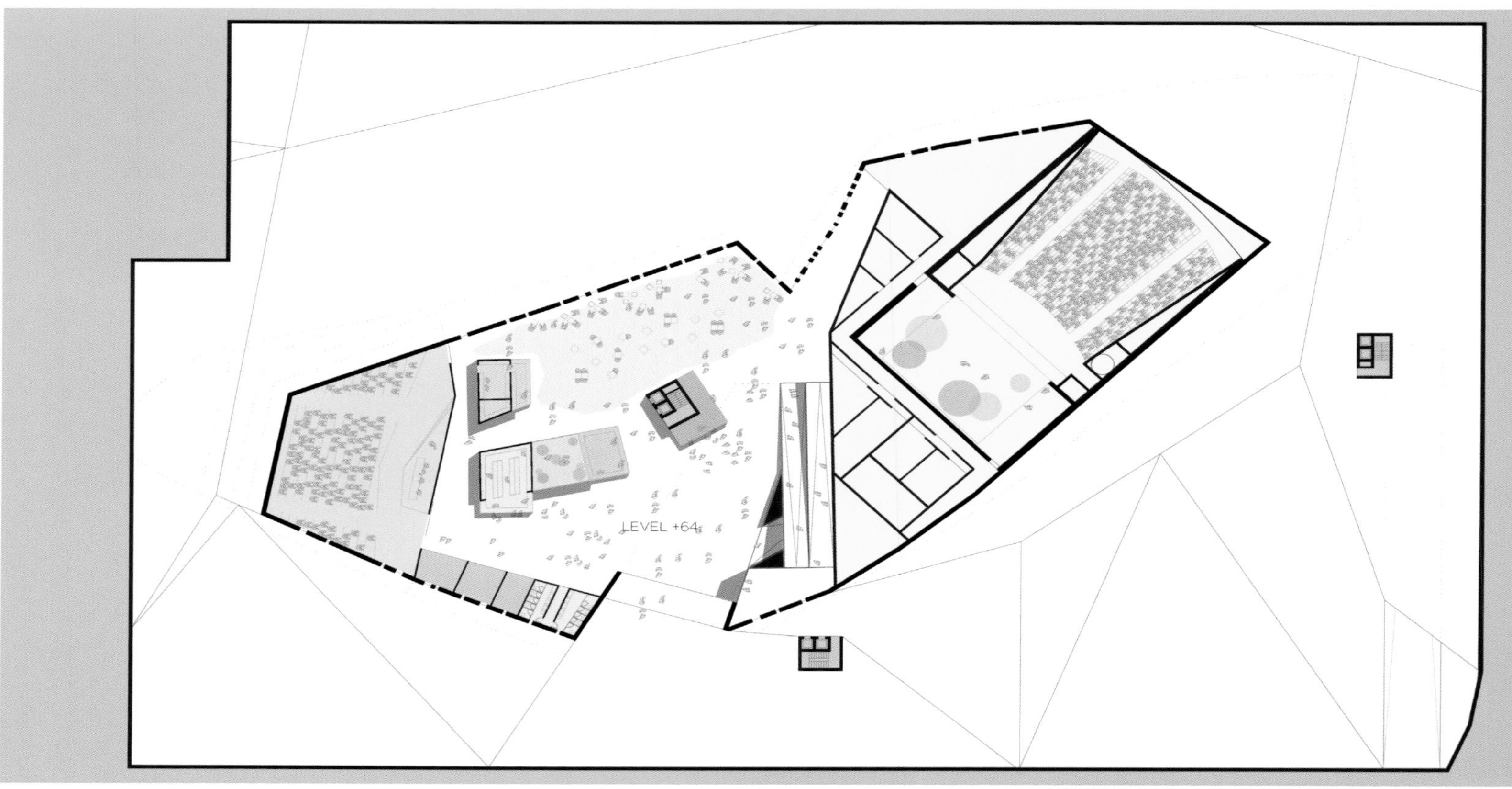

首层平面图 +64
图例 LEGEND

文化 / 剧场	Cultural_Theatre	2000 m2
文化 / 会议室	Cultural_Conference Room	500 m2
服务 / 买票站	Service_Tickets Office	33 m2
服务 / 服务台	Service_Reception	35 m2
服务 / 礼品店	Service_Gift Shop	45 m2
服务 / 衣帽间	Service_Cloak Room	60 m2
厨房和咖啡店	Kitchen & Coffee Shop	567 m2 (厨房：60 m2 + 其它：507 m2)
办公室	Office	68 m2
洗手间	Toilet	48 m2
坡道	Ramp	256 m2
分布空间	Distribution Space	1113 m2
首层总面积	TOTAL GROUND FLOOR	4725 m2

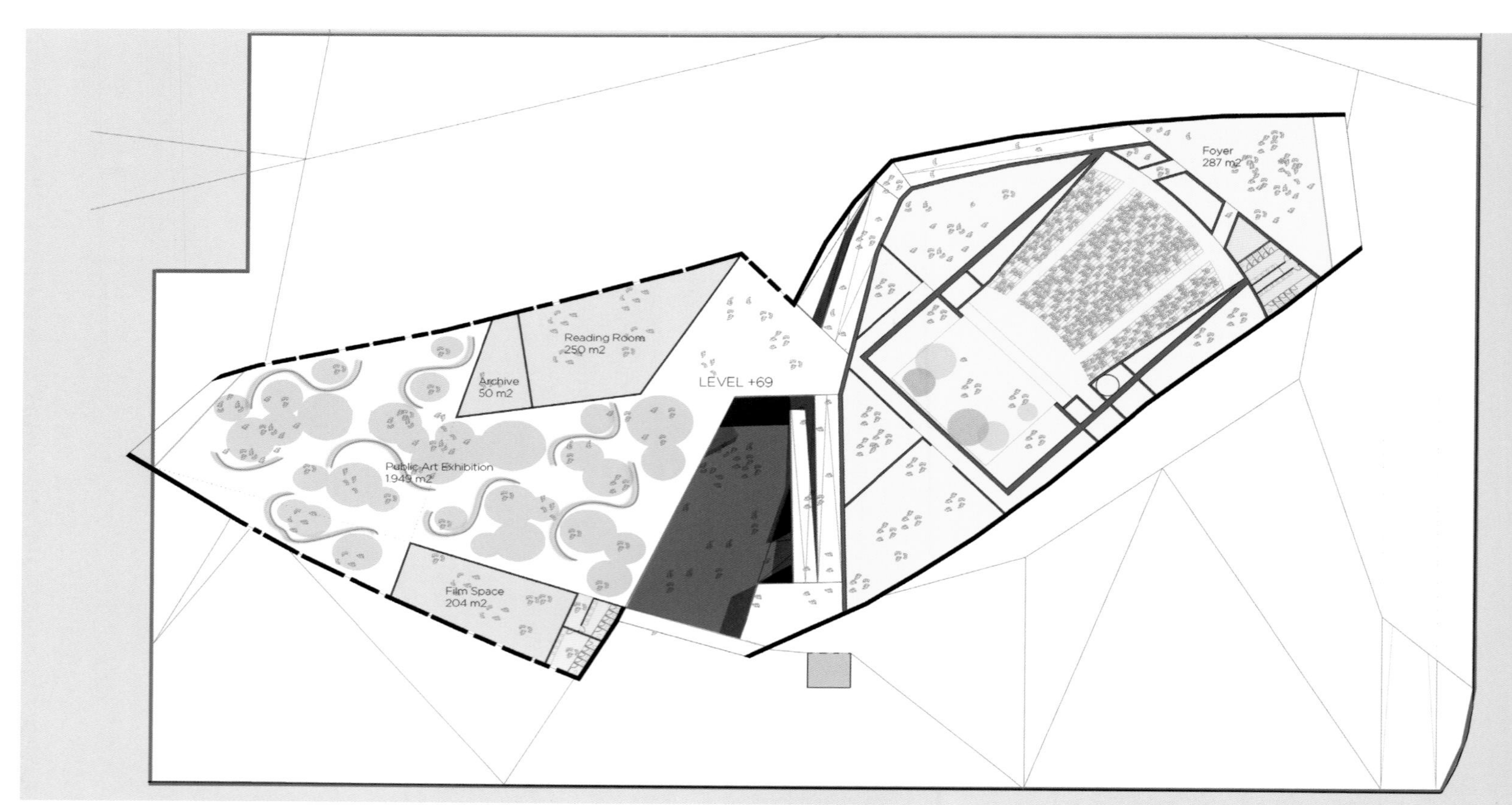

二层平面图 SECOND FLOOR LEVEL +69
图例 LEGEND

文化/剧场	Cultural_Theatre	Area accounted in the 1st Floor (Level +64)
大堂	Foyer	287 m2
文化/其它	Cultural_Others	2453 m2
洗手间	Toilet	55 m2
二层总面积	TOTAL 2ND FLOOR	2795 m2

Profile

The High-tech Science and Technology Cultural Center is located in the hi-tech zone. The latest technology is used in the design to create the shape of the 3D model. Modern expression modes are utilized to present cultural characteristics and traditional charm of this region.

Design Concept

Jinan is the intersection of Qi Culture and Lu Culture and one of the most important birthplaces of Chinese civilization. It has profound cultural connotation and rich artistic atmosphere which provide potential advantages and development opportunities for the project. Designers believe that these buildings need to reflect the cultural flavor of Jinan. They made a unique exclusive design, showing the soul and all its modern characters by software language.

Design Feature

The project is constituted by two buildings and a square. Both buildings are independent but connected. The two buildings, like dancing stones are placed in an urban stage that is the square. The organization of the parts is symmetric; the dancing stones are located in each side of an axis. Both parts are linked by the square. Like the city layout of Jinan, the square symbolizes the city while the two buildings represent the east and the west sides.

The square was renewed in order to emphasize its role as an urban stage. All stairs were removed except the one in the north-center, where the stairs work as a stage box. The rest of the square's circulation works by using ramps; ramps that allow a free movement without obstacles.

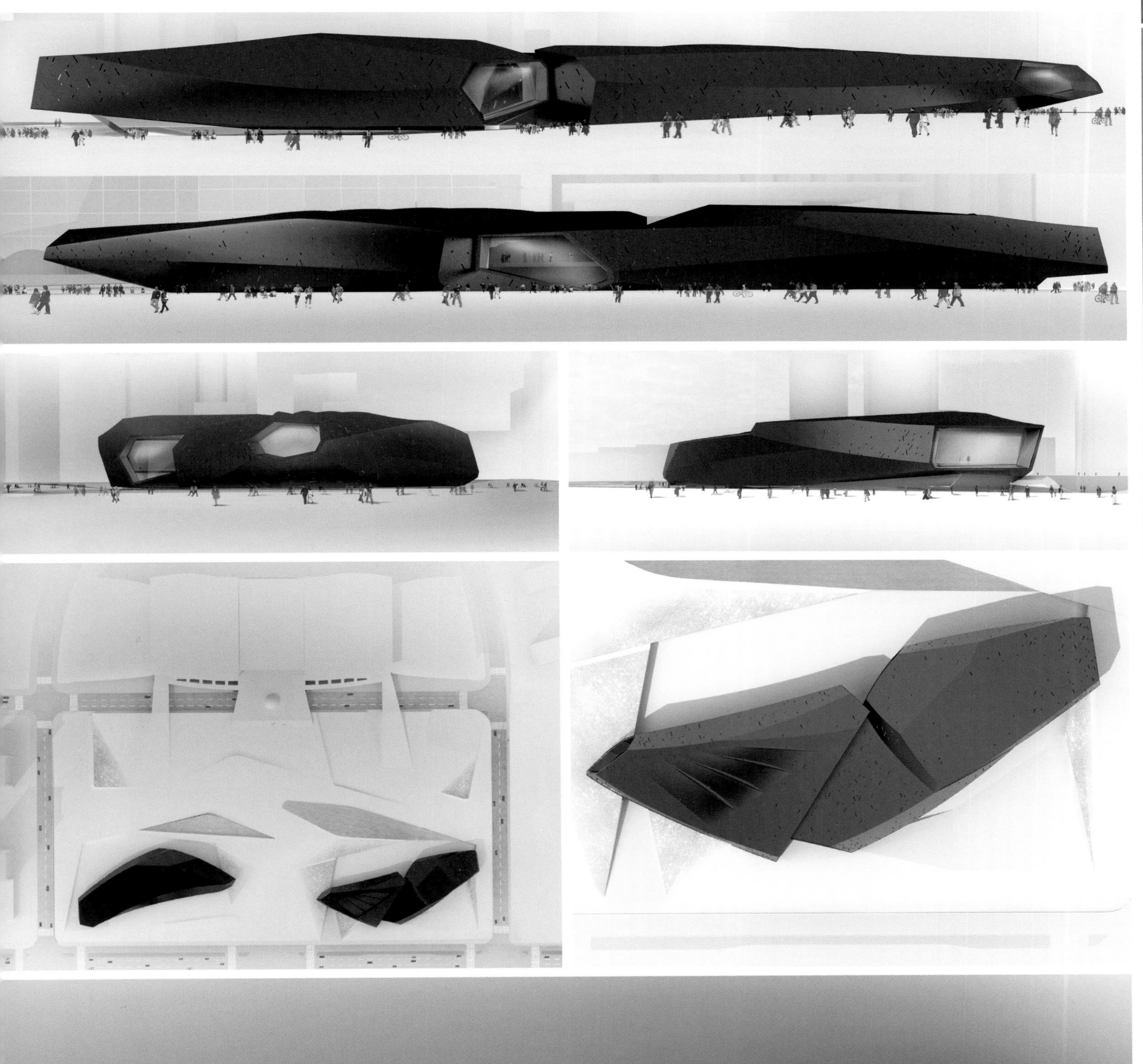

千灯艺体中心

China Qiandeng Art & Sports Center

设计单位：美国 KDG 建筑设计有限公司

Designed by: Kalarch Design Group

项目概况

千灯艺体中心位于江南水乡昆山市，这个高科技含量的建筑不仅考虑了大型聚集场所的公众需求，而且打破了公共体育项目成为政府的财政包袱、利用率低、维护经营惨淡的普遍现象，创造了可持续发展、自负盈亏、繁荣城市的新型都市核心。

设计构思

通过对国内外体育中心项目的研究，设计师在考虑客户需求的基础上，通过对功能分区的合理布局，提升项目的经济价值。设计采用了篮球馆和游泳馆组合的方式，两者体量接近，功能互补，同时，通过商业区的合理布置，不仅将体育设施与城市有机地联系在一起，而且有利于项目的经营与管理。

设计特色

这是一个古朴厚重的体育馆与精致迷离的游泳体育综合馆的二重奏：一个尺度宏大、古朴粗犷，似在追忆往昔；一个玲珑剔透、扑朔迷离，将人引入令人遐想的未来。这一厚重一明快的建筑体既形成鲜明的对比，又

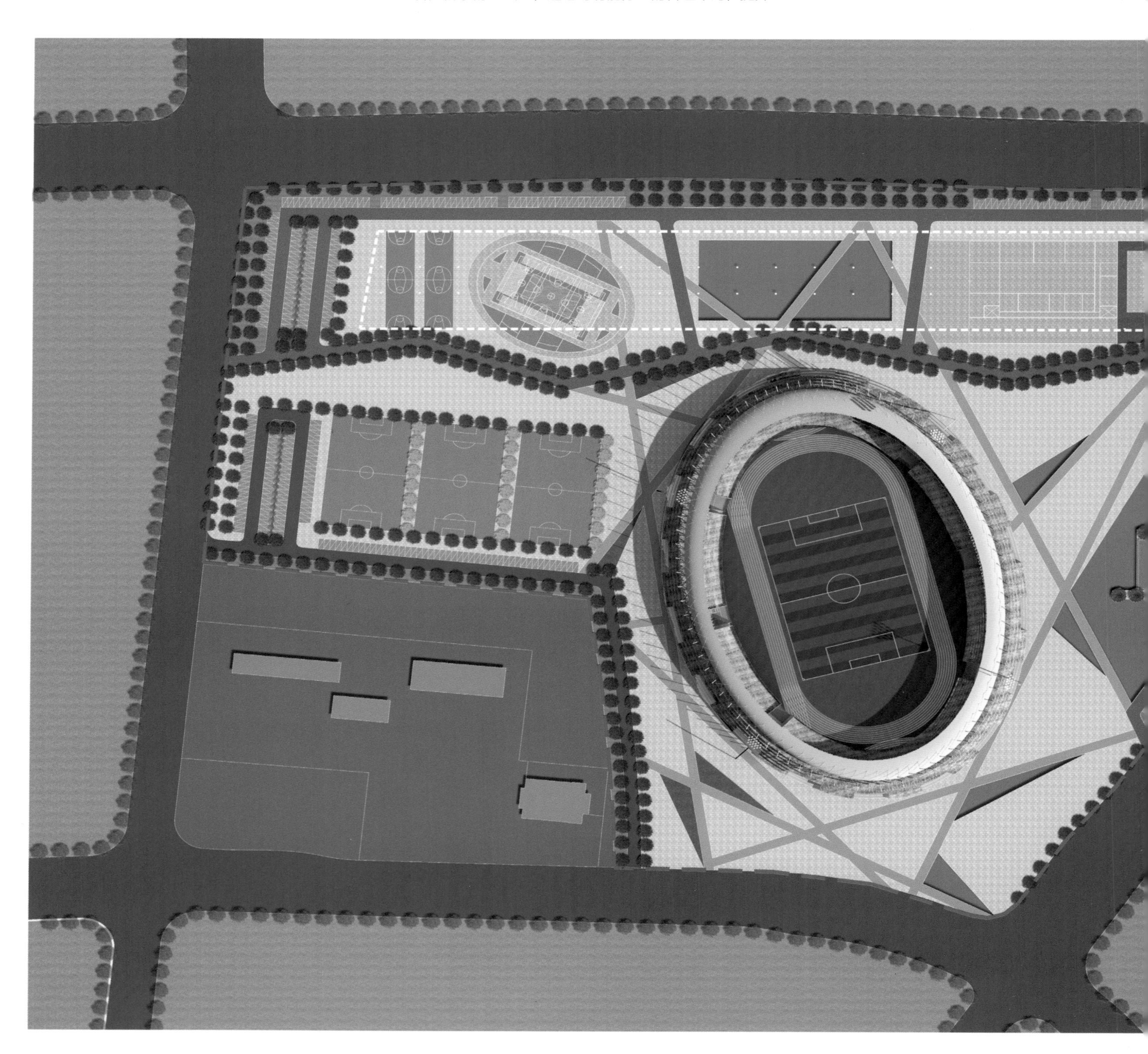

完美地融合在一起，将历史与未来、现实与想象联系在一起，置身其间，给人穿越时空般的梦幻体验。

综合建筑的功能需求和建筑投资等因素，设计将篮球馆和游泳馆分别设置。建筑形体规整，入口空间简洁而气派，两馆仅以一结构简单的通廊相连。明确清晰的结构系统，不仅是建筑魅力的源泉，而且为合理使用结构杆件来降低造价提供了有利的基础。

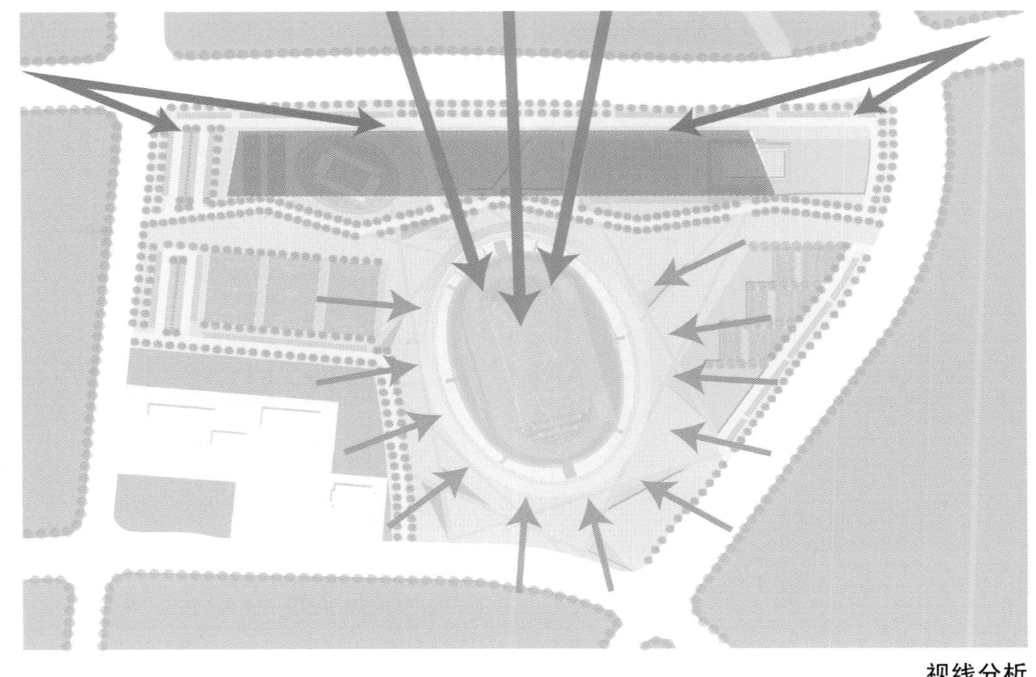

视线分析

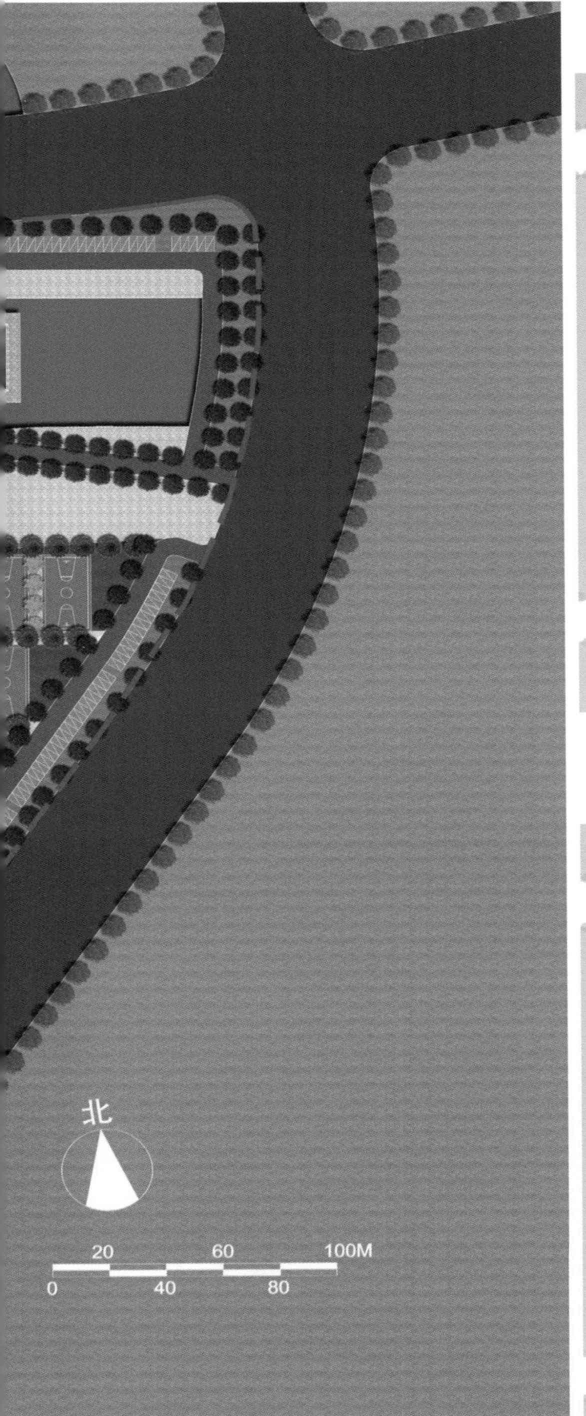

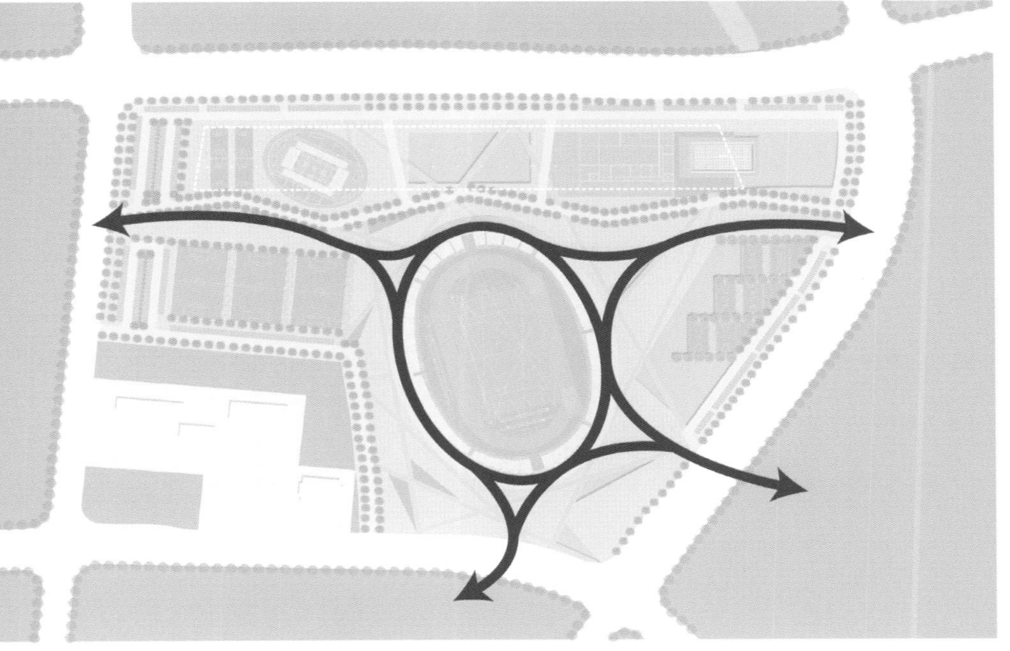

流线分析

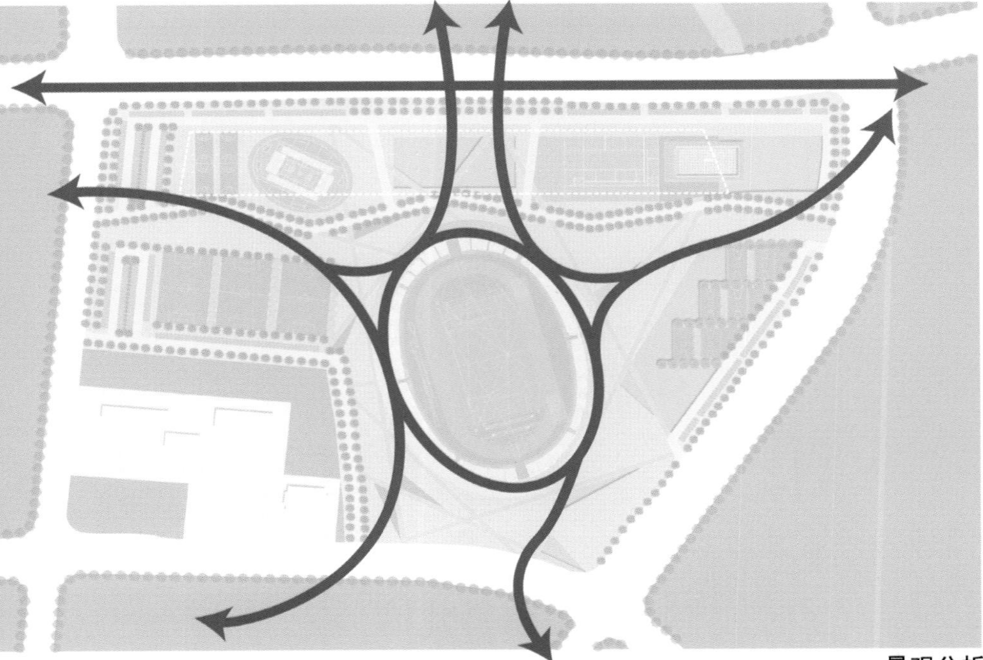

景观分析

Profile

Qiandeng Art & Sports Center is located in Jiangnan Watertown, Kunshan. This hi-tech building not only takes public demands for a large gathering place into account, but also has changed the old status—public sports events, with low utilization ratio and low profit operation, were financial burden of the government. It creates a new urban core which is sustainable, self-financing and prosperous.

Design Concept

Through research on domestic and overseas sports centers and based on customers' demands, designers have properly arranged functional subareas by which economic value of this project is greatly enhanced. Basketball Gymnasium and Natatorium are combined in this design for their similar volumes and complementary functions. Proper layout of commercial district which combines sports facilities with the city has facilitated project's operation and management.

Design Feature

The project has double identity as an antique shocking gymnasium and an exquisite puzzling natatorium: one is grand, antique and robust; the other is exquisitely carved, puzzling and enchanting. These two volumes are sharply contrasted while perfectly integrated. They link history with future and combines reality with imagination. People here shall enjoy dreamlike experience.

For the consideration of functional demands and architecture investments of this building complex, Basketball Gymnasium and Natatorium are respectively set. The complex has regular shape, simple but grand entrance space. The two venues are linked by a simple connecting corridor. Clear and explicit structural system is not only the source of architectural charm, but also uses structural components to reduce construction cost.

天津生态城

Tianjin Ecocity

设计单位：斯蒂文·霍尔建筑师事务所
项目地址：中国天津市

Designed by: Steven Holl Architects
Location: Tianjin, China

项目概况

中国的多彩建筑有着深厚而悠久的历史，天津生态城的设计参照了中国传统建筑的设计手法，将古代的彩饰运用到了21世纪的公共空间体验当中。

设计特色

七座新的文化建筑宛若环绕着古道河的七彩项圈，形成古道河边缘的一个独特的城市中心。连接这七座文化建筑的步行道路铺装着特殊的金色石材，这一人行步道的设计将人们独特的视觉体验置于首位。

每一栋建筑在颜色上和形态上都对应着一种珍贵的矿物：图书馆和档案馆大楼因带有北向采光的窗户和反光的黄铜色外壳，宛若黄色的水晶；老年人中心像是立方体的黄铁矿；文化和艺术博物馆则覆盖着天青石的独特瓷砖；生态博物馆的形态酷似公园中的绿色孔雀石；科学博物馆则由黑曜石打造成“贝壳状破裂”的曲线形态；规划博物馆的多孔形态仿若螺硫银；工人文化会馆则呈现出沙金石的色彩和质感。

生态设计

设计师在场地内设想了一个大型的地热井，可为整个综合体内的所有建筑供热和供冷。所有建筑中的太阳能光伏电池以及公共花园中的太阳能光伏绿廊将为这一城市片区提供25%-50%的电能，同时，中央商务区的圆形大厦安装了太阳能跟踪光伏遮阳屏，既能够起到遮阳的作用，也有助于实现大厦的自然供电。

Profile

The polychrome architecture of China has a deep and ancient history. The Tianjin Ecocity, adopted Chinese traditional architectural design techniques, brings the polychromy of Ancient China into the experience of public space in the 21st century.

Design Feature

Seven new cultural buildings shaped like a necklace of polychrome buildings form a unique city center at the Gudao River edge. Connecting all seven of the cultura buildings, the pedestrian path is paved with special go colored stones. This path puts the perspective of th human experience in the primary position of importanc for this master plan.

Each one of these buildings corresponds to a preciou mineral in color and morphology: the Library an Archives has the form of yellow crystal with specia north lights and a faceted brass shell; the Senior Cent is formed like pyrite in cubic clustering; the Cultur and Art Museum is sheathed in special tiles of lap lazuli; the Ecology Museum is formed like park gree malachite; the Science Museum is shaped by th curvilinear “concoidal fractures” that are characterist of black obsidian; the Planning Museum's porou morphology is like the mineral acanthine; the Worker Culture Hall is the texture and color of aventurine.

PARK
公园

TUNNEL ENTRANCE

CORNICHE
滨海路

HARBOR MASTER

Parking

Fuelling Berth

Public Quay

SHADE

Pedestrian Bridge Envelope

plaza dashed above

−6.0 +5.0 +5.5 +8.5 +4.5 +1.5 +10.0 +8.0 +7.75 +7.5 +7.25 +7.0 +6.75 +6.5 +6.25 +6.0 +5.5 +5.0 +4.75 +4.5 +4.25 +4.85

D.P. 1418 1354 1353 1352 296 9-3 1421 1401 1422 1368 1369 1400

LEGEND
图例

WATER 水体	
SHADE 遮阳	
SOFT LANDSCAPE 软质景观	
NATURAL ROCKS 天然石	
ART PIECES LOCATION 艺术品的位置	

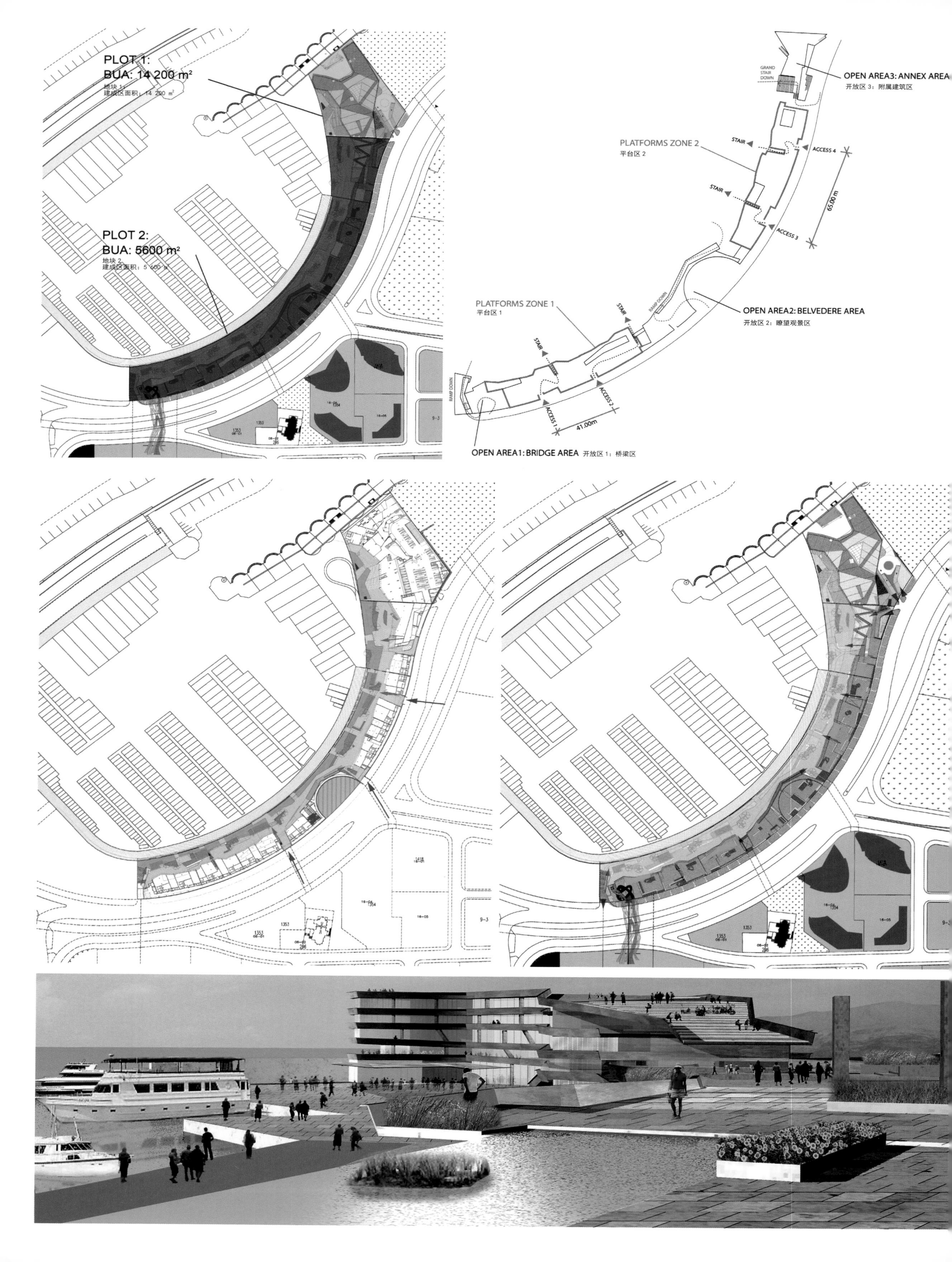
PLOT 1:
BUA: 14 200 m²
地块 1：
建成区面积：14 200 m²
PLOT 2:
BUA: 5600 m²
地块 2
建成区面积：5 600 m²
OPEN AREA3: ANNEX AREA
开放区 3：附属建筑区
GRAND STAIR DOWN
PLATFORMS ZONE 2
平台区 2
STAIR
ACCESS 4
65,00 m
ACCESS 3
OPEN AREA2: BELVEDERE AREA
开放区 2：瞭望观景区
RAMP DOWN
PLATFORMS ZONE 1
平台区 1
ACCESS 1
ACCESS 2
41,00m
OPEN AREA1: BRIDGE AREA 开放区 1：桥梁区

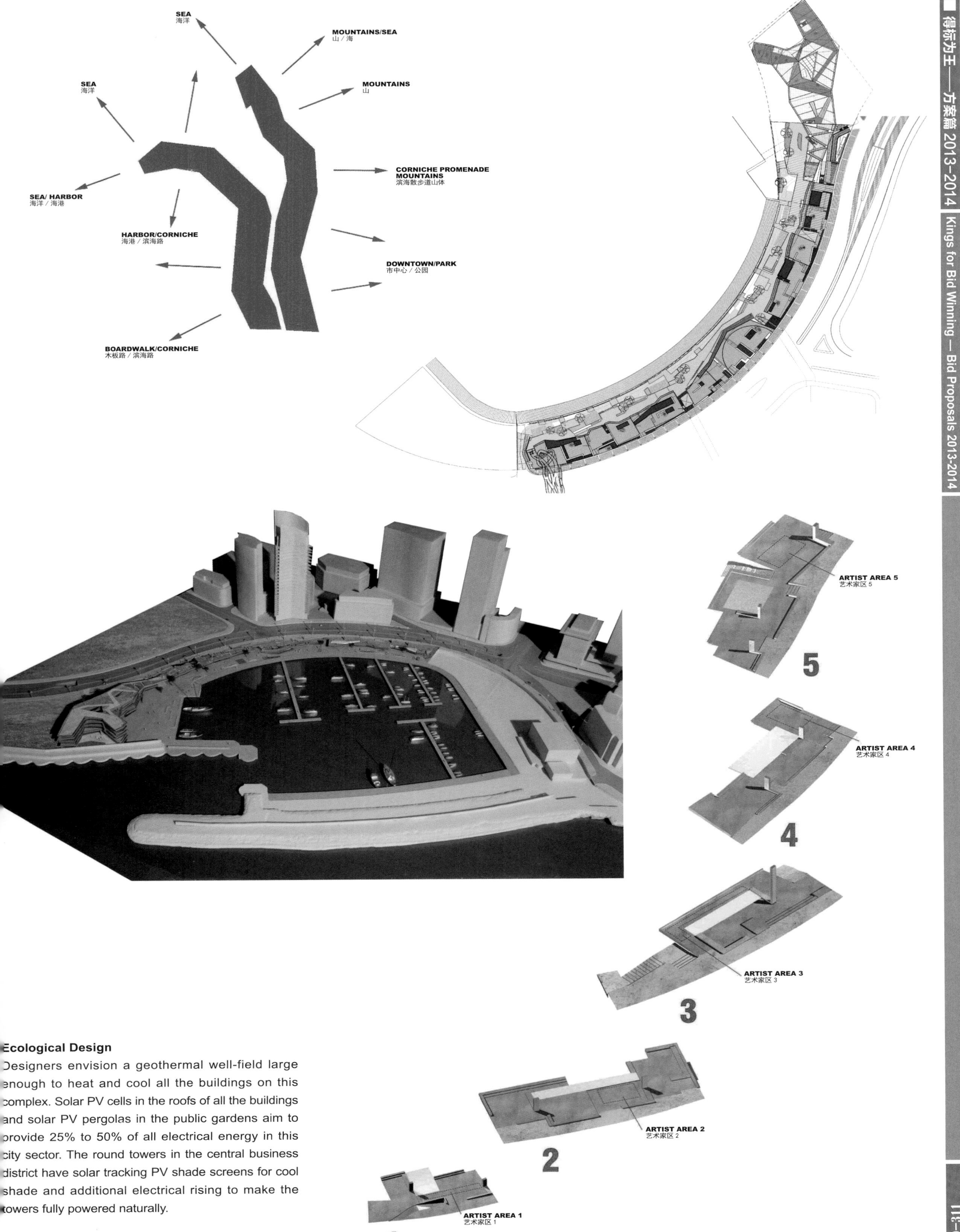

Ecological Design

Designers envision a geothermal well-field large enough to heat and cool all the buildings on this complex. Solar PV cells in the roofs of all the buildings and solar PV pergolas in the public gardens aim to provide 25% to 50% of all electrical energy in this city sector. The round towers in the central business district have solar tracking PV shade screens for cool shade and additional electrical rising to make the towers fully powered naturally.

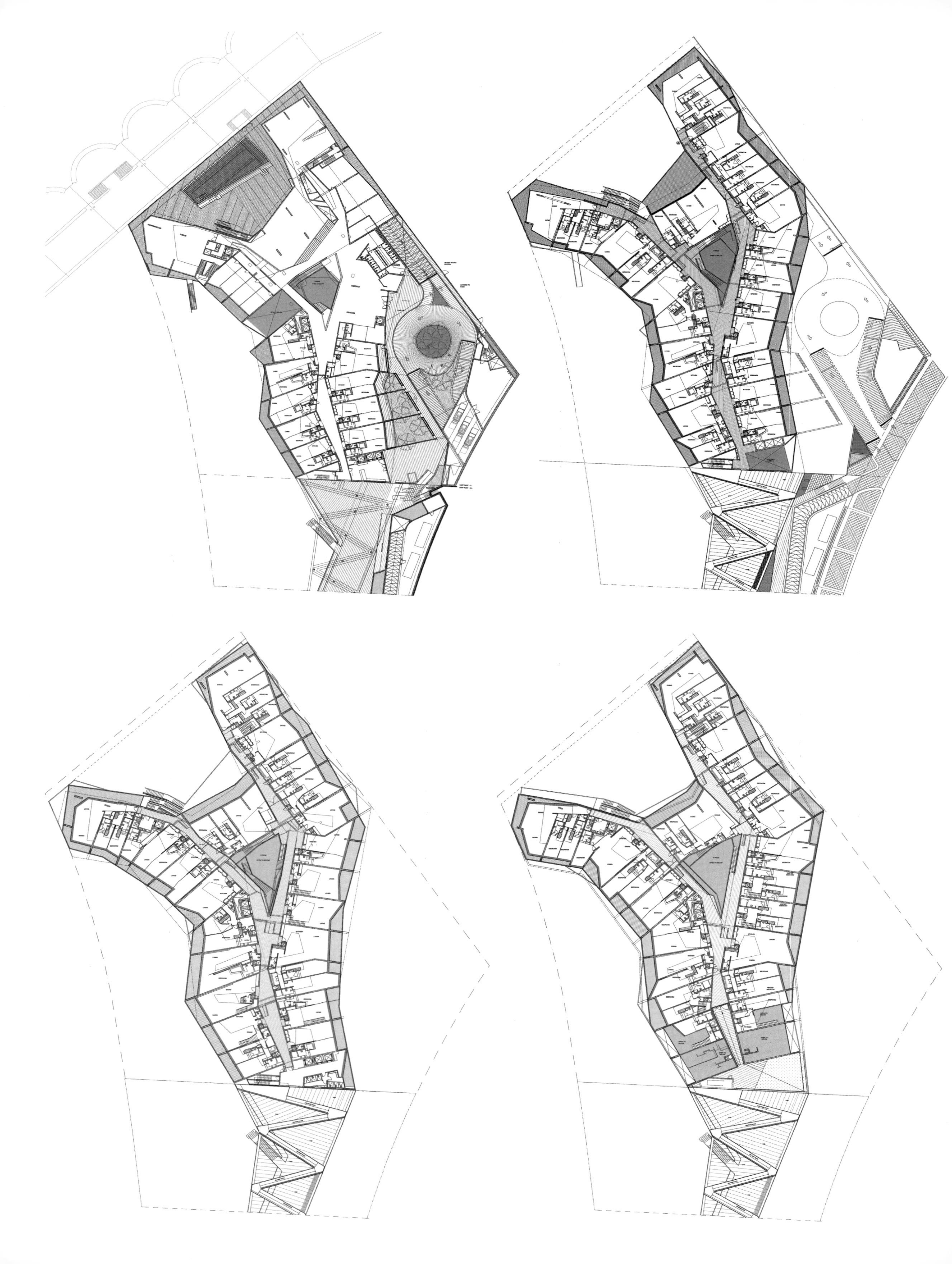

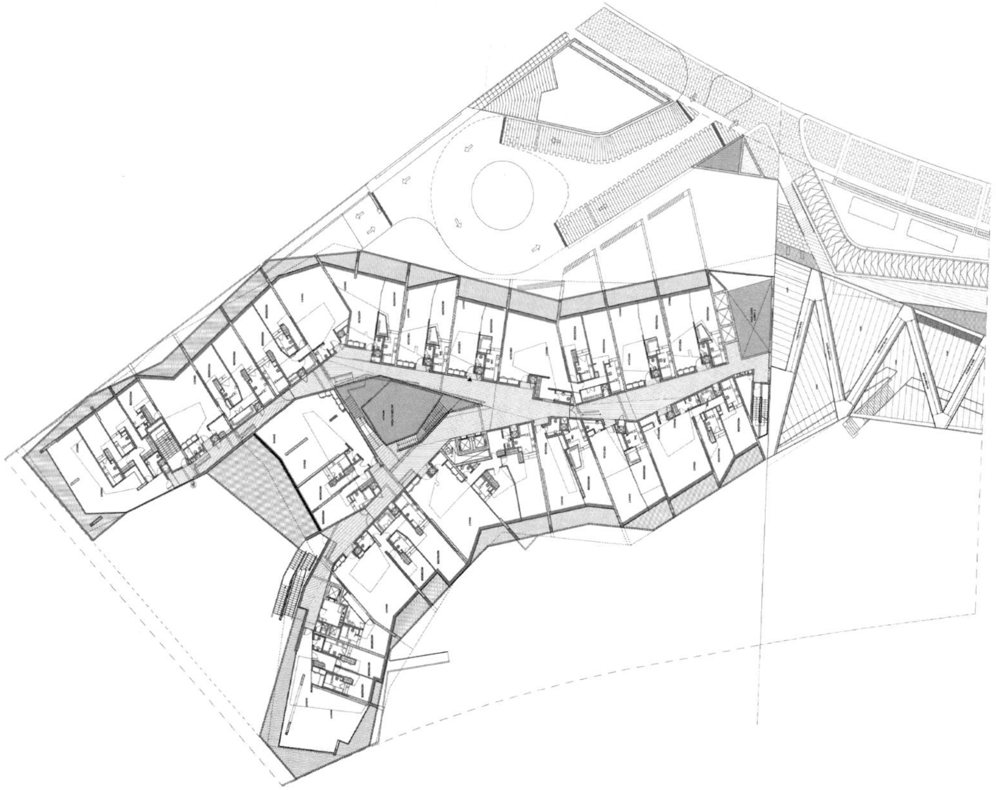

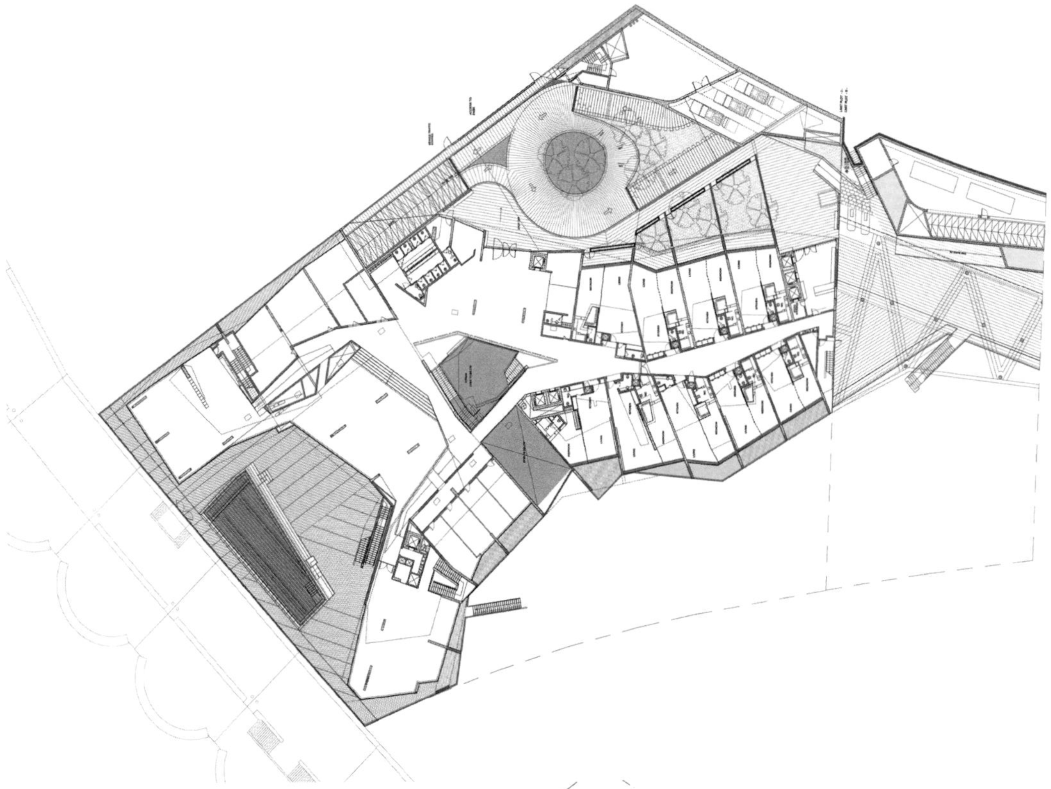

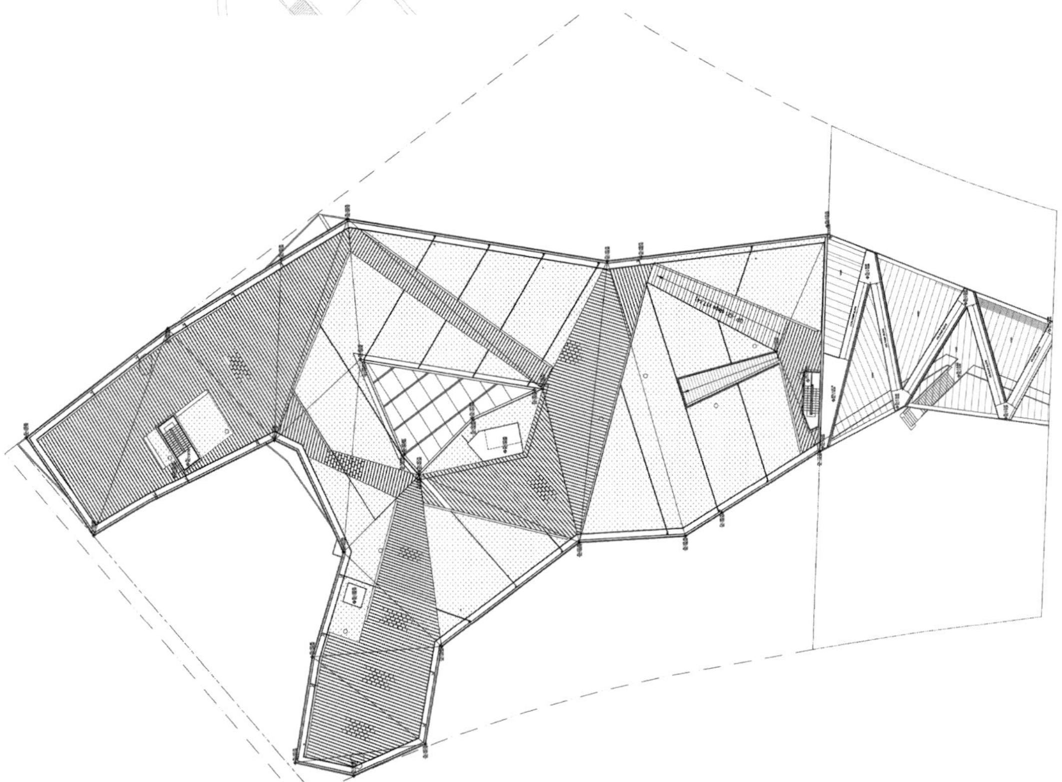

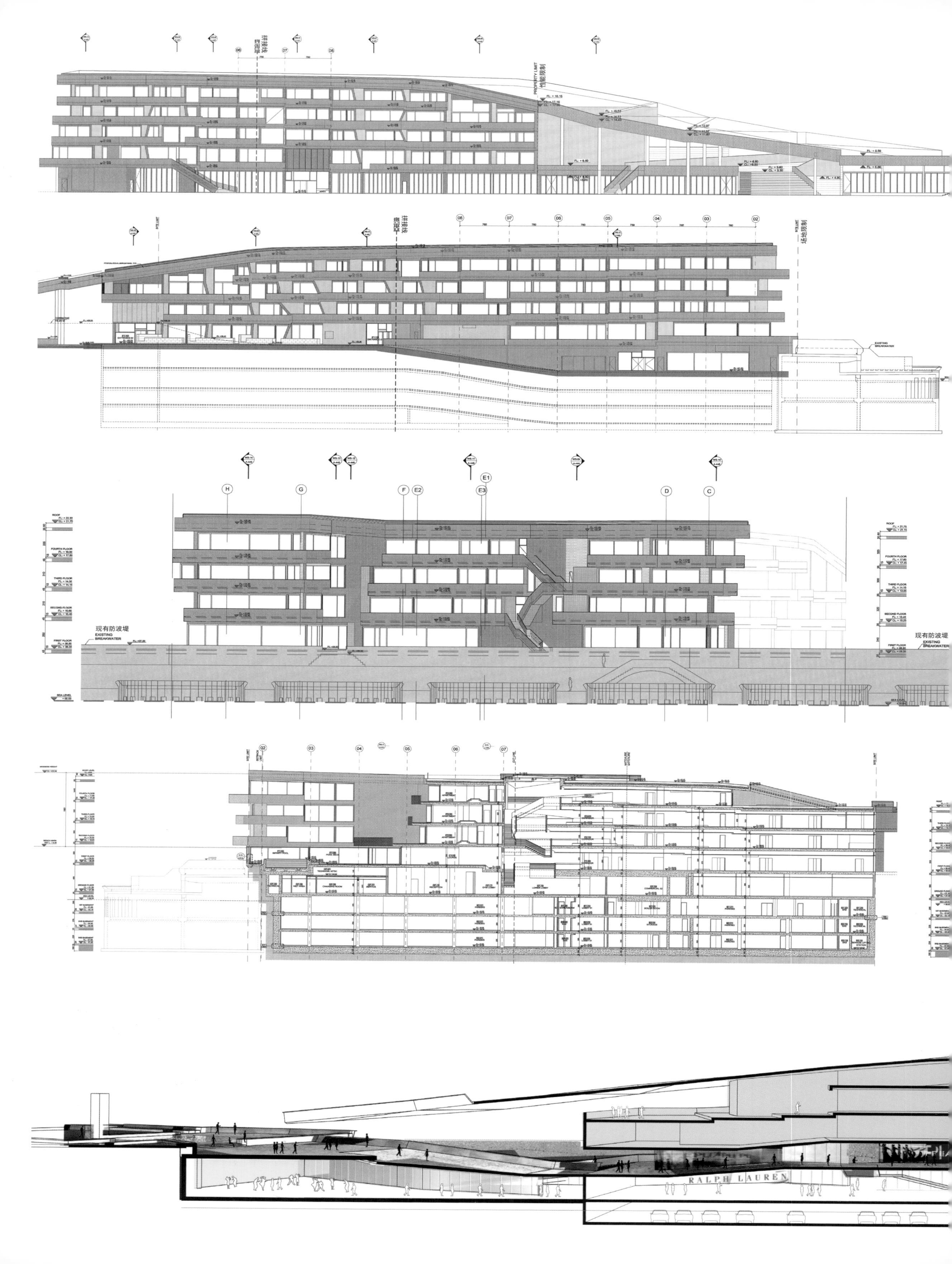

拼接线
PROPERTY LIMIT
性能限制
场地限制
EXISTING BREAKWATER
现有防波堤
EXISTING BREAKWATER
现有防波堤
EXISTING BREAKWATER
ROOF
FOURTH FLOOR
THIRD FLOOR
SECOND FLOOR
FIRST FLOOR
SEA LEVEL
RALPH LAUREN

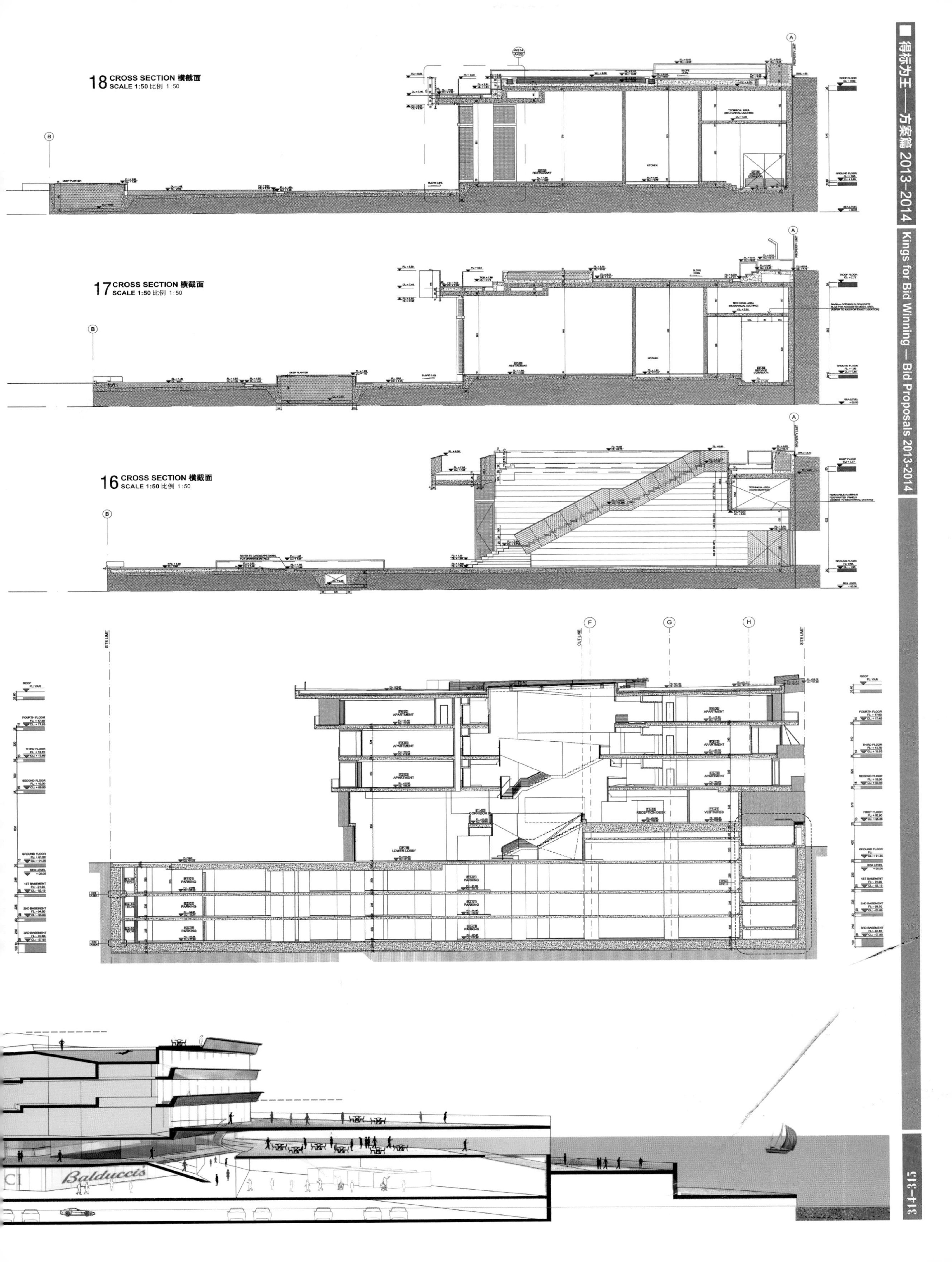
18 CROSS SECTION 横截面
SCALE 1:50 比例 1:50
17 CROSS SECTION 横截面
SCALE 1:50 比例 1:50
16 CROSS SECTION 横截面
SCALE 1:50 比例 1:50
Balduccis

福建福州平潭海峡论坛及中央商务区

Cross Strait Forum and CBD Development

设计单位：10 DESIGN（拾稼设计）
开发单位：平潭综合实验区规划局
项目地址：中国福建省福州市
占地面积：930 000 ㎡
建筑面积：518 000 ㎡
设计团队：Gordon Affleck　Brian Fok
Francisco Fajardo　Frisly Colop Morales
Laura Rusconi Clerici　Lukasz Wawrzenczyk
Maciej Setniewski　Mike Kwok
Ryan Leong（建筑）　Shane Dale（多媒体）
Ewa Koter　Fabio Pang（景观）

Designed by: 10 DESIGN
Client: Pingtan Comprehensive Experiment Planning Bureau
Location: Fuzhou, Fujian, China
Site Area: 930,000 m²
Gross Floor Area: 518,000 m²
Design Team: Gordon Affleck, Brian Fok, Francisco Fajardo, Frisly Colop Morales, Laura Rusconi Clerici, Lukasz Wawrzenczyk, Maciej Setniewski, Mike Kwok, Ryan Leong (Architecture Design); Shane Dale (CGI) Ewa Koter, Fabio Pang (Landscape Design)

项目概况

平潭是一个新兴的、推动中国大陆和台湾之间沟通和商贸往来的商业中心。平潭中央商务区总体规划将包括 330 万平方米的城市规划，而平潭海峡论坛中心将作为整个项目第一阶段的设计发展目标。

建筑设计

平潭海峡论坛中心项目是包含了剧院、会展中心、展览中心及配套商业和文化设施的设计。建筑设计结合了景观及滨海的自然元素，创造了一个流动的、开放的、与建筑融为一体的公共场所，充分体现了建筑的透明度和与周围环境互动的特征。

为使车辆对行人道路的影响降至最低，设计规划了一条以中央公园为起点，穿过湖泊和一系列的休闲零售服务场所，直至滨海自由畅通的线路，道路等要素被设计安排成景观梯田的组合。

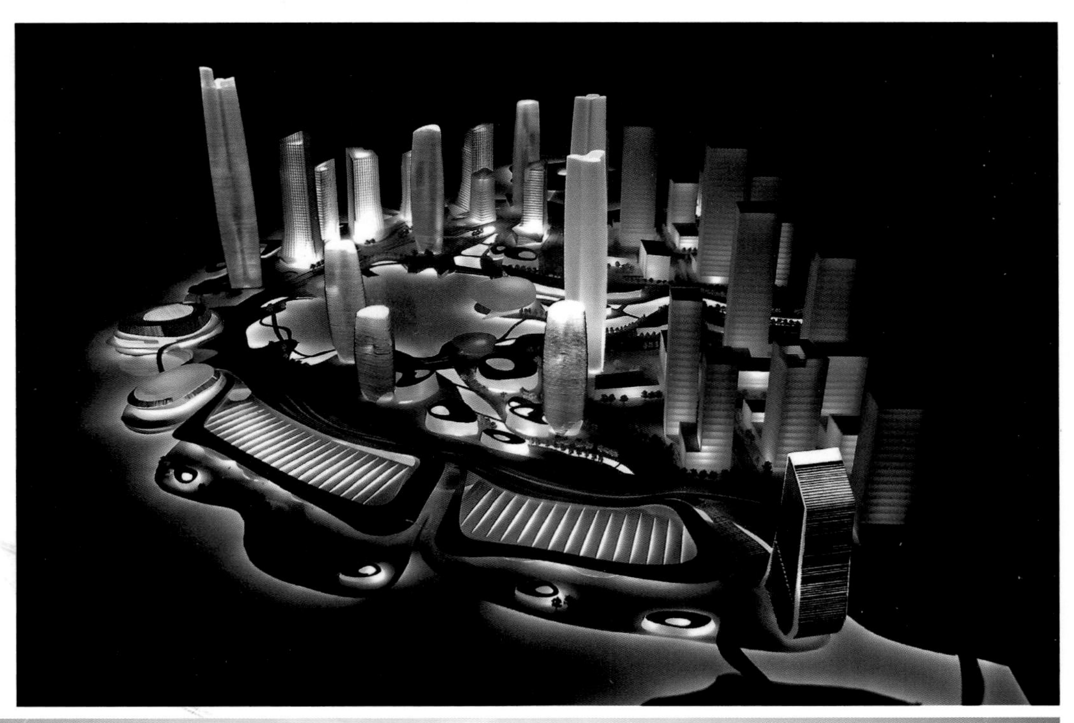

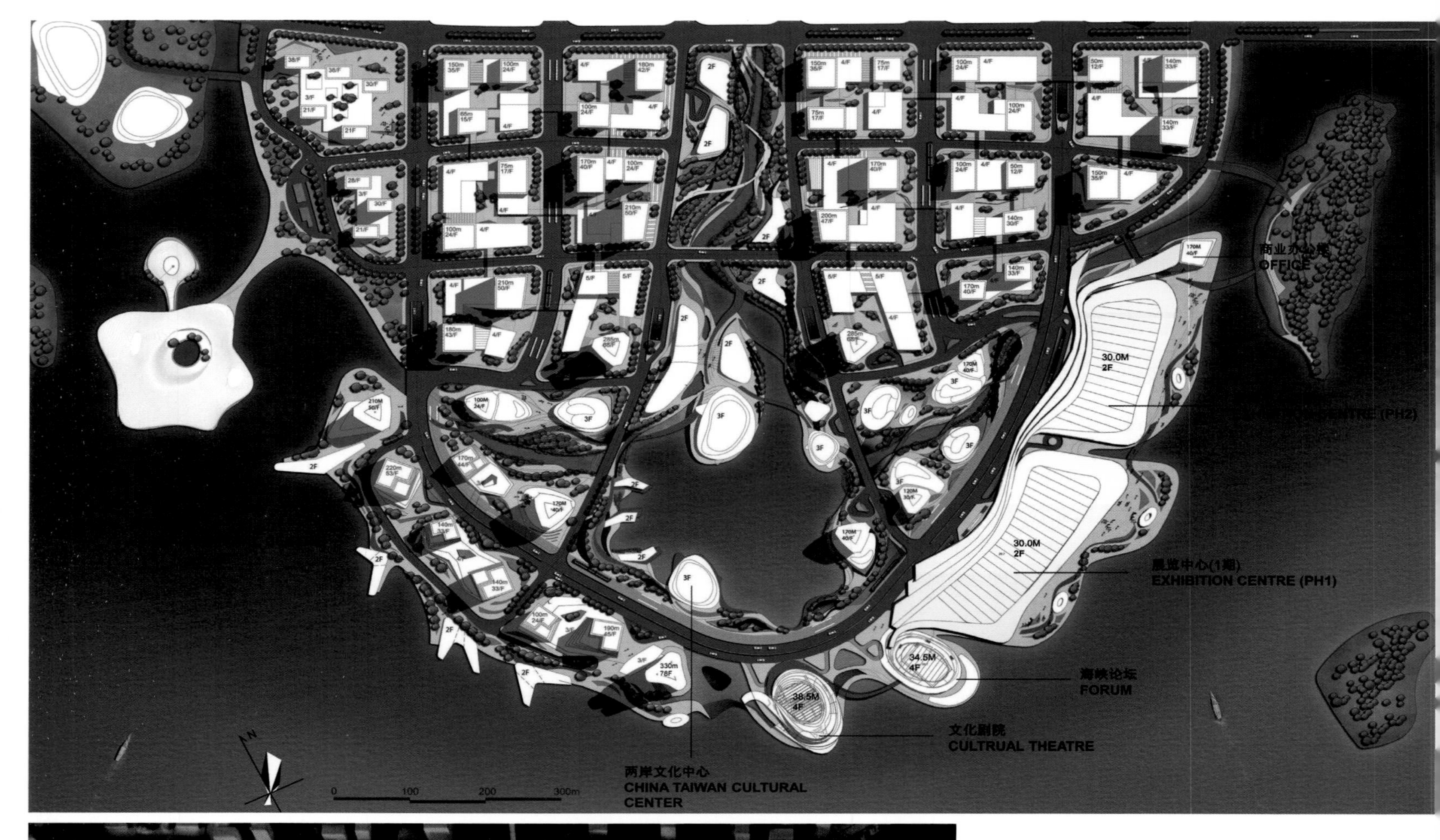

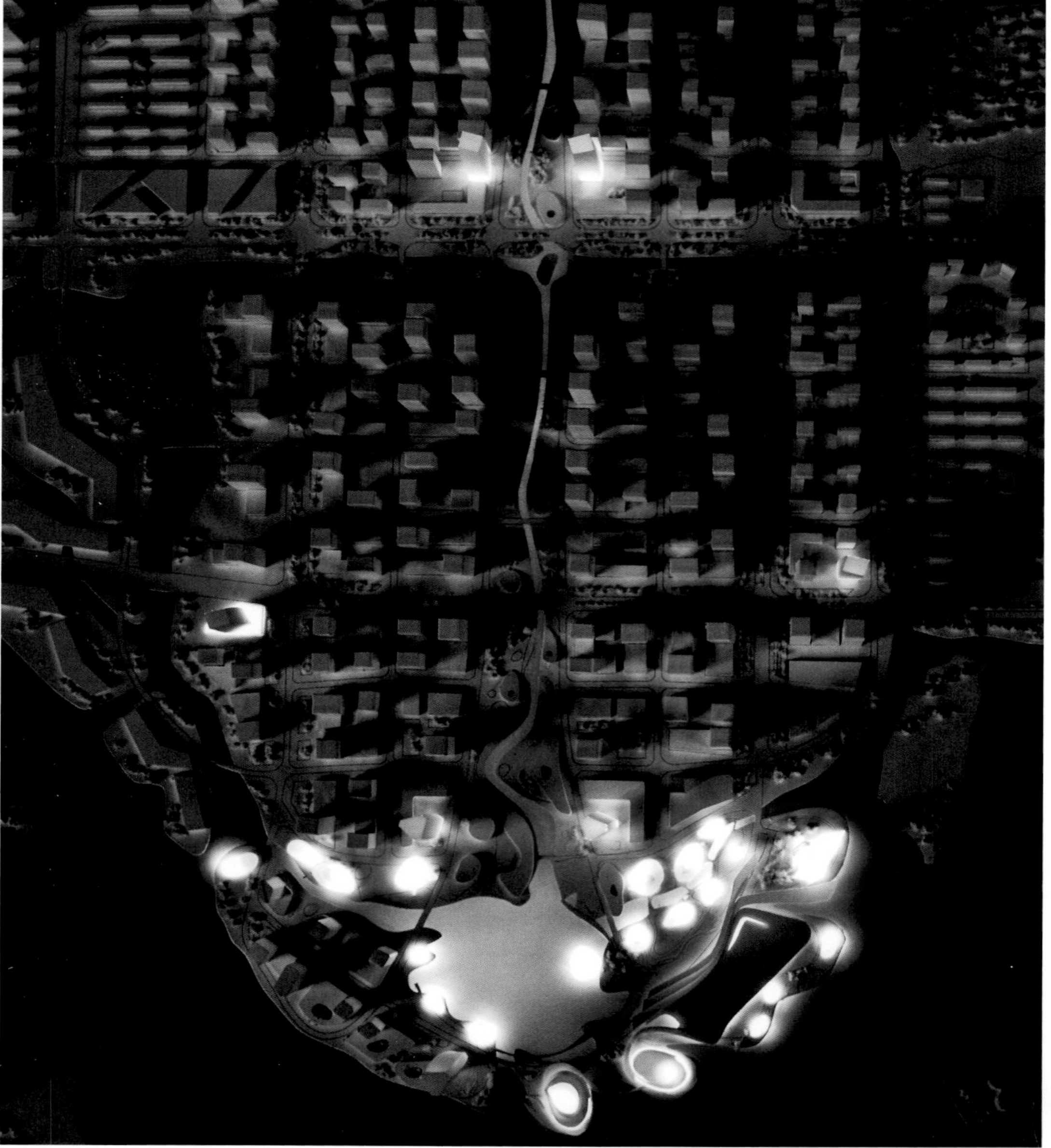

Profile

Pingtan is planned as a new commercial hub to drive communication and commercial trade between Chinese Mainland and Taiwan. The masterplan caters for some 3.3 million square meters of urban development, while the Cross Straits Forum would be in the first phase of development.

Architectural Design

The design of a new Cross Straits Forum includes theatre, convention, exhibition and auxiliary commercial and cultural facilities. To reflect the aspiration of transparency and dialogue the buildings are formed by converging elements that combine with the landscape and waterfront to create a fluid and open series of public spaces that meld into the buildings themselves. Service traffic, roads and trams are integrated into a series of terraced landscape levels to minimize impact of car traffic on pedestrian circulation routes and to create free access from the central axial park canal through the lake towards the waterfront through a series of leisure and retail lined canals.

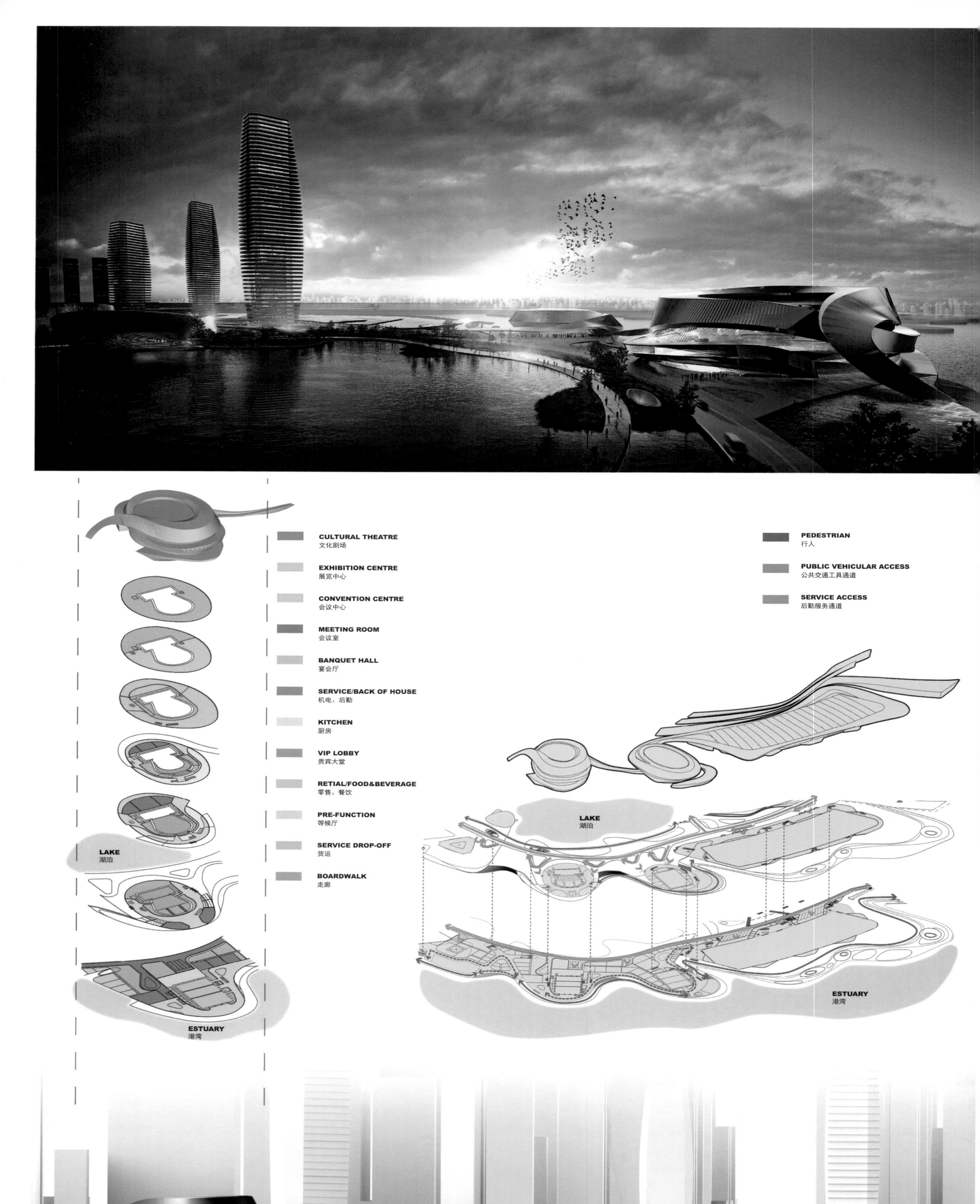
CULTURAL THEATRE
文化剧场
EXHIBITION CENTRE
展览中心
CONVENTION CENTRE
会议中心
MEETING ROOM
会议室
BANQUET HALL
宴会厅
SERVICE/BACK OF HOUSE
机电、后勤
KITCHEN
厨房
VIP LOBBY
贵宾大堂
RETIAL/FOOD&BEVERAGE
零售、餐饮
PRE-FUNCTION
等候厅
SERVICE DROP-OFF
货运
BOARDWALK
走廊
PEDESTRIAN
行人
PUBLIC VEHICULAR ACCESS
公共交通工具通道
SERVICE ACCESS
后勤服务通道
LAKE
湖泊
ESTUARY
港湾
LAKE
湖泊
ESTUARY
港湾